Rosalia A. Bowen
Rosalia A. Bowen

Calculus with Analytic Geometry

Calculus with Analytic Geometry

Paul K. Rees
Professor of Mathematics, Emeritus
Louisiana State University

Fred W. Sparks
Professor of Mathematics, Emeritus
Texas Technological College

McGraw-Hill Book Company
New York   St. Louis   San Francisco   Toronto
London   Sydney

Calculus with Analytic Geometry

Preface

As noted in the title, this book includes analytic geometry as well as
calculus. The analytic geometry is continued in chapters that are
separate from the calculus and is presented from time to time as
needed in the development of the calculus. This is done so that the
book may be used either as a text for a combined course in calculus
and analytic geometry, or as a text for calculus alone. We give
somewhat more analytic geometry than do most of the recent books
on the combined subjects, but less than is given in books on analytic
geometry. Differential and integral calculus of polynomials is presented along with the needed analytic geometry before any work on
transcendental functions is given.

Considerable effort has been expended to make the book teach-
able. The work tends toward the traditional, but we have not lost
sight of recent trends and have used modern terminology and concepts when they seemed appropriate. We hope that any lack of
sophistication is offset by the readability of the book from the student's
point of view. Besides believing that the student can understand the
book, we anticipate that he can apply the principles that are presented.

We have seriously attempted to include sufficient worked-out examples to illustrate the text material they follow and the problems they precede. We hope that the discussions in connection with the examples are such that the student will understand and become interested in them.

Exercises have been placed a lesson apart, for ease of assignment each day. Many of the concepts and techniques are of such a nature that more than one day is required for their mastery. We think that enough problems have been included for this purpose. There are about 3,700 problems in 116 exercises, so that more than one day can be spent on a considerable number of them. The problems are in groups of four of about the same order of difficulty and requiring essentially the same concepts and techniques; the order of difficulty increases from group to group. With this arrangement a good assignment could consist of each fourth problem in an exercise.

We wish to express our appreciation to CUPM and to many of our colleagues whose recommendations we have considered in deciding what topics to include and in deciding how to treat them. But in the final analysis, the selection of topics and method of treatment are ours and have been determined in the light of our years of teaching collegiate mathematics.

Paul K. Rees
Fred W. Sparks

Contents

Calculus with Analytic Geometry

1
Topics from algebra

1.1   Sets

One of the basic and useful concepts of mathematics is denoted by
the word "set." A *set* is a collection of well-defined objects or symbols
called *elements* or members of the set. By "well defined" we mean
that there is a criterion that enables us to decide whether an object
or symbol is or is not a member of the set. For example, suppose that
S is the set of all bicycles that are green. We can conclude: first,
a green bicycle is an element of S; second, a tricycle is not an element
of S; third, a bicycle that is not green is not a member of S.

As implied above, capital letters are frequently employed to
designate sets. Lowercase letters and numbers are often used to
designate the elements of a set. A set is also denoted by enclosing the
elements in braces { }. For example, if $A = \{a, b, c, d\}$, then A is a
set whose elements are a, b, c, and d. Furthermore, the notation
$B = \{1, 2, 3, \ldots, 99\}$ means that B is the set of natural numbers, or
the numbers used in counting, that are less than 100. Note that the
three dots between 3 and 99 indicate that the natural numbers

set

elements

between 3 and 99 are included in the set B. Another notation for a set that is frequently employed is illustrated by the following:

$$A = \{x \mid x \text{ is an even natural number less than } 5\}$$

The vertical bar is read "such that." It follows at once that $A = \{2, 4\}$. This is often referred to as set-building notation.

The notation $a \in A$ means that a is an element of the set A and is sometimes read "a belongs to A."

subset

proper subset

If each element of the set A is a member of the set B, then A is a *subset* of B, and we write $A \subseteq B$. If the set B contains at least one element that is not a member of A, then A is a *proper subset* of B, and this situation is indicated by $A \subset B$. For example, if $A = \{a, b, c, d\}$ and $B = \{a, b, c, d, e, f\}$ then $A \subset B$. If, however, $B = \{a, b, c, d\}$, then $A \subseteq B$.

identical sets

Two sets A and B are *identical* if and only if each is a subset of the other. For example, the sets $A = \{1, 2, 3, 4\}$ and $B = \{4, 2, 3, 1\}$ are identical. This situation is indicated by writing $A = B$.

null set

A set that contains no elements is called the *empty* or *null* set and is indicated by the symbol $\varnothing$.

Example 1 $\{x \mid x \text{ is a woman who has been president of the United States}\} = \varnothing$

Example 2 $\{x \mid x \text{ is a two-digit natural number less than } 10\} = \varnothing$

It frequently happens that the same set of elements belongs to each of two sets A and B. This set is called the intersection of A and B and is designated by $A \cap B$. More precisely we define the intersection below.

intersection of two sets

The *intersection* of two sets, $A \cap B$, is the set $\{x \mid x \in A \text{ and } x \in B\}$.

For example, if $A = \{x \mid x \text{ is a natural number less than } 10\}$ and $B = \{x \mid x \text{ is a natural number divisible by } 2\}$, then $A \cap B = \{2, 4, 6, 8\}$.

Obviously if two sets have no elements in common, their intersection is the null set $\varnothing$. For example, since no former governor of Texas has been a governor of California, then $\{x \mid x \text{ is a former governor of Texas}\} \cap \{x \mid x \text{ is a former governor of California}\} = \varnothing$.

disjoint sets

Two sets A and B are *disjoint* if $A \cap B = \varnothing$.

The concept of the intersection of sets can be extended to three or more sets. For example

$$A \cap B \cap C = \{x \mid x \in A \text{ and } x \in B \text{ and } x \in C\}$$

union

The *union* of two sets is denoted by $A \cup B$ and is defined to be the set of elements that belong to A or to B or to both A and B. In the symbolism of sets,

$$A \cup B = \{x \mid x \in A \text{ or } x \in B\}$$

Similarly,

$$A \cup B \cup C = \{x \mid x \in A \text{ or } x \in B \text{ or } x \in C\}$$

For example, if $A = \{1, 3, 5\}$, $B = \{2, 3, 6\}$, and $C = \{4, 5, 7, 9\}$, then $A \cup B = \{1, 2, 3, 5, 6\}$ and $A \cup B \cup C = \{1, 2, 3, 4, 5, 6, 7, 9\}$.

universal set   The totality of elements that are involved in any specific situation or discussion is called the *universal set* and is designated by U. For example, each of the various clubs, athletic teams, academic classes, and any other group whose members are students at a given college are subsets of the universal set composed of the entire student body of the college.

A method for picturing sets and certain relations between them was devised by an Englishman, John Venn (1834–1923). The fundamental idea is to represent a set by a simple plane figure. In order to illustrate the method, we shall use circles. We shall represent the universal set U by a circle C and shall define U to be the set of all points within and on the circumference of C. We shall represent the various subsets of U by circles wholly within the circle C. Figure 1.1 illustrates the device.

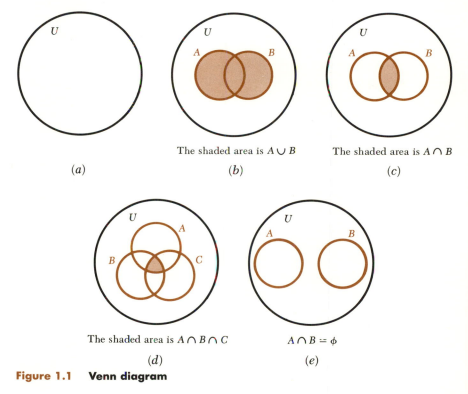

(a)

The shaded area is $A \cup B$

(b)

The shaded area is $A \cap B$

(c)

The shaded area is $A \cap B \cap C$

(d)

$A \cap B = \phi$

(e)

Figure 1.1   **Venn diagram**

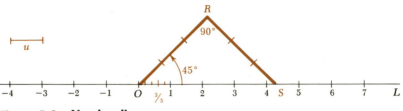

Figure 1.2   Number line

1.2   Real numbers

In elementary calculus we use the real number system almost exclusively. This system is defined in college algebra, and the numbers are interpreted by use of the number line such as line L shown in Fig. 1.2. In this figure the unit length u is laid off successively to the right and to the left of the point O on L. The positive integers 1, 2, 3, 4, 5, 6, . . . are associated with the successive right extremities of the intervals to the right of O and the negative integers -1, -2, -3, -4, -5, -6, . . . are associated with the left extremities of the intervals to the left of O. In order to obtain the point associated with $\frac{3}{5}$, we divide the interval from O to 1 into 5 equal parts and then associate $\frac{3}{5}$ with the right extremity of the third of these subintervals. We say that $\frac{3}{5}$ is the quotient of 3 and 5 and call it a *rational number*.

rational number

If we construct the right triangle ORS as indicated in the figure, the length of the line segment OS is $\sqrt{3^2 + 3^2} = \sqrt{18}$. Hence the point S is associated with $\sqrt{18}$. Furthermore, $\sqrt{18}$ cannot be expressed as the quotient of two integers.° We can, however, obtain a decimal representation of $\sqrt{18}$ by a repeated application of the square-root process of arithmetic. The process never terminates and the decimal fraction never becomes periodic. Thus we say that $\sqrt{18}$ is a nonterminating, nonperiodic decimal fraction. We call such numbers *irrational*.

irrational number

Each point on the line L is associated with one and only one number that is either an integer, a rational number, or an irrational number. Furthermore, each number is associated with one and only one point on the line. Since the number a is associated with only one point on L, we shall frequently refer to the number a as the point a.

As indicated in Fig. 1.3, we define the sum $a + b$ as the number associated with the point on L that is a distance of b to the right of the point a, and the sum $a + (-b)$ as the number associated with the point that is a distance of b to the left of a. It can be verified that the point $a + b$ is the same as $b + a$. We shall assume that this is true

° For proof of this statement see Rees and Sparks, College Algebra, 5th ed., McGraw-Hill Book Company, New York, 1967.

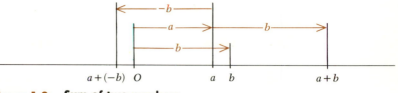

Figure 1.3   **Sum of two numbers**

for any two numbers; thus we have the axiom

$$a + b = b + a \tag{1}$$

The positive numbers are the numbers associated with the points to the right of O on L, and the negative numbers are those associated with the points to the left of O.

We say that $a > b$ if the point a is to the right of the point b on L and that $c < d$ if the point c is to the left of the point d.

By the above interpretation the point associated with the sum $a + (-a)$ is the point O. Furthermore the point $O + a$ is the point a. Heretofore, we have not called O a number, but have used the symbol to designate the reference point on L. We now define the number *zero* to be the number such that

$$\text{Zero} + a = a \tag{2}$$

and shall designate it with the symbol 0. It is consistent with the above reasoning to assign the number zero to the point O.

negative of a number   We shall now define the *negative* of the number a to be the number $-a$ such that

$$a + (-a) = 0 \tag{3}$$

As implied by the above discussion, the positive integers are the natural numbers, or the numbers used in counting. The negative integers are the negatives of the natural numbers, and 0 is the number such that $0 + a = a$. We are now in a position to define *the set of integers I* as follows:

the set of integers

$$I = \{p \mid p \text{ is a natural number}\} \cup \{0\} \cup \{n \mid n \text{ is the negative}$$
of a natural number$\}$

rational number   A *rational number* is a number that can be expressed as the quotient of two integers. Since the integer $n = n/1$, any integer is also a rational number. Hence, if J is the set of rational numbers, then $I \subset J$.

irrational number   An *irrational* number is a number whose decimal representation is a nonterminating, nonperiodic decimal. Such numbers cannot be expressed as the quotient of two integers. We shall represent the set of irrational numbers by K.

We now define the set of R of real numbers to be the union of the set of irrational numbers and the set of rational numbers. Hence we have

$$R = J \cup K$$

1.3 Inequalities

At times it is not sufficient to say that two numbers a and b are unequal. We can tell something about the way in which they are unequal by saying that a is greater than b or that a is less than b. This is done symbolically by writing $a > b$ or $a < b$. If a and b are real numbers, we define the inequalities $a > b$ and $a < b$ by saying that

$$a > b \qquad \text{if and only if} \qquad a - b \text{ is positive}$$

and $\qquad a < b \qquad$ if and only if $\qquad a - b$ is negative

sense of an inequality Two inequalities have the *same sense* or *opposite senses* according as the inequality signs in them point in the same or in opposite directions.

There are several laws or rules for use in work with inequalities. They include:

1. If $a > b$ and $b > c$, then $a > c$

2. If $a > b$, then $a + c > b + c$ for c any real number

3. If $a > b$, then $ac > bc$ for c any positive number

4. If $a > b$, then $ac < bc$ for c any negative number

Similar laws hold for operations with inequalities if the sense of each inequality in (1), (2), (3), and (4) is reversed. Thus, corresponding to (2), we have

5. If $a < b$, then $a + c < b + c$ for c any real number

Example 1 Prove that if $a < b$, then $a + c < b + c$ for c any real number.

Solution: The symbol $a < b$ means that

$$a - b < 0$$

Hence, $\qquad a - b + c + (-c) < 0 \qquad$ *since $c + (-c) = 0$*

and $\qquad (a + c) - (b + c) < 0 \qquad$ *rearranging and inserting parentheses*

Consequently, $\qquad a + c < b + c$

The other stated and indicated laws can be proved in a similar manner and some of the proofs will be called for in the next exercise.

open interval   The set of all numbers between two given numbers is called an *open interval*. Thus, all x such that $a < x < b$ is an open interval and consists of all the real numbers between a and b but does not include the numbers a and b. At times, we use

$$\{x \mid a < x < b\}$$

to indicate the open interval and we may use the more compact notation (a, b). If the numbers a and b are included as part of the interval, we write

$$\{x \mid a \leq x \leq b\} \qquad \text{or merely} \qquad [a, b]$$

closed interval   where $\leq$ is a symbol for "less than or equal to," and say that the interval is *closed*. An interval may be open at one end and closed at the other. The symbol $(a, b]$ is used to indicate that b is in the interval and a is not; furthermore, the symbol $[a, b)$ indicates that the interval is closed on the left and open on the right. The following symbols will also be used as defined:

$$(a, \infty) = \{x \mid x > a\}$$
$$[a, \infty) = \{x \mid x \geq a\}$$
$$(-\infty, a) = \{x \mid x < a\}$$
$$(-\infty, a] = \{x \mid x \leq a\}$$
$$(-\infty, \infty) = \{x \mid x \text{ is real}\}$$

solution set of an inequality   The *solution set of an inequality* consists of the set of numbers for which the inequality is a true statement. If two inequalities have the same solution set, they are said to be equivalent.

Example 2   Find the solution set of $3x - 8 < x - 2$.

Solution: The procedure here will be very similar to that followed in solving a linear equation. We first obtain an equivalent inequality in which the unknown and its coefficient are in one member and the other terms are in the other by adding $8 - x$ to each member as justified by (2). Thus, the equivalent inequality is

$$3x - 8 - x + 8 < x - 2 - x + 8$$

$$2x < 6 \qquad\qquad \textit{collecting}$$

$$x < 3 \qquad\qquad \textit{multiplying by } \tfrac{1}{2}$$

Consequently, the solution set is $\{x \mid x < 3\}$.

Example 3 Find the set of real numbers such that $\sqrt{12 - 3x}$ is a real number.

Solution: The square root of a real number is real if and only if the radicand is greater than or equal to zero. Therefore, we want the set of numbers x such that

$$12 - 3x \geq 0$$

We solve this inequality as follows:

$$12 - 3x - 12 \geq -12 \qquad \text{\textit{by (2)}}$$

$$-3x \geq -12 \qquad \text{\textit{collecting}}$$

$$x \leq 4 \qquad \text{\textit{by (4) with } } c = -\tfrac{1}{3}$$

Consequently, $\sqrt{12 - 3x}$ is a real number if and only if x is in $(-\infty, 4]$.

1.4 Inequalities that involve absolute values

In this section, we shall consider inequalities that involve absolute values. Consequently, we must recall what is meant by the absolute value of a real number. *If N is positive or zero, its absolute value is N; if N is negative, its absolute value is $-N$.* It is customary to denote the absolute value of N by $|N|$. Therefore, $|7| = 7$ and $|-5| = -(-5) = 5$.

If we use the definition of the absolute value of a number, we see that an inequality of the type

$$|ax + b| < c \qquad c > 0 \tag{1}$$

requires that $ax + b$ shall be a number that is between $-c$ and c; hence, (1) is satisfied if $ax + b > -c$ and simultaneously $ax + b < c$. Thus, (1) is satisfied by all x in the intersection of the solution sets of $ax + b > -c$ and $ax + b < c$; that is, by $\{x \mid ax + b > -c\} \cap \{x \mid ax + b < c\}$.

Example 1 Solve $|3x - 2| < 10$.

Solution: The given inequality is satisfied if and only if both

$$3x - 2 > -10 \qquad \text{and} \qquad 3x - 2 < 10$$

are satisfied. By adding 2 to each member of each of these inequalities and dividing by 3, we get

$$x > -\tfrac{8}{3} \qquad \text{and} \qquad x < 4$$

Therefore the given inequality is satisfied by all x such that $-\tfrac{8}{3} < x < 4$. Hence, the solution set is all x in $(-\tfrac{8}{3}, 4)$.

By use of the definition of the absolute value of a number, an inequality of the type

$$|rx + s| > c \qquad c > 0$$

is satisfied by all x such that $rx + s > c$ and also by all x such that $rx + s < -c$; hence, by each element in the union of these sets.

Example 2 Solve $|2x - 5| > 3$.

Solution: The given inequality is satisfied if $2x - 5 > 3$ and also if $2x - 5 < -3$. Now adding 5 to each member of each of these inequalities and dividing by 2, we see that the given inequality is satisfied if $x > 4$ and if $x < 1$; hence, the solution set is $\{x \mid x > 4\} \cup \{x \mid x < 1\}$.

1.5 Bounds

If there is a number b that is greater than or equal to every member of a set, then the set is said to be *bounded above* and b is called an *upper bound*. The set $\{x \mid 0 < x \le 1\}$ is bounded above and 1 is an upper bound; furthermore, any number greater than 1 is an upper bound of the given set since it is greater than any element of the set. If there is a number U that is an upper bound of a set and if no number less than U is an upper bound, then U is called the *least upper bound* of the set, which is abbreviated as l.u.b. It is the smallest number that is an upper bound of the set. If x is a number such that $0 < x \le 1$, then 1 is not only an upper bound but the least upper bound.

If there is a number a that is less than or equal to every member of a set, then the set is said to be *bounded below* and a is called a *lower bound*. Furthermore, if there is a number L that is a lower bound of a set and if no number greater than L is a lower bound, then L is called the *greatest lower bound* of the set, which is abbreviated as g.l.b. We shall assume that *every set that has a lower bound has a g.l.b.* and that *every set that has an upper bound has a l.u.b.* These two assumptions are called the *completeness axiom.*

upper bound

least upper bound

lower bound

greatest lower bound

completeness axiom

EXERCISE 1.1

1 Show that $A = \{x \mid x$ is a positive integer divisible by 7$\}$ is a proper subset of $B = \{x \mid x$ is a nonnegative integer divisible by 7$\}$.

2 Show that $A = \{x \mid x$ is an integer divisible by 7$\}$ is a proper subset of $B = \{x \mid x$ is an integer$\}$.

3   Show that $A = \{x \mid x$ is an integer divisible by 5$\}$ is not a subset of $B = \{x \mid x$ is divisible by 2$\}$.

4   Show that $A = \{x \mid x$ is divisible by 3$\}$ and $B = \{x \mid$ the sum of the digits in x is divisible by 3$\}$ are identical.

In Problems 5 to 8, give a set B that has A as a subset.

5 $A = \{x \mid x$ is a positive rational number$\}$

6 $A = \{x \mid x$ is an odd integer$\}$

7 $A = \{x \mid x$ is divisible by 4$\}$

8 $A = \{x \mid x$ is an integral power of 2$\}$

Find the union and intersection of sets A and B in each of Problems 9 to 12.

9 $A = \{1, 2, 3, 5\}, B = \{3, 4, 5\}$

10 $A = \{2, 4, 6, 7\}, B = \{7, 8, 13\}$

11 $A = \varnothing, B = \{x \mid x$ is rational$\}$

12 $A = \{0\}, B = \{x \mid x < 4,$ positive, and integral$\}$

If $A = \{-3, -1, 0, 2\}, B = \{-1, 1, 4\},$ and $C = \{-1, 2\},$ find the set called for in each of Problems 13 to 16.

13 $A \cup B \cup C$ **14** $(A \cap B) \cap C$

15 $(A \cap B) \cup C$ **16** $(A \cup B) \cap C$

17   Prove that if $a < b$ and $b < c$, then $a < c$.

18   Prove that if $a < b$, then $a + c < b + c$ for c any real number.

19   Prove that if $a < b$, then $ac < bc$ for c any positive number.

20   Prove that if $a < b$, then $a/c > b/c$ for c any negative number.

Solve the inequalities given in Problems 21 to 28.

21 $3x + 5 > x - 3$ **22** $7 - 4x \geq 2x - 5$

23 $5x - 9 < 6 + 2x$ **24** $11 - 6x \leq x - 3$

25 $|2x - 1| \leq 5$ **26** $|3 - 4x| < 9$

27 $|3x + 2| > 4$ **28** $|5x - 6| > 1$

Find the set of values of x for which the indicated root in each of Problems 29 to 32 is a real number.

29 $\sqrt{3x - 5}$ **30** $\sqrt{2 - 5x}$

31 $\sqrt{x^2 - 9}$ **32** $\sqrt{4 - x^2}$

Prove the statements concerning inequalities given in Problems 33 to 40.

33   If $y \neq x$ and $xy > 0$, then $x/y + y/x > 2$.

34   If $0 < x < y$, then $(x + 1)/(y + 1) > x/y$.

35   If $x > 0, y > 0,$ and $x \neq y$, then $(x + y)/2 > (xy)^{1/2}$. The last inequality in this problem states that the arithmetic mean of two unequal positive numbers is greater than their geometric mean.

36   If a and b are unequal positive numbers and n is a positive integer, then a^n and b^n are unequal in the same sense as a and b; furthermore, $\sqrt[n]{a}$ and $\sqrt[n]{b}$ are unequal in that sense.

37   If x is a rational approximation to $\sqrt{2}$, then $(x + 2)/(x + 1)$ is a better one. *Hint:* Show that $(x + 2)/(x + 1) = 1 + 1/(x + 1)$; then show that $x < 1 + 1/(x + 1)$ if $x < \sqrt{2}$ and that $x > (x + 2)/(x + 1)$ if $x > \sqrt{2}$.

38   If $0 < x < y$, then $x < \sqrt{xy} < y$.

39   If r, s, and t are distinct positive numbers, then $r + s + t > \sqrt{3(rs + rt + st)}$. *Hint:* Square each member, collect, replace s by $r + p$ and t by $r + kp$, $p > 0$, $k > 1$.

40   If n is a positive integer, then $n! \geq n^{n/2}$.

Prove each of the following theorems.

41 $|a + b| \leq |a| + |b|$

42 $|a - b| \geq |a| - |b|$

43 $|ab| = |a||b|$

44 $\left|\dfrac{a}{b}\right| = \dfrac{|a|}{|b|}$

1.6   Constants and variables

In this section, we shall consider fixed numbers, variable numbers, and relations between numbers. Symbols are used to represent numbers. A symbol that represents only one number throughout a discussion is *constant* called a *constant*. Thus, 7, 3.48, and 2π are constants. Furthermore, a symbol such as a, c, k, x, and y that is used to represent a number *may* be a constant and is one if it represents a definite, even though unspecified number. If a symbol is regarded as being replaceable by *variable* any one of a set of numbers, it is called a *variable*. If $V = \pi r^2 h$ represents the volume of water in a right circular cylinder where r denotes the cylinder's radius and h denotes the height of the water, then for a given cylinder, r is a constant and V and h are variables. However, since there are cylinders of varying radius, we think of r as a variable when considering all cylinders. Thus, we see that a symbol may be a constant in one discussion and a variable in another.

1.7   Functions

We shall now consider the function concept. Its essential characteristics are a set D of numbers x and a rule which determines exactly one *domain* corresponding number y for each x in D. The set D is called the *domain* of definition of the function, and the set R of all the y's is referred to *range* as the *range*. The range is determined if the domain and rule are given. We shall define the function concept by saying that:

1.   If D is a set of numbers

2. If for each x in D, there is a rule that specifies exactly one number y

3. If R is the set of all of the numbers y

function

then the set of all ordered pairs (x, y) such that $x \in D$ and y is the corresponding element in R is called a *function* with domain D and range R.

We frequently say "y is a function of x" to express the fact that there exists a rule by which y is determined uniquely when x is specified. Thus we say that the area of a circle is a function of its radius.

We often use either the letter y or the letter F as the name of a function, and we use the symbol $f(x)$ to denote the numbers paired with x. Thus, if $y = 2x + 3$ is the defining relation (that is, the rule) for the function, we write $f(x) = 2x + 3$ or $y = f(x) = 2x + 3$. Consequently, $f(1) = 2(1) + 3 = 5$, $f(2) = 2(2) + 3 = 7$ and $(1, 5)$ and $(2, 7)$ are two of the ordered pairs that belong to the function. The function consists of all ordered pairs $(x, f(x))$ such that $f(x) = 2x + 3$ and $x \in D$.

It has been customary for some time to write $y = f(x)$ as an abbreviation of the statement "y is a function of x" and of $\{(x, y) \mid y = f(x)\}$, and we shall use this notation from time to time.

Example 1

If D is the set of all real numbers x such that $1 \leq x \leq 10$ and if $y = f(x) = 3x + 2$, then corresponding to $x = 1$ we have $y = f(1) = 5$, corresponding to $x = \frac{5}{3}$ we have $y = f(\frac{5}{3}) = 7$, and corresponding to any x such that $1 \leq x \leq 10$ we have $y = f(x) = 3x + 2$. Hence the set of all ordered pairs (x, y) is called a function with domain D and range the set of all corresponding values of $y = f(x) = 3x + 2$. The range is readily seen to be all numbers y such that $5 \leq y \leq 32$ or $\{y \mid 5 \leq y \leq 32\}$.

Example 2

It may be observed that if D is the set of all numbers x such that $0 \leq x \leq 5$ and the rule is $x^2 + y^2 = 25$, there are two y's to pair with each x in D. Thus, corresponding to $x = 3$ we have $y = 4$ or -4. The resulting set of ordered pairs is sometimes called a "double-

relation

valued function," but most writers now prefer to call it a *relation* and to reserve the word function for the case in which there is exactly one y corresponding to each x. In the present example, either $y = +\sqrt{25 - x^2}$ or $y = -\sqrt{25 - x^2}$ would yield a function. If $y = +\sqrt{25 - x^2}$ is used, the function is

$$\{(x, y) \mid y = \sqrt{25 - x^2}, -5 \leq x \leq 5\}$$

Although a function is not completely determined unless its domain is specified, we may omit this specification if the domain D is all real

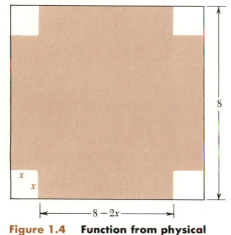

Figure 1.4   Function from physical considerations

numbers x for which the rule yields exactly one real number y for each x in D. Thus, it will be understood that the domain of "the function $y = \sqrt{4 - x^2}$" is $-2 \leq x \leq 2$ if nothing to the contrary is specified. If the rule were $y = 3x + 2$, it would similarly be understood that x may be any real number.

When a function arises from a physical problem, its domain is often determined by the physical conditions involved. Suppose, for example, a box is made from an 8 by 8 in. sheet of tin by cutting an x by x inch square from each corner and turning up the sides as in Fig. 1.4, and suppose we wish to investigate the way in which the volume v of the box obtained varies with x. Since $v = $ (area of base) (height), the rule giving v for any permissible value of x is $v = x(8 - 2x)^2$, and that x must be a number between 0 and 4. Hence, the function is $\{(x, v) \mid v = x(8 - 2x)^2\}$, and the domain is the set of all numbers x in $(0, 4)$. The range is the set of corresponding values of v.

1.8   The inverse of a function

By definition, no two ordered pairs in the function $f = \{(x, f(x))\}$ have the same first element. If the function is such that no two of the ordered pairs have the same second element, then the function has an inverse. Before defining the inverse of a function we shall illustrate the ideas involved.

We shall consider the functions $f = \{(x, 3x + 2)\}$ and $g = \{(x, (x - 2)/3)\}$ that are determined by the equations $y = f(x) = 3x + 2$ and $y = g(x) = (x - 2)/3$, respectively. Now if a is an element in the domain of f, then $(a, 3a + 2)$ is an ordered pair of numbers in the function f. Furthermore, if we replace x in g by $3a + 2$, we obtain $\left(3a + 2, \dfrac{3a + 2 - 2}{3}\right) = (3a + 2, a)$. Hence, if we interchange the elements in an ordered pair in f, we obtain an ordered pair in g. Moreover, $f[g(x)] = 3(x - 2)/3 + 2 = x$. We call attention to the fact that the equation $y = g(x)$ is obtained by solving $y = f(x)$ for x in terms of y, and then interchanging x and y. The function g is *inverse of a function* called the *inverse* of the function f, and it illustrates the following definition:

If no two ordered pairs in the function $\{x, f(x)\}$ have the same second element and if there exists a function $\{x, g(x)\}$ such that $f[g(x)] = x$, then g is the *inverse* of the function f.

We shall now show that this definition leads to the conclusion that each of the ordered pairs in g is an ordered pair in f with the elements interchanged. Now if a is an element in the domain of g and $g(a) = b$, then (a, b) is an ordered pair in g. Furthermore, if we replace x by b in $\{x, f(x)\}$, we have

$$\{x, f(x)\} = \{b, f(b)\}$$

$$= \{b, f(g(a))\} \qquad \textit{since } b = g(a)$$

$$= (b, a) \qquad \textit{since } f(g(a)) = a$$

Hence if (a, b) is an ordered pair in g, then (b, a) is an ordered pair in f. It follows that the domain of f is the range of g and the range of f is the domain of g.

It is customary to designate the inverse of f by f^{-1}.

Example 1   If the function f is the set of ordered pairs

$$\{(1, 5), (2, 6), (3, 7), (4, 8)\}$$

then its domain is $\{1, 2, 3, 4\}$ and its range is $\{5, 6, 7, 8\}$. Furthermore, the inverse f^{-1} is

$$\{(5, 1), (6, 2), (7, 3), (8, 4)\}$$

since this is the set obtained by interchanging first and second elements in f.

Example 2   Find the inverse of the function

$$\{(x, y) \mid y = f(x) = x^2 - 4, x \geq 0\} \tag{1}$$

Solution: For this function, the domain D is $\{x \mid x \geq 0\}$ and the range R is $\{y \mid y \geq -4\}$. To find the inverse, we begin by solving $y = x^2 - 4$ for x. We obtain $x = \pm\sqrt{y + 4}$ but we use only $x = +\sqrt{y + 4}$ since $x \geq 0$. We next interchange x and y and have

$y = \sqrt{x + 4}$. Consequently, the inverse of the given function is

$$\{(x, y) \mid y = f^{-1}(x) = \sqrt{x + 4}, x \geq -4\}$$

We can see that this is the inverse of the given function since

$$f[f^{-1}(x)] = f(\sqrt{x + 4}) = x + 4 - 4 = x$$

and $\qquad f^{-1}[f(x)] = f^{-1}(x^2 - 4) = \sqrt{x^2 - 4 + 4} = x$

Furthermore, the domain and range of f are the range and domain, respectively, of f^{-1}. We can show similarly that

$$\{(x, y) \mid y = G(x) = -\sqrt{x + 4}, x \geq -4\}$$

is the inverse of

$$\{(x, y) \mid y = F(x) = x^2 - 4, x \leq 0\}$$

EXERCISE 1.2

In each of Problems 1 to 12, find the domain and range.

1   $y = f(x) = 2x - 3$ 2 $y = f(x) = 3x + 1$
3   $y = f(x) = -x + 4$ 4 $y = f(x) = -2x - 5$
5   $y = f(x) = (x - 1)^2$ 6 $y = f(x) = (x + 2)^2$
7   $y = f(x) = x^2 - 2x + 3 = (x - 1)^2 + 2$
8   $y = f(x) = x^2 - 6x + 5$ 9 $y = f(x) = \sqrt{x}$
10 $y = f(x) = \sqrt{2x - 1}$ 11 $y = f(x) = \sqrt{2 - x}$
12 $y = f(x) = -\sqrt{5 - 2x}$

Find the inverse of the function in each of Problems 13 to 16, and find the domain and range of the inverse.

13 $\{(x, y) \mid y = f(x) = 3x - 6\}$
14 $\{(x, y) \mid y = f(x) = x^2 + 1, x \geq 0\}$
15 $\{(x, y) \mid y = f(x) = \sqrt{x + 3}, x \geq -3\}$
16 $\{(x, y) \mid y = f(x) = x^2 + 2x + 5, x \geq -1\}$

17   The local taxi company charges at the rate of 40 cents per mile and has a minimum charge of 50 cents. Express the fare as a function of the distance.

18   A TV repair man has a service charge of $5 per call and a charge of $2.75 per tube. Express his charge as a function of the number of tubes replaced.

19   The minimum monthly water bill in a college town is $3, and 6,000 gal is allowed for this charge. There is an extra charge of 30 cents for each 1,000 gal used beyond the minimum. Express the amount of the bill in dollars as a function of the number of gallons used beyond the minimum.

20   The monthly rate for electricity in a town is:

5 cents per kw-hr for the first 30 kw-hr with a minimum charge of $1.20
   3 cents per kw-hr for the next 170 kw-hr
   1 cent per kw-hr for all beyond 200 kw-hr

Express the amount of the bill as a function of the number of kilowatt-hours used.

21   If n is an integer greater than 1, let $f(n)$ represent the largest prime factor of n. Does this relation satisfy the requirements of a function?

22   If n is a positive integer, let $\phi(n)$ represent the number of positive integers that are not greater than n and that are prime relative to n. Does this relation satisfy the requirements of the definition of a function?

23   Let x represent any real number, and let $f(x)$ be the largest integer that is not greater than x. Does this satisfy the definition of a function?

24   Let x represent any real number, and let $f(x)$ be the smallest integer that is greater than x. Does this relation satisfy the definition of a function?

25   A box is to be made from a 6 by 8 in. rectangular piece of cardboard by cutting squares of equal size from each corner and then turning up the edges. Find the volume of the box as a function of the edge of the square, and give the domain.

26   A box is to be made from an 8 by 10 in. rectangular piece of tin by cutting a $4x$ by $5x$ in. rectangle from each corner with the $5x$ in. along the longer side. Express the volume of the box as a function of x, and find the domain.

27   A rectangular piece of tin that is 12 in. wide is folded down the middle so as to form a triangular cross section that is $2x$ in. across the base. Express the area of the cross section as a function of x, and find the domain.

28   A rectangular gutter without top is to be made from a piece of metal that is 10 in. wide by turning up the sides. Express the cross-sectional area as a function of the height h of the gutter, and find the domain.

2 Lines, circles, rational functions

2.1 Directed line segments

No distinction was made in plane geometry between the line segments *AB* and *BA* since we were then interested only in the length of a segment. It, however, is desirable at times to consider both length and direction.

If a positive direction has been selected on a segment, we shall call the segment a *directed line segment* and shall call its length a directed distance. The positive direction is often indicated by placing an arrow at one end of the segment as in Fig. 2.1, but it may also be indicated by the order in which the letters are written. In Fig. 2.1, the arbitrarily chosen positive direction is from left to right. Consequently, from right to left is the negative direction, since the direction opposite to the chosen positive one is considered to be negative. Therefore, we write $BA = -AB$.

A ———————————————→ B

Figure 2.1   Directed line segment

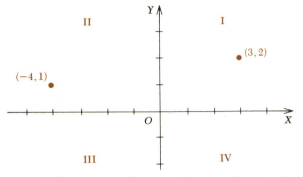

Figure 2.2   Rectangular cartesian-coordinate system

2.2   The cartesian-coordinate system

coordinate axes

quadrant

abscissa

ordinate

ordered pair

We use a pair of perpendicular directed lines as the frame of reference for the rectangular cartesian-coordinate system. These lines, shown in Fig. 2.2, are called the *coordinate axes*. The horizontal line is the *X axis*, the vertical line is the *Y axis*, and their intersection is the *origin*. The plane in which the coordinate axes are drawn is called a cartesian plane. The four parts into which the plane is divided by the coordinate axes are called the four *quadrants*. They are numbered from I to IV as in Fig. 2.2. The location of a point is determined by giving its distance and direction from each of the coordinate axes. The directed distance from the *Y* axis to the point is called the *abscissa* of the point and the directed distance from the *X* axis is known as the *ordinate* of the point. These two numbers are called the *coordinates* of the point; they are written in parentheses, separated by a comma, and the abscissa is always written first. Consequently, the coordinates are an *ordered pair* of numbers. If a point is to the right of the *Y* axis, its abscissa is positive; if to the left of the *Y* axis, its abscissa is negative. Furthermore, the ordinate of a point is positive or negative according as the point is above or below the *X* axis. The point $(3, 2)$ is 3 units to the right of the *Y* axis and 2 units above the *X* axis. It is shown in Fig. 2.2, as is the point $(-4, 1)$.

2.3   The distance between two points

There will be many occasions on which we shall need to know the distance between two points, and we shall now develop a formula for use at such times. In doing so, we shall designate the points by $P_1(x_1, y_1)$ and $P_2(x_2, y_2)$ as indicated in Fig. 2.3. The point $P_3(x_1, y_2)$ is the intersection of the line through P_1 parallel to the *Y* axis and the

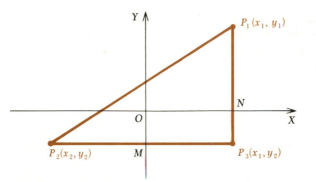

Figure 2.3   The distance between two points

line through P_2 parallel to the X axis; hence, its coordinates are as shown. We first notice that

$$P_2P_3 = P_2M + MP_3 = MP_3 - MP_2 = x_1 - x_2$$

and $\qquad P_3P_1 = P_3N + NP_1 = NP_1 - NP_3 = y_1 - y_2$

Using the Pythagorean theorem, we have

$$(P_1P_2)^2 = (P_2P_3)^2 + (P_3P_1)^2 = (x_1 - x_2)^2 + (y_1 - y_2)^2$$

Therefore, $P_1P_2 = \sqrt{(x_1 - x_2)^2 + (y_1 - y_2)^2}$

For other relative positions of P_1 and P_2, we obtain the same formula for P_1P_2 as the one above even though $x_1 - x_2$ or $y_1 - y_2$ or both are replaced by their negatives, because the square of a number *distance* and of its negative are equal. We may now say that *the length of the* *formula* *segment which connects* (x_1, y_1) *and* (x_2, y_2) *is*

$$P_1P_2 = \sqrt{(x_1 - x_2)^2 + (y_1 - y_2)^2} \qquad (2.1)$$

Example   Find the distance between $(5, -1)$ and $(2, 3)$.

Solution: If we use $(5, -1)$ as P_1, we get

$$P_1P_2 = \sqrt{(5 - 2)^2 + (-1 - 3)^2} = 5$$

2.4   Division of a line segment into a given ratio

We shall now find the coordinates of the point $P(x, y)$ on the segment that connects $P_1(x_1, y_1)$ and $P_2(x_2, y_2)$ such that

$$\frac{P_1P}{PP_2} = \frac{r_1}{r_2} \qquad (1)$$

where r_1 and r_2 are any two real numbers different from zero. In order to find the value of x, we shall draw a line segment through P_1 and P_2, locate $P(x, y)$ on it, and drop perpendiculars P_1Q_1, PQ, and P_2Q_2 to the X axis. The coordinates of the feet of these perpendiculars are as shown in Fig. 2.4. Then

$$\frac{r_1}{r_2} = \frac{P_1P}{PP_2} = \frac{Q_1Q}{QQ_2} \tag{2}$$

since if three parallels are cut by transversals, the segments into which the transversals are divided are proportional. Since Q_1Q and QQ_2 are on the X axis, we have, $Q_1Q = x - x_1$ and $QQ_2 = x_2 - x$. Consequently,

$$\frac{Q_1Q}{QQ_2} = \frac{x - x_1}{x_2 - x}$$

and, by (2)

$$\frac{x - x_1}{x_2 - x} = \frac{r_1}{r_2}$$

Finally, solving this equation for x, we have

point of division
$$x = \frac{r_2x_1 + r_1x_2}{r_1 + r_2} \qquad r_1 + r_2 \neq 0 \tag{2.2a}$$

Similarly, by drawing perpendiculars to the Y axis, we can obtain

$$y = \frac{r_2y_1 + r_1y_2}{r_1 + r_2} \qquad r_1 + r_2 \neq 0 \tag{2.2b}$$

We derived these formulas for P between P_1 and P_2, but we would obtain the same formulas for x and y with P_1 between P and P_2 or with P_2 between P and P_1. In these two cases r_1 and r_2 have opposite signs, that is, $r_1/r_2 < 0$.

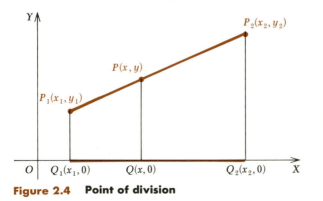

Figure 2.4   Point of division

There is a special case of the point-of-division formula that is often needed. It is the formula obtained if $r_1 = r_2$ and is called the *midpoint formula*. If we let $r_2 = r_1$ in (2.2a) and (2.2b) and take out the common factor r_1, we find that

$$x = \frac{x_1 + x_2}{2} \quad \text{and} \quad y = \frac{y_1 + y_2}{2} \quad (2.2')$$

midpoint *are the coordinates of the midpoint of the segment that connects* *formula* $P_1(x_1, y_1)$ *and* $P_2(x_2, y_2)$.

Example 1   Find the coordinates of the point $P(x, y)$ on the segment that connects $P_1(3, -4)$ and $P_2(10, 1)$ such that $P_1P/PP_2 = \frac{5}{2}$.

Solution: Since the ratio of P_1P to PP_2 is to be 5 to 2, we use $r_1 = 5$ and $r_2 = 2$ in the point-of-division formula. Thus, we get

$$x = \frac{2(3) + 5(10)}{2 + 5} = \frac{6 + 50}{7} = 8$$

and

$$y = \frac{2(-4) + 5(1)}{7} = -\frac{3}{7}$$

Consequently, the desired point is $(8, -\frac{3}{7})$.

Example 2   Find the midpoint of the segment that joins $P_1(4, 5)$ and $P_2(-1, 9)$.

Solution: If we substitute the given coordinates in the midpoint formulas, we get

$$x = \frac{4 + (-1)}{2} = \frac{3}{2} \quad \text{and} \quad y = \frac{5 + 9}{2} = 7$$

Hence, $(\frac{3}{2}, 7)$ is the midpoint of the segment.

EXERCISE 2.1

1   Find the directed distance from A to B and from B to A if the coordinates of A and B are $(3, 4)$ and $(3, y)$, respectively.

2   If we have $A(4, 7)$ and $B(x, 7)$ with x negative, is the directed distance BA equal to $x + 4$, $4 - x$, or $x - 4$?

3   If we have $C(2, y)$, $y > 0$, and $D(2, -3)$, is the directed distance CD equal to $y + 3$, $y - 3$, or $-3 - y$?

4   Which of the following represent the undirected distance between $A(3, 7)$ and $B(x, 7)$ if $x > 3$: $x - 3$, $3 - x$, $|x - 3|$, $\sqrt{(3 - x)^2}$.

5   Show that the triangle with vertices at $(2, 2)$, $(4, -2)$, and $(-4, 6)$ is a right triangle.

6   Show that the triangle with vertices at $(2, -2)$, $(7, 0)$, and $(5, 5)$ is a right triangle.

7   Is the quadrilateral with vertices at $A(1, 2)$, $B(7, 6)$, $C(11, 0)$, and $D(5, -4)$ a rectangle?

8   Is the quadrilateral with vertices at $A(-2, 1)$, $B(4, 5)$, $C(8, -1)$, and $D(2, -5)$ a rectangle?

9   Show that the triangle with vertices at $A(1, 1)$, $B(9, -7)$, and $C(0, -8)$ is isosceles.

10   Show that the circle that passes through $(-2, 0)$, $(-3, 7)$, and $(-2, 8)$ has its center at $(1, 4)$.

11   Show that $(-3, -7)$ is the center of a circle that passes through $(9, -2)$, $(2, 5)$, and $(-3, 6)$.

12   Find the center of the circle that is circumscribed about the triangle with vertices at $(5, 3)$, $(5, -7)$, and $(3, 7)$.

13   Find the coordinates of the third vertex of an equilateral triangle that has vertices at $(-1, 4)$ and $(9, -6)$.

14   Find the point on the X axis that is equidistant from $(-2, 9)$ and $(16, 1)$.

15   Find x so that (x, x) is equidistant from $(-3, 6)$ and $(5, -7)$.

16   Find x so that $(x, 5)$ is equidistant from $(2, -1)$ and $(-4, -2)$.

17   Determine x so that $(x, -3)$ is twice as far from $(6, 1)$ as from $(0, -1)$.

18   Find x so that $(x, 2x)$ is equidistant from $(-1, 2)$ and $(10, -1)$.

19   Find x so that $(x, x + 4)$ is equidistant from $(0, 5)$ and $(8, -1)$.

20   Find y so that $(y - 1, y)$ is twice as far from $(2, 3)$ as from $(-4, 0)$.

In each of Problems 21 to 28, find the coordinates of $P(x, y)$ such that $P_1P/PP_2 = r_1/r_2$.

21    $P_1(-4, 1)$, $P_2(6, -9)$, $r_1 = 2$, $r_2 = 3$
22    $P_1(8, -3)$, $P_2(-4, 5)$, $r_1 = 3$, $r_2 = 4$
23    $P_1(7, 1)$, $P_2(-9, 7)$, $r_1 = 3$, $r_2 = 5$
24    $P_1(13, -6)$, $P_2(-2, 4)$, $r_1 = 5$, $r_2 = 4$
25    $P_1(4, -3)$, $P_2(10, 5)$, $r_1 = 2$, $r_2 = -1$
26    $P_1(-9, 2)$, $P_2(3, 1)$, $r_1 = 4$, $r_2 = -3$
27    $P_1(5, 6)$, $P_2(7, -1)$, $r_1 = -2$, $r_2 = 3$
28    $P_1(8, -7)$, $P_2(-1, 2)$, $r_1 = -5$, $r_2 = 2$

Find the midpoint of the segment that connects the two points in each of Problems 29 to 32.

29    $(3, 4)$, $(7, 0)$ 30 $(5, -2)$, $(-3, 4)$
31    $(2, 3)$, $(5, -1)$ 32 $(4, -7)$, $(-1, 6)$

33   Find the coordinates of the points that divide the segment between $P_1(-6, -3)$ and $P_2(3, 0)$ into three equal parts.

34   Find the coordinates of the points that divide the segment that connects $P_1(8, -3)$ and $P_2(-4, 7)$ into four equal parts.

35   Find the coordinates of the point on each median of a triangle that is two-thirds of the way from the vertex to the midpoint of the opposite side if the vertices are at $P_1(-3, -4)$, $P_2(6, -2)$, and $P_3(3, 8)$.

36   Show that the point on a median of a triangle with vertices at (x_1, y_1), (x_2, y_2), and (x_3, y_3) that is two-thirds of the way from the vertex to the midpoint of the opposite side is

$$\left(\frac{x_1 + x_2 + x_3}{3}, \frac{y_1 + y_2 + y_3}{3}\right)$$

2.5   Inclination and slope

If a line l is not parallel to the X axis, it intersects that axis at some point Q as indicated in Fig. 2.5. The direction of the line relative to the X axis is determined by the magnitude of smallest counterclockwise angle α through which QX must be rotated so as to bring it into coincidence with the line l. This angle is called the *inclination* of l. The inclination of a line parallel to the X axis is zero by definition. The tangent of the angle of inclination of a line is called the *slope* of the line and is designated by m; hence $m = \tan \alpha$, $\alpha \neq 90°$.

inclination

slope

We shall now find an expression for the slope of a line in terms of the coordinates of two points on it. If the points are $P_1(x_1, y_1)$ and $P_2(x_2, y_2)$ as shown in Fig. 2.6a and b, then by the definition of the tangent of an angle, we get

$$\tan \alpha = \frac{MP_2}{P_1M} = \frac{y_2 - y_1}{x_2 - x_1} \qquad x_2 \neq x_1$$

Therefore, *if $P_1(x_1, y_1)$ and $P_2(x_2, y_2)$ are any two points on a line not perpendicular to the X axis, then the slope of the line is*

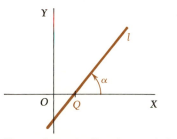

Figure 2.5   Inclination and slope

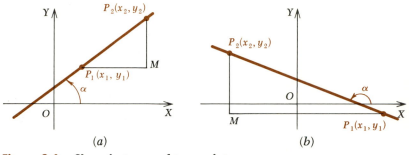

(a)   (b)

Figure 2.6   Slope in terms of two points

slope formula

$$m = \tan \alpha = \frac{y_2 - y_1}{x_2 - x_1} \tag{2.3}$$

If the line is perpendicular to the X axis, its inclination is 90° and there is no number to represent its slope since $\tan 90°$ does not exist.

2.6   Parallel and perpendicular lines

We shall consider two lines, neither of them parallel to the Y axis. If the lines are parallel as in Fig. 2.7, their inclinations α_1 and α_2 are equal; hence, $\tan \alpha_1 = \tan \alpha_2$. Conversely, if the slopes are equal, *parallel* then $\alpha_1 = \alpha_2$ and the lines are parallel. Thus, we see that *two* *lines* *lines that have slopes are parallel if their slopes are equal, and conversely.*

If the lines l_1 and l_2 are perpendicular, their inclinations differ by 90°; that is, $\alpha_1 = \alpha_2 + 90°$ or $\alpha_2 = \alpha_1 + 90°$ as in Fig. 2.8.

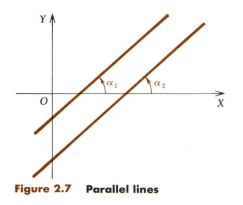

Figure 2.7   Parallel lines

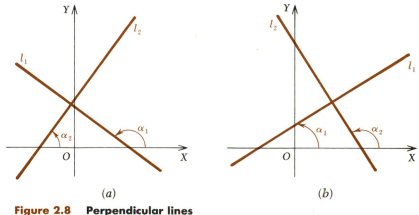

(a)   (b)

Figure 2.8   Perpendicular lines

In each case tan α_1 = tan $(\alpha_2 \pm 90°)$ = $-\cot \alpha_2$ = $-1/\tan \alpha_2$; hence,

$$m_1 = -\frac{1}{m_2} \quad \text{or} \quad m_1 m_2 = -1$$

since tan $\alpha_1 = m_1$ and tan $\alpha_2 = m_2$.

Conversely, if $m_1 = -1/m_2$,

$$\tan \alpha_1 = -\cot \alpha_2 \quad \text{and} \quad \alpha_1 = \alpha_2 \pm 90°$$

perpendicular lines   and it follows that the lines are perpendicular. Consequently, *two lines with slopes m_1 and m_2 are mutually perpendicular if and only if*

$$m_1 = -\frac{1}{m_2} \quad \text{or} \quad m_1 m_2 = -1 \tag{2.4}$$

The only case not covered by the theorem is that of two lines parallel to the X and Y axes, respectively. They are of course mutually perpendicular; the slope of the X axis is zero, and the Y axis has no slope.

2.7   Angle from one line to another

angle from one line to another   We shall let l_1 and l_2 be two lines that intersect at P and have slopes $m_1 = \tan \alpha_1$ and $m_2 = \tan \alpha_2$, respectively. The smallest counterclockwise angle θ through which l_1 must be rotated about P to make it coincide with l_2 is called the angle from l_1 to l_2. This is one of the two angles of intersection of l_1 and l_2, and we wish to find it, or its tangent, in terms of the slopes of the two lines. There are two cases to be considered.

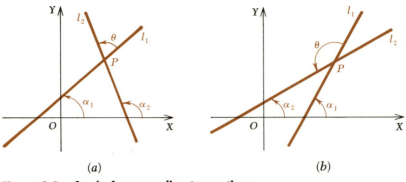

(a)   (b)

Figure 2.9   Angle from one line to another

Case 1. If, as in Fig. 2.9a, $\alpha_1 < \alpha_2$, then $\alpha_2 = \alpha_1 + \theta$ and $\theta = \alpha_2 - \alpha_1$. Hence,

$$\tan \theta = \tan (\alpha_2 - \alpha_1) = \frac{\tan \alpha_2 - \tan \alpha_1}{1 + \tan \alpha_2 \tan \alpha_1} = \frac{m_2 - m_1}{1 + m_2 m_1}$$

Case 2. If, as in Fig. 2.9b, $\alpha_1 > \alpha_2$, then $\alpha_1 = \alpha_2 + (180° - \theta)$ and

$$\theta = \alpha_2 - \alpha_1 + 180°$$

Hence,

$$\tan \theta = \tan (\alpha_2 - \alpha_1 + 180°) = \tan (\alpha_2 - \alpha_1) = \frac{m_2 - m_1}{1 + m_2 m_1}$$

The result is the same in both cases, and it follows that *if two lines l_1 and l_2 have slopes m_1 and m_2, respectively, then*

tan θ
$$\tan \theta = \frac{m_2 - m_1}{1 + m_2 m_1} \tag{2.5}$$

where θ is the angle from l_1 to l_2.

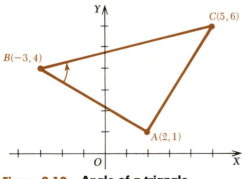

Figure 2.10   Angle of a triangle

Example   Find the angle B of the triangle with vertices at $A(2, 1)$, $B(-3, 4)$, and $C(5, 6)$.

Solution: The triangle is shown in Fig. 2.10 and ABC is the angle we want; hence, we need the slopes of BC and AB. They are

$$m(BC) = \frac{4 - 6}{-3 - 5} = \frac{1}{4}$$

and

$$m(AB) = \frac{4 - 1}{-3 - 2} = -\frac{3}{5}$$

Therefore, we have

$$\tan B = \frac{\frac{1}{4} - (-\frac{3}{5})}{1 + (\frac{1}{4})(-\frac{3}{5})} = \frac{5 + 12}{20 - 3} = 1$$

Consequently, $B = \tan^{-1} 1 = 45°$.

The other angles can be found in a similar manner.

EXERCISE 2.2

1   What is the slope of a line if its inclination is 60°?

2   What is the slope of a line if its inclination is 135°?

3   Find the inclination of a line if its slope is $-\sqrt{3}$.

4   Find the inclination of a line with slope $\sqrt{3}/3$.

In each of Problems 5 to 8, find the slope of the line through the given points.

5   $(2, 3), (5, -1)$

6   $(4, -2), (3, 0)$

7   $(-2, -7), (4, -1)$

8   $(-3, 6), (2, 1)$

9   Find the slope of all lines perpendicular to one with slope 3.

10   If a line has slope $-\frac{2}{5}$, what is the slope of a line perpendicular to it?

11   If a line perpendicular to AB has slope -6, what is the slope of AB?

12   What is the slope of BC if a perpendicular to it has slope $\frac{3}{4}$?

13   Without using the distance formula, show that $P(-8, 1)$, $Q(9, 0)$, $R(6, -6)$ is a right triangle.

14   Without using the distance formula, show that $P(8, 6)$, $Q(9, -6)$, $R(-4, 5)$ is a right triangle.

15   Show by use of slopes that the segment through $(2, 3)$ and $(8, -1)$ is parallel to the one through $(2, 3)$ and $(-1, 5)$.

16   Show that $(-2, -3)$, $(4, 6)$, $(5, 1)$, and $(3, -2)$ are the vertices of an isosceles trapezoid.

In each of Problems 17 to 20, find the angle from the line through A and B to the one through C and D.

17 $A(2, 1)$, $B(4, -2)$, $C(3, 8)$, $D(-1, 4)$

18 $A(5, 4)$, $B(2, 1)$, $C(-2, -1)$, $D(-5, -4)$

19 $A(3, 7)$, $B(2, 3)$, $C(5, -4)$, $D(1, -5)$

20 $A(0, 6)$, $B(5, -3)$, $C(-2, 4)$, $D(7, 0)$

In Problems 21 to 24, find the smallest angle of the triangle with vertices at P, Q, and R.

21 $P(2, 9)$, $Q(4, 1)$, $R(-3, -2)$ 22 $P(-1, 4)$, $Q(2, 0)$, $R(5, 3)$

23 $P(3, 6)$, $Q(2, -1)$, $R(-4, 5)$ 24 $P(1, -7)$, $Q(2, 3)$, $R(5, -4)$

2.8   Forms of the equation of a line

The equation of a line may be expressed in several forms, and we shall discuss four of these forms in this section. We shall first consider lines that are parallel to the Y axis. For such a line, x is a constant regardless of the value of y; hence, *the equation of a line parallel to the Y axis is $x = x_1$ where x_1 is the directed distance of the line from the Y axis.* Similarly, $y = y_1$ is the equation of a line parallel to the X axis.

line parallel to the Y axis

A line is determined if two points on it are known. If the two points are $P_1(x_1, y_1)$ and $P_2(x_2, y_2)$, then the slope of the line is

$$m = \frac{y_2 - y_1}{x_2 - x_1}$$

Furthermore, the point $P(x, y)$ is on the line if and only if the slope $(y - y_1)/(x - x_1)$ of PP_1 is equal to the slope of P_1P_2. Therefore,

$$\frac{y - y_1}{x - x_1} = \frac{y_2 - y_1}{x_2 - x_1} \qquad x_2 \neq x_1$$

(2.6)

or $$y - y_1 = \frac{y_2 - y_1}{x_2 - x_1}(x - x_1)$$

two-point form

is the equation of the line through $P_1(x_1, y_1)$ and $P_2(x_2, y_2)$. This is called the *two-point form* of the equation.

Example 1   Find the equation of the line through $(2, 5)$ and $(6, -3)$.

Solution: If $P(x, y)$ is any point on the line, then

$$\frac{y - 5}{x - 2} = \frac{-3 - 5}{6 - 2}$$

is the equation of the line since each member is an expression for the slope if and only if (x, y) is on the line. This equation can be written in the form $y - 5 = -2(x - 2)$ by reducing the right member to lowest terms and multiplying by $x - 2$.

If we know the slope m of a line and that $P_1(x_1, y_1)$ is on the line, then

$$\frac{y - y_1}{x - x_1} = m$$

if and only if $P(x, y)$ is on the line. Consequently, *the equation of the line through $P_1(x_1, y_1)$ and with slope m is*

$$\frac{y - y_1}{x - x_1} = m \qquad \text{or} \qquad y - y_1 = m(x - x_1) \qquad (2.7)$$

point-slope form

This is called the *point-slope form* of the equation.

Example 2   Find the equation of the line through $(4, 1)$ and with slope $-\frac{2}{5}$.

Solution: If we substitute the given values in the point-slope form of the equation, we find that

$$\frac{y - 1}{x - 4} = -\frac{2}{5}$$

is the desired equation. If we simplify this equation, we get $2x + 5y = 13$.

The x coordinate of the point of intersection of a line and the X axis is called the *x intercept* of the line. Similarly, if a line intersects the Y axis at $(0, b)$, then b is called the *y intercept*.

If we take $(0, b)$ as the point in the point-slope form of the equation, we have $(y - b)/x = m$. Now, multiplying by x and adding b to each member of the resulting equation, we get

slope–y-intercept form

$$y = mx + b \qquad (2.8)$$

This is called the *slope–y-intercept* form of the equation.

Example 3   Express the equation $2x + 3y = 12$ in the slope–y-intercept form and draw the line.

Solution: If we solve this equation for y, we find that $y = -2x/3 + 4$ is the slope–y-intercept form; hence, the slope is $-\frac{2}{3}$ and the y intercept is 4. Therefore, the graph goes through $(0, 4)$. We could use this fact and the slope to draw the line but we shall use the two intercepts. As seen by substituting $y = 0$ in the given equation and solving for x, the x intercept is 6. Consequently, the line is the one through $(6, 0)$ and $(0, 4)$ as shown in Fig. 2.11.

We shall now consider the equation $Ax + By + C = 0$. If $B = 0$, the equation becomes $Ax + C = 0$ or $x = -C/A$ and represents a

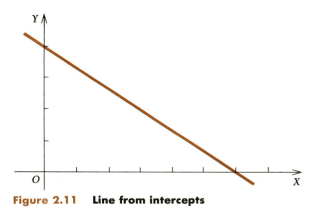

Figure 2.11   Line from intercepts

line parallel to the Y axis. If $B \neq 0$, we can solve for y and have

$$y = -\frac{A}{B}x - \frac{C}{B}$$

The graph of this equation is a line with slope $m = -A/B$ and y intercept equal to $-C/B$. Furthermore, we found earlier in this section that any line parallel to the Y axis has an equation of the form $x = x_1$ and any line not parallel to this axis has an equation of the form $y = mx + b$. Each of these can be put in the form $Ax + By + C = 0$. Consequently, we know that *the graph of $Ax + By + C = 0$ is always a straight line and that a straight line always has an equation of that form.* For this reason, a first-degree equation in two variables is often referred to as a *linear* equation.

2.9   Directed distance from a line to a point

We shall now derive a formula for the directed distance from a line l whose equation is

$$Ax + By + C = 0 \qquad A > 0, B > 0$$

to a point $P_1(x_1, y_1)$ which is above the line. Since $\tan \alpha = -A/B$ and we have specified that A and B are positive, it follows that α is between $90°$ and $180°$. We now drop a perpendicular EP_1 from P_1 to l, and from P_1 we draw a parallel DP_1 to the Y axis as shown in Fig. 2.12; hence, the angle EP_1D is $180° - \alpha$. Furthermore from the triangle EP_1D, we see that

$$\cos (180° - \alpha) = \frac{d}{DP_1} = \frac{d}{y_1 - y_2}$$

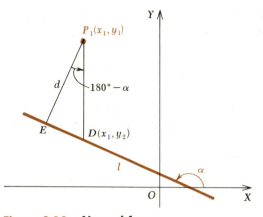

Figure 2.12   Normal form

Consequently,

$$d = (y_1 - y_2) \cos (180° - \alpha) = (y_1 - y_2)(-\cos \alpha) \qquad (1)$$

This directed distance is positive since $y_1 - y_2$ and $-\cos \alpha$ are positive; furthermore,

$$\cos \alpha = \frac{1}{\sec \alpha}$$

$$= -\frac{1}{\sqrt{1 + \tan^2 \alpha}} \qquad \cos \alpha < 0$$

$$= -\frac{1}{\sqrt{1 + (-A/B)^2}}$$

so since $\tan \alpha = -A/B$

$$\cos \alpha = -\frac{B}{\sqrt{A^2 + B^2}} \qquad B > 0 \qquad (2)$$

We now make use of the fact that $D(x_1, y_2)$ is on l; hence, its coordinates satisfy the equation of l and we have $Ax_1 + By_2 + C = 0$. Therefore, $y_2 = -(Ax_1 + C)/B$ and

$$y_1 - y_2 = y_1 + \frac{Ax_1 + C}{B} = \frac{Ax_1 + By_1 + C}{B} \qquad (3)$$

If we substitute the expressions for $\cos \alpha$ and $y_1 - y_2$ from (2) and (3) in (1), we find that

$$d = \frac{Ax_1 + By_1 + C}{\sqrt{A^2 + B^2}}$$

is the directed distance from l to P_1.

By use of a similar argument, we can show that the distance from the line whose equation is $Ax - By + C = 0$, $A > 0$, $B > 0$, to $P_1(x_1, y_1)$ is

$$d = \frac{Ax_1 - By_1 + C}{-\sqrt{A^2 + B^2}}$$

Furthermore, in each case it can be proved that d is positive if P_1 is above the line and negative if P_1 is below the line. Consequently, we conclude that *the directed distance d from the line whose equation is $Ax \pm By + C = 0$, $A > 0$, $B > 0$, to the point $P_1(x_1, y_1)$ is*

$$d = \frac{Ax_1 \pm By_1 + C}{\pm\sqrt{A^2 + B^2}} \tag{2.9}$$

where the signs preceding $B > 0$ and $\sqrt{A^2 + B^2}$ are the same; furthermore, d is positive or negative according as P_1 is above or below the line.

Example 1   Find the directed distance from the line whose equation is $5x - 12y + 6 = 0$ to $P_1(2, -3)$.

Solution: Since $B = -12 < 0$, we substitute the coordinates of P_1 in

$$\frac{5x - 12y + 6}{-\sqrt{5^2 + (-12)^2}} = \frac{5x - 12y + 6}{-13}$$

and get $$d = \frac{5(2) - 12(-3) + 6}{-13} = -4$$

Consequently, P_1 is 4 units below the given line.

Example 2   If l_1 and l_2 are the graphs of the equations

$$2x + y = 4 \tag{1}$$

and $$4x - 2y = 7 \tag{2}$$

respectively, (*a*) find the equation of the bisector of the obtuse angle between the lines, and (*b*) find the equation of the bisector of the acute angles between the lines.

Solution: (*a*) The graphs of the two lines are shown in Fig. 2.13, and we see that all points, to the right of A, on the bisector of the obtuse angles between the lines is above one of the lines and below the other, and vice versa to the left of A. Furthermore, each point on the bisector of an angle is equidistant from the sides. Hence, if $P(x, y)$ is on the bisector of the obtuse angles, we have $|d_1| = |d_2|$. However, since P is above one line and below the other, the directed distances

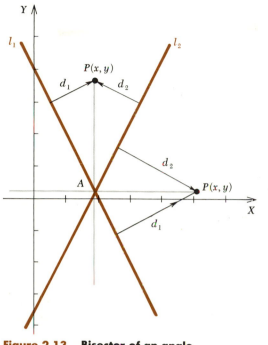

Figure 2.13   Bisector of an angle

differ in sign. Hence we have $d_1 = -d_2$. By use of (2.9), we have

$$d_1 = \frac{2x + y - 4}{\sqrt{5}} \qquad \text{and} \qquad d_2 = \frac{4x - 2y - 7}{-2\sqrt{5}}$$

Note that the sign that precedes the radical is the same as the sign of the coefficient of y. Therefore

$$\frac{2x + y - 4}{\sqrt{5}} = -\frac{4x - 2y - 7}{-2\sqrt{5}}$$

This is one form of the required equation, and if we simplify it, we get $4y = 1$.

(b) All points above A on the bisector of the acute angle are above each line, and those on the bisector below A are below each line. Hence we have $d_1 = d_2$; thus, one form of the required equation is

$$\frac{2x + y - 4}{\sqrt{5}} = \frac{4x - 2y - 7}{-2\sqrt{5}}$$

The simplified form is

$$8x = 15$$

EXERCISE 2.3

Find the equation of the line that satisfies the conditions given in each of Problems 1 to 20.

1   Through $(3, 2)$ and $(5, 7)$
2   Through $(4, 1)$ and $(7, -2)$
3   Through $(2, 6)$ and $(4, 1)$
4   Through $(-5, 4)$ and $(1, 9)$
5   Through $(-3, 4)$, $m = 2$
6   Through $(7, 5)$, $m = -3$
7   Through $(6, -1)$, $m = \frac{2}{3}$
8   Through $(-1, -2)$, $m = -\frac{3}{4}$
9   $m = -3$, $b = 5$ 10 $m = 5$, $b = -2$
11   $m = -\frac{2}{7}$, $b = -1$ 12 $m = \frac{1}{5}$, $b = \frac{3}{5}$
13   $\alpha = \pi/4$, $b = 0$ 14 $\alpha = \pi/3$, $b = \frac{1}{2}$
15   $\alpha = 5\pi/6$, $b = -2/\sqrt{3}$ 16 $\alpha = 2\pi/3$, $b = -\frac{2}{5}$
17   Through $(2, 4)$, parallel to $y = 2x - 7$
18   Through $(3, -1)$, parallel to $y = 3x + 5$
19   Through $(-6, 5)$, perpendicular to $y = -x/2 - 4$
20   Through $(-2, -5)$, perpendicular to $y = 2x/3 - 7$

Change the equation in each of Problems 21 to 24 to slope–y–intercept form.

21   $2x + 5y = 10$ 22 $3x - 4y = 12$
23   $6x - 2y = 9$ 24 $9x + 3y = 14$
25   Find the value of c so that $(2, 3)$ is on $3x - 5y + c = 0$.
26   Determine a so that $(-1, 4)$ is on $ax + 2y - 6 = 0$.
27   Show that $Ax + By = Ax_1 + By_1$ passes through $P_1(x_1, y_1)$ and is parallel to $Ax + By + C = 0$.
28   Show that $Bx - Ay = Bx_1 - Ay_1$ passes through $P_1(x_1, y_1)$ and is perpendicular to $Ax + By + C = 0$.

Find the directed distance from the line to the point in each of Problems 29 to 32.

29   $4x - 3y - 5 = 0$, $(7, 6)$
30   $8x + 15y - 22 = 0$, $(2, -3)$
31   $-12x + 5y - 18 = 0$, $(-1, -4)$
32   $24x - 7y + 30 = 0$, $(2, 4)$

Find the equation of the bisector of the acute angles between the lines in Problems 33 and 34.

33   $3x + 4y = 10$ 34 $12x - 5y = -39$
     $5x - 12y = 26$ $-3x + 4y = -20$

Find the equation of the bisector of the obtuse angles between the lines in Problems 35 and 36.

35 $8x - 15y = 34$
         $7x + 24y = -75$

36 $5x + 12y = 52$
         $24x - 7y = 50$

2.10   The circle

circle

We shall derive and discuss the equation of a circle in this and the next few sections. We shall begin by stating that *a circle consists of the set of points in a plane such that the distance of each from a fixed point of the plane is the same nonnegative constant.* The fixed point and given distance are called the *center* and *radius,* respectively, of the circle.

center

radius

We shall now derive the equation of a circle with center at $C(h, k)$ and radius r as shown in Fig. 2.14. If $P(x, y)$ is any point on the circle, then $CP = r$. Since CP and r are both expressions for the radius, $CP - r = 0$. This is a symbolic statement of the equation and we shall translate it into an equation that involves x, y, and appropriate constants.

If we multiply each member of $CP - r = 0$ by $CP + r$, we get $(CP)^2 - r^2 = 0$. This equation and $CP - r = 0$ are satisfied by the same set of points since $CP + r$ cannot be zero inasmuch as CP and r are both undirected segments. Now, by use of the distance formula, $(CP)^2 = r^2$ becomes

$$(x - h)^2 + (y - k)^2 = r^2$$

This equation is satisfied by all points (x, y) that are r units from (h, k), and it is not satisfied by any other point. Therefore, we know that *the equation of the circle with center at $C(h, k)$ and radius r is*

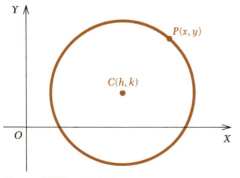

Figure 2.14   **A circle**

equation of a circle

$$(x - h)^2 + (y - k)^2 = r^2 \tag{2.10}$$

If the center (h, k) is at the origin, the equation of the circle takes the form $x^2 + y^2 = r^2$.

standard form

Equation (2.10) is called the *standard form* of the equation of a circle. The circle is real if $r^2 > 0$; it reduces to a point if $r^2 = 0$, and it is called an imaginary circle if $r^2 < 0$.

Example

Find the equation of the circle with center at $(3, -2)$ and radius 5.

Solution: Since we are given that $h = 3$, $k = -2$, and $r = 5$, the equation of the circle can be obtained by substituting in (2.10) and is

$$(x - 3)^2 + (y + 2)^2 = 5^2$$

If we expand the binomials and combine constants, the equation becomes

$$x^2 + y^2 - 6x + 4y - 12 = 0$$

2.11   The general form

If we expand the binomials in Eq. (2.10), we obtain

$$x^2 - 2hx + h^2 + y^2 - 2ky + k^2 = r^2$$

Since this equation contains x^2, y^2, a constant times x, a constant times y, and the constants h^2, k^2 and r^2, it can be expressed in the form

$$x^2 + y^2 + Dx + Ey + F = 0 \tag{2.11}$$

general form

Equation (2.11) is known as the *general form* of the equation of a circle, and the equation of any circle can be put in this form. If (2.11) is expressed in standard form by completing the squares of the quadratics in x and y, we have

$$\left(x + \frac{D}{2}\right)^2 + \left(y + \frac{E}{2}\right)^2 = \frac{D^2 + E^2}{4} - F = \frac{D^2 + E^2 - 4F}{4}$$

point circle

imaginary circle

If the constant $D^2 + E^2 - 4F = 4r^2 > 0$, then (2.11) represents a circle; if $4r^2 = 0$, Eq. (2.11) represents the single point $(-D/2, -E/2)$ and is called a *point circle;* finally, if $4r^2 < 0$, Eq. (2.11) does not represent a circle but is called an *imaginary circle* since its form is that of a circle except for r^2 being negative.

If we compare (2.10) and (2.11), we see that $h = -D/2$, $k = -E/2$, and $r = \sqrt{D^2 + E^2 - 4F}/2$.

Example

Change each of the following equations to standard form

(a) $x^2 + y^2 - 4x + 2y - 4 = 0$
(b) $x^2 + y^2 - 4x + 2y + 5 = 0$
(c) $x^2 + y^2 - 4x + 2y + 6 = 0$

and find the center and radius for each real circle.

Solution: We shall begin by writing (a) in the form

$$x^2 - 4x + y^2 + 2y = 4$$

If we now complete the square of each quadratic, we get

$$x^2 - 4x + 4 + y^2 + 2y + 1 = 4 + 4 + 1$$
$$(x - 2)^2 + (y + 1)^2 = 3^2$$

Consequently, (a) represents a circle of radius 3 with center at $(2, -1)$. Similarly (b) can be expressed in the form

$$(x - 2)^2 + (y + 1)^2 = 0$$

It represents the single point $(2, -1)$ and is thus a point circle. Finally, (c) can be expressed in the form

$$(x - 2)^2 + (y + 1)^2 = -1$$

It is not satisfied by any pair of real numbers and is therefore an imaginary circle.

EXERCISE 2.4

Find the equation of the circle described in each of Problems 1 to 16.
1 Center at $(2, 5)$, radius 3
2 Center at $(1, 4)$, radius 5
3 Center at $(4, -2)$, radius 6
4 Center at $(-3, 4)$, radius 5
5 Center at $(2, -1)$, through $(5, 3)$
6 Center at $(-3, 5)$, through $(2, 7)$
7 Center at $(-4, -3)$, through the origin
8 Center at $(0, 7)$, through $(24, 0)$
9 Center at $(5, 4)$, tangent to $3x - 4y + 6 = 0$
10 Center at $(2, -6)$, tangent to $x + 2y + 5 = 0$
11 Center at $(-3, 5)$, tangent to $7x + 4y + 1 = 0$
12 Center at $(-7, -4)$, tangent to $4x - 3y - 14 = 0$
13 With $(2, 3)$ and $(6, -5)$ as ends of a diameter
14 Ends of a diameter at $(-3, 2)$ and $(12, -6)$
15 Ends of a diameter at $(2, 4)$ and the intersection of $x - 2y = 5$ and $3x + 8y = 15$

16   One end of a diameter at $(5, -6)$ and the other at the intersection of $x - 5y = -2$ and $2x - 15y = 1$

Find the radius and coordinates of the center of each circle in Problems 17 to 24.

17 $x^2 + y^2 - 4x - 2y = 4$
18 $x^2 + y^2 - 8x - 10y = -41$
19 $x^2 + y^2 + 6x - 4y = 36$
20 $x^2 + y^2 + 2x + 6y = 6$
21 $4x^2 + 4y^2 - 4x - 12y = 15$
22 $4x^2 + 4y^2 + 12x - 20y = -25$
23 $9x^2 + 9y^2 - 36x + 24y = -36$
24 $25x^2 + 25y^2 + 40x - 80y = 20$

Determine the value or set of values of k for which the equation in each of Problems 25 to 28 represents a real circle, a point circle, and an imaginary circle.

25 $x^2 + y^2 - 2x - 4y = k$
26 $x^2 + y^2 - 6x + 8y = k$
27 $x^2 + y^2 + 8x - 12y = k + 3$
28 $x^2 + y^2 - 3x + 5y = k - 9.5$

29   Show that the equation $x^2 - y^2 + 2Dx + 2Ey + F = 0$ represents a circle with radius $\sqrt{D^2 + E^2 - F}$ and center at $(-D, -E)$ for $D^2 + E^2 - F > 0$ and that it represents a point circle if $D^2 + E^2 - F = 0$.

30   Prove that any angle inscribed in a semicircle is a right angle.

31   Find the equation satisfied by the set of points such that each is k times as far from (a, b) as from (c, d).

32   Find the equation that is satisfied by the set of points such that each is the midpoint of a segment from the circle $x^2 + y^2 = 4r^2$ to $(2a, 0)$.

2.12   The circle and three conditions

By examining the standard and general forms of the equation of a circle, we see that each contains three arbitrary constants. They are h, k, and r for the standard form and D, E, and F for the general form. Consequently, if we are given three conditions that determine a circle, we can find its equation by setting up and solving the corresponding three equations in h, k, and r, or in D, E, and F. The decision whether to use the standard or general form should be made only after the data are studied.

Example 1   Find the equation of the circle that passes through $P_1(4, -2)$, $P_2(5, 5)$, and $P_3(-3, -1)$.

Solution: We shall use the general form $x^2 + y^2 + Dx + Ey + F = 0$ since by so doing we can obtain a system of three linear equations in three unknowns. If we substitute the coordinates of P_1, of P_2, and then of P_3 in the general form of the equation, we get

$$16 + 4 + 4D - 2E + F = 0$$
$$25 + 25 + 5D + 5E + F = 0$$
$$9 + 1 - 3D - E + F = 0$$

This system of equations can be solved by one of several methods. The solutions are found to be $D = -2$, $E = -4$, and $F = -20$. Consequently, the equation of the desired circle is $x^2 + y^2 - 2x - 4y - 20 = 0$. If we express it in standard form, it becomes $(x - 1)^2 + (y - 2)^2 = 5^2$ and represents the circle with center at $(1, 2)$ and radius equal to 5.

Example 2 Find the equation of the circle with center on $x + y = 4$ and on $5x + 2y = -1$ and radius 3.

Solution: The radius is known; hence, we shall use the standard form

$$(x - h)^2 + (y - k)^2 = r^2$$
$$= 9 \qquad r = 3$$

Since the center (h, k) is on $x + y = 4$ and on $5x + 2y = -1$, its coordinates satisfy those equations. Thus, we have

$$h + k = 4 \qquad \text{and} \qquad 5h + 2k = -1$$

By solving the first of these equations for k in terms of h and substituting in the second, we have

$$5h + 2(4 - h) = -1$$
$$5h + 8 - 2h = -1$$
$$3h = -9$$
$$h = -3$$

Consequently, $-3 + k = 4$ and $k = 7$. Therefore, the circle is of radius 3, its center is at $(-3, 7)$ and its equation is $(x + 3)^2 + (y - 7)^2 = 3^2$.

2.13 Families of circles

We pointed out in the last section that there are three essential constants in the equation of a circle. Consequently, if two of them are given or are determined by given conditions, the third may be assigned a value arbitrarily provided that the condition which determines it is independent of, and does not contradict, the conditions that determine the other two. Given two conditions, there is a different *family* circle for each value assigned to the arbitrary (third) constant, and we *of circles* call the set of circles thus obtained a *family of circles*.

Example   Find the equation of the family of circles with center at $(2, -3)$. Select the member with radius 5.

Solution: If we use the standard form of the equation of a circle, we see that all circles with center at $(2, -3)$ have the equations

$$(x - 2)^2 + (y + 3)^2 = r^2$$

where the domain of r is all nonnegative real numbers.

The member with radius 5 is obtained by replacing r by 5 in the equation of the family. Thus, we get

$$(x - 2)^2 + (y + 3)^2 = 5^2$$

as the desired member.

EXERCISE 2.5

Find the equation of each circle described in Problems 1 to 20.

1   Through $(7, 9)$, $(0, 2)$, and $(6, 10)$
2   Through $(3, -7)$, $(10, 0)$, and $(-14, 10)$
3   Circumscribed about the triangle with vertices at the intersections of $4x - 3y + 11 = 0$, $x - y + 2 = 0$, and $x + y - 6 = 0$
4   Circumscribed about the triangle with vertices at the intersections of $4x - y = 5$, $7x - 4y = 11$, and $x - y = -1$
5   Center at $(1, 2)$ and tangent to $5x - 12y = 46$
6   Center at $(4, -3)$ and tangent to $3x - 4y + 1 = 0$
7   Concentric with $x^2 - 8x + y^2 + 6y = 10$ and through $(-2, 5)$
8   Concentric with $x^2 + y^2 - 10x + 4y = 7$ and with the same radius as $x^2 + y^2 - 2x - 6y = 15$
9   Through $(4, 6)$ and $(-3, 5)$ with center on $3x - y = 1$
10   Through $(4, 4)$ and $(-3, -3)$ with center on $2x + y + 7 = 0$
11   Through $(1, 0)$ and $(2, -1)$ and tangent to the Y axis
12   Through $(-1, 2)$ and $(6, 1)$ and tangent to the X axis
13   Tangent to $3x + 4y = 35$ and $4x - 3y = -20$ and through $(6, -2)$
14   Tangent to $4x + 3y = 16$ and $5x - 12y = 38$ and through $(-6, -3)$
15   Tangent to $3x - 4y + 12 = 0$, center on $x + 4y = -1$, radius 5
16   Tangent to $12x - 5y = 40$, center on $2x + y = -5$, radius 13
17   Tangent to $4x - 3y = 18$ at $(6, 2)$ and passing through $(-1, 1)$

18   Tangent to $5x - 12y = -21$ at $(3, 3)$ and passing through $(-4, -14)$

19   Tangent to $15x - 8y = 182$ at $(10, -4)$, center on $x + y + 1 = 0$

20   Tangent to $5x - 12y = -31$ at $(1, 3)$, center on $2x + y = 3$

21   What equation represents the family of circles through the origin and with center on the X axis?

22   What equation represents the family of circles with center on $y = x$ and passing through the origin?

23   What is the equation of the family of circles with center on $x + y = 0$ and tangent to both axes?

24   What is the equation of the family of circles with center on $y = 2x$ and passing through the origin?

25   Let the equations of two intersecting circles be

$$x^2 + y^2 + D_1x + E_1y + F_1 = 0$$

and

$$x^2 + y^2 + D_2x + E_2y + F_2 = 0$$

Show that for all real k, except -1, the equation

$$x^2 + y^2 + D_1x + E_1y + F_1 + k(x^2 + y^2 + D_2x + E_2y + F_2) = 0 \tag{1}$$

represents a circle that passes through the intersections of the given circles. Show that (1) represents a line through the intersections of the given circles if $k = -1$. This line is called the *radical axis*.

26   By use of the equation in Problem 25, find the equation of the family of circles through the intersections of $x^2 + y^2 + 2x - 4y = 4$ and $x^2 + y^2 - 6x + 2y = 6$. Find the equation of the radical axis.

27   Find the equation of the family of circles with center on $3x - y = 4$ and radius 5.

28   Find the equation of the family of circles that is tangent to $x + y = 3$ at $(-2, 5)$.

Select the members of the family in Problem 26 that satisfy the conditions given in Problems 29 to 32 by determining the proper values of k.

29   Radius equal to 2.5   **30**   Center on $x = 2y$

31   Through $(2, 2)$ **32**   Through $(9, -1)$

2.14   Intercepts

By the *intercepts* of a graph on the X axis, we mean the abscissas of the points where the graph touches or crosses the X axis. These numbers can be found by setting $y = 0$ in the equation of the curve and solving for x. By the intercepts on the Y axis, we mean the ordinates of the points where the graph touches or crosses the Y axis. They can be found by setting $x = 0$ and solving for y.

Example   If $f(x, y) = xy - 2x - y - 1 = 0$, then $f(x, 0) = -2x - 1 = 0$. It follows that $x = -\frac{1}{2}$ is the x intercept. Similarly, the y intercept is $y = -1$ since $f(0, y) = -y - 1 = 0$ yields $y = -1$.

2.15   Symmetry

We shall discuss a concept in this section which simplifies the sketching of many graphs. Most of us have an intuitive understanding of the meaning of symmetry, but we shall formalize it by means of several definitions.

symmetry relative to a line   Two points are said to be *symmetric with respect to a line* if the line is the perpendicular bisector of the line segment that joins the points. A curve is said to be *symmetric with respect to a line* if for each point on the curve, the point symmetrically located with respect to the line is also on the curve. For example, a circle is symmetric with respect to any diameter.

symmetry relative to a point   Two points are *symmetric with respect to a third point* if the third point is the midpoint of the segment that joins them. A curve is *symmetric with respect to a point* if for each point on the curve, the point symmetrically located with respect to the given point is also on the curve. For example, a circle is symmetric with respect to its center.

It is not an easy matter to determine whether a line or point of symmetry exists for a given curve. We shall be interested in symmetry with respect to the coordinate axis and the origin; we shall now give appropriate tests and prove them later in this section.

symmetry relative to the X axis   *A curve whose equation is $f(x, y) = 0$ is symmetric with respect to the X axis if and only if $f(x, -y) = \pm f(x, y)$.*

Example 1   Test $f(x, y) = x^3 - xy^2 + 2y^4 - 5 = 0$ for symmetry with respect to the X axis.

Solution: We shall indicate that y is replaced by $-y$ in the equation by use of $f(x, -y)$. We find that

$$f(x, -y) = x^3 - x(-y)^2 + 2(-y)^4 - 5 = x^3 - xy^2 + 2y^4 - 5 = f(x, y)$$

Hence, the curve represented by $f(x, y) = 0$ is symmetric with respect to the X axis.

We shall now give two more tests.
A curve whose equation is $f(x, y) = 0$ is symmetric with respect to

the Y axis if and only if $f(-x, y) = \pm f(x, y)$. Furthermore, *it is symmetric with respect to the origin if and only if* $f(-x, -y) = \pm f(x, y)$.

Example 2   Test $f(x, y) = x^3 - xy^2 - y = 0$ for symmetry with respect to the Y axis and the origin.

Solution: We begin by finding that

$$f(-x, y) = (-x)^3 - (-x)y^2 - y = -x^3 + xy^2 - y \neq \pm f(x, y)$$

Hence, $f(x, y) = 0$ is not symmetric with respect to the Y axis. We shall now test for symmetry with respect to the origin.

$$f(-x, -y) = (-x)^3 - (-x)(-y)^2 - (-y) = -x^3 + xy^2 + y = -f(x, y)$$

Therefore, the curve represented by $f(x, y) = 0$ is symmetric with respect to the origin.

We shall now prove that the test given for symmetry with respect to the X axis is valid. The points $P_1(x, y)$ and $P_2(x, -y)$ are symmetrically located with respect to the X axis since that axis is the perpendicular bisector of the segment P_1P_2. Therefore, if the equation obtained by replacing y by $-y$ is identical with the equation of a given curve, then corresponding to each point $P_1(x, y)$ on the curve the symmetrically located point $P_2(x, -y)$ is also on it and the curve is symmetric with respect to the X axis. Furthermore, since the graphs of $f(x, y) = 0$ and $-f(x, y) = 0$ are the same, the curve is symmetric with respect to the X axis if $f(x, -y) = -f(x, y)$.

The proofs of the tests for symmetry with respect to the Y axis and the origin can be given similarly.

2.16   Asymptotes

The concept that we shall discuss in this section is very useful in visualizing and constructing the graphs of some types of equations in two variables. We shall illustrate the principles involved with an example.

If we solve the equation $xy - y - 2x = 0$ for y in terms of x, we get

$$y = \frac{2x}{x - 1} \tag{1}$$

Now if we replace x by 1 in Eq. (1), we see that no corresponding value of y exists since $2/0$ is not a number. If, however, we let x approach 1 from the right, the denominator is positive and approaches zero. Hence y increases indefinitely and becomes greater than any

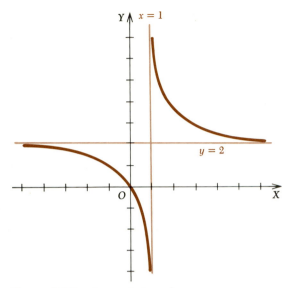

Figure 2.15   Asymptotes of a curve

preassigned quantity. Consequently, as indicated in Fig. 2.15 the graph of (1) recedes infinitely far from the X axis as it approaches the vertical line through $(1, 0)$. We call this line a vertical asymptote of the curve.

Similarly, if we solve $xy - y - 2x = 0$ for x, we have

$$x = \frac{y}{y - 2} \tag{2}$$

and by an argument of the same type as that above, we can show that the graph of $y = 2$ is a horizontal asymptote of the graph of Eq. (2).

In general, we define the vertical and the horizontal asymptotes of a graph, if either exists, as follows:

If $f(x, y) = 0$ and if, as x approaches a from either direction, $|y|$ increases indefinitely, then the vertical line through $(a, 0)$ is a *vertical asymptote* of the graph of $f(x, y) = 0$. Furthermore, if, as y approaches b from either direction, $|x|$ increases indefinitely, then the horizontal line through $(0, b)$ is a *horizontal asymptote* of the graph.

asymptotes defined

We shall now consider the case in which the solution of $f(x, y) = 0$ for y in terms of x is in the form $y = g(x)/h(x)$ where $g(x)$ and $h(x)$ are polynomials. If $h(a) = 0$, then by the factor theorem $h(x) = (x - a)^n k(x)$ where $n \geq 1$ and $k(a) \neq 0$. Now as x approaches a, $|x - a|$ becomes and remains less than any preassigned number. Consequently if $g(a) \neq 0$, $|y|$ increases indefinitely. Hence the graph

tests for asymptotes

of $x = a$ is a vertical asymptote of the graph of $f(x, y) = 0$. Similarly if the solution of $f(x, y) = 0$ is $x = r(y)/t(y)$ and if $t(b) = 0$ while $r(b) \neq 0$, then the graph of $y = b$ is a horizontal asymptote.

Example 1   Find the horizontal and vertical asymptotes of

$$f(x, y) = xy - 3x - y - 2 = 0$$

Solution: If we solve this equation for y, we get

$$y = \frac{3x + 2}{x - 1}$$

Consequently, $x - 1 = 0$ is a vertical asymptote. Similarly, solving for x, we have

$$x = \frac{y + 2}{y - 3}$$

Consequently, $y - 3 = 0$ is a horizontal asymptote.

Example 2   Sketch the graph of $f(x, y) = xy - 3x - y - 2 = 0$.

Solution: We first shall find the intercepts of the curve represented by this equation, as discussed in Sec. 2.14. They are $x = -\tfrac{2}{3}$ and

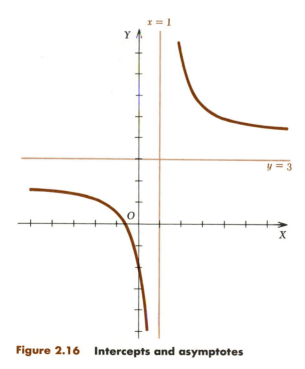

Figure 2.16   Intercepts and asymptotes

$y = -2$. The asymptotes were found in Example 1 to be $x = 1$ and $y = 3$. The usual tests will show that the curve is not symmetric with respect to either axis or the origin. The intercepts, asymptotes, and the points $(2, 7)$, $(3, 5)$, and $(5, 4)$ were used in sketching the graph of the function which is shown in Fig. 2.16.

EXERCISE 2.6

Find the intercepts and asymptotes of the curves represented by the equations below. Test each for symmetry and sketch the graph.

1 $y = 2/3x$

2 $y = 9/(x - 3)$

3 $y = 6/(2x + 3)$

4 $y = 2/(x + 4)$

5 $y = (x + 3)/2x$

6 $y = (5 - 2x)/3x$

7 $y = (x + 2)/(2x - 3)$

8 $y = (x - 3)/(3x - 1)$

9 $y = 8/(x^2 + 3)$

10 $y = 5/(x^2 + 2)$

11 $y = 4x/(x^2 + 4)$

12 $y = 6x/(x^2 + 3)$

13 $y = 6x/(x^2 - 4)$

14 $y = 8x/(x^2 - 5)$

15 $y = 3x/(2x^2 - 9)$

16 $y = 2x/(x^2 - 9)$

17 $y = 2x/(x + 2)^2$

18 $y = 3x/(x - 1)^2$

19 $y = 6/(x^2 - 4)$

20 $y = 4/(x^2 - 1)$

21 $y = (9 - x^2)/(x^2 - 5)$

22 $y = (x^2 + 4)/(x^2 - 4)$

23 $y = (x^2 - 4)/(x^2 + 4)$

24 $y = x^2/(x^2 - 12)$

25 $y = x^3/(x^2 - 3)$

26 $y = x^2/(x - 2)$

27 $y = x^3/(3x + 6)$

28 $y = x^3/(x - 2)$

29 $3xy - 6y + 2x - 9 = 0$

30 $2xy - 4y + 3x - 8 = 0$

31 $xy + 2x + y - 4 = 0$

32 $xy + 2y - 4x = 0$

33 $2xy + 4y = x^2 + 4$

34 $xy = x^2 - 4$

35 $x^2y + 3y = 2x^2 + 7$

36 $x^2y + 2x^2 = 8$

37 $x^2y + 4x = 4y$

38 $x^2y + xy = 3x - y + 6$

39 $x^2y - 4x + y = 3$

40 $x^2y - 9y = x^2 + 4x$

Limits

3.1   Increments

increment   If a variable assumes two values, the difference between them is often called an *increment*. In fact any change in a variable is called an increment. We shall be interested in variables that assume all values in an interval that may be open, closed or semiopen. The increment of the variable x is often called *delta x* and denoted by Δx. This symbol is not to be considered a product but merely a symbol for the change in x, which change may be positive or negative.

In the graph of the function $y = \sqrt{x}$ as shown in Fig. 3.1, x and y have definite values at A. The change in x from A to B is represented by Δx, and the corresponding change in the value of y is represented by Δy. Since the square root of a number increases as the number increases, the value of Δy is positive for Δx positive. As we shall see later, there are functions for which Δy is negative for Δx positive.

Example   If $y = f(x) = x^2 - 2$, find Δy for $x = 4$ and $\Delta x = 1$.

Solution: If we substitute 4 for x in $f(x)$, we find that $f(4) = 4^2 - 2 = 14$; furthermore $x + \Delta x = 4 + 1 = 5$ and $f(5) = 5^2 - 2 = 23$. Therefore, $\Delta y = f(x + \Delta x) - f(x) = f(5) - f(4) = 23 - 14 = 9$.

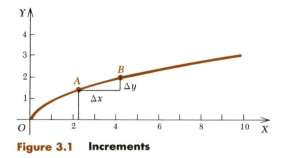

Figure 3.1   Increments

3.2   The limit of a function

The concept of the limit of a function was encountered in plane geometry in connection with the area of a circle; it is one of the essential ideas used in calculus. We shall now develop the concept involved in the limit of a function $f(x)$ as the independent variable x approaches a constant a through an interval of values. To proceed with the development, we need to introduce the concept of a neighborhood of a point.

neighborhood   Any open interval $(a - b_1, a + b_2)$ that contains a is called a *neighbor-*
deleted   *hood* of a; furthermore, if a is removed from the open interval $(a - b_1,$
neighborhood $a + b_2)$, we have a *deleted*, or *punctured*, *neighborhood*. Thus, the interval $(3, 6)$ with 5 removed is a deleted neighborhood of 5.

Before defining the concept of a limit, we shall illustrate the ideas involved by considering the function

$$y = f(x) = \frac{1 - \cos x}{x}$$

If $x = 0$, we have

$$f(0) = \frac{1 - \cos 0}{0} = \frac{1 - 1}{0} = \frac{0}{0}$$

Since $0/0$ is not a number, $f(x)$ has no value for $x = 0$, and we say that $f(0)$ does not exist. If, however, we let $x = 0.3, 0.2$ and 0.1, we find that $f(x)$ is 0.15, 0.10, and 0.05, respectively. Consequently, it appears that as x is replaced by values that approach zero, the value of $f(x)$ gets closer and closer to zero. This is, in fact, the situation as we shall prove when we study indeterminate forms.

We shall indicate the limit of $f(x)$ as x approaches a by writing

$$\lim_{x \to a} f(x)$$

Furthermore, we define this limit L, if it exists, by the statement that *if for every neighborhood N of L there exists a deleted neighborhood D of a in the domain of f such that $f(x)$ is in N for every x in D, then*

$$\lim_{x \to a} f(x) = L$$

Example 1 $\lim\limits_{x \to 2} 3x = 6$ since $3x$ can be made arbitrarily near 6 by taking x sufficiently near 2.

Example 2 $\lim\limits_{x \to -2} (5x + 4) = -6$ since $5x + 4$ can be made arbitrarily near -6 by taking x sufficiently near -2.

Example 3 $\lim\limits_{x \to 3} [(3x^2 + 1)/(x + 1)] = {}^{28}\!/_{4} = 7$ since the numerator can be made arbitrarily near 28 and the denominator arbitrarily 4 by taking x sufficiently near 3.

The analogy between the present situation and that referred to in the first paragraph of this section should be observed. The *limit* of the area of the inscribed polygon as each of its sides approaches zero is, by definition, the area of the circle. We are not concerned with what happens if each side *is* zero.

3.3   Theorems on limits

When we stated in the preceding section that

$$\lim_{x \to -2} (5x + 4) = -6 \qquad \text{and} \qquad \lim_{x \to 3} \frac{3x^2 + 1}{x + 1} = 7$$

we were using, without realizing it, the following basic theorems:

If $\lim\limits_{x \to a} g(x) = L$ *and* $\lim\limits_{x \to a} h(x) = M$, *then*

theorems on limits

$$\lim_{x \to a} [g(x) + h(x)] = L + M \tag{1}$$

$$\lim_{x \to a} [g(x)h(x)] = LM \tag{2}$$

$$\lim_{x \to a} \frac{g(x)}{h(x)} = \frac{L}{M} \qquad \text{if } M \neq 0 \tag{3}$$

Proofs of these statements will be given after we have discussed some of the methods for evaluating limits. As a special case of (2), we mention that, for any positive integer p,

$$\lim_{x \to a} x^p = a^p \qquad since \ \lim_{x \to a} x = a$$

It is also true that

$$\lim_{x \to a} \sqrt[n]{x} = \sqrt[n]{a} \qquad n \text{ an odd integer}$$

and

$$\lim_{x \to a} \sqrt[n]{x} = \sqrt[n]{a} \qquad n \text{ an even integer, } a > 0$$

The student will probably regard all of these things as intuitively evident. He will not at this stage see any need, for example, to prove that $\lim\limits_{x \to 4} \sqrt{x} = \sqrt{4}$.

3.4   Division by zero

The student should recall that the symbol $\dfrac{N}{D}$ (or N/D), where N and

zero in division

D are numbers, is defined to denote the unique number q, if it exists, such that $qD = N$. Thus $^{16}\!/_2 = 8$ and $^0\!/_4 = 0$. If $N \neq 0$ and $D = 0$, then since $qD = q \cdot 0 = 0$, there is no replacement for q such that $qD = N$. Therefore, N/D is not a number if $N \neq 0$ and $D = 0$. For example, there is no number b such that $7/0 = b$, since $0 \cdot b = 0$ for all numbers b. Therefore, division of $N \neq 0$ by 0 is not a permissible operation. Similarly, if $0/0 = a$, then a must be such that $0 \cdot a = 0$. This last condition is satisfied by all numbers, and since we require a unique result, the division of 0 by 0 is excluded from consideration.

3.5 $\lim\limits_{x \to a} (N/D)$ **if** $\lim\limits_{x \to a} N = \lim\limits_{x \to a} D = 0$

We have already observed that if $\lim\limits_{x \to a} g(x) = L$ and $\lim\limits_{x \to a} h(x) = M$, then

$$\lim_{x \to a} \frac{g(x)}{h(x)} = \frac{L}{M} \qquad \text{if } M \neq 0$$

We shall frequently be concerned with the case in which M is zero and usually with the case in which L and M are *both* zero. In many such cases we can find the limit, if there is one, by using the methods of the following examples.

Example 1   Find
$$\lim_{x \to 3} \frac{2x^2 - 5x - 3}{2x - 6}$$

Solution: The numerator and denominator are polynomials in x and both are zero if $x = 3$. It follows from the factor theorem of algebra that $x - 3$ is a factor of each. In fact

$$\frac{2x^2 - 5x - 3}{2x - 6} = \frac{(2x + 1)(x - 3)}{2(x - 3)} = \frac{2x + 1}{2} \qquad \text{if } x \neq 3$$

Thus for all values of x as near 3 as we please, and in fact for *all* values of x other than 3 itself (with which we are not concerned), the value of the original fraction is precisely the same as that of $(2x + 1)/2$.

Consequently the original fraction and $(2x + 1)/2$ must have the same limit as $x \to 3$, and this limit is obviously 7/2 or 3.5. We may write out our solution briefly as follows:

$$\lim_{x \to 3} \frac{2x^2 - 5x - 3}{2x - 6} = \lim_{x \to 3} \frac{(2x + 1)(x - 3)}{2(x - 3)} = \lim_{x \to 3} \frac{2x + 1}{2} = \frac{7}{2} = 3.5$$

The student must keep in mind the meaning of our result. It is that *the value of the fraction*

$$\frac{2x^2 - 5x - 3}{2x - 6}$$

is arbitrarily near 3.5 for all values of x that are sufficiently near 3, the value 3 itself for x being excluded from consideration. Note the similarity between this and the statement that the area of the inscribed regular polygon is arbitrarily near that of the circle if the length of each side is sufficiently *near* zero, the value zero itself being excluded.

Example 2   Find
$$\lim_{h \to 0} \frac{\sqrt{5 + h} - \sqrt{5}}{h}$$

Solution: As in Example 1, we are dealing with a situation in which the limits of numerator and denominator are both zero. In order to evaluate the limit, we shall multiply both numerator and denominator by $\sqrt{5 + h} + \sqrt{5}$. Thus we have

$$\lim_{h \to 0} \frac{\sqrt{5 + h} - \sqrt{5}}{h} = \lim_{h \to 0} \frac{\sqrt{5 + h} - \sqrt{5}}{h} \frac{\sqrt{5 + h} + \sqrt{5}}{\sqrt{5 + h} + \sqrt{5}}$$

$$= \lim_{h \to 0} \frac{5 + h - 5}{h(\sqrt{5 + h} + \sqrt{5})}$$

$$= \lim_{h \to 0} \frac{1}{\sqrt{5 + h} + \sqrt{5}}$$

$$= \frac{1}{\sqrt{5} + \sqrt{5}} = \frac{1}{2\sqrt{5}}$$

In each of the above examples we dealt with a fraction in which the limits of both numerator and denominator were zero. The limit of the fraction was 3.5 in Example 1 and $1/2\sqrt{5}$ in Example 2. The reader should consider the following fractions and decide in each case what happens to the value of the fraction as x approaches zero. In each case both numerator and denominator approach zero.

$$\lim_{x \to 0} \frac{6x^2}{x}, \ \lim_{x \to 0} \frac{6x}{x^2}, \ \lim_{x \to 0} \frac{6x}{x}, \ \lim_{x \to 0} \frac{x^2}{6,000,000x}$$

Finally, it should be observed that if the limit of the denominator is zero and that of the numerator is not zero, then the fraction has no limit. Consider, for example,

$$\lim_{x \to 0} \frac{5 - 2x}{5x}$$

We can readily verify that for x near zero the absolute value of the fraction is very large. Thus

If $x = 0.01$ then $\dfrac{5 - 2x}{5x} = \dfrac{4.98}{0.05} = 99.6$

If $x = 0.001$ then $\dfrac{5 - 2x}{5x} = \dfrac{4.998}{0.005} = 999.6$

If $x = -0.001$ then $\dfrac{5 - 2x}{5x} = \dfrac{5.002}{-0.005} = -1000.4$

Furthermore we can make the value of the fraction numerically as large as we please by taking x near enough to zero. The function value is positive if x is positive and negative if x is negative. We describe the situation by saying that as x approaches zero from the right, the value of the fraction increases beyond bound or, more conveniently, that it approaches infinity, and we write

$$\lim_{x \to 0^+} \frac{5 - 2x}{5x} = \infty$$

We similarly describe the situation that exists when x approaches zero from the left by writing

$$\lim_{x \to 0^-} \frac{5 - 2x}{5x} = -\infty$$

3.6   Infinity as a limit

If the value of $f(x)$ approaches some number L as x increases beyond any selected number, we say that the limit of $f(x)$ as x approaches infinity is L and write

$$\lim_{x \to \infty} f(x) = L$$

If $f(x)$ increases beyond any selected number as x approaches infinity, we say that the limit of $f(x)$ as x approaches infinity is infinite and

we write
$$\lim_{x \to \infty} f(x) = \infty$$

Example 1
$$\lim_{x \to \infty} \frac{2x+1}{x-3} = \lim_{x \to \infty} \frac{2+1/x}{1-3/x}$$ *dividing numerator and denominator by x*

$$= \frac{2+0}{1-0} = 2$$

Example 2   We call attention to the fact that

$$\lim_{x \to \infty} \frac{5-2x}{5x} = \frac{5-2(\infty)}{5(\infty)} = -\frac{\infty}{\infty}$$

and that since $-\infty/\infty$ is not a well-defined number, we cannot tell what the limit is in this form. If, however, $x \neq 0$, we have

$$\frac{5-2x}{5x} = \frac{5/x-2}{5}$$

Consequently, $\lim_{x \to \infty} \dfrac{5-2x}{5x} = \lim_{x \to \infty} \dfrac{5/x-2}{5} = -0.4$

Therefore, the Y axis and $y = -0.4$ are asymptotes of the curve as indicated in the graph of $y = (5 - 2x)/5x$, shown in Fig. 3.2.

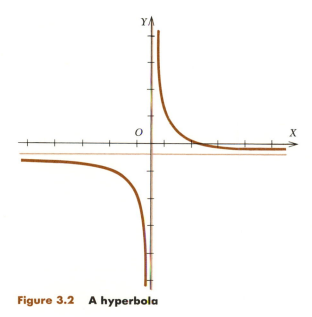

Figure 3.2   A hyperbola

3.7 Continuity

If the limit of a function f as x → a exists and is a number L and if
continuity
defined
f(a) is also L, then f(x) is said to be continuous for x = a. This may
be expressed more briefly by saying that f(x) is continuous for x = a
if $\lim_{x \to a} f(x) = f(a).$

The limit of x^2 as $x \to 2$ is 4 and its value for x equal to 2 is $f(2) = 4$;
hence, the function is continuous at $x = 2$. The situation is shown
graphically in Fig. 3.3. As $x \to 2$ from either side, the lengths of the
ordinates approach 4. *At* $x = 2$ the ordinate is 4. If a function is not
discontinuous continuous for a specified value of x, it is said to be *discontinuous* at
that value.

If a function is continuous at every point of an interval, it is said to
be *continuous over the interval*. The function x^2 is continuous over
any interval of the variable x since if a is any replacement for x, then
$\lim_{x \to a} x^2 = a^2.$

The function whose graph is shown in Fig. 3.2 is discontinuous at
$x = 0$, but is continuous over any interval that does not include $x = 0$.

It can be verified that

$$\lim_{x \to 0^+} 2^{1/x} = \infty \qquad \text{and} \qquad \lim_{x \to 0^-} 2^{1/x} = 0$$

and that, consequently, $y = 2^{1/x}$ is discontinuous for $x = 0$.

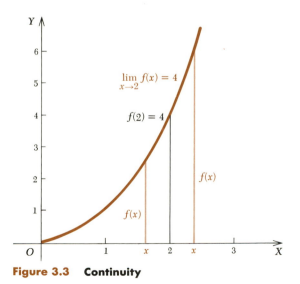

Figure 3.3 **Continuity**

EXERCISE 3.1

In Problems 1 to 4, find Δy for x and Δx as given for each function.
1   $y(x) = 2x - 3;\ x = 2,\ \Delta x = 1;\ x = 5,\ \Delta x = 1$
2   $y(x) = -3x + 1;\ x = -1,\ \Delta x = 2;\ x = 7,\ \Delta x = 2$
3   $y(x) = x^2 - x + 1;\ x = -2,\ \Delta x = 1;\ x = 3,\ \Delta x = 1$
4   $y(x) = x^2 + 3x - 5;\ x = -5,\ \Delta x = 2;\ x = 6,\ \Delta x = 2$

5   If x is near 3, what number is the value of the function $3x^2 - 2$ near? Is it true that the values of the function are arbitrarily near this number for all values of x that are sufficiently near 3? Is this number the limit of the function as $x \to 3$?

6   What number is the value of the function $x^2 - x - 1$ near if x is near 5? Are the values of the function arbitrarily near this number for all replacements for x that are sufficiently near 5? Is this number the limit of the function as $x \to 5$?

7   What number is the value of the function $(x^2 + 6)/(x - 4)$ near if x is near 6? Is the value of the function arbitrarily near this number for all replacements for x that are sufficiently near 6? Is this number the limit of the function as $x \to 6$?

8   What number is the value of the function $(x^3 - 5)/(x + 2)$ near if x is near 2? Is it true that the value of this function is arbitrarily near this number if x is sufficiently near 2? Is this number the limit of the function as $x \to 2$?

9   Is there a number L such that the values of the function $(x^2 + 1)/(x - 2)$ are near L for all values of x near 2? What statement can you make about the limit of this function as $x \to 2$?

10   Is there a number L such that the values of the function $(x^2 + 7x + 10)/(x^2 - 9)$ are near L for all values of x near 3? What statement can you make about the limit of this function as $x \to 3$?

11   Find the value of the function $(x^2 - 9)/(x - 3)$ for $x = 2.9$, 2.99, 3.1, 3.01. These results indicate that a number L may be the limit of this function as $x \to 3$. What is this number? Prove that this number is actually the limit.

12   Find the values of the function $(x^2 - 3x - 10)/(x - 5)$ for $x = 4.9, 4.99, 5.1, 5.01$. These results indicate that what number may be the limit of this function as $x \to 5$? Prove that this number is actually the limit.

In each of Problems 13 to 16, find the limit of the given function as $h \to 0$.

13   $\dfrac{(3 + h)^2 - 9}{h}$

14   $\dfrac{(2 + h)^3 - 8}{h}$

15 $\dfrac{1/(2 + h) - \frac{1}{2}}{h}$

16 $\dfrac{\sqrt{4 + h} - 2}{h}$

17 Evaluate $\lim\limits_{x \to 2} 4x/(3x - 5)$. Find the value of this function corresponding to $x = 2$. Is the function continuous for $x = 2$? Is there a value of x for which the function is not continuous? Sketch the graph.

18 Evaluate $\lim\limits_{x \to 1} 3x/(x^2 + 1)$. Find the value of this function corresponding to $x = 1$. Is the function continuous at $x = 1$? Is there any value of x for which it is not continuous? Sketch its graph.

19 Evaluate $\lim\limits_{x \to -3} (4x + 1)/(2 + x^2)$. Find the value of the function for $x = -3$. Is the function continuous for $x = -3$? Is there any value of x for which it is not continuous? Sketch the graph.

20 Evaluate $\lim\limits_{x \to -1} (x^2 - x + 2)/(x - 1)$. Find the value of the function for $x = -1$. Is the function continuous for $x = -1$? Is it continuous for $x = 1$? Sketch the graph.

Find each of the following limits if it exists.

21 $\lim\limits_{x \to 2} \dfrac{x^3 - 8}{x - 2}$

22 $\lim\limits_{x \to -1} \dfrac{x^3 + 1}{x + 1}$

23 $\lim\limits_{x \to \frac{1}{2}} \dfrac{2x^2 + 7x - 4}{4x - 2}$

24 $\lim\limits_{x \to -5} \dfrac{3x + 15}{x^2 - 25}$

25 $\lim\limits_{x \to 2} \dfrac{2 - x}{x + 2}$

26 $\lim\limits_{h \to 0} \dfrac{h^2 + 3h}{h - 1}$

27 $\lim\limits_{h \to 0} \dfrac{1}{h}[(6 + h)^2 - 36]$

28 $\lim\limits_{x \to -a} \dfrac{x^2 - a^2}{x + a}$

29 $\lim\limits_{x \to 2} \dfrac{x^3 - 4x}{x^3 - 2x^2}$

30 $\lim\limits_{h \to 0} \dfrac{h^3 - 4h}{h^3 - 2h^2}$

31 $\lim\limits_{x \to x_1} \dfrac{4x^2 - 4x_1{}^2}{x - x_1}$

32 $\lim\limits_{x \to x_1} \dfrac{x^3 - x_1{}^3}{x - x_1}$

33 $\lim\limits_{h \to 0} \dfrac{(x + h)^2 - x^2}{h}$

34 $\lim\limits_{h \to 0} \dfrac{(x + h)^3 - x^3}{h}$

35 $\lim\limits_{h \to 0} \dfrac{1}{h}\left[\dfrac{1}{\sqrt{x + h}} - \dfrac{1}{\sqrt{x}}\right]$

36 $\lim\limits_{h \to 0} \dfrac{\sqrt{x + h} - \sqrt{x}}{h}$

3.8   Limits and continuity in terms of ε and δ

In Sec. 3.2, we said that if the value of a function $f(x)$ is arbitrarily near a number L for all values of x that are sufficiently near a given number a, then L is the limit of $f(x)$ as x approaches a. This was and

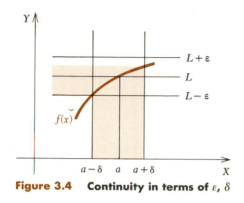

Figure 3.4 **Continuity in terms of ε, δ**

is stated symbolically by writing

$$\lim_{x \to a} f(x) = L$$

This statement can be shown pictorially as in Fig. 3.4. In it $f(x)$ is within a distance ε of L provided that x is within a distance δ of a. This states that $f(x)$ is between $L - \varepsilon$ and $L + \varepsilon$ if x is between $a - \delta$ and $a + \delta$. This introduction of ε and δ enables us to give the definition of a limit in the following form.

ε, δ in limits **Definition** If given any $\varepsilon > 0$, there exists a $\delta > 0$ such that

$$|f(x) - L| < \varepsilon \qquad \text{whenever } 0 < |x - a| < \delta$$

then $f(x)$ approaches the limit L as x approaches a.

Example 1 Show that 25 is the limit of x^2 as x approaches 5.

Solution: In keeping with the definition of a limit of a function as the independent variable approaches a constant, we must show that given any $\varepsilon > 0$, there exists a number $\delta > 0$ such that $|x^2 - 25| < \varepsilon$ whenever $|x - 5| < \delta$ since $f(x)$, L, and a are given as x^2, 25, and 5, respectively. To do this, we notice that $|x^2 - 25| < \varepsilon$ is implied by

$$|(x - 5)(x + 5)| < \varepsilon \qquad \textit{factoring}$$

$$|x - 5|\,|x + 5| < \varepsilon$$

We now assume that x is in $(4, 6)$ since it is approaching 5 and use the fact that $|x + 5| < 11$ for $|x - 5| < 1$. Therefore,

$$|x - 5|\,|x + 5| < |x - 5|11 \qquad \text{if } |x - 5| < 1$$

Consequently, $|x^2 - 25| < \varepsilon$ if $11|x - 5| < \varepsilon$ and $|x - 5| < 1$, hence, if $|x - 5| < \varepsilon/11$ and $|x - 5| < 1$. Therefore,

$$|x^2 - 25| < \varepsilon \qquad \text{if } |x - 5| < \delta$$

where δ is the smaller of 1 and $\varepsilon/11$. Therefore, we have shown that 25 is the limit of x^2 as x approaches 5.

If $\varepsilon = 0.11$, then $\delta = 0.01$ in this example since it is the smaller of 1 and $\varepsilon/11 = 0.01$.

We shall now state the definition of continuity in terms of ε and δ.

ε, δ in
continuity

Definition   If given any $\varepsilon > 0$, there exists a $\delta > 0$ such that

$$|f(x) - f(a)| < \varepsilon \qquad \text{whenever } |x - a| < \delta$$

then $f(x)$ is continuous at $x = a$.

This definition may be stated in the form: *$f(x)$ is continuous at $x = a$ if $f(a)$ is defined,* $\lim\limits_{x \to a} f(x)$ *exists, and* $\lim\limits_{x \to a} f(x) = f(a)$.

We are now able to prove that certain functions are continuous at a value of x or over an interval of values.

Example 2   Show that $f(x) = x^2$ is continuous at $x = 5$.

Solution: In Example 1, we showed that $\lim\limits_{x \to 5} x^2 = 25$; hence, we need only show $f(5) = 25$. This is seen to be true by substitution. Hence, $f(x) = x^2$ is continuous for $x = 5$.

Example 3   Prove that $f(x) = x^2$ is continuous at $x = a$ if a is any positive number.

Solution: Since, by substitution, $f(a) = a^2$, we know that $f(a)$ is defined. We must show that given $\varepsilon > 0$, there exists $\delta > 0$ such that

$$|x^2 - a^2| < \varepsilon \qquad \text{whenever } |x - a| < \delta$$

We begin by observing that $|x^2 - a^2| < \varepsilon$ if

$$|(x - a)(x + a)| < \varepsilon \qquad \text{\small since } x^2 - a^2 = (x - a)(x + a)$$

Now if $|x - a| < 1$, then $-1 < x - a < 1$ and it follows that $2a - 1 < x + a < 2a + 1$. Hence $|x + a| < 2a + 1$. Therefore

$$|x - a|\,|x + a| < (2a + 1)|x - a| \qquad \text{if } |x - a| < 1$$

Consequently, $|x^2 - a^2| < \varepsilon$ if $(2a + 1)|x - a| < \varepsilon$ and $|x - a| < 1$; hence, if $|x - a| < \varepsilon/(2a + 1)$ and $|x - a| < 1$. Finally

$$|x^2 - a^2| < \varepsilon \qquad \text{whenever } |x - a| < \delta$$

where δ is the smaller of $\varepsilon/(2a + 1)$ and 1. Thus, if $\varepsilon = a/10$, then $\delta = a/(20a + 10)$ since it is the smaller of $a/(20a + 10)$ and 1.

3.9   Proofs of the limit theorems

We shall now prove two of the three theorems on limits that are given in Sec. 3.3. In each of them, we use

$$\lim_{x \to a} g(x) = L \qquad \text{and} \qquad \lim_{x \to a} h(x) = M$$

The first of the theorems states that the limit of a sum is equal to the sum of the limits. To prove this, we recall that $\lim_{x \to a} g(x) = L$ means that given any $\varepsilon_1 > 0$, there exists $\delta_1 > 0$ such that $|f(x) - L| < \varepsilon_1$ for $|x - a| < \delta_1$ and that $\lim_{x \to a} h(x) = M$ means that given $\varepsilon_2 > 0$, there exists $\delta_2 > 0$ such that $|h(x) - M| < \varepsilon_2$ for $|x - a| < \delta_2$. Consequently, if we choose δ to be the smaller of δ_1 and δ_2 and $\varepsilon/2$ to be the larger of ε_1 and ε_2, we have

$$|g(x) - L| < \frac{\varepsilon}{2} \qquad \text{and} \qquad |h(x) - M| < \frac{\varepsilon}{2} \qquad \text{if } |x - a| < \delta$$

Therefore, adding corresponding members of the two inequalities that involve $\varepsilon/2$, we see that

$$|g(x) - L| + |h(x) - M| < \varepsilon \qquad \text{if } |x - a| < \delta$$

hence, since the absolute value of the sum of two numbers is less than or equal to the sum of the absolute values, we have

$$|g(x) - L + h(x) - M| < \varepsilon \qquad \text{if } |x - a| < \delta$$

and

$$|g(x) + h(x) - (L + M)| < \varepsilon \qquad \text{if } |x - a| < \delta$$

Consequently,

$$\lim_{x \to a} [g(x) + h(x)] = L + M \tag{1}$$

This theorem can be extended to any finite number of functions. If the domains of $g(x)$ and $h(x)$ are G and H, respectively, then the domain of $g(x) + h(x)$ is $G \cap H$.

In proving that the limit of a product is the product of the limits, we shall make use of

$$\lim_{x \to a} kf(x) = k \lim_{x \to a} f(x) \tag{2}$$

We shall begin by proving this statement for $k > 0$. If $\lim_{x \to a} f(x) = F$, then given $\varepsilon_1 > 0$, there exists $\delta > 0$ such that $|f(x) - F| < \varepsilon_1$ if $|x - a| < \delta$.

Multiplying each member of $|f(x) - F| < \varepsilon_1$ by $k > 0$, we have

$$k|f(x) - F| < k\varepsilon_1 \qquad \text{if } |x - a| < \delta$$
$$|kf(x) - kF| < \varepsilon \qquad \text{if } |x - a| < \delta, \text{ and } \varepsilon = k\varepsilon_1$$

Therefore, $\qquad \lim_{x \to a} kf(x) = kF = k \lim_{x \to a} f(x)$

We shall now prove that the limit of a product is equal to the product of the limits. Since $\lim\limits_{x \to a} g(x) = L$ and $\lim\limits_{x \to a} h(x) = M$, we know that given any $\varepsilon_1 > 0$, there exists $\delta_1 > 0$ such that

$$|g(x) - L| < \varepsilon_1 \qquad \text{for } |x - a| < \delta_1$$

and given any $\varepsilon_2 > 0$, there exists $\delta_2 > 0$ such that

$$|h(x) - M| < \varepsilon_2 \qquad \text{for } |x - a| < \delta_2$$

If we designate the smaller of δ_1 and δ_2 by δ and the larger of ε_1 and ε_2 by $\sqrt{\varepsilon}$, we have

$$|g(x) - L| < \sqrt{\varepsilon} \qquad \text{and} \qquad |h(x) - M| < \sqrt{\varepsilon} \qquad \text{if } |x - a| < \delta$$

Therefore, $\qquad |g(x) - L|\,|h(x) - M| < \varepsilon \qquad \text{if } |x - a| < \delta$

and it follows that

$$|g(x)h(x) - Mg(x) - Lh(x) + LM| < \varepsilon \qquad \text{if } |x - a| < \delta$$

Hence, $\quad \lim\limits_{x \to a} g(x)h(x) - \lim\limits_{x \to a} Mg(x) - \lim\limits_{x \to a} Lh(x) + LM = 0$

and $\qquad \lim\limits_{x \to a} g(x)h(x) - ML - LM + LM = 0 \qquad \textcolor{brown}{by\ (2)}$

Now combining $-LM$ and LM and adding ML to each member, we have

$$\lim\limits_{x \to a} g(x)h(x) = ML$$

as we wanted to prove.

This theorem can also be extended to any finite number of functions. The domain of $g(x)h(x)$ is the $G \cap H$ if G and H are the domains of $g(x)$ and $h(x)$.

EXERCISE 3.2

1   If $\lim\limits_{x \to a} f(x) = 0$ and $\lim\limits_{x \to a} g(x) = k \neq 0$, what can be said about $\lim\limits_{x \to a} f(x)/g(x)$?

2   If $\lim\limits_{x \to a} f(x) \neq 0$ and $\lim\limits_{x \to a} g(x) = 0$, what can be said about $\lim\limits_{x \to a} f(x)/g(x)$?

3   If $\lim\limits_{x \to a} f(x) = h$ and $\lim\limits_{x \to a} g(x) = k$, what can be said about $\lim\limits_{x \to a} f(x)g(x)$?

4   If $\lim\limits_{x \to a} f(x) = A$ and $\lim\limits_{x \to a} g(x) = A$, what can be said about $\lim\limits_{x \to a} [f(x) - g(x)]$?

5   Sketch the graph of the function $4^{-1/x}$. Discuss the continuity of the function.

6   Sketch the graph of the function $1/(1 + 2^{1/x})$. Discuss the continuity of the function.

7   Sketch the graph of the function $2^{1/x}/(1 + 2^{1/x})$. Discuss the continuity of the function.

8   If $\lim_{x \to a} f(x) = L$ and $\lim_{x \to a} g(x) = M \neq 0$, prove that $\lim_{x \to a} [f(x)/g(x)] = L/M$.

In Problems 9 to 16 prove by use of the ε, δ definition that each function is continuous for the given value of x.

9 $f(x) = x^2 + 2$, $x = 3$ **10** $f(x) = 3x + 5$, $x = a$

11 $f(x) = x^3$, $x = 2$ **12** $f(x) = x^3 - x$, $x = 4$

13 $f(x) = \sqrt{2x - 1}$, $x = 5$ **14** $f(x) = \sqrt{3x + 1}$, $x = 5$

15 $f(x) = \dfrac{x + 2}{x + 1}$, $x = a \neq -1$

16 $f(x) = \dfrac{3x - 1}{2x + 3}$, $x = a \neq -1.5$

In Problems 17 to 20, assume that f(x) and g(x) are continuous for x = a.

17   Prove that the sum of two continuous functions is continuous.

18   Prove that the product of a constant and a continuous function is continuous.

19   Prove that the product of two continuous functions is continuous.

20   Prove that $f(x)/g(x)$ is continuous for $x = a$ unless $g(a) = 0$.

4

The derivative of a function

4.1   Introduction

The fundamental problem of differential calculus is that of finding the rate of change of the value of a function $f(x)$ relative to x. We shall now consider in detail some simple examples in order to make the basic idea clear.

Example 1   If the side of the square shown on the left in Fig. 4.1 is x_1 in., where x_1 is any positive number, then the corresponding area is given by the relation $A_1 = x_1^2$. We shall assume that the side is increasing and shall find the rate at which the area is increasing *relative to the side*. Thus, we want to find the number of square inches that are added to the area for each inch added to the side.

For this purpose we observe that if the side is increased to x in., the area increases to a new value $A = x^2$ sq in. The increase in area is

$$A - A_1 = (x^2 - x_1^2) \text{ sq in.}$$

and this increase results from an increase of $(x - x_1)$ in. in the side. We shall now consider the fraction

$$\frac{\Delta A}{\Delta x} = \frac{A - A_1}{x - x_1} = \frac{x^2 - x_1^2}{x - x_1}$$

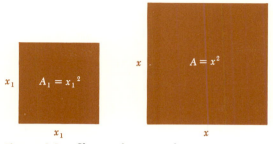

Figure 4.1   Change in area of a square

This fraction is obtained by dividing the number of square inches added to the area by the corresponding number of inches added to the side; it gives the *average* number of square inches added to the area per inch added to the side. It is called the *average rate of change of the area relative to the side* over the interval $[x_1, x]$.

average rate of change

As x is taken progressively nearer to x_1, the above quotient gives the corresponding average rate of change over smaller intervals, all of which have as one end point the fixed value x_1. Hence, we define the *instantaneous* rate of change of the area with respect to the side as

instantaneous rate of change

$$\lim_{x \to x_1} \frac{A - A_1}{x - x_1} = \lim_{x \to x_1} \frac{x^2 - x_1{}^2}{x - x_1} = \lim_{x \to x_1} \frac{(x + x_1)(x - x_1)}{x - x_1}$$

$$= \lim_{x \to x_1} (x + x_1) = 2x_1$$

This result shows that if the side of a square is x_1 in. long, the rate of change of the area relative to the side is $2x_1$ sq in. per in. Any other unit of measurement may be used instead of inches. Thus, for $x_1 = 3$ ft the area is increasing at 6 sq ft per ft, and for $x_1 = 9$ miles the area is increasing at 18 sq miles per mile.

This result has the following nongeometric interpretation: As x changes, the value of the function x^2 also changes. The rate at which x^2 changes *relative to* x is always equal to $2x$. Thus if $x = 8$, then x^2 is changing 16 times as fast as x.

Example 2   Find the rate of change of the volume of a sphere relative to its radius.

Solution: The volume V_1 is expressed in terms of a specified radius r_1 by the formula $V_1 = 4\pi r_1{}^3/3$. If a sphere has initial radius r_1 and changes to a new value r, we see that the volume changes from an initial value V_1 to a new value V, where

$$V = \frac{4\pi r^3}{3} \qquad \text{and} \qquad V_1 = \frac{4\pi r_1{}^3}{3}$$

The change in volume is

$$\Delta V = V - V_1 = \frac{4\pi r^3}{3} - \frac{4\pi r_1{}^3}{3}$$

The *average rate of change* of the volume relative to the radius is obtained by dividing the change in volume by the corresponding change in radius as indicated by

$$\frac{\Delta V}{\Delta r} = \frac{V - V_1}{r - r_1} = \frac{4\pi r^3/3 - 4\pi r_1{}^3/3}{r - r_1} = \frac{4\pi(r^3 - r_1{}^3)/3}{r - r_1}$$

The required instantaneous rate is the limit of this fraction as r approaches r_1. Thus,

$$\lim_{r \to r_1} \frac{V - V_1}{r - r_1} = \lim_{r \to r_1} \frac{4\pi}{3} \frac{r^3 - r_1{}^3}{r - r_1} = \lim_{r \to r_1} \frac{4\pi}{3}(r^2 + rr_1 + r_1{}^2)$$

$$= \frac{4\pi}{3}(r_1{}^2 + r_1{}^2 + r_1{}^2) = 4\pi r_1{}^2$$

Consequently, the rate of change of the volume relative to the radius is equal to $4\pi r^2$ for any value of r. If the radius is 3 ft and is increasing, then at this instant the volume is increasing at a rate of $4\pi(3)^2 = 36\pi$ or about 113 cu ft per ft change in radius.

Example 3 If a rock is dropped, its distance below the starting point at the end of t_1 sec is given approximately by the formula $S_1 = 16t_1{}^2$, where S_1 is in feet. Find the rate of change of distance relative to time and interpret this result.

Solution: At the end of t_1 sec the rock is S_1 ft below the starting point, and at the end of t sec it is S ft below the starting point, where

$$S_1 = 16t_1{}^2 \qquad \text{and} \qquad S = 16t^2$$

as indicated in Fig. 4.2. The distance traveled by the rock in the time interval from t_1 to t is

$$S - S_1 = 16t^2 - 16t_1{}^2$$

If we divide this distance by the corresponding time interval $t - t_1$, we obtain the *average number of feet traveled by the rock per second* over this interval. The *average speed* of the rock over this interval is

$$\frac{\Delta S}{\Delta t} = \frac{S - S_1}{t - t_1} = \frac{16t^2 - 16t_1{}^2}{t - t_1}$$

As we take t closer and closer to t_1, this fraction gives the average speed over smaller and smaller time intervals, all of which have t_1 as one end point. The *instantaneous speed* at the time t_1 is defined as

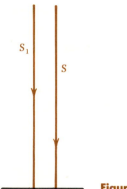

Figure 4.2 Speed of a falling body

$$\lim_{t \to t_1} \frac{S - S_1}{t - t_1} = \lim_{t \to t_1} \frac{16t^2 - 16t_1{}^2}{t - t_1} = \lim_{t \to t_1} 16(t + t_1) = 32t_1$$

The rate of change of S relative to t is thus equal to $32t$ ft per sec at any time t and may be interpreted as the speed of the rock.

EXERCISE 4.1

1 In Fig. 4.1 assume that $x = 4$ in. so that $A = 16$ sq in. Go through the steps of Example 1 using 4 and x_1 instead of x and x_1, and thus find the rate of change of the area relative to x for the instant that $x = 4$ in. Compute the average rate of change of the area with respect to x over the interval from $x = 4$ to 4.5 in.

2 If the edge of a cube is 2 in., its surface area is $6(2)^2$ or 24 sq in. If the edge is increased to x in., the new surface area will be $6x^2$ sq in. Find the value of the fraction

$$\frac{6x^2 - 24}{x - 2}$$

if $x = 2.5$ and interpret the result. Find the limit of the fraction as $x \to 2$ and interpret this result.

3 Find the average rate of change of the volume of a cube relative to its edge x over the interval from $x = 2.4$ to 2.6 in. Find the instantaneous rate if $x = 2.4$ in.

4 Find the average rate of change of the surface area of a cube relative to its edge x over the interval from $x = 4$ to 4.5 in. Find the instantaneous rate if $x = 4$ in.

5 Find the rate of change of the area of a circle relative to its radius.

6   Find the rate of change of the surface area of a sphere relative to its radius.

7   Find the rate of change of the volume of a cube relative to its edge.

8   Find the rate of change of the area of an equilateral triangle relative to a side.

9   A rectangular box has a square base of side x ft and its height is $4x$ ft. Find the rate of change of its volume relative to x.

10   A right pyramid has a square base of side x in. and its height is $6x$ in. Find the rate of change of its volume relative to x.

11   Find the rate of change of the volume of a cube relative to its surface area. *Hint:* If the edge is x_1 then the volume is $x_1{}^3$ and the surface area is $6x_1{}^2$. If the edge changes to a new value x, the volume becomes x^3 and the surface area becomes $6x^2$. Consequently,

$$\frac{V - V_1}{S - S_1} = \frac{x^3 - x_1{}^3}{6x^2 - 6x_1{}^2}$$

12   Find the rate of change of the volume of a sphere relative to the area πr^2 of a great circle.

13   Find the rate of change of the volume of a right circular cylinder of radius r and height h relative to the cross-sectional area.

14   Find the rate of change of the volume of a sphere relative to its surface area.

15   The radius r of a right circular cylinder is increasing and its height h is constant. Find the rate at which its total surface area increases relative to r.

16   Find the rate of change of the volume of a right circular cone relative to its radius r if its height h is a constant.

17   Show that when $x = 25$, the square root of x is increasing one-tenth as fast as x.

18   Find, for any positive value of x, the rate of change of the square root of x relative to x.

19   Find the average rate of change of the value of the function $x^2 - 4x$ relative to x over the interval from $x = 7$ to 9. Find the instantaneous rate when $x = 7$.

20   Find the average rate of change of the value of the function $x^3 + 6$ relative to x over the interval from $x = 5$ to 7. Find the instantaneous rate when $x = 6$.

21   Find the rate of change of the value of the function $1/x$ relative to x. What is the significance of the negative sign?

22   Find the rate of change of the value of the function $(x + 1)/(x - 1)$ relative to x. In particular, show that if x has the value 3 and is increasing, the value of this function is decreasing one-half as fast as x is increasing.

23 Show that for $x = 1$ the value of the function $x^2 + \sqrt{x}$ increases 2½ times as fast as x.

24 Show that for $x = 4$ the value of the function $1/\sqrt{x}$ decreases one-sixteenth as fast as x increases.

4.2 The derivative of a function

We may apply the idea of the preceding section to any function $y = f(x)$ in order to investigate the rate of change of $f(x)$ relative to x.

If x_1 and $f(x_1)$ are a corresponding pair of numbers and if x and $f(x)$ are another corresponding pair, then the fraction

$$\frac{f(x) - f(x_1)}{x - x_1}$$

is called the *average rate of change* of the function value relative to x over the interval from x_1 to x. The *limit* of this fraction as $x \to x_1$ is, by definition, the corresponding instantaneous rate of change of $f(x)$ at $x = x_1$ in keeping with our discussion in Sec. 4.1. This limit is *the derivative* also called the *derivative* with respect to x of the given function evaluated at $x = x_1$. This definition will now be put in a more formal manner by stating that

If $$\lim_{x \to x_1} \frac{f(x) - f(x_1)}{x - x_1}$$

exists, then the function $f(x)$ is said to be differentiable for the number x_1, and the limit is called the derivative of $f(x)$ with respect to x evaluated at $x = x_1$.

We shall frequently use the symbol $f'(x)$ to denote the derivative. In terms of this notation,

$$f'(x_1) = \lim_{x \to x_1} \frac{f(x) - f(x_1)}{x - x_1} \tag{1}$$

The graphical significance of the derivative may be seen from Fig. 4.3. Since $f(x_1) = MP_1$ and $f(x) = NP$, it follows that $f(x) - f(x_1) = LP$. Also $x_1 = OM$ and $x = ON$; hence $x - x_1 = MN = P_1L$. Therefore,

$$\frac{f(x) - f(x_1)}{x - x_1} = \frac{LP}{P_1L} = \tan \phi$$

Consequently, the average rate of change of the value of the function relative to x over the interval from x_1 to x is equal to the slope of the chord P_1P.

Now as x assumes values nearer to x_1, P moves along the curve and approaches P_1. The limiting position, if any, of the secant line P_1P as

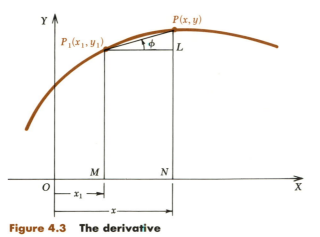

Figure 4.3   The derivative

geometric interpretation

$P \rightarrow P_1$ is, by definition, the *tangent line* to the curve at P_1. Hence, *the value of the derivative $f'(x)$, for any value of x for which it exists, is equal to the slope of the tangent line to the graph of the function at the corresponding point of the curve.*

The definition of the derivative can be stated in a form that is often desirable by letting the points be $P(x, y)$ and $Q(x + \Delta x, y + \Delta y)$ as in Fig. 4.4. If this is done, the definition of the derivative becomes:

$$\text{If} \qquad \lim_{\Delta x \to 0} \frac{f(x + \Delta x) - f(x)}{\Delta x} \qquad (2)$$

exists, then the function $y = f(x)$ is differentiable for the number x and the limit is the derivative of $f(x)$ with respect to x.

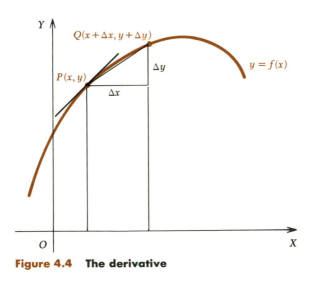

Figure 4.4   The derivative

Example   Find the derivative with respect to x of the function $6x^2 + 4$.

Solution: We shall let $f(x) = 6x^2 + 4$ and denote the required derivative by $f'(x)$. Then, if we use form (1) of the definition, we have

$$f'(x) = \lim_{x \to x_1} \frac{f(x) - f(x_1)}{x - x_1} = \lim_{x \to x_1} \frac{(6x^2 + 4) - (6x_1^2 + 4)}{x - x_1}$$

$$= \lim_{x \to x_1} \frac{6(x^2 - x_1^2)}{x - x_1} = \lim_{x \to x_1} 6(x + x_1) = 12x_1$$

We obtain the same result by using form (2) of the definition of the derivative. Thus,

$$f'(x) = \lim_{\Delta x \to 0} \frac{f(x + \Delta x) - f(x)}{\Delta x} = \lim_{\Delta x \to 0} \frac{[6(x - \Delta x)^2 + 4] - (6x^2 + 4)}{\Delta x}$$

$$= \lim_{\Delta x \to 0} \frac{6[x^2 + 2x\,\Delta x + (\Delta x)^2 - x^2]}{\Delta x} = \lim_{\Delta x \to 0} 6(2x + \Delta x) = 12x$$

4.3 Notation

We have just found that for any value of x, the derivative with respect to x of the function $6x^2 + 4$ is $12x$. If we write $f(x) = 6x^2 + 4$, then we can write $f'(x) = 12x$.

If we denote the function values by y and write $y = 6x^2 + 4$, then we may denote the corresponding values of the derivative by y' and write $y' = 12x$. Several notations are used to denote the derivative of $y = f(x)$ with respect to x. They include

$$D_x f, \; D_x y, \; \frac{df}{dx}, \; \text{and} \; \frac{dy}{dx}$$

Hence, if $y = f(x) = 6x^2 + 4$, we can write

$$D_x y = D_x(6x^2 + 4) = 12x \qquad \text{or} \qquad \frac{d}{dx}(6x^2 + 4) = 12x$$

The symbol dy/dx may be regarded as arising in the manner pictured in Fig. 4.4. In the figure, $\Delta x = x + \Delta x - x$ and $f(x + \Delta x) - f(x) = y + \Delta y - y = \Delta y$. Hence,

$$\lim_{\Delta x \to 0} \frac{f(x + \Delta x) - f(x)}{\Delta x} \qquad \text{becomes} \qquad \lim_{\Delta x \to 0} \frac{\Delta y}{\Delta x}$$

and we let dy/dx denote the limit of the fraction $\Delta y/\Delta x$ as Δx approaches zero. We shall use the symbol $\dfrac{dy}{dx}\bigg|_{x=a} = \dfrac{dy}{dx}(a)$, or $D_x y \bigg|_{x=a} = D_x y(a)$, to denote the value of $D_x y$ at the point for which $x = a$.

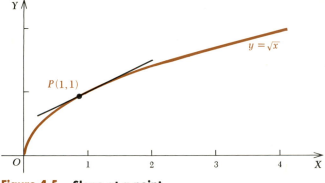

Figure 4.5   Slope at a point

Example

If $y = \sqrt{x}$ then

$$D_x y \Big|_{x=x_1} = \lim_{x \to x_1} \frac{\sqrt{x} - \sqrt{x_1}}{x - x_1} = \lim_{x \to x_1} \frac{\sqrt{x} - \sqrt{x_1}}{x - x_1} \frac{\sqrt{x} + \sqrt{x_1}}{\sqrt{x} + \sqrt{x_1}}$$

$$= \lim_{x \to x_1} \frac{x - x_1}{x - x_1} \frac{1}{\sqrt{x} + \sqrt{x_1}}$$

Since, if $x \neq x_1$, then $(x - x_1)/(x - x_1) = 1$

$$D_x y \Big|_{x=x_1} = \lim_{x \to x_1} \frac{1}{\sqrt{x} + \sqrt{x_1}} = \frac{1}{2\sqrt{x_1}} \qquad \text{if } x_1 \neq 0$$

If $x_1 = 0$, the limit is $1/2\sqrt{0}$, which is not a number since division by zero is not a permissible operation. Hence, we see that $\sqrt{x}$ is not differentiable for $x = 0$. Also, since we wish to confine our attention to the field of real numbers, we require that x must not be negative. For any $x > 0$

$$\frac{d}{dx}(\sqrt{x}) = D_x(\sqrt{x}) = \frac{1}{2\sqrt{x}}$$

In particular, if $x = 1$, the value of this derivative is ½. This means that the slope of the tangent line to the graph at the point $(1, 1)$ is ½ as shown in Fig. 4.5. It also means that if $x = 1$, $\sqrt{x}$ changes one-half as fast as x.

EXERCISE 4.2

In each of Problems 1 to 20 find the derivative with respect to x of the given function using form (2) of the definition.

1 $3x^2 + 2$	**2** $4x^2 + 3x$	**3** $x^2 - 5x$
4 $x^3/2$	**5** $x^3 - 2x$	**6** $2x^3 + 3x$

7 $2x^2 - 7x$	8 $x^4 - 3x^2$	9 $3/x$
10 $4/(x + 2)$	11 $x/(2x + 1)$	12 $(x + 2)/(x + 3)$
13 $8/(x^2 + 4)$	14 $1/\sqrt{x}$	15 $3/(x^2 - 2)$
16 $x/(x^2 + 2)$	17 $8x/(x^2 - 16)$	18 $x^2/(x^2 + 4)$
19 $(x^2 - 1)/(x^2 + 1)$	20 $(x^2 + 4)/(x^2 - 4)$	

21   Given $y = 4t^3 + 3$, find $D_t y$.
22   Given $w = u^2 - 3u$, find $D_u w$.
23   Given $S = 4\sqrt{w}$, find ds/dw.
24   Given $A = 4\pi r^2$, find dA/dr.
25   Given $y = \sqrt{x + 2}$, find $D_x y$.
26   Given $Q = t(2t + 1)$, find dQ/dt.
27   Given $S = (t - 1)/(t^2 + 4)$, find $D_t S$.
28   Given $T = (u - 3)/(u^2 + 2)$, find $D_u T$.

In each of Problems 29 to 36 sketch the graph of the given equation, and find the slope of the tangent line to the graph at the points indicated. Draw the corresponding tangent lines.

29   $y = x^2/2 + x$; $(-2, 0)$, $(3, 2.5)$
30   $y = 2x/3 + 1$; $(-1, \frac{1}{3})$, $(3, 3)$
31   $y = 2\sqrt{x}$; $(1, 2)$, $(4, 4)$
32   $y = x^3/2$; $(1, \frac{1}{2})$, $(2, 4)$
33   $y = 4/x$; $(2, 2)$, $(8, \frac{1}{2})$
34   $y = (x + 2)/(x - 1)$; $(0, -2)$, $(-2, 0)$
35   $y = (3x - 6)/(x + 2)$; $(2, 0)$, $(-5, 7)$
36   $y = 4x/(x + 4)$; $(-2, -4)$, $(0, 0)$

37   If $Q = 6/(t + 1)$, what is the instantaneous rate of change of Q relative to t at $t = 3$?
38   If $R = 3/(2t - 1)$, find the instantaneous rate of change of R relative to t at $t = 2$.
39   Find the instantaneous rate of change of y with respect to x at $x = 4$ if $y = 2x/(x + 2)$.
40   What is the instantaneous rate of change of y with respect to x at $x = 3$ if $y = x^2/(2x - 3)$?

4.4   Concerning the existence of the derivative

In order to make our examples as simple as possible, we have usually thought of x as being greater than x_1 so that $\Delta x = x - x_1$ is positive. It must be remarked that x may be either greater or less than the initial value x_1. In fact when we say that

$$\lim_{x \to x_1} \frac{f(x) - f(x_1)}{x - x_1}$$

exists, we mean that there is a certain number such that the values of the fraction are arbitrarily near this number for *all* values of x that are sufficiently near to x_1, regardless of whether $x > x_1$ or $x < x_1$.

right-hand and left-hand derivatives

If x is restricted to values *greater* than x_1 in the above limit, the corresponding derivative is called the *right-hand derivative*. If $x < x_1$, it is called the *left-hand derivative*. Both of these limits must exist and they must be equal, if the function is to have a derivative at the point in question.

It should be noted that when $x \to x_1$, the denominator of the fraction approaches zero. The fraction cannot then have a limit unless the numerator also approaches zero. This requires that

$$\lim_{x \to x_1} f(x) = f(x_1)$$

necessary condition for a derivative

which means that the function must be *continuous* at a point if it is to have a derivative at that point. Since the limit need not exist even if the numerator *does* approach zero, a continuous function need not have a derivative. Thus, continuity is a *necessary* but not a *sufficient* condition for differentiability.

Example

The function $1 + \sqrt[3]{x^2}$, whose graph is shown in Fig. 4.6, is continuous for all values of x. We can show that it does not have a derivative at $x = 0$ by considering the quotient

$$\frac{f(x) - f(0)}{x - 0} = \frac{(1 + \sqrt[3]{x^2}) - 1}{x} = \frac{1}{\sqrt[3]{x}}$$

This fraction has no limit as $x \to 0$. It turns out that the function does have a derivative for all other values of x.

Most of the functions with which we shall be concerned will be continuous over the interval in which they are defined, and they will have derivatives except possibly at certain isolated points.

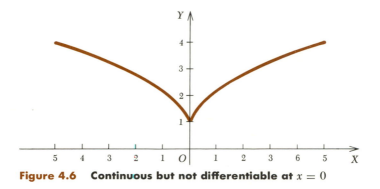

Figure 4.6   Continuous but not differentiable at $x = 0$

4.5 Increasing and decreasing functions

The function $y = f(x)$ is said to be *increasing* at $x = x_1$ if there is an interval I enclosing x_1 in which $f(x) \leq f(x_1)$ for all values of x in I to the left of x_1 and $f(x) \geq f(x_1)$ for all values of x in I to the right of x_1. If $y = f(x)$ is increasing for all x in an interval, then $y = f(x)$ is said to be increasing in the interval. A decreasing function is defined in a corresponding manner.

Example 1   The function $\sqrt{x}$ is an increasing function for all positive values of x, since if a and b are any two positive numbers with $b > a$, then $\sqrt{b} > \sqrt{a}$. The graph, shown in Fig. 4.5, rises from left to right.

Let x_1 and x be two values of x in an interval in which $f(x)$ is increasing. If $x > x_1$, then $f(x) \geq f(x_1)$, and if $x < x_1$, then $f(x) \leq f(x_1)$. Thus the numerator and denominator of the fraction

$$\frac{f(x) - f(x_1)}{x - x_1}$$

are both positive or both negative, and the value of the fraction is positive, or the numerator and the fraction are zero. Its limit as $x \to x_1$, if it exists, must then be a positive number or zero. Therefore we know that:

derivative of an increasing function   *If $f(x)$ is increasing over an interval, the value of its derivative at any given point of the interval must be either a positive number or zero.*

derivative of a decreasing function   Similarly, *if $f(x)$ is decreasing, its derivative is either negative or zero.*

Conversely, it can be shown that if $f'(x)$ is positive for a certain value of x, then the function $f(x)$ is increasing over at least a small interval about that point and if $f'(x)$ is negative for a certain value of x then $f(x)$ is decreasing in the immediate neighborhood of that point.

Example 2   If $y = (x^3 - 12x)/4$, over what set of values of x does y decrease as x increases?

Solution: By using the usual procedure we can find that

$$D_x y = \frac{3(x^2 - 4)}{4}$$

This derivative is negative if and only if x is between -2 and $+2$, and it is therefore in this interval that y decreases as x increases.

Observe that $D_x y = 0$ if $x = -2$ and if $x = 2$ and that the tangent line to the graph is horizontal and has zero slope at the corresponding points A and B in Fig. 4.7. For $x = 1$, $D_x y = -2\frac{1}{4}$ and this is the

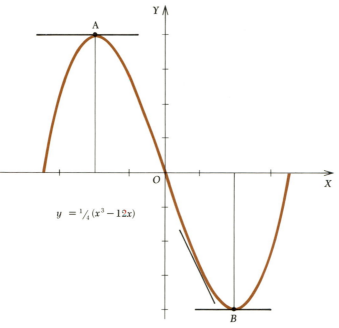

$$y = \tfrac{1}{4}(x^3 - 12x)$$

Figure 4.7   Increasing and decreasing intervals

slope of the corresponding tangent line. Thus, if $x = 1$, the value of $y = (x^3 - 12x)/4$ is *decreasing* at an instantaneous rate of 2¼ units per unit of increase in x.

EXERCISE 4.3

1   Give an opinion as to the truth of each of the following assertions:
   (*a*) If $f(x)$ is a constant for $a \le x \le b$, then the average rate of change of $f(x)$ over this interval is zero.
   (*b*) If the average rate of change of $f(x)$ over the interval $a \le x \le b$ is zero, then $f(x)$ is constant over this interval.
2   Show that the average rate of change of the function $f(x) = 10x - x^2$ is zero over the interval $2 \le x \le 8$. Over what part of the interval is $f(x)$ increasing and over what part is it decreasing? What is the largest value of $f(x)$ in the interval?
3   Show that the function $\phi(x) = \sqrt{x}$ does not have a right-hand derivative at $x = 0$. *Hint:* Consider the fraction

$$\frac{\phi(x) - \phi(0)}{x - 0} = \frac{\sqrt{x} - 0}{x - 0} \qquad x > 0$$

Show that it does not have a limit as $x \to 0^+$.

4   Show that the function $f(x) = \sqrt{1 - x^2}$ does not have a left-hand derivative at the point for which $x = 1$. *Hint:* Consider the fraction

$$\frac{f(x) - f(1)}{x - 1} = \frac{\sqrt{1 - x^2} - 0}{x - 1} \qquad 0 < x < 1$$

Show that it does not have a limit as $x \to 1^-$.

In each of Problems 5 to 10 determine the x interval or intervals over which the value of y is increasing.

5   $y = x^2 - 8x$ 6 $y = 4x - x^2$

7   $y = 2/x$ 8 $y = -5/(x + 1)$

9   $y = x^3 - 6x^2$ 10 $y = 3x - x^3$

11   Show that the function $(x + 1)/(x - 1)$ is a decreasing function over any interval that does not include the point $x = 1$.

12   Let $y = 1/(x^2 + 1)$. Show that y decreases as x increases if $x > 0$ and that y increases as x increases if $x < 0$.

13   If $y = 4x/(x^2 + 4)$, show that y increases as x increases over the interval $-2 < x < 2$. Draw the corresponding graph. What slope does the curve have at the origin? At what points is the slope equal to zero?

14   Sketch the graph of the equation $y = 8/(x^2 - 4)$. Use $D_x y$ to show that y increases as x increases if $x < -2$ and if $-2 < x < 0$. What is the slope of the tangent line to the graph at the point $(4, \frac{2}{3})$?

15   If $y = x^3$, find the value of $D_x y$ for $x = 2$ by evaluating

$$\lim_{x \to 2} \frac{x^3 - 2^3}{x - 2} \qquad \text{or} \qquad \lim_{\Delta x \to 0} \frac{(2 + \Delta x)^3 - 2^3}{\Delta x}$$

16   If $y = 4/x$, find the value of $D_x y$ at $x = 2$ by evaluating

$$\lim_{x \to 2} \frac{4/x - 4/2}{x - 2} \qquad \text{or} \qquad \lim_{\Delta x \to 0} \frac{4/(2 + \Delta x) - 4/2}{\Delta x}$$

17   If $y = \sqrt{x + 2}$, find $D_x y$ in terms of x by evaluating

$$\lim_{x \to x_1} \frac{\sqrt{x + 2} - \sqrt{x_1 + 2}}{x - x_1}$$

Check by writing down the corresponding fraction involving Δx and finding its limit as $\Delta x \to 0$. For what values of x is the result valid in the field of real numbers?

18   If $y = x^4$, find $D_x y$ in terms of x by evaluating

$$\lim_{x \to x_1} \frac{x^4 - x_1{}^4}{x - x_1}$$

Check by writing down the corresponding fraction involving Δx and finding its limit as $\Delta x \to 0$.

19 A right pyramid has a rectangular base that is x ft wide and $2x$ ft long. The height of the pyramid is $3x$ ft. Express its volume V as a function of x. Find the value of $D_x V$ for $x = 2\frac{1}{2}$ ft. In what units is this rate expressed?

20 The number N of grams of a substance in solution varies with the time t in minutes that the substance has been in contact with the solvent in accordance with the formula $N = 16/(t + 5)$. Find the value of $D_t N$ if $t = 3$. In what units is this rate expressed?

21 If an object is dropped from the top of a cliff and falls under the action of gravity alone, its distance S in feet below the top of the cliff at the end of t sec is given approximately by the formula $S = 16t^2$. Find the value of $D_t S$ when $t = 4$, and explain the meaning of the result. In what units is this rate expressed?

22 A cylinder contains 200 cu in. of air at a pressure of 15 lb per sq in. If the air is now compressed by moving a piston in the cylinder and if this is done under a condition of constant temperature, the pressure will increase as the volume decreases, the relation between them being $pv = 3{,}000$. Find the value of $D_v p$ for $v = 50$ cu in. In what units is this rate expressed?

23 A certain quantity Q varies with the time t in accordance with the formula $Q = 8t^2 - t^3$, $0 \le t \le 8$. Over what part of this time interval is Q increasing?

24 If r in. is the radius of a sphere, then the number of cubic inches in its volume and the number of square inches in its surface area are given, respectively, by the functions $4\pi r^3/3$ and $4\pi r^2$. Evaluate

$$\lim_{r \to r_1} \frac{4\pi r^3/3 - 4\pi r_1^3/3}{4\pi r^2 - 4\pi r_1^2}$$

What physical meaning can be attached to the result?

5 Differentiation of algebraic functions

5.1 Introduction

Heretofore, we have obtained derivatives by use of the limiting process. We shall now use this fundamental procedure to obtain differentiation formulas. In this chapter, we shall derive only those formulas that are needed in differentiating algebraic functions. Consequently, we shall begin by developing formulas for the derivative of a constant, of a constant times a function, of a sum, and of a power. We shall then be able to differentiate a polynomial. The remainder of the chapter will be devoted to the derivation and use of formulas for the derivatives of a quotient, a function of a function, and an implicit function.

5.2 The derivative of a polynomial

We shall represent functions that have derivatives for the values of x under consideration by $u(x)$ and $v(x)$ or, in this chapter, merely by u and v. We shall represent the increments of u and v that correspond to the increment Δx of x by Δu and Δv, respectively. It follows that Δu and Δv approach zero as Δx does since u and v are differentiable functions of x. Finally, we shall represent the limits of $\Delta u/\Delta x$ and $\Delta v/\Delta x$ as Δx approaches zero by du/dx and dv/dx, respectively.

We shall now find the derivative of a constant by use of the limit process. If

$$y = c$$

then $\qquad y + \Delta y = c \qquad$ for c a constant

$$\Delta y = c - c = 0 \qquad \frac{\Delta y}{\Delta x} = \frac{0}{\Delta x} = 0$$

Consequently, $\qquad \lim_{\Delta x \to 0} \frac{\Delta y}{\Delta x} = 0$

derivative of a constant and $\qquad D_x c = 0 \qquad$ for c a constant $\qquad$ (5.1)

This is as we should expect since a constant does not change and the derivative is the rate of change of one quantity relative to another.

We shall now consider the product of a constant and a differentiable function and obtain a formula for its derivative. If Δy and Δu are the changes in y and u, respectively, due to a change Δx in x and if

$$y = cu$$

then $\qquad y + \Delta y = c(u + \Delta u)$

$$\Delta y = c\,\Delta u \qquad \frac{\Delta y}{\Delta x} = c\,\frac{\Delta u}{\Delta x}$$

Therefore,

$$\lim_{\Delta x \to 0} \frac{\Delta y}{\Delta x} = \lim_{\Delta x \to 0} c\,\frac{\Delta u}{\Delta x} = c \lim_{\Delta x \to 0} \frac{\Delta u}{\Delta x} = c\,\frac{du}{dx}$$

Consequently,

$$D_x cu = c D_x u \qquad (5.2)$$

derivative of a constant times a function This formula can be expressed in words by saying that *the derivative of a constant times a differentiable function is the constant times the derivative of the function.*

We shall now consider the sum of two differentiable functions. If

$$y = u + v$$

then $\qquad y + \Delta y = u + \Delta u + v + \Delta v$

$$\Delta y = \Delta u + \Delta v \qquad \frac{\Delta y}{\Delta x} = \frac{\Delta u}{\Delta x} + \frac{\Delta v}{\Delta x}$$

Therefore,

$$\lim_{\Delta x \to 0} \frac{\Delta y}{\Delta x} = \lim_{\Delta x \to 0} \left(\frac{\Delta u}{\Delta x} + \frac{\Delta v}{\Delta x} \right) = \lim_{\Delta x \to 0} \frac{\Delta u}{\Delta x} + \lim_{\Delta x \to 0} \frac{\Delta v}{\Delta x} = \frac{du}{dx} + \frac{dv}{dx}$$

Consequently,

$$D_x(u + v) = D_xu + D_xv \tag{5.3}$$

derivative of a sum and it follows that *the derivative of the sum of two differentiable functions is the sum of their derivatives.* By repeating this procedure, we can readily show that *the derivative of the sum of a finite number of differentiable functions is the sum of their derivatives.*

As a fourth step in the procedure for obtaining the derivative of a polynomial, we shall consider u^n with n a positive integer. If

$$y = u^n$$

then

$$y + \Delta y = (u + \Delta u)^n$$

$$= u^n + \frac{nu^{n-1}}{1} \Delta u + \frac{n(n-1)}{1 \cdot 2} u^{n-2}(\Delta u)^2 + \cdots + (\Delta u)^n$$

$$\Delta y = nu^{n-1}\Delta u + \frac{n(n-1)}{2} u^{n-2}(\Delta u)^2 + \cdots + (\Delta u)^n$$

Therefore, dividing by Δx and indicating the limit, we see that

$$\lim_{\Delta x \to 0} \frac{\Delta y}{\Delta x}$$

$$= \lim_{\Delta x \to 0} \left[nu^{n-1}\frac{\Delta u}{\Delta x} + \frac{n(n-1)}{2} u^{n-2}\frac{\Delta u}{\Delta x} \Delta u + \cdots + \frac{\Delta u}{\Delta x}(\Delta u)^{n-1} \right]$$

hence, $$\frac{dy}{dx} = nu^{n-1}\frac{du}{dx}$$

since the limit of a sum is the sum of the limits and since Δu approaches zero as Δx does. Consequently, if n is a positive integer,

$$D_xu^n = nu^{n-1} D_xu \tag{5.4}$$

derivative of a power of a function and we can say that *the derivative of a constant power of a differentiable function is the exponent times the function to a power one less than the given power times the derivative of the function.*

We have proved (5.4) only for positive integral exponents; however, it is true for all constant exponents. We shall so use it even though the proof will not be given until the chapter on logarithms.

If $u = x$, then $u + \Delta u = x + \Delta x$, $\Delta u = \Delta x$, $\Delta u/\Delta x = 1$, $D_xu = 1$; hence, $D_xx = 1$. We differentiate x^n by use of (5.4) and have

$$D_xx^n = nx^{n-1} \tag{5.4'}$$

Since a polynomial is the sum of a finite number of terms of the form cx^n, we can obtain the derivative of a polynomial by use of (5.1) to (5.4'). Thus,

$$D_x(a_0x^n + a_1x^{n-1} + \cdots + a_{n-1}x + a_n)$$
$$= na_0x^{n-1} + (n-1)a_1x^{n-2} + \cdots + a_{n-1}$$

Example 1 $\qquad\qquad\qquad\qquad D_x7 = 0 \qquad$ *by (5.1)*

Example 2 $\qquad\qquad\qquad\qquad D_x5x^3 = 5D_xx^3 \qquad$ *by (5.2)*
$$= 5(3x^2) \qquad \textit{by (5.4')}$$

Example 3 $\qquad\qquad D(x^4 + x^{-2}) = 4x^3 + (-2)x^{-3} \qquad$ *by (5.3) and (5.4')*

Example 4 $\quad D_x(3x^2 + 1)^{1.5} = 1.5(3x^2 + 1)^{0.5}\,D_x(3x^2 + 1) \qquad$ *by (5.4)*
$$= 1.5(3x^2 + 1)^{0.5}(D_x3x^2 + D_x1) \qquad \textit{by (5.3)}$$
$$= 1.5(3x^2 + 1)^{0.5}(6x) \qquad \textit{by (5.2), (5.4'), and (5.1)}$$
$$= 9x(3x^2 + 1)^{0.5}$$

Example 5 Find the slope of $y = x^3 - 2x^2 + 6x$ for $x = 2$.

Solution: Since the slope of the curve is a geometric interpretation of the derivative, we shall find the derivative and then evaluate it for $x = 2$. Thus, using (5.3), (5.2), and (5.4'), we have

$$D_xy(x) = 3x^2 - 4x + 6$$

Hence, $\qquad\qquad y'(2) = 3(2)^2 - 4(2) + 6 = 10$

EXERCISE 5.1

Find the derivative of the function in each of Problems 1 to 16.

1	$y = x^2 + 3x - 2$	2	$y = x^3 + 3x^2 + 7$
3	$y = 2x^5 + 7x^2 + 5x$	4	$y = 3x^4 - 4x^3 + 6x$
5	$y = x^3 + x^{-2}$	6	$y = 5x^2 + x^{-3}$
7	$y = x^{-4} + 3/x$	8	$y = 2x^{-1} - 2/x$
9	$y = 2x^3 + \sqrt{x}$	10	$y = \sqrt[3]{x} + 2/\sqrt{x}$
11	$y = \sqrt{2x^3 + x}$	12	$y = \sqrt[3]{x^3 + 3x}$
13	$y = (4x^3 + 6x)^{3/2} + x^{-1}$	14	$y = \sqrt{x^4 - 4x^5} - 2x^{-3}$
15	$y = \sqrt{2x - 3} - 1/\sqrt{3x - 2}$		
16	$y = \sqrt[3]{x^6 - 3x^4}$		

Find the derivative with respect to x of the function in each of Problems 17 to 20. Evaluate the derivative for the given value of x.

17 $\quad y = x^3 + 5x - 3\sqrt{x},\ x = 4$

18 $\quad y = x^4 - 3\sqrt[3]{2x},\ x = 4$

19 $\quad y = \sqrt{2x - 1} + 1/\sqrt{x + 4},\ x = 5$

20 $\quad y = (x^2 + x + 2)^{-1} + 1/\sqrt{4 - x},\ x = 3$

Find the slope of the curve at the specified point in each of Problems 21 to 24.

21   $y = \sqrt[3]{x - 1} - 6\sqrt{x}, \ (9, -16)$

22   $y = 4/\sqrt{3x + 4} - \sqrt{2x + 1}, \ (4, -2)$

23   $y = (2x^3 - x - 4)^{-1}, \ (2, 0.1)$

24   $y = (x^2 - 3x + 1)^{-2} + 1/\sqrt{3x - 5}, \ (2, 2)$

Find the value or values of x for which the derivative is zero in each of Problems 25 to 32. Find the replacement sets for x for which the slope is negative and for which it is positive.

25   $y = x^2 - x + 3$ 26 $y = 3x^2 - 5x + 1$

27   $y = 2x^2 + 3x - 7$ 28 $y = x^2 + 2x - 5$

29   $y = 2x^3 - 7x^2 + 8x - 1$ 30 $y = x^3 + 2x^2 - 4x + 6$

31   $y = 4x^3 + 5x^2 - 8x - 5$ 32 $y = x^3 + 5x^2 + 3x - 2$

5.3   The derivative of a product

We shall let $u(x)$ and $v(x)$ represent differentiable functions and obtain a formula for the derivative of their product by use of the limiting process. If

$$y = uv$$

and Δy, Δu, and Δv are the changes in y, u, and v, respectively, due to the change Δx in x, then

$$y + \Delta y = (u + \Delta u)(v + \Delta v)$$

and

$$\Delta y = (u + \Delta u)(v + \Delta v) - uv = uv + u\,\Delta v + v\,\Delta u + \Delta u\,\Delta v - uv$$

$$= u\,\Delta v + v\,\Delta u + \Delta u\,\Delta v$$

Now dividing by Δx, indicating the limit as Δx approaches zero, and using the theorems on the limit of a sum and of a product, we have

$$\lim_{\Delta x \to 0} \frac{\Delta y}{\Delta x} = \lim_{\Delta x \to 0} u \frac{\Delta v}{\Delta x} + \lim_{\Delta x \to 0} v \frac{\Delta u}{\Delta x} + \lim_{\Delta x \to 0} \frac{\Delta u}{\Delta x} \lim_{\Delta x \to 0} \Delta v$$

Therefore, $\dfrac{dy}{dx} = u \dfrac{dv}{dx} + v \dfrac{du}{dx} + \dfrac{du}{dx} \cdot 0$

since u and v are differentiable functions of x and $\lim\limits_{\Delta x \to 0} \Delta v = 0$. Therefore,

$$D_x uv = u\,D_x v + v\,D_x u \tag{5.5}$$

derivative of a product   and we can say that *the derivative of the product of two differentiable functions is the first function times the derivative of the second plus the second function times the derivative of the first.*

Example   Find the derivative with respect to x of $x^2(3x - 1)^{1/2}$.

Solution: Since the given function is a product, we must use (5.5). Thus, we get

$$D_x x^2(3x - 1)^{1/2} = x^2 \, D_x(3x - 1)^{1/2} + (3x - 1)^{1/2} \, D_x x^2$$

$$\text{by (5.5)}$$

$$= x^2(\tfrac{1}{2})(3x - 1)^{-1/2} \, D_x(3x - 1) + (3x - 1)^{1/2}(2x)$$

$$\text{by (5.4) and (5.4')}$$

$$= \frac{x^2(3x - 1)^{-1/2}(3)}{2} + (3x - 1)^{1/2}(2x)$$

$$\text{by (5.2) and (5.3)}$$

The part of this problem that involves calculus is now completed but some change in algebraic form may be desirable. If we decide to eliminate the negative exponent and to get a common denominator, we multiply numerator and denominator of the first fraction by $(3x - 1)^{1/2}$ and those of the second fraction by $2(3x - 1)^{1/2}$. Thus,

$$D_x x^2(3x - 1)^{1/2}$$

$$= \frac{3x^2(3x - 1)^{-1/2}}{2} \frac{(3x - 1)^{1/2}}{(3x - 1)^{1/2}} + (3x - 1)^{1/2}(2x) \frac{(3x - 1)^{1/2}}{(3x - 1)^{1/2}} \left(\frac{2}{2}\right)$$

$$= \frac{3x^2 + 4x(3x - 1)}{2(3x - 1)^{1/2}} = \frac{15x^2 - 4x}{2(3x - 1)^{1/2}}$$

5.4   The derivative of a quotient

We shall consider two differentiable functions $u(x)$ and $v(x)$ and obtain a formula for the derivative of their quotient.

$$\text{If} \quad y = \frac{u}{v} \quad \text{then} \quad y + \Delta y = \frac{u + \Delta u}{v + \Delta v}$$

and

$$\Delta y = \frac{u + \Delta u}{v + \Delta v} - \frac{u}{v} = \frac{uv + v \, \Delta u - uv - u \, \Delta v}{(v + \Delta v)v} = \frac{v \, \Delta u - u \, \Delta v}{(v + \Delta v)v}$$

Now, dividing by Δx, we have

$$\lim_{\Delta x \to 0} \frac{\Delta y}{\Delta x} = \lim_{\Delta x \to 0} \frac{v \dfrac{\Delta u}{\Delta x} - u \dfrac{\Delta v}{\Delta x}}{(v + \Delta v)v}$$

Therefore,

$$D_x \frac{u}{v} = \frac{v \, D_x u - u \, D_x v}{v^2} \tag{5.6}$$

derivative of a quotient   and we can say that *the derivative of the quotient of two differentiable functions is the denominator times the derivative of the numerator minus the numerator times the derivative of the denominator all divided by the square of the denominator.*

Example   Find the derivative of $(x^3 - 2x)/(x^2 + 1)$.

Solution: Since the given function is a quotient, we shall use (5.6) and obtain

$$D_x \frac{x^3 - 2x}{x^2 + 1} = \frac{(x^2 + 1)\, D_x(x^3 - 2x) - (x^3 - 2x)\, D_x(x^2 + 1)}{(x^2 + 1)^2}$$

$$= \frac{(x^2 + 1)(3x^2 - 2) - (x^3 - 2x)(2x)}{(x^2 + 1)^2} \qquad \text{\textit{by} (5.3), (5.2), \textit{and} (5.4')}$$

$$= \frac{3x^4 + 3x^2 - 2x^2 - 2 - 2x^4 + 4x^2}{(x^2 + 1)^2} \qquad \text{\textit{expanding}}$$

$$= \frac{x^4 + 5x^2 - 2}{(x^2 + 1)^2} \qquad \text{\textit{collecting}}$$

EXERCISE 5.2

Find the derivative of the function in each of Problems 1 to 32.

1   $(x^2 - 2)(3x - 1)$ 2 $(2x^2 - 3x)(4x + 1)$
3   $(2x^2 - 3x + 4)(x^2 + 3)$ 4 $(x^2 - x - 5)(2x + 7)$
5   $(x^2 + 5)/(2x - 3)$ 6 $(x^2 + 7)/(3x + 2)$
7   $(x^2 - 4x + 2)/(3x + 1)$ 8 $(3x^2 - 5x + 2)/(2x - 5)$
9   $(x^2 + 1)^2(2x + 5)$ 10 $(x^2 - 2)^3(3x + 1)$
11 $(x^2 + 2x)^3(5x - 2)^2$ 12 $(2x^2 - 5x + 1)^2(4x + 3)^3$
13 $(x^2 + 2)^2/(3x + 1)$ 14 $(x^2 + x)^3/(2x - 3)$
15 $(x^2 + 3x + 1)^3/(4x + 3)$ 16 $(x^2 - 2x + 3)^2/(5x - 1)$
17 $(2x - 1)\sqrt{x^2 - x}$ 18 $(3x - 2)\sqrt{3x^2 - 4x}$
19 $(x^2 - 7)\sqrt{x^2 - 5}$ 20 $(x + 1)\sqrt{x^2 + 2x}$
21 $(x^2 + 3x)/\sqrt{2x + 3}$ 22 $(2x - 5)/\sqrt{x^2 - 5x}$
23 $(3x - 4)/\sqrt{3x^2 - 8x}$ 24 $(x^2 - 6x)/\sqrt{x - 3}$

25 $\left(\dfrac{2x - 3}{x + 2}\right)^2$ 26 $\left(\dfrac{3x - 1}{x + 3}\right)^3$

27 $\left(\dfrac{2x + 5}{3x + 2}\right)^4$ 28 $\left(\dfrac{4x - 3}{3x + 4}\right)^3$

29 $\dfrac{x}{\sqrt{x^2 - a^2}}$ 30 $\dfrac{\sqrt{x^2 - a^2}}{x}$

31 $\dfrac{x - a}{\sqrt{x^2 - 2ax}}$ 32 $\left(\dfrac{a - x}{a + x}\right)^{1/2}$

33   Derive the formula for the derivative of a product by evaluating

$$\lim_{x \to x_1} \frac{f(x) - f(x_1)}{x - x_1}$$

for $f(x) = u(x)v(x)$.

34   Derive the formula for the derivative of a quotient by evaluating the limit given in Problem 33 for $f(x) = u(x)/v(x)$.

35   The number N of grams of a certain substance in a solution decreases from an initial value of 14 in accordance with the law $N = 28/(0.5t^2 + 2)$ where t represents the elapsed time in minutes. At what rate is N decreasing at the end of 6 min?

36   If a compact object is dropped from the top of a cliff and falls under the force of gravity alone, its velocity v in feet per second at the end of t sec is given approximately by $v = 32t$. Find the value of $D_t v$ for $t = 2$ sec, and explain its physical meaning.

5.5   The chain rule

We shall now consider the situation in which y depends on u and u depends on x. Therefore, if x is given an increment Δx, then u takes on an increment Δu and this causes y to take on an increment Δy. Consequently, if $\Delta u \neq 0$, we have

$$\frac{\Delta y}{\Delta x} = \frac{\Delta y}{\Delta x}\frac{\Delta u}{\Delta u} \qquad \textit{multiplying by } \frac{\Delta u}{\Delta u}$$

$$= \frac{\Delta y}{\Delta u}\frac{\Delta u}{\Delta x} \qquad \textit{rearranging}$$

Hence, since the limit of a product is equal to the product of the limit, we see that

$$\lim_{\Delta x \to 0} \frac{\Delta y}{\Delta x} = \lim_{\Delta x \to 0}\frac{\Delta y}{\Delta u} \lim_{\Delta x \to 0} \frac{\Delta u}{\Delta x}$$

Furthermore, if u is a differentiable function of x, it follows that Δu approaches zero as Δx does, and we can write

$$\lim_{\Delta x \to 0} \frac{\Delta y}{\Delta x} = \lim_{\Delta u \to 0}\frac{\Delta y}{\Delta u} \lim_{\Delta x \to 0} \frac{\Delta u}{\Delta x}$$

$$= D_u y \, D_x u$$

provided that y is a differentiable function of u. Therefore, we know that

chain *rule*   *If u is a differentiable function of x for certain values of x and y is a differentiable function of u for the corresponding values of u, then*

$$\frac{dy}{dx} = \frac{dy}{du}\frac{du}{dx} = D_u y\, D_x u \qquad (5.7)$$

This is called the *chain rule*.

Example If $y = u^3$ and $u = x^2 - 3x + 1$, find $D_x y$.

Solution: Since $D_u y = 3u^2$ and $D_x u = 2x - 3$, it follows from (5.7) that

$$D_x y = (3u^2)(2x - 3)$$
$$= 3(x^2 - 3x + 1)^2(2x - 3) \qquad \textit{replacing u in terms of x}$$

5.6 Implicit functions

An equation which defines y as a function of x, but which is not solved for y in terms of x, is said to define y as an implicit function of x. Thus, the equation $x^2 - xy + y - 4 = 0$ defines the function

$$y = f(x) = \frac{4 - x^2}{1 - x}$$

whose domain is all real numbers except $x = 1$. Since this function is a quotient, we can find its derivative by use of (5.6). Frequently, however, it is difficult or even impossible to solve an equation in x and y for y in terms of x. In such cases, we use a process that is called implicit differentiation in order to find dy/dx.

If $F(x, y) = 0$ defines the function $y = f(x)$, then $F[x, f(x)] \equiv 0$ and $D_x F = 0$. Now if $F(x, y) = x^2 - xy + y - 4 = 0$, it follows from the formulas for the derivative of a sum that

$$D_x F = D_x x^2 - D_x xy + D_x y - D_x 4 = 0$$

hence
$$2x - x\, D_x y - y + D_x y = 0$$

and
$$D_x y = \frac{y - 2x}{1 - x}$$

This example illustrates the fact that if $F(x, y) = 0$ defines y as a differentiable function of x, we can find dy/dx by treating y as a differentiable function of x in obtaining $dF/dx = 0$ and then by solving $dF/dx = 0$ for dy/dx.

An equation $F(x, y) = 0$ may define two or more differentiable functions of x. Thus the equation

$$x^3 + 3x^2 + xy^2 - 2y^2 = 0 \qquad (1)$$

whose graph is shown in Fig. 5.1 defines y implicitly as two continuous

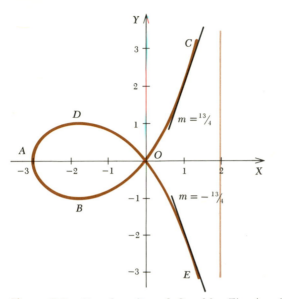

Figure 5.1 **Two functions defined by** $F(x, y) = 0$

functions of x over the interval $[-3, 2)$ as seen by solving (1) for y in terms of x. Thus,

$$y = x\sqrt{\frac{3 + x}{2 - x}} \quad \text{and} \quad y = -x\sqrt{\frac{3 + x}{2 - x}} \qquad (2)$$

We have solved (1) for y in terms of x to show that (1) defines two functions, but this is not necessary in order to find dy/dx.

Example 1 Find dy/dx from the equation $F(x, y) = x^3 + 3x^2 + xy^2 - 2y^2 = 0$ without solving for y in terms of x. Evaluate $D_x y(x, y)$ at $(1, 2)$ and $(1, -2)$.

Solution: We begin by differentiating the sum with respect to x treating y as a differentiable function of x with derivative dy/dx. Thus,

$$D_x F(x, y) = D_x x^3 + D_x 3x^2 + D_x xy^2 - D_x 2y^2 = 0$$

Hence
$$3x^2 + 6x + 2xy\, D_x y + y^2 - 4y\, D_x y = 0$$

and
$$D_x y = \frac{3x^2 + 6x + y^2}{4y - 2xy} \qquad (3)$$

Consequently, $D_x y(1, 2) = {}^{13}\!/_4$ and $D_x y(1, -2) = -{}^{13}\!/_4$.

Any pair of coordinates (x, y) satisfying (1) may be substituted in (3). If the right-hand member of (3) has a value, then this is the slope of the tangent line drawn to the graph of (1) at this point. For example, the point $(-2, -1)$ is on the curve (1), and we find by substitution that $(dy/dx)(-2, -1) = -\frac{1}{8}$. The point $(0, 0)$ is also on the graph of (1), but the right-hand member of (3) does not have a value at this point. The slope of either branch at $(0, 0)$ can be found by differentiating the corresponding one of Eqs. (2) and substituting $x = 0$. Thus, the slope of the branch $ABOC$ for $x = 0$ is $\sqrt{6}/2$ and that of $ADOE$ is $-\sqrt{6}/2$.

In using this procedure we must be careful to differentiate each term with respect to x. Since y is regarded as an unknown function of x, the term xy^2 is a product of two functions of x, and the product formula (5.5) must be used. In order to obtain the derivative of y^2 with respect to x, we must use formula (5.4); thus,

$$\frac{d}{dx}(y^2) = 2y\frac{d}{dx}(y) = 2y\frac{dy}{dx}$$

Here, since y is a function of x, y^2 is of the form u^n.

Finally, it should be mentioned that an equation in x and y may not define y as a function of x. No adequate discussion of the situation can be given here. The following examples will show that a certain amount of caution is necessary.

Example 2 The above procedure, applied to the equation $x^2 + y^2 = 0$, yields

$$\frac{dy}{dx} = -\frac{x}{y}$$

The given equation is satisfied only by $(0, 0)$, and for this pair of values the expression obtained for dy/dx is meaningless.

Example 3 The equation $x^2 + 2xy + (y - x)^2 = y^2 + 10$ does not define y as a function of x for any value of x. In fact, it can be reduced to $x^2 = 2$.

EXERCISE 5.3

Find the derivative of y with respect to x in terms of x in each of Problems 1 to 20.

1   $y = (3x^2 - 2)^5$ 2 $y = (2x^3 - x)^4$

3   $y = (x^4 - 3x)^3$ 4 $y = (x^2 - x + 2)^3$

5   $y = \left(\dfrac{x+1}{x-1}\right)^{1/2}$ 6 $y = \left(\dfrac{2x-1}{x-2}\right)^{1/3}$

7   $y = \left(\dfrac{3x+2}{2x-3}\right)^{1/3}$ 8 $y = \left(\dfrac{x-2a}{x+2a}\right)^{1/2}$

9   $y = u^2, u = x^2 + 3$ 10 $y = u^3, u = 2x - 1$

11 $y = u^2 - 3u,\ u = x^2 - 3x$

12 $y = u^2 + u + 1,\ u = x^2 - x - 1$

13 $y = u^2/(u^2 + 1),\ u = x + 2$

14 $y = u^2/(u^3 + 1),\ u = x/(x - 1)$

15 $y = u/(u^2 - 2),\ u = (x^2 - 2)/x$

16 $y = (u + 1)/(u - 1),\ u = (x - 1)/(x + 1)$

17 $y = u^2 + 2,\ u = \sqrt{x} + x$

18 $y = 4u/(u^2 + 1),\ u = x^2$

19 $y = u^3(\sqrt{u} + 2),\ u = \sqrt{x + 1}$

20 $y = u(\sqrt{u} - 1),\ u = x^2\sqrt{2x - 1}$

21 A point moves along the line $y = 3x - 5$ so that $x = 2t$. At what rate is y changing for $t = 2$?

22 A point moves along the curve $y = x^2 - 2x - 15$ so that $x = 2t$. What is the value of dy/dt for $t = 0,\ \frac{1}{2},\ 2$?

23 A point moves along $xy = 6$ so that $tx = 2$. Is y increasing or decreasing? What is the rate of change of dy/dt for $t = 2$?

24 A point moves along $x^2 + y^2 = 9$ so that $x = 3t$. Find dy/dt for $t = 0,\ \sqrt{3}/2$.

In each of Problems 25 to 36 find dy/dx, both with and without solving for y in terms of x.

25 $2x + 5y = 10$ 26 $ax + by + c = 0$

27 $x^2 + y^2 = a^2$ 28 $x^2 - y^2 = 4$

29 $b^2x^2 + a^2y^2 = a^2b^2$ 30 $x^2/4 + y^2/2 = 1$

31 $y^2 = 2px$ 32 $y^2 = 12x$

33 $x^2 + y^2 = 4x$ 34 $x^{1/2} + y^{1/2} = a^{1/2}$

35 $y^2 = (4 + x)/(4 - x)$ 36 $xy - y - 2x - 5 = 0$

Sketch the graph for the curve in each of Problems 37 to 48, and find the slope of the tangent line at the given points, both with and without solving for y in terms of x.

37 $xy - 2y = 8;\ (4, 4)$ 38 $xy - x^2 = 3;\ (1, 4)$

39 $x^2y + y - 4 = 0;\ (-1, 2)$ 40 $x^2y - 4y = 8;\ (4, \frac{2}{3})$.

41 $x^2y - 4y - 2x^2 - 10 = 0;\ (1, -4)$ and $(-4, \frac{1}{2})$

42 $x^2y + 4y - 4x = 0;\ (0, 0)$ and $(2, 1)$

43 $x^2 - y^2 = 9;\ (5, -4)$ 44 $9x^2 + 36y^2 = 324;\ (-2, \sqrt{8})$

45 $x^2 + y^2 = 10x;\ (8, -4)$ 46 $x^2 + y^2 + 26y = 0;\ (-12, -8)$

47 $x^4 + 2y^4 = 48;\ (-2, 2)$ 48 $y^2 = x(x - 4)(x - 6);\ (2, 4)$

In each of Problems 49 to 56, find the value of $D_x y$ at the given point.

49 $y^2 - 8x + 6y + 17 = 0;\ (3, -7)$

50 $x^2 + y^2 - 4x - 6y - 12 = 0;\ (6, 0)$

51 $x^2 + 4y^2 - 20x - 40y + 100 = 0;\ (4, 1)$

52 $x^2 - 2xy + 4y^2 - 4x = 0;\ (4, 0)$

53 $x^2 + 2xy + y^2 - 8x - 6y = 0$; (0,0)
54 $x^2y^2 - 3y^2 = 2x^2 + 6$; (3, 2)
55 $x^4 - 3xy^2 + 2y^3 = 0$; (−4, 8)
56 $x^2 - 2y^3 - 8y^2 = 0$; (−4, −2)

In each of the following, find dy/dx.

57 $x^2 + xy - 4y^2 = 1$ 58 $x^2 + y^2 - 8x - 24y = 10$
59 $2xy + 5y = 3x - 3$ 60 $x^2 + 4y^2 + 4y = 32$
61 $y^2 - 2x^2 = 4 + 3xy$ 62 $x^3 + y^3 = 3axy$
63 $x^3 + 4y^3 = 6 - 4xy^2$ 64 $x^3 = 4y^3 + 6xy$
65 $4x^2y + 2y^3 = 1 + xy^2$ 66 $x^4 - 2y^4 = x^2y^2 - 6$
67 $3x^2 + 8xy - 3y^2 - 4x - y = 6$
68 $x^4 + xy^3 + y - 2x + 4 = 0$
69 $x^4 - 5x^2y^2 + xy - 4 = 0$
70 $x^4 + 2xy^3 + y + 4x + 4 = 0$
71 $x^3 + y^3 + x^2 - 4xy - 3x + y = 4$
72 $x^3 + 4y^3 - x^2y = 3$
73 $y^2 = x^3/(x^2 + 4)$ 74 $y^2 = 16/(x^4 - 4x^2)$
75 $y^3 = (x + 2)/(x - 2)$ 76 $y^3 = (3x - 1)/(x + 3)$

The differential

6.1 The differential

We shall consider a variable x with given domain and let f be a differentiable function over it. We have found that it is valuable to associate with each number x of the domain not only the function value $f(x)$ but also the number $f'(x)$. We shall now introduce a new concept that is closely related to the derivative.

Corresponding to each number x of the domain, we take the number $f'(x)$ and multiply it by an arbitrary real number, which we denote by the symbol dx. This product is called the *differential* of the function and is denoted by $df(x)$. Symbolically,

$$df(x) = f'(x)\, dx$$

where dx is an arbitrarily chosen real number. We may designate the function by y; we then write $dy = f'(x)\, dx$.

Example 1   If $f(x) = x^3 + 5x$, then $df(x) = (3x^2 + 5)\, dx$.

Example 2   If $y = f(x) = \sqrt{x^2 - 4}$, $x > 2$, then $dy = (x/\sqrt{x^2 - 4})\, dx$.

Observe that the value of the differential of f is fixed when we choose

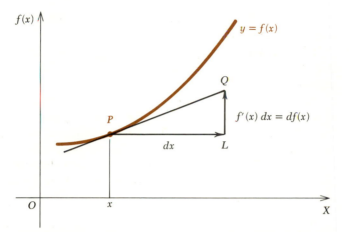

Figure 6.1   **Geometric interpretation of the differential**

the value of x and select the number dx. Thus if we choose $x = 4$ and $dx = 2$ in Example 1 above, we get $df(x) = [3(4^2) + 5]2 = 106$. Since the choice of dx is independent of the value of x, the differential is a function of these *two* independent variables. Under certain conditions we may wish to use the same number dx with every x in a certain interval; then, $df(x)$ will depend only on x.

A geometric interpretation of the differential is shown in Fig. 6.1. The curve is the graph of the function f, and P is the point on the graph with abscissa x. We draw a line segment PL of arbitrary length through P parallel to the X axis and agree to call its directed length dx. Thus, we let $PL = dx$. We then complete the right triangle PLQ by drawing PQ tangent to the curve at P and drawing LQ through L perpendicular to PL. Now, by referring to the right triangle PLQ, we see that the directed length LQ is equal to $f'(x) dx$ since

$$\frac{LQ}{PL} = f'(x)$$

and hence $\qquad LQ = f'(x) \cdot PL = f'(x) dx$

We may associate such a triangle with every point on the curve for which $f'(x)$ exists and is different from zero; for each triangle the directed length $PL = dx$ is any nonzero° real number that we care to choose. When we select a positive number for dx, L is to the right of P as in Fig. 6.1. The choice of a negative number for dx puts L to the left of P. Likewise, since the *directed* length LQ represents the value

---

° Corresponding to $dx = 0$, the triangle is a single point.

of $f'(x)\ dx$, Q will be *above* L when the differential is *positive* and *below* when it is *negative*.

If $f'(x) = 0$ for some particular number x, then $f'(x)\ dx = 0$ regardless of the number we choose as dx. In this case the triangle becomes degenerate in the sense that Q coincides with L.

When we introduced the symbol dy/dx to denote the derivative of y with respect to x, the separate symbols dy and dx were given no individual meanings, and the entire symbol was not regarded as a quotient. It is important to observe that we have now given meanings to the symbols dy and dx and have chosen these definitions so that the quotient obtained by dividing dy by dx, $dx \neq 0$, is the derivative of y with respect to x.

Example 3   If $y = x^3$, then $dy = 3x^2\ dx$, where dx is an arbitrary real number. For any $dx \neq 0$ we may divide dy by dx. The result is $3x^2$, which is the derivative of y with respect to x.

The distinction between the derivative dy/dx and the differential dy should be thoroughly understood. The derivative is the *rate* of change of y relative to x. The differential is the *amount y would change* if the rate of change remained what it was for the chosen x and if x changed by dx. Thus in Fig. 6.1, LQ is the amount by which the function value would change if we start at $P(x, y)$ and if we let the abscissa change by an amount dx, provided that we go along the tangent line instead of along the curve. The following physical example should help to clarify the situation.

Example 4   If a rock is dropped from the top of a cliff, its distance below the top at the end of t sec is given approximately by the formula

$$s = 16t^2$$

Therefore, $\dfrac{ds}{dt} = 32t$ and $ds = 32t\ dt$

The value of ds/dt for $t = 3$ sec is 96 ft per sec. This is the *speed* at which the rock is falling at this instant. The value of ds with $t = 3$ and $dt = 0.01$ is

$$ds = (32)(3)(0.1) = 9.6 \text{ ft}$$

This is the distance that the rock would fall in the next 0.1 sec if the speed remained what it was at the beginning of this interval, that is, when $t = 3$.

As an even simpler example, one might reflect on the relation between the statements "John gets $2.25 per hr" and "John would get \$9.00 if he works 4 hr." In the first, we are stating the rate of pay. In the second, we are stating the amount of pay that he would get if this rate was maintained for 4 hr.

6.2   Formulas for finding the differential

We can derive the formulas for the differential by use of the definition and the formulas for derivatives. For example, if $y = f(x)$ is a constant for $a \leq x \leq b$, then $f'(x) = 0$ for every x in this interval; consequently $dy = f'(x)\, dx = 0$. We can express this in words by saying that the differential of a constant is zero.

By using the definition of the differential, we can readily show that if u and v are differentiable functions of x, then

$$d(u + v) = du + dv$$

$$d(uv) = u\, dv + v\, du$$

differential formulas

$$d\left(\frac{u}{v}\right) = \frac{v\, du - u\, dv}{v^2}$$

$$d(u^n) = nu^{n-1}\, du$$

We shall give the proof of the second of these formulas and leave the others to the student.

If u and v are differentiable functions of x, then the product uv is also differentiable, and we know that

$$\frac{d}{dx}(uv) = u\frac{dv}{dx} + v\frac{du}{dx}$$

Now, by definition,

$$d(uv) = \frac{d}{dx}(uv)\, dx = \left(u\frac{dv}{dx} + v\frac{du}{dx}\right) dx$$

$$= u\frac{dv}{dx}\, dx + v\frac{du}{dx}\, dx = u\, dv + v\, du$$

In the last step we used the fact that, by definition, dv is the derivative of v with respect to x times dx and similar thinking for du/dx times dx.

EXERCISE 6.1

Prove the formulas given in Problems 1 to 4 for u and v differentiable functions of x.

1   $d(u + v) = du + dv$ 2 $d(u - v) = du - dv$
3   $d(u/v) = (v\, du - u\, dv)/v^2$ 4 $du^n = nu^{n-1}\, du$

Find the differential of the function in each of Problems 5 to 24.

5 $3x^4 + 2x + 1$ 6 $5x^3 - 3x^2 + 4$

7 $2x^2 - 5x + 6$ 8 $4x^5 - 5x^4 + 11x$

9 $x^3 + 2/x$ 10 $5x - 3/x^2$

11 $2x^4 - 7/x^3$ 12 $6x^2 + 3/x^3$

13 $3x^3 + \sqrt[3]{x^2}$ 14 $x^{-1} + 3x^{2/3}$

15 $5x^{3/5} + 2x^{-2}$ 16 $6x^{-2/3} + 2\sqrt{x}$

17 $x\sqrt{x^2 - 1}$ 18 $3x^2\sqrt{16 - x^2}$

19 $x^3\sqrt[3]{x^3 + 3x^2}$ 20 $x\sqrt[5]{x^5 - 5x}$

21 $x^2/\sqrt{x^2 + 4}$ 22 $(x - 1)/\sqrt{x^2 + x}$

23 $2x/\sqrt{4x^2 - 9}$ 24 $(x^2 - x)/\sqrt{2x - 1}$

In each of Problems 25 to 32, sketch the graph of the function, and find the value of dy for the given values of x and dx.

25 $y = x^2 + 4x - 1$, $x = 1$, $dx = 0.5$

26 $y = x^2 - 2x + 7$, $x = 3$, $dx = -0.5$

27 $y = x^3 - 1$, $x = 2$, $dx = -\frac{1}{3}$

28 $y = x^3 - 3x^2$, $x = 1$, $dx = \frac{1}{3}$

29 $y = \sqrt{x}$, $x = 4$, $dx = -1$

30 $y = \sqrt[3]{x}$, $x = 27$, $dx = \frac{2}{3}$

31 $y = 2x\sqrt{x}$, $x = 4$, $dx = -0.5$

32 $y = x/\sqrt{x + 1}$, $x = 8$, $dx = \frac{1}{4}$

33   A particle starts at the point $(3, 0)$ and moves to the right along the X axis in such a manner that its distance from the origin at the end of t min is given by the formula

$$x = 3 + \frac{\sqrt{t}}{2}$$

Find the value of dx/dt for $t = 16$, and compute dx corresponding to $t = 16$ and $dt = \frac{1}{2}$. Interpret each of these quantities physically, giving the units in which each is expressed.

34   When a ball is thrown vertically upward from the ground with an initial velocity of 60 ft per sec, its distance above the ground at the end of t sec is given approximately by the formula

$$y = 60t - 16t^2 \qquad 0 \le t \le 3.75$$

Find the value of dy/dt for $t = 2$. Compute the value of dy corresponding to $t = 2$ and $dt = 0.1$. Interpret each of these quantities physically, giving the units in which each is expressed.

35   A cylinder that is fitted with a piston contains 400 cu in. of air at a pressure of 15 lb per sq in. If the air is compressed by moving the piston and the temperature is held constant, the pressure increases

as the volume decreases, the relation between them being $pv = 6,000$. Compute the value of dp/dv for $v = 100$ cu in. Also compute the value of dp corresponding to $v = 100$ cu in. and $dv = -2$ cu in. Interpret each of these quantities physically, giving the units in which each is expressed.

36   The number of cubic centimeters of alcohol in a solution decreases with the time t in accordance with the formula

$$N = \frac{60}{t^2/2 + 4}$$

for t in hours. Find the value of dN/dt for $t = 4$ and the value of dN corresponding to $t = 4$ and $dt = 0.1$ hr. Interpret each of these quantities physically, giving the units in which each is expressed.

37   Sketch the graph of the equation $y = 12/(x^2 + 4)$. Show that for $x = 2$ we have $dy = -\frac{3}{4} dx$ where dx is an arbitrary real number. Exhibit on the graph the values of dy corresponding to $dx = 1, 2,$ and -1.

38   Sketch the graph of the equation $y = \sqrt{25 - x^2}$. Compute the values of dy corresponding to $x = -4$ and $dx = 1, 3,$ and -1. Show these on the graph.

39   Sketch the graph of the equation $x^2/64 + y^2/27 = 1$. Compute the values of dy at the point $(4, \frac{9}{2})$ corresponding to $dx = 1, 4,$ and -4. Show these on the graph.

40   Sketch the graph of the equation $y^2 = 12x$. Compute the values of dy at the point $(3, 6)$ corresponding to $dx = 0.5, 1,$ and 2. Show these on the graph.

6.3   The differential of implicit and composite functions

The differential has a property which makes it simpler to use than the derivative. The student will recall that when we use derivatives, we must be careful to distinguish between the derivative of a function $f(x)$ with respect to x and its derivative with respect to some other variable, say t. Thus, if $y = f(x)$ and $x = \phi(t)$, then $dy/dx = f'(x)$ but

$$\frac{dy}{dt} = \frac{dy}{dx}\frac{dx}{dt} = f'(x)\phi'(t)$$

When we deal with differentials, however, we need not distinguish between the case in which x is the independent variable and that in which x is, in turn, a function of some other variable. If we think of the equations $y = f(x)$ and $x = \phi(t)$ as defining y as a function of t by, say,

$$y = f[\phi(t)] = G(t)$$

then, by definition, dy is equal to the derivative of y with respect to t times an arbitrary real number dt. Thus,

$$dy = \left(\frac{dy}{dx}\frac{dx}{dt}\right) dt \qquad \textit{by the chain rule}$$

$$= f'(x)\phi'(t)dt$$

Now, in view of the relation $x = \phi(t)$, we can replace $\phi'(t)\ dt$ by dx. We thus get

$$dy = f'(x)\ dx$$

This is the expression that we have for dy if we disregard the relation $x = \phi(t)$ entirely and think of x as the independent variable. The only difference between the two cases lies in the amount of freedom that we have in selecting the numbers x and dx. If x is the independent variable, then x and dx are chosen at random from their respective domains. If x depends on the independent variable t, first t and dt are chosen and then x and dx are determined from the relations $x = \phi(t)$ and $dx = \phi'(t)\ dt$.

In view of the above result, we return to our original definition of the differential of a function $f(x)$ and agree to call dx the differential of x whether dx is an arbitrary real number or the differential of some function. This is consistent with the fact that for the function $f(x) = x$, we have $f'(x) = 1$ and $df(x) = 1\ dx = dx$.

We may state our basic result briefly by saying that the differential of a function $f(x)$ is always $f'(x)\ dx$ regardless of whether x is regarded as a function of some other variable.

Example 1   If $y = x^2 + 3x + 1$ and $x = t^2 + 2$, find dy in terms of x and dx, in terms of x, t, and dt, and in terms of t and dt. Find dy/dx and dy/dt.

Solution: In order to express dy in terms of x and dx, we need use only the equation $y = x^2 + 3x + 1$. Now, taking the differential of each member, we have

$$dy = 2x\ dx + 3dx$$

$$= (2x + 3)\ dx \qquad\qquad (1)$$

In order to express dy in terms of x, t, and dt, we replace dx by $2t\ dt$ as obtained from $x = t^2 + 2$ and write

$$dy = (2x + 3) \cdot 2t\ dt$$

Finally to express dy in terms of t and dt, we merely replace x by $t^2 + 2$ as given and get

$$dy = [2(t^2 + 2) + 3] \cdot 2t\ dt$$

$$= (4t^3 + 14t)\ dt \qquad\qquad (2)$$

This same result could have been obtained by replacing x by $t^2 + 2$ in $y = x^2 + 3x + 1$ and then taking the differential. Thus,

$$y = (t^2 + 2)^2 + 3(t^2 + 2) + 1$$
$$= t^4 + 7t^2 + 11$$

and
$$dy = (4t^3 + 14t)\, dt$$

We can find dy/dx from (1) and dy/dt from (2). They are $dy/dx = 2x + 3$ and $dy/dt = 4t^3 + 14t$.

Example 2   If x and y are subject to the relation $x^2 + y^2 = 25$, then they cannot both be independent variables. The equation may be regarded as defining y in terms of x or x in terms of y, or both x and y may be regarded as functions of some third variable t. In this case the equation has the form $F(t) = 25$ for a suitable domain for t. In any case the differential of x^2 is $2x\, dx$, that of y^2 is $2y\, dy$, and consequently

$$d(x^2 + y^2) = 2x\, dx + 2y\, dy$$

Now, if we wish to regard x and y as functions of t, so that we have an equation of the form $F(t) = 25$, we see that $F'(t)$, and consequently dF, is zero for every t in its domain. We thus have immediately the relation $2x\, dx + 2y\, dy = 0$, or

$$x\, dx + y\, dy = 0$$

This relation may be solved for dy/dx or dx/dy to obtain these derivatives at all points at which they exist. The relation has the geometric interpretation shown in Fig. 6.2. If we start at any point $P(x, y)$ on the circle $x^2 + y^2 = 25$ and move along the tangent line until the abscissa has changed by an amount dx, the change dy in the

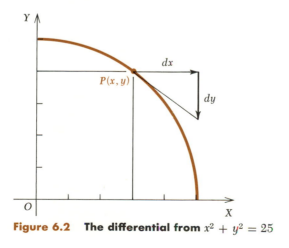

Figure 6.2   The differential from $x^2 + y^2 = 25$

ordinate will be such that $x\,dx + y\,dy = 0$. Thus if $x = 3$, $y = 4$, and $dx = 2$, then $dy = -\tfrac{3}{2}$.

Example 3   We shall think of a fixed quantity of gas that is confined in a cylinder and is being compressed at constant temperature by a piston that moves in the cylinder. The pressure and volume will change in accordance with the law $pv = k$, where k is a constant. If we wish, we may think of both p and v as changing with the time t so that our equation is $F(t) = k$. The differential of pv is $p\,dv + v\,dp$ in any case; hence, we may write

$$p\,dv + v\,dp = 0 \qquad (1)$$

We can solve this relation for dp/dv or for dv/dp, or use it in the form (1).

Example 4   In order to find $D_x y$ from the equation

$$y^3 + 4x^2 y^2 = 8$$

we may proceed as follows by taking the differential. Thus,

$$3y^2\,dy + 4(x^2 \cdot 2y\,dy + y^2 \cdot 2x\,dx) = 0$$

$$(3y^2 + 8x^2 y)\,dy = -8xy^2\,dx$$

$$\frac{dy}{dx} = -\frac{8xy^2}{3y^2 + 8x^2 y} = -\frac{8xy}{3y + 8x^2}$$

6.4   The differential dy as an approximation to Δy

We shall see in this section that if $y = f(x)$ is a differentiable function of x, then the differential dy is an approximation to the increment Δy. Symbolically, this is:

If $\Delta y = f(x + dx) - f(x)$, then $dy = f'(x)\,dx \doteq \Delta y$ where $\doteq$ is a symbol for "approximately equal to."

In proving this statement, we shall replace the arbitrary number dx by Δx and have $\Delta y = f(x + \Delta x) - f(x)$ as we had when deriving differentiation formulas. Now subtracting $dy = f'(x)\,dx = f'(x)\,\Delta x$ from Δy, dividing by Δx, and taking the limit as Δx approaches zero, we get

$$\lim_{\Delta x \to 0}\frac{\Delta y - dy}{\Delta x} = \lim_{\Delta x \to 0}\frac{\Delta y - f'(x)\,\Delta x}{\Delta x} = \lim_{\Delta x \to 0}\left(\frac{\Delta y}{\Delta x} - f'(x)\right) = 0$$

$dy \doteq \Delta y$ by the definition of the derivative. Consequently, we see that as Δx approaches zero, the differential dy approaches the increment Δy.

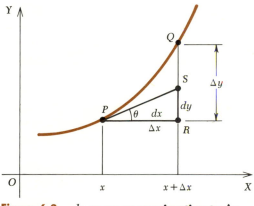

Figure 6.3 dy **as an approximation to** Δy

The situation is shown graphically in Fig. 6.3. In it, we let $P(x, y)$ be the point on the curve $y = f(x)$ with abscissa x and let $Q(x + \Delta x,$ $y + \Delta y)$ be the point with abscissa $x + \Delta x$. Then Δy is the directed distance RQ since that is the change in y due to the change Δx in x. Furthermore, dy is the directed distance RS since $RS = \tan \theta\, dx =$ $f'(x)\, dx = dy$. This is the amount y would have changed if the rate of change of y with respect to x at P had been maintained throughout the interval from P to R.

Example   About how much material is used in putting a cover 0.02 in. thick on a spherical ball that is 6 in. in radius?

Solution: In solving this problem we make use of the fact that the volume of a sphere is $v = 4\pi r^3/3$. An approximation to the change in volume due to a change of dr in the radius is $dv = 4\pi r^2\, dr$. Putting in the given values of r and dr, we see that the approximate number of cubic inches of material used is

$$dv = 4\pi(6)^2(0.02) = 2.88\pi = 9.04$$

If x_1 and y_1 are the true values of the variables in $y = f(x)$ and if $x = x_1 + dx$ is the measured or observed value of x for which $f(x_1 + dx) = y_1 + \Delta y$, then

$$\frac{dy}{y} \doteq \frac{\Delta y}{y_1}$$

relative change   is called the *relative change* in y due to the change dx in x. Furthermore, if dy/y is expressed as a percent, it is called the *percent of change*. *percent of change*   Therefore, the relative change in the example is $dv/v = 2.88\pi/$ $288\pi = 1/100$ and the percent of change is 1.

EXERCISE 6.2

In each of Problems 1 to 8 a relation between x and y is given. Find the corresponding relation between their differentials, and then find both dy/dx and dx/dy.

1   $3xy + y^2 = 6x$ 2 $xy^2 + y = 8x$
3   $2x^2 - 3y^2 = 5xy$ 4 $x^2 + 3y^2 + x - y = 4$
5   $y^2x + x^2y = 5$ 6 $x^3 + y^3 - 3xy = 0$
7   $6x - 5x^2y + y^2 = 17$ 8 $x^2y^2 = 6x - 2y$

In each of Problems 9 to 16, y is given as a function of x, and x as a function of t. By use of differentials, find dy/dx in terms of x, and dy/dt in terms of x or t or of t and x.

9    $y = x^2 + 2x - 1, x = 3t - 2$
10   $y = x^2 - 5x + 2, x = 2t + 1$
11   $y = 3x - 4, x = t^2 - 7t - 1$
12   $y = 2x + 7, x = 2t^2 - 5$
13   $y = x\sqrt{x^2 - 1}, x = \sqrt{t^2 + 2}, |x| > 1$
14   $y = x^2\sqrt{4 - x^2}, -2 \le x \le 2, x = \sqrt{2t - 1}, t \ge \frac{1}{2}$
15   $y = x/\sqrt{x^2 + 2}, x = \sqrt{t - 1}/t, t \ge 1$
16   $y = \sqrt{x^2 - x}/x, x > 1, x = t/\sqrt{t^2 - t}, t > 1$

17   The radius of a circle is 32 in. By approximately how much does the area increase if the radius increases by $\frac{1}{16}$ in.?
18   If the edge of a cube is 8 cm and increases by 0.1 cm, what is the approximate change in volume?
19   The radius of a right circular cylinder of height 16 in. and radius 4 in. decreases by $\frac{1}{4}$ in. What is the approximate change in volume?
20   About how much does the area of a triangle of base 10 in. and altitude 8 in. change if the base and altitude each change by $\frac{1}{2}$ in.?

Sketch the graph of the equation given in each of Problems 21 to 24, compute Δy and dy for the given conditions, and show them on the graph.

21   $y = x^2/4, x = 3, dx = 0.5$
22   $y = 8 - x^3/8, x = 2, dx = \frac{1}{3}$
23   $y = x^3 - 3x - 2, x = 2, dx = 1$
24   $y = 10 - x^2, x = 3, dx = 1$

Approximate the numbers in Problems 25 to 32 by use of differentials.

25   $\sqrt{6.35}$; use $6.35 = 6.25 + 0.10$
26   $\sqrt{101}$ 27 $\sqrt[3]{1.704}$ 28 $\sqrt[3]{9}$
29   $(3.02)^3$ 30 $(1.98)^4$ 31 99^3
32   $(10.2)^2$

33   Two boys are supposed to lay out a square with each side 32 ft. What is approximately the greatest error in the area if each side may be off as much as 0.5 ft?

34   What is the allowable error in the edge of a cubic box that is intended to contain 1,000 cu in. if the error in volume is not to exceed 2 cu in.?

35   The radius of a circle is to be measured and its area computed. The error in the measurement of the radius does not exceed 0.005 in., and the error in the area must not exceed 0.1 sq in. What is the radius of the largest circle for which these specifications are met?

36   The radius of a sphere is to be measured and its volume computed. If the possible error in the radius is 0.001 in. and the allowable error in volume is 0.5 cu in., what is the radius of the largest sphere for which these specifications hold?

37   For what range of values of x can $\sqrt[3]{x + 3}$ be replaced by $\sqrt[3]{x}$ if an error of not more than 0.01 is permissible?

38   Show that an error of 1 percent in measuring the radius of a sphere will result in an error of about 3 percent in the computed volume.

39   Find the approximate relative error in the computed volume and surface area of a cube due to an error of 1 percent in the edge.

40   Show that the relative error in the computed value of kx^n due to a small error in x is approximately n times the relative error in x.

41   Find the derivative of the volume of a cube with respect to its surface area by dividing dV by dS.

42   Find the derivative of the volume of a sphere with respect to its surface area by dividing dV by dS.

43   Find the derivative of the area of a circle with respect to the circumference by division of appropriate differentials.

44   Find the derivative of the area of a square with respect to the perimeter by division of appropriate differentials.

7
The indefinite integral

7.1 Introduction

In earlier chapters of this book, we considered the problem of finding the derivative of a given function and studied some of the applications of the derivative. This chapter and part of the later ones will be devoted to the problem of finding the function when its derivative is given. The process of finding the function whose derivative is given is called *integration;* it is the inverse of differentiation. We shall see in later chapters that, by means of integration, we are able to find the length of a curve, areas, volumes, work done by a variable force, liquid pressure, and the center of gravity and to solve other problems.

7.2 The indefinite integral

If we let $f(x)$ be a function that is defined in an interval $a < x < b$ and let $F(x)$ be another function such that $dF(x) = f(x)\,dx$, we then say that *an integral* of $f(x)$ in the interval is $F(x)$ and write

definition of $\int f(x)\,dx$

$$\int f(x)\,dx = F(x) + C \tag{1}$$

The inclusion of the additive constant is justified, since the derivative of $F(x) + C$ is the same as the derivative of $F(x)$ in as much as the derivative of a constant is zero. The definition of the indefinite integral is often given by saying that $F(x)$ *is an integral of* $f(x)$ *if and only if* $dF(x) = f(x)\,dx$, *that is, if and only if* $F'(x) = f(x)$. By use of this definition, we have

$$\frac{d}{dx}\left[\int f(x)\,dx\right] = \frac{dF(x)}{dx} = f(x) \tag{2}$$

integrand   In (1), $f(x)$ is called the *integrand*.

Example

$$\int (2x + 3)\,dx = x^2 + 3x + C$$

where C is an arbitrary constant, since $D_x(x^2 + 3x + C) = 2x + 3$.

We shall need to make use of the fact that *if two functions $F(x)$ and $G(x)$ have the same derivative for all values of x in the interval $a < x < b$, then $F(x)$ and $G(x)$ differ by a constant,* that is, $F(x) - G(x) = C$. This will be proved by use of the law of the mean in Sec. 14.2. We shall now let

$$H(x) = F(x) - G(x) \qquad \text{for all } x \text{ such that } a < x < b$$

Hence $H'(x) = F'(x) - G'(x) = 0$ for x between a and b since $F(x)$ and $G(x)$ have the same derivative. Therefore, $H(x)$ is a constant. Consequently, if $F(x)$ is a value of $\int f(x)\,dx$, then any other value, say, $G(x)$ differs from $F(x)$ by a constant since $F(x)$ and $G(x)$ have the same derivative $f(x)$. Consequently, *if $F(x)$ is an integral of $f(x)$ in $a < x < b$, then any other integral of $f(x)$ in the interval is equal to $F(x) + C$, and we have*

$$\int f(x)\,dx = F(x) + C \qquad \text{for } C \text{ an arbitrary constant}$$

7.3   Three integration formulas

The first of the integration formulas that we shall consider states that *the integral of the sum of several functions is equal to the sum of the integrals of the separate functions.* This can be stated in symbols by writing

$$\int (u + v + \cdots + w)\,dx = \int u\,dx + \int v\,dx + \cdots + \int w\,dx \tag{7.1}$$

By use of (2) of Sec. 7.2 and the fact that the differential of the sum of several functions is the sum of their differentials, we see that the derivative of each member of (7.1) is $u + v + \cdots + w$. Consequently, (7.1) is true.

The second integration formula that we shall consider states that *the integral of the product of a constant and a function is equal to the product of the constant and the integral of the function. Symbolically, this is*

$$\int au \, dx = a\int u \, dx \qquad (7.2)$$

Now by use of (2) of Sec. 7.2 and the fact that the differential of a constant times a function is the constant times the differential of the function, we see that the derivative of each member of (7.2) is au. Consequently, (7.2) is true.

As a third integration formula, we shall prove that

$$\int u^n \, du = \frac{u^{n+1}}{n+1} + C \qquad n \neq -1 \qquad (7.3)$$

This is true since

$$\frac{d}{du} \frac{u^{n+1}}{n+1} = u^n \qquad \text{and} \qquad \frac{dC}{dx} = 0$$

We can now use (7.1), (7.2), and (7.3) to find an integral of any polynomial and some other functions.

Example 1
$$\int x^2 \, dx = \frac{x^3}{3} + C \qquad \text{by use of (7.3)}$$

Example 2
$$\int 5x^3 \, dx = 5\int x^3 \, dx \qquad \text{by use of (7.2)}$$
$$= \frac{5x^4}{4} + C \qquad \text{by use of (7.3)}$$

Example 3
$$\int (3x^2 + 5x^4 + 2) \, dx = \int 3x^2 \, dx + \int 5x^4 \, dx + \int 2 \, dx$$
$$\text{by use of (7.1)}$$
$$= 3\int x^2 \, dx + 5\int x^4 \, dx + 2\int dx$$
$$\text{by use of (7.2)}$$
$$= \frac{3x^3}{3} + C_1 + \frac{5x^5}{5} + C_2 + 2x + C_3$$
$$\text{by use of (7.3)}$$
$$= x^3 + x^5 + 2x + C$$
$$\text{simplifying and replacing } C_1 + C_2 + C_3 \text{ by } C$$

Example 4
$$\int (x^2 + 3x)^4 (2x + 3) \, dx = \frac{(x^2 + 3x)^5}{5} + C \qquad \text{by use of (7.3)}$$

This example calls for a word of explanation. If we think of $x^2 + 3x$ as a function u, then u is raised to a power 4 and is multiplied by the differential $(2x + 3) \, dx$ of the function u. Consequently, $\int (x^2 + 3x)^4 (2x + 3) \, dx$ is of the form $\int u^n \, du$; hence, the integral is obtained by use of (7.3).

Example 5 $\int (x^2 + 2x + 3)^{2/3}(x + 1)\, dx$ is not of the form $\int u^n\, du$ since $d(x^2 + 2x + 3) = (2x + 2)\, dx = 2(x + 1)\, dx$. We, however, can get a function of the form of (7.3) by multiplying the integrand by 2 and we offset this by multiplying the integral by ½. If these operations are performed, we have

$$\int (x^2 + 2x + 3)^{2/3}(x + 1)\, dx = ½\int (x^2 + 2x + 3)^{2/3} \cdot 2(x + 1)\, dx$$
$$= ½\int (x^2 + 2x + 3)^{2/3}\, d(x^2 + 2x + 3)$$
$$= ½(x^2 + 2x + 3)^{5/3} \div \tfrac{5}{3} + C$$
$$= 0.3(x^2 + 2x + 3)^{5/3} + C \qquad ½ \div \tfrac{5}{3} = 0.3$$

In using (7.3), we must be certain that we have not only u^n but also du. As in Example 5, we can often get du by multiplying the integrand by a constant; we then must multiply the integral by the reciprocal of that constant. This is in effect multiplying by 1 and it changes the form without changing the value of the function.

7.4   Integration by substitution

Quite often, the process of integration can be simplified by use of a substitution or change of variable. Differentiation is used in obtaining a desirable change of variable. If $f(u)$ and $F(u)$ are related by the equation

$$\frac{d}{du} F(u) = f(u)$$

then $\qquad\qquad \int f(u)\, dx = F(u)$

and if u is a function of x, say $u(x)$, then

$$\frac{dF[u(x)]}{dx} = \frac{dF(u)}{du}\frac{du}{dx} \qquad \text{\textit{by the chain rule}}$$
$$= f[u(x)]u'(x) \qquad u'(x) = \frac{du}{dx}$$

Therefore, $dF[u(x)] = f[u(x)]u'(x)\, dx$ and

$$\int f[u(x)]u'(x)\, dx = F(u) = \int f(u)\, du \qquad (7.4)$$

This formula is quite useful if the integrand is a function of x that can be written as the product of a function of a function of x, say $u(x)$, and the derivative $u'(x)$ of u with respect to x.

Example 1   By use of (7.4), we shall perform the integration indicated by

$$\int (x^2 + 3x)^4(2x + 3)\, dx$$

Solution: If we let $u(x) = x^2 + 3x$, then the function of u is $(x^2 + 3x)^4 = f(u)$ and $u'(x) = 2x + 3$; hence, the integrand in the given integral is the product of $f(u)$ and $u'(x)$. Therefore,

$$\int (x^2 + 3x)^4 (2x + 3) \, dx = \int u^4 \, du \qquad u = x^2 + 3x$$

$$= \frac{u^5}{5} + C$$

$$= \frac{(x^2 + 3x)^5}{5} + C$$

as was obtained when we worked this problem before in Example 4 of Sec. 7.3.

Example 2   By use of (7.4), we shall perform the integration indicated by

$$\int \frac{(2x^2 + 1) \, dx}{\sqrt[3]{(2x^3 + 3x + 1)^2}}$$

Solution: If we let $u(x) = 2x^3 + 3x + 1$

then $$du = u'(x) \, dx = (6x^2 + 3) \, dx = 3(2x^2 + 1) \, dx$$

Consequently, $$(2x^2 + 1) \, dx = \frac{du}{3}$$

and we have

$$\int \frac{(2x^2 + 1) \, dx}{\sqrt[3]{(2x^3 + 3x + 1)^2}} = \int \frac{du/3}{\sqrt[3]{[u(x)]^2}}$$

$$= \frac{1}{3} \int [u(x)]^{-2/3} \, du \qquad \text{by (7.2)}$$

$$= \frac{1}{3} \left[\frac{[u(x)]^{1/3}}{\frac{1}{3}} \right] + C \qquad \text{by (7.3)}$$

$$= [u(x)]^{1/3} + C$$

$$= (2x^3 + 3x + 1)^{1/3} + C$$

EXERCISE 7.1

In Problems 1 to 36 perform the indicated integrations by use of (7.1), (7.2), (7.3), and (7.4).

1 $\int x^2 \, dx$

2 $\int x^{-3} \, dx$

3 $\int x^{-1/2} \, dx$

4 $\int x^4 \, dx$

5 $\int 2x^{-3} \, dx$

6 $\int 3x^5 \, dx$

7 $\int 3x^2 \, dx$

8 $\int 7x^{-8} \, dx$

9   $\int(x^2 + 3x + 5)\ dx$ 10 $\int(x^3 - x - 4)\ dx$

11   $\int(x + 1/x^2)\ dx$ 12 $\int(x^2 - 2/x^3)\ dx$

13   $\int(x^2 - 4x + 7)^3(2x - 4)\ dx$

14   $\int(2x^2 - 3x + 1)^2(4x - 3)\ dx$

15   $\int(x^3 - 3x)^5(3x^2 - 3)\ dx$

16   $\int(x^4 - 5x^2 - 7x)^3(4x^3 - 10x - 7)\ dx$

17   $\int(x^2 + 4x - 1)^4(x + 2)\ dx$

18   $\int(x^3 + 6x)^2(6x^2 + 12)\ dx$

19   $\int(x^4 + 5x^2)^5(2x^3 + 5x)\ dx$

20   $\int(x^5 + 5x^3 + 10x)^4(x^4 + 3x^2 + 2)\ dx$

21   $\int 3x\sqrt{x}\ dx$ 22 $\int 9x^3\sqrt{x}\ dx$

23   $\int \sqrt{a^2 - x^2}\ x\ dx$ 24 $\int \sqrt{x^2 + a^2}\ x\ dx$

25   $\int \dfrac{x^2 - 2}{\sqrt{x}}\ dx$ 26 $\int \dfrac{x^3 - 4x}{\sqrt{x}}\ dx$

27   $\int \dfrac{x\sqrt{x^2 - 9}}{x^2 - 9}\ dx$ 28 $\int \dfrac{3x\sqrt{4 + x^2}}{4 + x^2}\ dx$

29   $\int \left(\dfrac{1}{x^2} - 3\sqrt{x} + \sqrt[3]{8x}\right)\ dx$

30   $\int \left(\dfrac{3}{x^2} - \dfrac{1}{\sqrt{x}} + \sqrt[3]{x}\right)\ dx$

31   $\int \left(\sqrt{x}\ \sqrt[3]{x} + \dfrac{4}{x^3}\right)\ dx$ 32 $\int \left(\dfrac{\sqrt{x}}{\sqrt[3]{x}} + \dfrac{x^3 + 2}{x^2}\right)\ dx$

33   $\int(x + 4)(x - 3)\ dx$ 34 $\int(2x - 1)(x + 2)\ dx$

35   $\int(x^2 + 2x + 4)(2x + 1)\ dx$

36   $\int(x^2 - 3x + 1)(x - 3)\ dx$

37   Evaluate $\int(x + 1)^2\ dx$ with and without expanding. Compare the constants of integration.

38   Evaluate $\int(2x - 1)^3\ dx$ with and without expanding. Compare the constants of integration.

39   Explain why (7.3) cannot be used to evaluate $\int(x^2 + 3)^{1/2}\ dx$.

40   Explain why (7.3) cannot be used to evaluate $\int \dfrac{2x - 5}{x^2 - 5x + 1}\ dx$.

7.5   Some applications

In Sec. 7.2, we defined the indefinite integral by saying that

$$\int f(x)\ dx = F(x) + C \qquad C \text{ an arbitrary constant} \qquad (1)$$

if and only if $dF(x) = f(x)\ dx$ for $a < x < b$. If we let $y = \int f(x)\ dx =$

$F(x) + C$, then we have a family of curves and can obtain a member of the family by assigning a value to the arbitrary constant C. The value to be assigned to C can be determined by requiring that the member of the family of curves will pass through a specified point.

Example 1   Find the equation of the family of curves whose slope is $4x - 5$. Select the member through $(3, 1)$.

Solution: Since dy/dx is the slope, we have $dy/dx = 4x - 5$. Therefore $dy = (4x - 5) \, dx$ and

$$y = \int (4x - 5) \, dx = 2x^2 - 5x + C$$

This is the equation of the family of parabolas with the given slope, and we get a member of the family for each value assigned to C. In order to find the value of C for the member that passes through $(3, 1)$, we make use of the fact that the coordinates of a point on a curve satisfy the equation of the curve. Thus, substituting $(3, 1)$ in the equation $y = 2x^2 - 5x + C$ of the family, we have

$$1 = 2(3)^2 - 5(3) + C = 3 + C$$

Hence, $C = -2$ for the member through $(3, 1)$, and the equation of the member is $y = 2x^2 - 5x - 2$. It and two other members of the family are shown in Fig. 7.1.

At times we can find the equation of the curve whose slope is a function of the variables of x and y and which passes through a given point. This can be done provided that the function is such that we can change the given equation

$$\frac{dy}{dx} = G(x, y)$$

into one in which dy is multiplied by a function of y alone and dx is multiplied by a function of x alone and provided that we can integrate the resulting equation.

Example 2   Find the equation of the family of curves for which the slope is $(5 - x)/(y - 3)$. Select the member that passes through $(2, -1)$.

Solution: We are given that the slope is $(5 - x)/(y - 3)$; hence

$$\frac{dy}{dx} = \frac{5 - x}{y - 3}$$

Now, multiplying each member by $(y - 3) \, dx$, we have

$$(y - 3) \, dy = (5 - x) \, dx$$

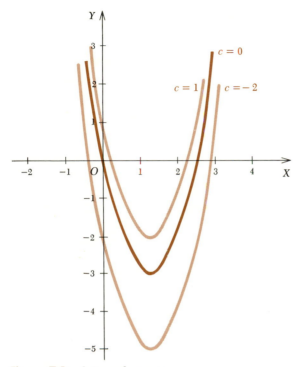

Figure 7.1 Integral curves

and the variables are separated. Equating the integrals of the two members and putting the constant of integration on the right, we find that

$$\frac{y^2}{2} - 3y = 5x - \frac{x^2}{2} + C$$

Multiplying by 2, adding $x^2 - 10x$ to each member, and completing the squares, we have

$$(x - 5)^2 + (y - 3)^2 = 2C + 34 \qquad (1)$$

This is a family of circles. They are real circles if $C > -17$ and imaginary circles if $C < -17$; furthermore, the circle degenerates to the point $(5, 3)$ if $C = -17$.

In order to obtain the equation of the circle that passes through $(2, -1)$, we replace (x, y) by $(2, -1)$ in (1) and solve for C. Thus, we get

$$(-3)^2 + (-4)^2 = 2C + 34$$
$$2C = 9 + 16 - 34 = -9$$

Consequently, the required circle is

$$(x - 5)^2 + (y - 3)^2 = 5^2$$

7.6   Motion of falling bodies

It is known that an acceleration in the direction of the force is produced if a force of F lb acts on a particle whose weight is W lb. It has been found experimentally that the acceleration in feet per second per second is given by the equation

$$F = \frac{W}{g} a \qquad (1)$$

where the constant g is the acceleration of gravity whose value is approximately 32.2 ft per sec per sec. If F and W are known, (1) can be solved for a. Consequently, velocity, in terms of time, can be found by an integration since $a = dv/dt$. After the velocity is obtained, the displacement or distance s, in terms of time, can be found by means of another integration since $v = ds/dt$.

Example   A compact ball is thrown vertically upward with an initial velocity of 75 ft per sec from a point that is 40 ft above the ground. Find its velocity and distance from the ground at the end of t sec.

Solution: We shall consider upward as the positive direction. Consequently,

$$F = \frac{W}{g} a = -W$$

since the only force, other than air resistance (which we shall neglect), acting on the ball after it is in the air is the acceleration of gravity, and it is acting in the negative direction. Consequently, we have $a = -g$; hence

$$\frac{dv}{dt} = -g \qquad \textit{since } a = \frac{dv}{dt}$$

$$v = \int -g \, dt = -gt + c_1 \qquad (2)$$

Therefore, since $v = ds/dt$, we have

$$s = \int (-gt + c_1) \, dt$$

$$s = \frac{-gt^2}{2} + c_1 t + c_2 \qquad (3)$$

We must now make use of the initial conditions in order to evaluate c_1 and c_2. Since the ball is given an initial velocity of 75 ft per sec, we know that $v = 75$ for $t = 0$. Substituting these values in (2), we have

$$75 = -g \cdot 0 + c_1$$

Hence $c_1 = 75$ and (2) becomes

$$v = -gt + 75$$

Since the ball is thrown from 40 ft in the air, we know that $s = 40$ when $t = 0$. Now, substituting these values along with $c_1 = 75$ in (3) yields

$$40 = \frac{-g}{2} \cdot 0 + 75 \cdot 0 + c_2$$

Therefore, $c_2 = 40$ and (3) becomes

$$s = \frac{-gt^2}{2} + 75t + 40$$

EXERCISE 7.2

In each of Problems 1 to 12, find the equation of the family of curves whose slope is given, and find C for the member through the given point.

1 $y' = 2x + 6$, $(1, 8)$ 2 $y' = 2x - 7$, $(2, -10)$

3 $y' = 4x - 4$, $(1, -1)$ 4 $y' = 6x + 5$, $(-1, 4)$

5 $y' = 3x^2 - 3$, $(2, 2)$ 6 $y' = 3x^2 - 12x - 3$, $(3, -4)$

7 $y' = 6x^2 - 10x + 7$, $(0, 5)$

8 $y' = 2x^3 + 8x^2 + 5x + 3$, $(-3, 6)$

9 $dy/dx = (x + 1)/(y - 1)$, $(1, 1)$

10 $dy/dx = (x - 2)/(4 - y)$, $(2, 0)$

11 $dy/dx = (2x - 1)/(3 - 4y)$, $(-2, 3)$

12 $dy/dx = 2/(6y - 5)$, $(5, -3)$

In Problems 13 to 20, $a = dv/dt$ is the acceleration at any time t, v_0 is the velocity for $t = 0$, and y_0 is the distance of the moving body from a fixed point on the line of motion for $t = 0$, y is the distance at time t. Find y in each problem, and find v unless it is given. These problems are not for motion under gravity, except for Problem 13, since acceleration is not a constant.

13 $v = 2t + 3$, $y_0 = 0$ 14 $v = (t + 1)^2$, $y_0 = 25/3$

15 $v = (t + 2)^{-2}$, $y_0 = 15$ 16 $v = 8\sqrt{t}$, $y_0 = 20$

17 $a = 2t$, $v_0 = 0$, $y_0 = 0$

18 $a = t^{-1/2}$, $v_0 = 3$, $y_0 = 6$

19 $a = 15\sqrt{t + 1}$, $v_0 = 7$, $y_0 = 9$

20 $a = 48(2t + 1)^2$, $v_0 = 8$, $y_0 = 5$

21   From the top of a building 50 ft high, a stone is thrown vertically upward with an initial velocity of 80 ft per sec. Express its velocity and its distance above the ground at any time t in terms of t

by beginning with $a = -32$ and appropriate values of v_0 and y_0. Find the greatest height reached by the projectile.

22   From a point 700 ft above the ground, a small heavy particle is thrown vertically downward with initial velocity of 60 ft per sec. Derive expressions for its velocity and its distance from the ground at any time t. With what velocity does it strike the ground?

23   Derive expressions for velocity and position at any time t of a body dropped from a height of h ft above the earth. Show that its velocity when it strikes the earth will be $\sqrt{2gh}$ if air resistance is neglected.

24   Find the velocity with which a heavy small ball hits the ground if dropped from a height of 81 ft above the ground.

8

The definite integral

8.1   Introduction

In the preceding chapter we considered the indefinite integral $\int f(x)\,dx = F(x) + C$ where $F'(x) = f(x)$. In this case $F(x)$ varies with x and C is an arbitrary constant.

An integral of the type $\int_a^b f(x)\,dx$ is a *definite integral*. We evaluate a definite integral as follows:

$$\text{If} \qquad F'(x) = f(x) \qquad \text{then} \qquad \int_a^b f(x)\,dx = F(b) - F(a) \qquad (1)$$

The function $f(x)$ is the integrand and the constants a and b are the lower and upper limits, respectively.

If we replace x by t and b by x in (1), we have

$$\int_a^x f(t)\,dt = F(x) - F(a)$$

Similarly, $\qquad \int_a^x f(y)\,dy = \int_a^x f(z)\,dz = F(x) - F(a)$

Consequently, if the lower limit is a constant and the upper limit is a variable, the integral is a function of the upper limit. The letter that appears in the integrand is merely a catalytic agent and has no effect on the result obtained.

In subsequent sections we shall define the definite integral as the limit of a certain sum as the number of addends approaches infinity, and we shall show that this definition leads to the above procedure.

We shall now introduce a symbol for a certain type of sum. Thus, in place of

$$u_1 + u_2 + u_3 + \cdots + u_n$$

we write $\displaystyle\sum_{i=1}^{n} u_i$ and read "the sum from $i = 1$ to n of u_i." In keeping with this, we have

$$\sum_{i=1}^{7} a_i = a_1 + a_2 + a_3 + a_4 + a_5 + a_6 + a_7$$

partition

norm

We shall also divide an interval $[a, b]$ into n subintervals that may be equal but need not be in our study of the definite integral. Such a division into subintervals is called a partition of the interval or merely a *partition* and is symbolized by σ. We shall use d_σ for the maximum of the lengths of the subintervals and call it the *norm* of the partition σ.

8.2   The definite integral

$\displaystyle\int_a^b f(x)\,dx$

We shall consider a function $y = f(x)$ whose domain includes the closed interval $[a, b]$. We partition the interval by selecting any numbers $a = x_0 < x_1 < x_2 < \cdots < x_{i-1} < x_i < \cdots < x_{n-1} < x_n = b$ as the points of subdivision as in Fig. 8.1 and represent the lengths of the subintervals by $\Delta x_1 = x_1 - x_0,\ \Delta x_2 = x_2 - x_1, \ldots, \Delta x_i = x_i - x_{i-1}, \ldots, \Delta x_n = x_n - x_{n-1}$. We let d_σ be the norm of the partition. Furthermore, we select z_i arbitrarily on the ith subinterval. Now we form the product $f(z_i)\,\Delta x_i$ for each subinterval; then if $\displaystyle\lim \sum_{i=1}^{n} f(z_i)\,\Delta x_i$ exists as d_σ approaches zero, we call this limit "*the integral of $f(x)$ from $x = a$ to $x = b$*" and write

$$\int_a^b f(x)\,dx = \lim_{d_\sigma \to 0} \sum_{i=1}^{n} f(z_i)\,\Delta x_i \tag{1}$$

We call attention to the fact that all $\Delta x_i \to 0$ with d_σ. The

Figure 8.1   Partition

Riemann sum
sum $\sum_{i=1}^{n} f(z_i) \, \Delta x_i$ is called a *Riemann sum* and the integral (1) is called

Riemann integral
a *Riemann integral* in honor of Georg Friedrich Bernhard Riemann who lived from 1826 to 1866 and clarified the concept of an integral.

The above definition of the definite integral is sometimes altered by requiring that all subintervals be of the same length. If this is done, then the summation in (1) is replaced by $\lim_{n \to \infty} \sum_{i=1}^{n} f(z_i) \, \Delta x_i$ since automatically $d_\sigma \to 0$ if all Δx_i are the same size and the number of subintervals approaches infinity.

terminology
In (1), the function $f(x)$ is called the *integrand;* the numbers a and b are called the *lower limit* and *upper limit*, respectively; the variable x is the *variable of integration.* $\sum_{i=1}^{n} f(z_i) \, \Delta x_i$ is referred to as the *approximating sum,* and $f(z_i) \, \Delta x_i$ is an *element* or *term* of this approximation.

If we notice that the definite integral in (1) does not depend on x but is a symbol for the limit of the approximating sum, we see that the variable of integration is a filler and that

$$\int_a^b f(x) \, dx = \int_a^b f(y) \, dy = \int_a^b f(t) \, dt = \int_a^b f(w) \, dw = \cdots$$

We can approximate the integral (1) by finding $f(z_i) \, \Delta x_i$ and evaluating the Riemann approximating sum $r_n = \sum_{i=1}^{n} f(z_i) \, \Delta x_i$. Furthermore, we can sometimes evaluate the integral by finding the limit of r_n as n approaches infinity if all subintervals are of equal length or if the length of the longest subinterval approaches zero. If all subintervals are of equal length, we say the partition is *regular*.

regular partition

Example 1
Approximate $\int_0^2 x \, dx$ by using four subintervals of equal length and taking z_i to be the midpoint of the ith subinterval.

Solution: Since the length of the interval is $2 - 0 = 2$ and there are to be four subintervals of equal size, each subinterval is of length $\frac{2}{4} = \frac{1}{2}$. The points of division are as shown in Fig. 8.2. The values of z_i are as shown in the figure since each is to be the midpoint of the subinterval. Therefore,

$$\sum_{i=1}^{4} f(z_i) \, \Delta x_i = (0.25 + 0.75 + 1.25 + 1.75)0.5 = 2$$

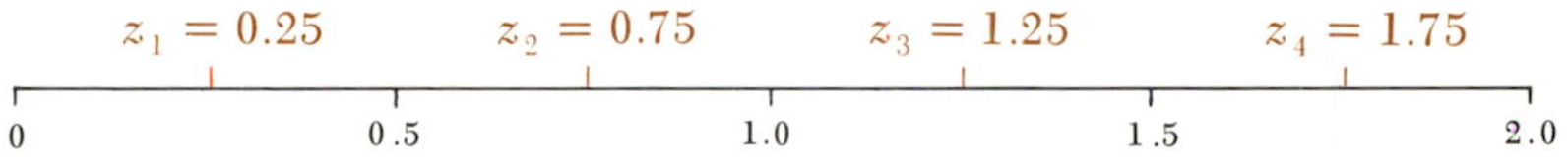

$z_1 = 0.25 \qquad z_2 = 0.75 \qquad z_3 = 1.25 \qquad z_4 = 1.75$

| 0 | 0.5 | 1.0 | 1.5 | 2.0 |

Figure 8.2

The following summation formulas can be proved by mathematical induction and will be helpful in approximating and evaluating definite integrals.

$$\sum_{i=1}^{n} i = 1 + 2 + \cdots + n = \frac{n(n + 1)}{2} \tag{2}$$

$$\sum_{i=1}^{n} i^2 = 1^2 + 2^2 + \cdots + n^2 = \frac{n(n + 1)(2n + 1)}{6} \tag{3}$$

$$\sum_{i=1}^{n} i^3 = 1^3 + 2^3 + \cdots + n^3 = \left[\frac{n(n + 1)}{2} \right]^2 \tag{4}$$

Example 2   Evaluate $\int_0^3 x^3 \, dx$ by use of (1) with a regular partition.

Solution: Since the length of the interval is $3 - 0 = 3$ and we are to use a regular partition of n subintervals, each subinterval is of length $\Delta x = 3/n$. Therefore, the points of subdivision are

$$0, \Delta x, 2\Delta x, 3\Delta x, \ldots, (n - 1) \Delta x, n \, \Delta x$$

We shall choose $z_1 = \Delta x$, $z_2 = 2\Delta x, \ldots, z_i = i \, \Delta x, \ldots, z_n = n \, \Delta x$. Thus we have

$$\int_0^3 x^3 \, dx = \lim_{n \to \infty} \sum_{i=1}^{n} f(z_i) \, \Delta x$$

$$= \lim_{n \to \infty} \sum_{i=1}^{n} f(i \, \Delta x) \, \Delta x$$

$$= \lim_{n \to \infty} \sum_{i=1}^{n} \left(i \frac{3}{n} \right)^3 \frac{3}{n}$$

$$= \lim_{n \to \infty} \frac{81}{n^4} \sum_{i=1}^{n} i^3$$

$$= \lim_{n \to \infty} \frac{81}{n^4} \left[\frac{n(n + 1)}{2} \right]^2 \qquad \textit{by (4)}$$

$$= \lim_{n \to \infty} \frac{81}{n^4} \frac{n^4 + 2n^3 + n^2}{4} \qquad \textit{expanding}$$

$$= \lim_{n \to \infty} \frac{81}{4} \left(1 + \frac{2}{n} + \frac{1}{n^2} \right)$$

$$= \frac{81}{4} \qquad \textit{since } \lim_{n \to \infty} \frac{2}{n} = \lim_{n \to \infty} \frac{1}{n^2} = 0$$

EXERCISE 8.1

1   Show that $\displaystyle\sum_{i=1}^{n} ci = cn(n+1)/2$.

2   Show that $\displaystyle\sum_{i=1}^{n} ca_i = c \sum_{i=1}^{n} a_i$.

3   Show that $\displaystyle\sum_{i=1}^{n} (a_i + b_i) = \sum_{i=1}^{n} a_i + \sum_{i=1}^{n} b_i$.

4   Show that $\displaystyle\sum_{i=1}^{n} a_i b_i \neq \sum_{i=1}^{n} a_i \sum_{i=1}^{n} b_i$.

Evaluate the sum in each of Problems 5 to 12.

5   $\displaystyle\sum_{i=1}^{n} 6i^2$

6   $\displaystyle\sum_{i=1}^{n} (8i^3 + 6i^2)$

7   $\displaystyle\sum_{i=1}^{n} (12i^2 - 2i) - 3n(n+1)$

8   $\displaystyle\sum_{i=1}^{n} (4i^3 - 6i^2 + 2i)$

9   $\displaystyle\sum_{i=1}^{n} 6i(i-1)$

10   $\displaystyle\sum_{i=1}^{n} 2i(3i+5)$

11   $\displaystyle\sum_{i=1}^{n} 12i(i+1)(i-1)$

12   $\displaystyle\sum_{i=1}^{n} 2i(i+1)(2i+1)$

Find an approximation to the integral in each of Problems 13 to 20 by use of a Riemann sum, a regular partition with n as given, and z_i at the right end of the subinterval.

13   $\displaystyle\int_{0}^{2} x^2 \, dx, \, n = 4$

14   $\displaystyle\int_{1}^{7} (x+1) \, dx, \, n = 6$

15   $\displaystyle\int_{1}^{4} x^3 \, dx, \, n = 3$

16   $\displaystyle\int_{3}^{8} (x^2 - x) \, dx, \, n = 5$

17   $\displaystyle\int_{1}^{5} \frac{dx}{x}, \, n = 4$

18   $\displaystyle\int_{4}^{7} \frac{x}{x+1} \, dx, \, n = 3$

19   $\displaystyle\int_{1}^{3} \frac{dx}{x^2}, \, n = 4$

20   $\displaystyle\int_{2}^{5} \frac{x+1}{x^2+1} \, dx, \, n = 3$

Evaluate each integral in Problems 21 to 28 by means of a Riemann sum corresponding to a regular partition of the interval. Compare Problems 21 to 24 with 13 to 16.

21   $\displaystyle\int_{0}^{2} x^2 \, dx$

22   $\displaystyle\int_{1}^{7} (x+1) \, dx$

23   $\displaystyle\int_{1}^{4} x^3 \, dx$

24   $\displaystyle\int_{3}^{8} (x^2 - x) \, dx$

25   $\displaystyle\int_{1}^{5} 4x^2 \, dx$

26   $\displaystyle\int_{0}^{1} (8x^3 + 6x^2 + 1) \, dx$

27   $\displaystyle\int_{2}^{3} 6x(x-1) \, dx$

28   $\displaystyle\int_{0}^{2} 2x(3x+5) \, dx$

8.3   Some properties of integrals

We shall begin by stating without proof that *if a function f is con-*

existence of
$\int_a^b f(x)\,dx$

tinuous over [a, b], *then* $\int_a^b f(x)\,dx$ *exists. In this case we say that f is integrable on* [a, b]. A proof of this statement can be found in most advanced calculus books.

We shall now give several theorems that are useful in connection with definite integrals. If we partition [a, b] from b to a instead of from a to b with the same points of subdivision and the same z_i in the two cases, the effect is to replace Δx_i by $-\Delta x_i$; hence

$$\int_b^a f(x)\,dx = -\int_a^b f(x)\,dx \qquad (8.1)$$

*some
properties
of integrals*

If $b = a$, Eq. (8.1) becomes

$$\int_a^a f(x)\,dx = -\int_a^a f(x)\,dx$$

Consequently,

$$\int_a^a f(x)\,dx = 0 \qquad (8.2)$$

We shall assume that f is continuous and hence, for a regular partition, that

$$\lim_{n\to\infty} \sum_{i=1}^{n} f(z_i)\,\Delta x = \int_a^b f(x)\,dx$$

exists. We shall further assume that $a \le c \le b$ and $c = x_k$. Then,

$$\sum_{i=1}^{n} f(z_i)\,\Delta x = \sum_{i=1}^{k} f(z_i)\,\Delta x + \sum_{i=k+1}^{n} f(z_i)\,\Delta x$$

Now taking the limit of each sum as the number of subdivisions becomes infinite and replacing each infinite sum by the appropriate integral, we have

$$\int_a^b f(x)\,dx = \int_a^c f(x)\,dx + \int_c^b f(x)\,dx \qquad (8.3)$$

We also find that

$$\int_a^b Cf(x)\,dx = C\int_a^b f(x)\,dx \qquad (8.4)$$

*more
properties
of integrals*

as an immediate consequence of the fact that

$$\lim_{n\to\infty} \sum_{i=1}^{n} Cf(z_i)\,\Delta x = C\lim_{n\to\infty} \sum_{i=1}^{n} f(z_i)\,\Delta x$$

If we make use of the fact that the limit of a sum is equal to the sum of the limits, we find that

$$\int_a^b [f(x) + g(x)] \, dx = \int_a^b f(x) \, dx + \int_a^b g(x) \, dx \tag{8.5}$$

If $f(x) \geq g(x)$ for all x in $[a, b]$, then for $\Delta x > 0$, we have

$$\sum_{i=1}^n f(z_i) \, \Delta x \geq \sum_{i=1}^n g(z_i) \, \Delta x$$

Now taking the limit as $n \to \infty$ and replacing each infinite sum by the appropriate definite integral, we see that:

If $f(x) \geq g(x)$ for all x in $[a, b]$

then $\int_a^b f(x) \, dx \geq \int_a^b g(x) \, dx \tag{8.6}$

We shall assume that the function f is continuous and hence is bounded in $[a, b]$. If m is the greatest lower bound and M the least upper bound of $f(x)$ in $[a, b]$, then $m \leq f(x) \leq M$. Consequently, we see that

If m and M are the g.l.b. and l.u.b., respectively, of $f(x)$ in $[a, b]$, then

$$m \, \Delta x_i \leq f(z_i) \, \Delta x_i \leq M \, \Delta x_i$$

and using (8.6),

$$m(b - a) \leq \int_a^b f(x) \, dx \leq M(b - a) \tag{8.7}$$

since $\lim_{n \to \infty} \sum_{i=1}^n m \, \Delta x_i = m \lim_{n \to \infty} \Delta x_i = m(b - a)$

$$\lim_{n \to \infty} \sum_{i=1}^n f(z_i) \, \Delta x_i = \int_a^b f(x) \, dx$$

$$\lim_{n \to \infty} \sum_{i=1}^n M \, \Delta x_i = M \lim_{n \to \infty} \sum_{i=1}^n \Delta x_i = M(b - a)$$

From (8.7), we see that $\int_a^b f(x) \, dx = K(b - a)$ where $m \leq K \leq M$.

Furthermore, since $f(x)$ is continuous, it assumes the value K for some value of x in $[a, b]$. Therefore,

law of mean for integrals

If $f(x)$ is continuous in $[a, b]$, then there is at least one value x_0 of x in $[a, b]$ such that $f(x_0) = K$, and we have

$$\int_a^b f(x) \, dx = (b - a)f(x_0) \tag{8.8}$$

This is called *the law of the mean for integrals.*

We shall now prove that

If
$$F(x) = \int_a^x f(t)\, dt$$

where $f(t)$ is continuous for $a \le t \le b$ and x is in $[a, b]$, then F has a derivative F' where $F'(x) = f(x)$ for $a \le x \le b$.

We propose to prove, under the stated hypotheses, that

$$\frac{dF(x)}{dx} = \frac{d}{dx} \int_a^x f(t)\, dt = f(x)$$

We first note that by (8.8) $\int_a^x f(t)\, dt = (x - a)f(x_0)$ where $a \le x_0 \le x$ depends on x. Consequently the definite integral is a function of the upper limit. We shall next express ΔF as a definite integral as follows:

$$\Delta F = F(x + \Delta x) - F(x)$$

$$= \int_a^{x+\Delta x} f(t)\, dt - \int_a^x f(t)\, dt$$

$$= \int_a^x f(t)\, dt + \int_x^{x+\Delta x} f(t)\, dt - \int_a^x f(t)\, dt \qquad \text{by (8.3)}$$

$$= \int_x^{x+\Delta x} f(t)\, dt$$

Hence
$$\frac{\Delta F}{\Delta x} = \frac{1}{\Delta x} \int_x^{x+\Delta x} f(t)\, dt$$

Now if m and M are the g.l.b. and the l.u.b., respectively, of $f(t)$ in $[x, x + \Delta x]$, then by (8.7),

$$m \le \frac{1}{\Delta x} \int_x^{x+\Delta x} f(t)\, dt \le M \qquad \text{if } \Delta x > 0 \tag{1}$$

Consequently, since $f(t)$ is continuous for $x \le t \le x + \Delta x$, then m and M approach $f(x)$ as Δx approaches zero. Furthermore, since

$$\frac{\Delta F}{\Delta x} = \frac{1}{\Delta x} \int_x^{x+\Delta x} f(t)\, dt$$

is between m and M, then $\Delta F/\Delta x$ also approaches $f(x)$. Therefore,

$$F'(x) = \lim_{\Delta x \to 0} \frac{1}{\Delta x} \int_x^{x+\Delta x} f(t)\, dt = f(x)$$

If $\Delta x < 0$, the inequality signs in (1) would be reversed and the above argument would still be valid.

Since a function is continuous if its derivative exists, it follows that $\int_a^x f(t)\, dt$ *is continuous in $[a, b]$ if $f(x)$ is continuous for that interval of values of x.*

8.4   The fundamental theorem of integral calculus

We shall now state and prove the fundamental theorem of integral calculus.

fundamental theorem of integral calculus   *If F and f are functions whose domains include $[a, b]$, if f is continuous, and if $F'(x) = f(x)$, then*

$$\lim_{d_o \to 0} \sum_{i=1}^{n} f(z_i)\, \Delta x_i = \int_a^b f(x)\, dx = F(b) - F(a) \tag{8.9}$$

In order to prove this important theorem, we let

$$\phi(x) = \int_a^x f(t)\, dt \tag{1}$$

hence $\phi'(x) = f(x)$. We, however, have $F'(x) = f(x)$ by hypothesis. Now, since $\phi'(x) = F'(x)$, it follows that $\phi(x)$ and $F(x)$ differ by a constant for $a \leq x \leq b$. Consequently, we write

$$\phi(x) = F(x) + C$$

and $$\int_a^x f(t)\, dt = F(x) + C \qquad \textit{by use of (1)} \tag{2}$$

If we now let $x = a$ in (2), we have

$$0 = F(a) + C \qquad \textit{by use of (8.2)}$$

Therefore $$C = -F(a)$$

and (2) becomes

$$\int_a^x f(t)\, dt = F(x) - F(a)$$

Finally, if $x = b$, this becomes

$$\int_a^b f(t)\, dt = F(b) - F(a)$$

The first and second members of (8.9) are equal from the definition of the definite integral. Hence, (8.9) is true.

If $F'(x) = f(x)$, it is customary to write

$$\int_a^b f(x)\, dx = F(x)\Big|_a^b = F(b) - F(a)$$

and read "the integral from a to b of $f(x)\, dx$ is $F(x)$ evaluated from a to b is equal to $F(b) - F(a)$."

Example 1 $$\int_1^5 x^2\, dx = \frac{x^3}{3}\Big|_1^5 = \frac{(5^3 - 1^3)}{3} = \frac{124}{3}$$

Example 2
$$\int_0^2 (4x^3 - 9x^2)\, dx = \int_0^2 4x^3\, dx - \int_0^2 9x^2\, dx \qquad \textit{by (8.5)}$$

$$= 4\int_0^2 x^3\, dx - 9\int_0^2 x^2\, dx \qquad \textit{by (8.4)}$$

$$= \frac{4x^4}{4}\Big|_0^2 - \frac{9x^3}{3}\Big|_0^2 \qquad \textit{by (8.9)}$$

$$= (x^4 - 3x^3)\Big|_0^2 \qquad \textit{simplifying}$$

$$= 2^4 - 3(2)^3 = 16 - 24 = -8$$

Example 3 In order for $\int_0^{2\sqrt{2}} \sqrt{x^2 + 1}\, x\, dx$ to be of the form $u^n\, du$, we must introduce the constant factor 2 into the integrand. Doing this and offsetting it by multiplying by ½, we have

$$\int_0^{2\sqrt{2}} \sqrt{x^2 + 1}\, x\, dx = \frac{1}{2} \int_0^{2\sqrt{2}} \sqrt{x^2 + 1} \cdot 2x\, dx$$

$$= \frac{(x^2 + 1)^{3/2}}{2}\left(\frac{2}{3}\right)\Big|_0^{2\sqrt{2}}$$

$$= \frac{(8 + 1)^{3/2} - 1^{3/2}}{3} = \frac{(27 - 1)}{3} = \frac{26}{3}$$

EXERCISE 8.2

Find the value of the definite integral in each of Problems 1 to 16.

1 $\int_1^4 2x\, dx$

2 $\int_0^3 (3x^2 - 4x)\, dx$

3 $\int_1^3 (3x + 2)\, dx$

4 $\int_2^5 (4x^3 - 6x^2 - 2)\, dx$

5 $\int_0^1 (x^4 + 2x^2 - 3)\, dx$

6 $\int_{-2}^1 (6x^2 - 4x + 3)\, dx$

7 $\int_{1/2}^1 (x^2 + 1/x^2)\, dx$

8 $\int_{-2}^{-1} (x^3 + 1/x^3)\, dx$

9 $\int_0^3 (x^2 + 1)^2\, dx$

10 $\int_{-1}^1 (x^3 - 2)^2\, dx$

11 $\int_{-2}^0 (2x - 1)^3\, dx$

12 $\int_{-2}^{-1} (3x + 4)^3\, dx$

13 $\int_1^{\sqrt{6}} \sqrt{x^2 + 3}\, x\, dx$

14 $\int_{-2}^1 (3x - 2)^{2/3}\, dx$

15 $\int_0^3 (x^3 + 3x)^{1/2}(x^2 + 1)\, dx$

16 $\int_{-2}^2 (3x^2 + 4)^{1/2} x\, dx$

Prove that each of Problems 17 to 20 is true by use of one of the properties of integrals and by evaluating each integral.

17 $\int_1^3 (x^2 - 3x)\, dx = -\int_3^1 (x^2 - 3x)\, dx$

18 $\int_2^5 \sqrt{x}\, dx + \int_2^5 \sqrt{x-1}\, dx = \int_2^5 (\sqrt{x} + \sqrt{x-1})\, dx$

19 $\int_{-1}^3 (2x + 3)^{3/2}\, dx = \int_{-1}^1 (2x + 3)^{3/2}\, dx + \int_1^3 (2x + 3)^{3/2}\, dx$

20 $\int_{-1}^1 (5 - 4x)^{1/2}\, dx < \int_{-1}^1 (7 - 4x)^{1/2}\, dx$

Assume that $\int f(x)\, dx = F(x) + C$ and prove the statement in each of Problems 21 to 24 by evaluating the integrals or by use of the properties of integrals.

21 $\int_a^b f(x)\, dx + \int_b^c f(x)\, dx + \int_c^a f(x)\, dx = 0$

22 $\int_a^b f(x)\, dx - \int_a^c f(x)\, dx = \int_c^b f(x)\, dx$

23 $\int_a^b f(x)\, dx - \int_c^b f(x)\, dx = -\int_d^a f(x)\, dx + \int_d^c f(x)\, dx$

24 $\int_a^c f(x)\, dx - \int_b^c f(x)\, dx = \int_a^d f(x)\, dx + \int_d^b f(x)\, dx$

25 Prove that $\dfrac{d}{dx}\int_x^b f(x)\, dx = -f(x).$

26 Prove that $\int_{-a}^a f(x)\, dx = 0$ if $f(-x) = -f(x)$. Such a function is called an *odd function.*

27 Prove that $\int_{-a}^a f(x)\, dx = 2\int_0^a f(x)\, dx$ if $f(-x) = f(x)$. Such a function is called an *even function.*

28 Prove that $\int_{-a}^a [f(x) - f(-x)]\, dx = 0.$

Verify the equation in the appropriate one of Problems 21 to 28 by evaluating the integrals in each of Problems 29 to 36.

29 $\int_{-2}^2 (4x^3 - x)\, dx = 0$ **30** $\int_{-3}^3 (x^5 - x^3)\, dx = 0$

31 $\int_{-1}^{1} (x^2 + 3)\, dx = 2\int_{0}^{1} (x^2 + 3)\, dx$

32 $\int_{-2}^{2} (5x^4 - x^2 + 7)\, dx = 2\int_{0}^{2} (5x^4 - x^2 + 7)\, dx$

33 $\int_{-3}^{3} [(x^2 - x) - (x^2 + x)]\, dx = 0$

34 $\int_{-4}^{4} [(2 - 3x) - (2 + 3x)]\, dx = 0$

35 $\int_{-3}^{3} [f(x) - f(-x)]\, dx = 0$ for $f(x) = x^3 - x$

36 $\int_{-s}^{s} [f(x) - f(-x)]\, dx = 0$ for $f(x) = 2x^2 - 3x$

8.5   The area under a curve

We found in the last section that the fundamental theorem enables us to replace an infinite sum by a definite integral and then to evaluate that integral. We shall now discuss an application of the theorem.

We shall consider a continuous function $y = f(x)$ on the closed interval $a \leq x \leq b$ and assume that $f(x) \geq 0$ for each x in this interval. We want to find the area bounded by $y = f(x)$, the X axis, $x = a$, and $x = b$ as shown in Fig. 8.3. We begin by dividing the interval $[a, b]$ into n equal subintervals. The points of subdivision will be represented by $x_0 = a$, x_1, x_2, ..., x_{i-1}, x_i, ..., x_{n-1}, $x_n = b$ and the subintervals are of lengths $\Delta x_1 = x_1 - x_0, \ldots$, $\Delta x_i = x_i - x_{i-1}, \ldots$, $\Delta x_n = x_n - x_{n-1}$. The desired area is the sum of the areas above the subintervals into which $[a, b]$ is divided. The ith one of these is above the interval bounded by x_{i-1} and x_i, and we shall call it A_i as indicated in Fig. 8.3. If m_i and M_i are the minimum and maximum values, respectively, of $f(x)$ in this subinterval, then

$$m_i\, \Delta x_i \leq A_i \leq M_i\, \Delta x_i$$

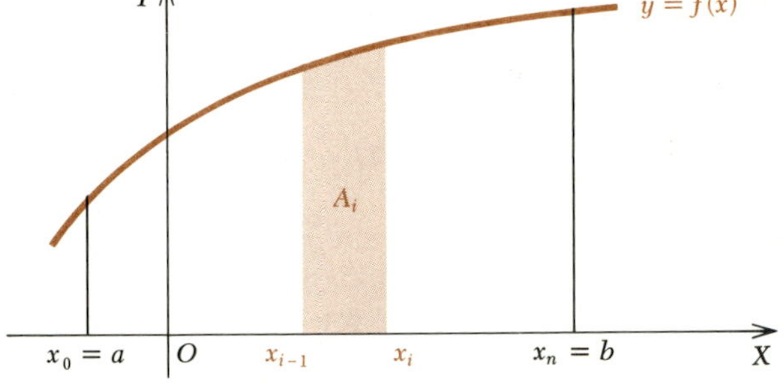

Figure 8.3   Area under a curve

It follows since $f(x)$ is continuous that there is a value x_i' of x in the ith subinterval such that $A_i = f(x_i') \Delta x_i$. The desired area A is the sum of the areas above the subintervals and is

$$A = A_1 + A_2 + \cdots + A_i + \cdots + A_n$$
$$= f(x_1') \Delta x_1 + f(x_2') \Delta x_2 + \cdots + f(x_i') \Delta x_i + \cdots + f(x_n') \Delta x_n$$
$$= \sum_{i=1}^{n} f(x_i') \Delta x_i$$

as long as each x_i' is the properly chosen value of x in the ith subinterval even if the number of subintervals increases indefinitely so that each Δx approaches zero. Furthermore,

$$\lim_{n \to \infty} \sum_{i=1}^{n} f(x_i') \Delta x_i = \int_{a}^{b} f(x)\, dx$$

Consequently, *if* $y = f(x)$ *is continuous and nonnegative in* $[a, b]$, *then the area bounded by* $y = f(x)$, $x = a$, *the* X *axis, and* $x = b$ *is*

area under a curve

$$A = \lim_{n \to \infty} \sum_{i=1}^{n} f(x_i') \Delta x_i = \int_{a}^{b} f(x)\, dx \qquad (8.10)$$

where Δx_i is a subinterval in a regular partition of $[a, b]$.

This area is often referred to as the *area under the curve* $y = f(x)$ and between a and b.

Example 1   Find the area bounded by $y = f(x) = 10 - x^2$, the X axis, $x = 1$, and $x = 3$.

Solution: The area called for is indicated in Fig. 8.4.

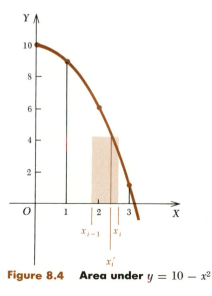

Figure 8.4   Area under $y = 10 - x^2$

We shall subdivide the interval from $x = 1$ to $x = 3$ into n equal subintervals. Then the portion of the desired area that is above the ith subinterval is $A_i = f(x_i') \, \Delta x_i = (10 - x_i'^2) \, \Delta x_i$ provided that x_i' is the properly chosen value of x. Therefore, the desired area is

$$A = \lim_{n \to \infty} \sum_{i=1}^{n} f(x_i') \, \Delta x_i = \lim_{n \to \infty} \sum_{i=1}^{n} (10 - x_i'^2) \, \Delta x_i = \int_{1}^{3} (10 - x^2) \, dx$$

$$= \left(10x - \frac{x^3}{3}\right)\Big|_{1}^{3} = 30 - \frac{27}{3} - \left(10 - \frac{1}{3}\right) = 11\tfrac{1}{3}$$

Example 2   Find the area bounded by $y = \sqrt{11 - x}$, the lines $x = 2$ and $x = 10$, and the X axis.

Solution: The desired area is indicated in Fig. 8.5. The area is

$$A = \int_{2}^{10} \sqrt{11 - x} \, dx = -\int_{2}^{10} (11 - x)^{1/2} (-dx)$$

$$= -(11 - x)^{3/2}\left(\frac{2}{3}\right)\Big|_{2}^{10} = -\frac{2[(11 - 10)^{3/2} - (11 - 2)^{3/2}]}{3}$$

$$= -\frac{2(1 - 27)}{3} = \frac{52}{3}$$

If the graph of the continuous function $y = f(x)$ is entirely below the X axis, then the value of each $f(x_i')$ is negative and if the interval $[a, b]$ is so divided that each Δx_i is positive, then each product in $\sum_{i=1}^{n} f(x_i') \, \Delta x_i$ is negative; consequently, the corresponding definite integral is negative. Therefore, the area bounded by the continuous, nonpositive function $y = f(x)$, the lines $x = a$ and $x = b$, and the X axis is $-\int_{a}^{b} f(x) \, dx$.

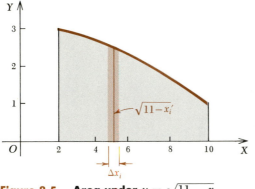

Figure 8.5   **Area under** $y = \sqrt{11 - x}$

If the continuous function $y = f(x)$ is positive for some values of x in $a \leq x \leq b$ and negative for others, then $\int_a^b f(x) \, dx$ is the difference

f(x) positive and negative in [a, b]

of the areas above and below the X axis. Consequently, if the continuous function $y = f(x)$ is nonnegative for all x in $[a, c]$ and nonpositive for all x in $[c, b]$, then the area bounded by $y = f(x)$, $x = a$, the X axis, and $x = b$ is

$$A = \int_a^c f(x) \, dx - \int_c^b f(x) \, dx$$

Example 3   Find the area bounded by $y = (x - 1)^3$, $x = -1$, the X axis, and $x = 2$.

Solution: The graph of $y = (x - 1)^3$ is shown in Fig. 8.6. We cannot find the desired area by evaluating the integral of $f(x) = (x - 1)^3$ from $x = -1$ to $x = 2$ since $f(x)$ is negative for x in a part of $-1 \leq x \leq 2$ and positive for x in another part. In particular $(x - 1)^3$ is negative for $x < 1$ and positive for $x > 1$. The desired area is

$$A = -\int_{-1}^1 (x - 1)^3 \, dx + \int_1^2 (x - 1)^3 \, dx$$

$$= \frac{-(x - 1)^4}{4} \bigg|_{-1}^1 + \frac{(x - 1)^4}{4} \bigg|_1^2$$

$$= \frac{-(1 - 1)^4}{4} + \frac{(-1 - 1)^4}{4} + \frac{(2 - 1)^4}{4} - \frac{(1 - 1)^4}{4}$$

$$= 0 + 4 + \frac{1}{4} - 0 = \frac{17}{4}$$

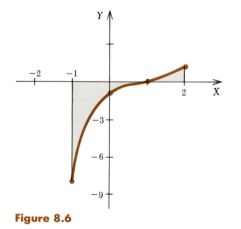

Figure 8.6

EXERCISE 8.3

Find the area under the curve and between the given lines in each of Problems 1 to 20.

1 $y = 2x + 1$, $x = 0$ to $x = 2$
2 $y = x + 9$, $x = -4$ to $x = 2$
3 $y = 3x - 1$, $x = 1$ to $x = 4$
4 $y = 4x + 7$, $x = -1$ to $x = 5$
5 $y = x^2 + 1$, $x = 0$ to $x = 3$
6 $y = 2x^2 - x$, $x = -3$ to $x = 2$
7 $y = 5x - x^2$, $x = 0$ to $x = 5$
8 $y = 3 + 2x - x^2$, $x = -1$ to $x = 2$
9 $y = x^3 + 2x$, $x = 1$ to $x = 3$
10 $y = x^3 + 3x^2 - 1$, $x = -2$ to $x = 2$
11 $y = x^3 - x + 4$, $x = -1$ to $x = 3$
12 $y = -x^3 + x + 2$, $x = -3$ to $x = 0$
13 $y = -2x + 3$, $x = -5$ to $x = 4$
14 $y = 2x - x^2$, $x = 1$ to $x = 3$
15 $y = x^2 - 4x + 3$, $x = -1$ to $x = 2$
16 $y = 2 - x - 2x^2 + x^3$, $x = 0$ to $x = 3$
17 $y = \sqrt{x + 2}$, $x = -1$ to $x = 7$
18 $y = \sqrt{6 - x}$, $x = -3$ to $x = 2$
19 $y = \sqrt{3x + 7}$, $x = -1$ to $x = 3$
20 $y = \sqrt{9 - 2x}$, $x = -8$ to $x = 4$

21 Evaluate

$$\int_0^a \sqrt{a^2 - x^2}\, dx$$

by identifying it with the area of a part of a circle.

22 Find the total area enclosed by the ellipse $b^2x^2 + a^2y^2 = a^2b^2$ by using the result of Problem 21.

23 Find the area bounded by $y^2 = 4x$, the X axis, $x = 0$, and $x = 9$.

24 Find the area bounded by $xy^2 = 9$, the X axis, $x = 1$, and $x = 9$.

8.6   The area between two curves

area between two curves

We shall now discuss the procedure for finding the area bounded by two curves and the ordinates $x = a$ and $x = b$ (see Fig. 8.7). If $y_1 = f(x)$, $y_2 = g(x)$, and $f(x) \geq g(x)$ for $a \leq x \leq b$, then the area between the curves is the area under $y_1 = f(x)$ decreased by the area under $y_2 = g(x)$. This is given by

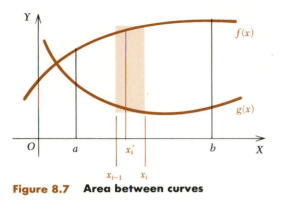

Figure 8.7   Area between curves

$$A = \lim_{n \to \infty} \sum_{i=1}^{n} [f(x_i') - g(x_i')]\, \Delta x \qquad \text{for a regular partition}$$

$$= \int_{a}^{b} [f(x) - g(x)]\, dx$$

$$= \int_{a}^{b} f(x)\, dx - \int_{a}^{b} g(x)\, dx$$

Example 1   Find the area bounded by $f(x) = 1 + 5x - x^2$, $g(x) = 4x - x^2$, $x = 1$, and $x = 3$.

Solution: The bounding curves and an element of the area are shown in Fig. 8.8. If a regular partition of the interval is used, we can indicate the area by

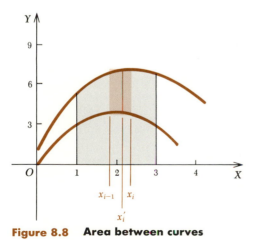

Figure 8.8   Area between curves

$$A = \lim_{n \to \infty} \sum_{i=1}^{n} [f(x_i') - g(x_i')]\, \Delta x$$

$$= \lim_{n \to \infty} \sum_{i=1}^{n} [(1 + 5x_i' - x_i'^2) - (4x_i' - x_i'^2)]\, \Delta x$$

$$= \int_{1}^{3} [(1 + 5x - x^2) - (4x - x^2)]\, dx$$

$$= \int_{1}^{3} (1 + x)\, dx = \left(x + \frac{x^2}{2}\right)\Big|_{1}^{3} = 3 + \frac{9}{2} - \left(1 + \frac{1}{2}\right) = 6$$

The same procedure as outlined and illustrated above can be used to find the area between two curves that intersect at two or more points or the area bounded by an ordinate and two intersecting curves.

Example 2   Find the area bounded by $y = x^2 - 4x + 5$ and $y = 2x - 3$.

Solution: The limits of integration are determined by the intersections of the curves. The graphs are shown in Fig. 8.9, and we shall find the abscissas of the points of intersection by eliminating y between the given pair of equations. Thus we get

$$x^2 - 4x + 5 = 2x - 3$$
$$x^2 - 6x + 8 = 0 \qquad \text{\textit{adding} } -2x + 3 \text{ \textit{to each member}}$$
$$(x - 2)(x - 4) = 0$$
$$x = 2, 4$$

Therefore, since $2x - 3 > x^2 - 4x + 5$ if $2 \le x \le 4$ the desired area is

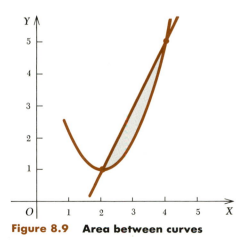

Figure 8.9   Area between curves

$$A = \int_2^4 [(2x - 3) - (x^2 - 4x + 5)]\, dx$$

$$= \int_2^4 (-x^2 + 6x - 8)\, dx$$

$$= \left(-\frac{x^3}{3} + 3x^2 - 8x \right) \Big|_2^4$$

$$= \left(-\frac{64}{3} + 48 - 32 \right) - \left(-\frac{8}{3} + 12 - 16 \right)$$

$$= -\frac{56}{3} + 20 = \frac{4}{3}$$

EXERCISE 8.4

Find the areas bounded by the curves whose equations are given in Problems 1 to 36.

1 $y = x^2 + 2x + 3$, $y = x + 1$, $x = -1$, $x = 3$
2 $y = x^2 - x + 4$, $y = 2x - 3$, $x = 1$, $x = 4$
3 $y = x^3$, $y = 4x + 1$, $x = 0$, $x = 2$
4 $y = x^3 + 3x + 1$, $y = 2x - 3$, $x = -1$, $x = 3$
5 $y = x^2 - 2x$, X axis   6 $y = x^2 - 3x$, X axis
7 $y = 9 - x^2$, X axis   8 $y = x^2 - 3x - 4$, X axis
9 $y = x^2 + 3x - 4$, $y = x^2 - 2x + 1$, $x = 3$
10 $y = x^2 - 4x + 3$, $y = x^2 + x - 2$, $x = 4$
11 $y = x^2 + 6x - 5$, $y = x^2 + 3x + 1$, $x = -2$
12 $y = 3 + 3x - x^2$, $y = -x^2 + 2x + 3$, $x = 3$
13 $y = 8x - x^2$, $y = 2x$ 14 $y = x^2 - 4x + 6$, $y = 3x$
15 $y = 2 + 2x - x^2$, $y = -2x - 3$
16 $y = -3 + 2x - x^2$, $y = x - 5$
17 $y = x^3 - 2x^2 + 5$, $y = 4x - 3$
18 $y = x^3 - x^2 - 4x$, $y = -3x - 1$
19 $y = -7 + 2x + 3x^2 - x^3$, $y = -7x + 20$
20 $y = 7 + 3x - 2x^2 - x^3$, $y = -x - 1$
21 $y = 3x^2 - 5x + 3$, $y = x^2$
22 $2y = x^2$, $y^2 = 16x$ 23 $x^2 = 6y$, $x^2 = 12y - 9$
24 $y = 8x - x^2$, $3y = x^2$ 25 $y = x^3 + 8$, $y = (x + 2)^2$
26 $y = (x - 1)^3$, $y = x^2 + 12x - 37$
27 $y = x^3 - 2x^2 - 3x + 4$, $y = (x - 1)^2$
28 $4y = x^2(5 - x)$, $y = (x - 2)^2$

Use the result of Problem 21, Exercise 8.3, to find the areas bounded by graphs of the equations in Problems 29 to 32.

29 $y^2 - 2xy + 2x^2 = 1$ 30 $y^2 - 4xy + 5x^2 = 1$
31 $y^2 - 6xy + 10x^2 = 4$ 32 $y^2 - 4xy + 5x^2 = 9$

In Problems 33 to 36, take the element of area parallel to the X axis.

33   $2x = 4y^2 - y^3$, the Y axis, $y = 1$, $y = 3$

34   $x = y^2 - 4y$, Y axis, $y = 0$, $y = 4$

35   $x = y^2(4 - y)$, $x = y(4 - y)$

36   $x = y(y + 3)$, $x = y + 3$

8.7   Improper integrals

Heretofore, we have discussed only those definite integrals that have finite limits and have considered only situations in which the integrand $f(x)$ is a continuous function of x for all x in its domain. Quite frequently, we encounter definite integrals that do not satisfy these conditions. The conditions are not satisfied if either or both limits of integration are infinite or if the integrand $f(x)$ is discontinuous or undefined for a value of x between the limits of integration or at one of them.

We shall now consider the situation in which at least one of the limits of integration is infinite. We shall give definite integrals with infinite limits a meaning by saying that

If a and b are fixed numbers and if f(x) is continuous on any domain involved for the variable x, then

*infinite
limits*

$$\int_a^\infty f(x)\,dx = \lim_{h \to \infty} \int_a^h f(x)\,dx \tag{8.11}$$

$$\int_{-\infty}^b f(x)\,dx = \lim_{h \to -\infty} \int_h^b f(x)\,dx \tag{8.12}$$

$$\int_{-\infty}^\infty f(x)\,dx = \lim_{h \to -\infty} \int_h^c f(x)\,dx + \lim_{h \to \infty} \int_c^h f(x)\,dx \tag{8.13}$$

where c is any number.

*convergent
and divergent
integrals*

If the limit exists, the improper integral is said to be *convergent*, and if the limit fails to exist, the improper integral is *divergent*.

Example 1   Evaluate $\displaystyle\int_1^\infty \frac{9\,dx}{(x+2)^2}$

Solution: We shall make use of the definition given by (8.11) and replace the given integral by

$$\lim_{h \to \infty} \int_1^h \frac{9\,dx}{(x+2)^2} = \lim_{h \to \infty} \frac{-9}{x+2}\bigg|_1^h = \lim_{h \to \infty} \left(\frac{-9}{h+2} + \frac{9}{3}\right) = 0 + 3 = 3$$

The graph of $y = 9/(x+2)^2$ for $1 \le x \le h$ is shown in Fig. 8.10. The area A under the curve from $x = 1$ to $x = h$ is

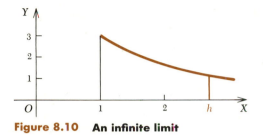

Figure 8.10 An infinite limit

$$A = -\frac{9}{h+2} + 3$$

As h moves indefinitely far to the right, this area continues to increase and approaches 3 as a limit.

Example 2   Evaluate $\int_2^{\infty} \frac{dx}{\sqrt{x-1}}$

Solution: We shall integrate from 2 to h and then take the limit of the resulting function of h as $h \to \infty$.

$$\int_2^h \frac{dx}{\sqrt{x-1}} = \int_2^h (x-1)^{-1/2}\, dx = 2\sqrt{x-1}\,\Big|_2^h$$

$$= 2(\sqrt{h-1} - \sqrt{2-1}) = 2(\sqrt{h-1} - 1)$$

We now evaluate this as $h \to \infty$ and find that

$$\lim_{h \to \infty} 2(\sqrt{h-1} - 1) = \infty$$

Consequently, the integral is divergent.

We shall now consider an improper integral in which the integrand is discontinuous or undefined at one of the limits of integration or for some value between them. In doing this we use the fact that "limit as $h \to b^-$" indicates that h approaches b through values less than b; similarly, $\lim\limits_{h \to a^+}$ means that h approaches a through values greater than a. We shall define the improper integrals mentioned above by saying that

If $f(x)$ is continuous for $a \le x < b$ but discontinuous or undefined for $x = b$, then

$$\int_a^b f(x)\, dx = \lim_{h \to b^-} \int_a^h f(x)\, dx = \lim_{\varepsilon \to 0^+} \int_a^{b-\varepsilon} f(x)\, dx \qquad (8.14)$$

discontinuous   If $f(x)$ is continuous for $(a, b]$ but discontinuous or undefined *integrand*   for $x = a$, then

$$\int_a^b f(x)\, dx = \lim_{h \to a^+} \int_h^b f(x)\, dx = \lim_{\varepsilon \to 0^+} \int_{a+\varepsilon}^b f(x)\, dx \qquad (8.15)$$

If f(x) is discontinuous or undefined for x = c, a < c < b, and continuous for all other x in [a, b], then

$$\int_a^b f(x)\, dx = \int_a^c f(x)\, dx + \int_c^b f(x)\, dx \qquad (8.16)$$

where the two integrals on the right are improper integrals of a type already considered in (8.14) and (8.15).

Example 3   Evaluate $\int_2^6 \dfrac{dx}{\sqrt{x-2}}$

Solution: A sketch of the graph of $y = \dfrac{1}{\sqrt{x-2}}$ is shown in Fig. 8.11. The integrand is not defined for the lower limit of integration; hence the integral is of the type in (8.15). Consequently, we write

$$\int_2^6 \frac{dx}{\sqrt{x-2}} = \lim_{\varepsilon \to 0^+} \int_{2+\varepsilon}^6 \frac{dx}{\sqrt{x-2}} = \lim_{\varepsilon \to 0^+} 2\sqrt{x-2}\,\Big|_{2+\varepsilon}^6$$

$$= \lim_{\varepsilon \to 0^+} (2\sqrt{6-2} - 2\sqrt{2+\varepsilon-2}) = 4 - 0 = 4$$

Example 4   Evaluate $\int_2^6 \dfrac{dx}{(x-3)^2}$

Solution: Since the integrand is undefined for the value 3 of x between the limits of integration, we make use of (8.16) and write

$$\int_2^6 \frac{dx}{(x-3)^2} = \int_2^3 \frac{dx}{(x-3)^2} + \int_3^6 \frac{dx}{(x-3)^2}$$

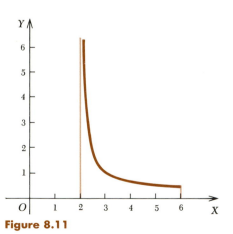

Figure 8.11

We now evaluate these improper integrals by use of (8.14) and (8.15) since each is undefined for a limit of integration. Thus,

$$\int_2^3 \frac{dx}{(x-3)^2} = \lim_{h \to 3^-} \int_2^h \frac{dx}{(x-3)^2} \qquad \text{by (8.14)}$$

$$= \lim_{h \to 3^-} \frac{-1}{x-3} \Big|_2^h$$

$$= \lim_{h \to 3^-} \left(\frac{-1}{h-3} + \frac{1}{2-3} \right)$$

$$= \infty$$

and

$$\int_3^6 \frac{dx}{(x-3)^2} = \lim_{h \to 3^+} \int_h^6 \frac{dx}{(x-3)^2} = \lim_{h \to 3^+} \frac{-1}{x-3} \Big|_h^6$$

$$= \lim_{h \to 3^+} \left(\frac{-1}{6-3} - \frac{-1}{h-3} \right) = \infty$$

Consequently, the integral is divergent.

It is true that if $\int f(x)\, dx = F(x)$ and $F(x)$ is continuous in $[a, b]$, then $\int_a^b f(x)\, dx$ can be evaluated without (8.16). We shall show it for the case in which $f(x)$ is continuous over $[a, b]$ except for a vertical asymptote at $x = c$ with c in (a, b). For this situation, we have

$$\int_a^b f(x)\, dx = \lim_{\varepsilon \to 0} \int_a^{c-\varepsilon} f(x)\, dx + \lim_{\varepsilon \to 0} \int_{c+\varepsilon}^b f(x)\, dx$$

$$= \lim_{\varepsilon \to 0} [F(c - \varepsilon) - F(a)] + \lim_{\varepsilon \to 0} [F(b) - F(c + \varepsilon)]$$

$$= F(b) - F(a)$$

since

$$\lim_{\varepsilon \to 0} F(c - \varepsilon) = \lim_{\varepsilon \to 0} F(c + \varepsilon) = F(c)$$

Example 5 Evaluate $\int_2^{11} \dfrac{dx}{\sqrt[3]{x-3}}$.

Solution: Since the integrand $f(x) = 1/\sqrt[3]{x-3}$ is continuous in $[2, 11]$ except for a vertical asymptote in $(2, 11)$ and since $F(x) = 1.5(x-3)^{2/3}$ is continuous in $[2, 11]$, we need not use (8.16) but can write

$$\int_2^{11} \frac{dx}{\sqrt[3]{x-3}} = 1.5(x-3)^{2/3} \Big|_2^{11}$$

$$= 1.5[8^{2/3} - (-1)^{2/3}] = 1.5(4 - 1) = 4.5$$

EXERCISE 8.5

In Problems 1 to 28 draw an appropriate figure, and evaluate each of the following integrals or show that it diverges.

1 $\displaystyle\int_2^\infty \frac{dx}{x^2}$ 2 $\displaystyle\int_1^\infty \frac{dx}{\sqrt{x}}$

3 $\displaystyle\int_0^\infty \frac{dx}{(2+x)^3}$ 4 $\displaystyle\int_5^\infty \frac{dx}{(x-4)^{4/3}}$

5 $\displaystyle\int_{-\infty}^1 \frac{dx}{x^4}$ 6 $\displaystyle\int_{-\infty}^2 \frac{dx}{(x-3)^2}$

7 $\displaystyle\int_{-\infty}^{-1} \frac{dx}{x^{1/3}}$ 8 $\displaystyle\int_{-\infty}^{-2} \frac{dx}{\sqrt{2-x}}$

9 $\displaystyle\int_{-\infty}^\infty \frac{2x\,dx}{(x^2+1)^3}$ 10 $\displaystyle\int_{-\infty}^\infty \frac{x^2\,dx}{(x^3+1)^2}$

11 $\displaystyle\int_{-\infty}^\infty x\sqrt{x^2+5}\,dx$ 12 $\displaystyle\int_{-\infty}^\infty x\sqrt[3]{x^2+7}\,dx$

13 $\displaystyle\int_1^5 \frac{dx}{(x-1)^2}$ 14 $\displaystyle\int_2^3 \frac{dx}{\sqrt{x-2}}$

15 $\displaystyle\int_2^{\sqrt{5}} \frac{x\,dx}{(x^2-4)^3}$ 16 $\displaystyle\int_1^5 \frac{dx}{\sqrt{x-1}}$

17 $\displaystyle\int_0^1 \frac{dx}{(x-1)^{1/3}}$ 18 $\displaystyle\int_{-5}^{-2} \frac{x\,dx}{(x^2-4)^2}$

19 $\displaystyle\int_{-29}^3 \frac{dx}{\sqrt[5]{x-3}}$ 20 $\displaystyle\int_3^5 \frac{dx}{(x-5)^2}$

21 $\displaystyle\int_1^9 \frac{dx}{(x-3)^2}$ 22 $\displaystyle\int_0^1 \frac{dx}{(2x-1)^3}$

23 $\displaystyle\int_0^2 \frac{dx}{\sqrt[3]{x-1}}$ 24 $\displaystyle\int_{-4}^{-2} \frac{dx}{\sqrt[5]{x+3}}$

25 $\displaystyle\int_{-\infty}^0 \frac{dx}{(2+x)^3}$ 26 $\displaystyle\int_0^\infty \frac{dx}{(x-4)^{2/3}}$

27 $\displaystyle\int_0^\infty \frac{x\,dx}{(x^2-1)^2}$ 28 $\displaystyle\int_{-\infty}^\infty \frac{dx}{\sqrt{2-x}}$

If it exists, find the specified area in each of Problems 29 to 32.

29   Between the X axis and $y = 1/x^2$ for $x \geq 1$
30   Between the X axis and $x^3 y = 1$ for $x \geq 2$
31   In the first quadrant under $y = (x+1)^{-1/2}$
32   In the first quadrant under $y = (x+1)^{-3/2}$

The conics

9.1 Introduction

The equation

$$Ax^2 + Bxy + Cy^2 + Dx + Ey + F = 0$$

conic
section
with A, B, C, D, E, and F constants and x and y variables is the general second-degree equation in two variables. The curve that corresponds to the general equation is called a *conic* or *conic section* and consists of the set of points whose coordinates satisfy the equation. The name conic or conic section is used since the curve is the intersection of a plane and a right circular conic surface.° If the plane does not pass through the vertex, the conic is:

parabola
1. A *parabola* if the plane is parallel to an element of the cone as in Fig. 9.1*a*

ellipse
2. An *ellipse* if the plane cuts entirely across one nappe of the conic surface as in Fig. 9.1*b*

° To generate a *right circular conic surface,* we choose a fixed point on a fixed line and rotate another line that passes through the fixed point about the fixed line in such a way that there is a constant angle between the two lines. The fixed point is called the *vertex* of the conic surface, any position of the rotating line is an *element,* and the part of the cone on either side of the vertex is called a *nappe.* The fixed line through the vertex that makes equal angles with all elements of the conic surface is called the *axis.*

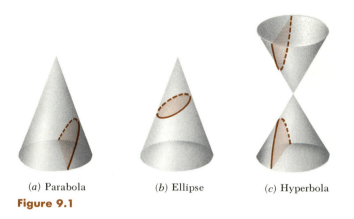

(*a*) Parabola   (*b*) Ellipse   (*c*) Hyperbola

Figure 9.1

hyperbola   **3.**   A *hyperbola* if the plane cuts both nappes of the conic surface as in Fig. 9.1*c*

The types of conics were studied and named by the Greeks who discovered many of their properties. The properties discovered by the Greeks include the ones that we shall use later in this chapter as definitions.

9.2   The parabola

We shall now study the parabola. It is defined by the statement: The *parabola* is the set of all points in a plane that are equidistant from a fixed point and a fixed line. The fixed point is called the *focus* and the fixed line is the *directrix*. The line through the focus and perpendicular to the directrix is called the *axis* and the intersection of the axis and the parabola is the *vertex*. The line segment through the focus, perpendicular to the axis, and intercepted by the parabola is called the *focal chord*, or *latus rectum*. These lines and points are shown in Fig. 9.2.

parabola defined

focus

directrix

vertex

focal chord

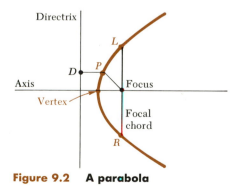

Figure 9.2   A parabola

We shall now derive the equation of a parabola with axis parallel to the X axis and with $x = h - a$ as the equation of the directrix, the point $F(h + a, k)$ as the focus, and $P(x, y)$ as any point of the parabola. Then, as shown in Fig. 9.3, the point D has the coordinates $(h - a, y)$ since it is on the directrix and is the same distance as P from the X axis. Therefore, the symbolical form $|FP| = |DP|$ of the definition, by use of the distance formula, becomes

$$\sqrt{(x - h - a)^2 + (y - k)^2} = \sqrt{(x - h + a)^2 + (y - y)^2}$$

Now considering $x - h$ as a single quantity and equating the squares of the members of the last equation, we have

$$[(x - h) - a]^2 + (y - k)^2 = [(x - h) + a]^2$$

Expanding the terms in brackets, we find that

$$(x - h)^2 - 2a(x - h) + a^2 + (y - k)^2 = (x - h)^2 + 2a(x - h) + a^2$$

Finally, adding $-(x - h)^2 + 2a(x - h) - a^2$ to each member and collecting like terms, we see that

$$(y - k)^2 = 4a(x - h) \tag{9.1}$$

is the equation of the parabola with $x = h - a$ as directrix and focus at $F(h + a, k)$. It is called the standard form.
The equation of the axis is

$$y = k \tag{9.2}$$

since the axis is through the focus and perpendicular to the directrix; furthermore, the vertex is at (h, k) since that point is the intersection of the axis and the parabola.

Thus, *the standard-form equation of the parabola with vertex at (h, k) and focus at $(h + a, k)$ is*

$$(y - k)^2 = 4a(x - h) \tag{9.1}$$

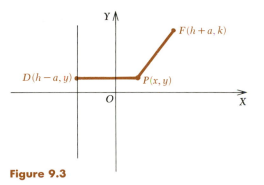

Figure 9.3

standard
forms   It can be shown in a similar manner that *the standard-form equation of the parabola with vertex at* (h, k) *and focus at* $(h, k + a)$ *is*

$$(x - h)^2 = 4a(y - k) \qquad (9.3)$$

vertex at
the origin   If the vertex (h, k) is at the origin, Eqs. (9.1) and (9.3) reduce to

$$y^2 = 4ax \qquad (9.1')$$

and
$$x^2 = 4ay \qquad (9.3')$$

The left member of (9.1) is positive or zero since it is the square of a real quantity; hence, the right member must also be nonnegative. Therefore, a and $x - h$ must be of the same sign or $x - h = 0$. Consequently, *if* $a > 0$, *then* $x > h$ *and the curve is to the right of the vertex;* furthermore, *if* $a < 0$, *then* $x < h$ *and the curve is to the left of the vertex.* It can be shown in a similar manner that the parabola (9.3) *extends upward from the vertex if* $a > 0$ *and extends downward from the vertex if* $a < 0$.

We can find the length of the focal chord by solving the equation of the line along which it lies simultaneously with that of the parabola and then using the distance formula. These equations are $x = h + a$ and $(y - k)^2 = 4a(x - h)$. If we substitute the expression for x from the first into the second, we have

$$(y - k)^2 = 4a(h + a - h) = 4a^2$$

Therefore, $y - k = \pm 2a$ and $y = k \pm 2a$. Hence, the ends of the focal chord are at $L(h + a, k - 2a)$ and $R(h + a, k + 2a)$ as shown in Fig. 9.4. The distance between them is $4a$ if $a > 0$ and $-4a$ if $a < 0$. Consequently, we can now say that *the length of the focal*
length of
focal
chord   *chord of the parabola* (9.1) *or* (9.3) *is* $4|a|$.

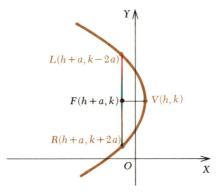

Figure 9.4   Focal chord

Example   Find the equation and length of the focal chord of the parabola with
vertex at $(1, 2)$ and focus at $(-3, 2)$.

Solution: In this problem, the vertex and focus are the same distance
from the X axis; hence, we must use (9.1). Furthermore, the focus is
4 units to the left of the vertex; consequently, $2a = -4$ and
$4a = -8$. Therefore, the equation of the parabola is

$$(y - 2)^2 = -8(x - 1)$$

The length of the focal chord is $4|a| = 8$.

EXERCISE 9.1

*Find the equation of each parabola described in Problems 1 to 20.
Sketch each one.*
   1   Vertex at $(4, 3)$, focus at $(2, 3)$
   2   Vertex at $(6, 2)$, focus at $(2, 2)$
   3   Vertex at $(5, 3)$, focus at $(5, 1)$
   4   Vertex at $(-2, -1)$, focus at $(-2, 3)$
   5   Vertex at $(7, 1)$, $y = 5$ as directrix
   6   Vertex at $(3, -2)$, $x = 11$ as directrix
   7   Vertex at $(-3, 4)$, $x = 1$ as directrix
   8   Vertex at $(2, -1)$, $y = 3$ as directrix
   9   Focus at $(5, 0)$, $x = 9$ as directrix
   10   Focus at $(4, 1)$, $y = -7$ as directrix
   11   Focus at $(-2, 3)$, $y = 5$ as directrix
   12   Focus at $(7, 2)$, $x = -1$ as directrix
   13   Vertex at $(3, 1)$, ends of focal chord at $(1, 5)$ and $(1, -3)$
   14   Vertex at $(2, 6)$, ends of focal chord at $(6, 8)$ and $(-2, 8)$
   15   Vertex at $(4, -2)$, ends of focal chord at $(-4, 2)$ and $(5, 2)$
   16   Vertex at $(-1, 3)$, ends of focal chord at $(-5, 11)$ and $(-5, -5)$
   17   Vertex at $(2, 1)$, axis parallel to the X axis, through $(6, 5)$
   18   Vertex at $(1, -3)$, axis parallel to the X axis, through $(-3, 1)$
   19   Vertex at $(2, 3)$, axis parallel to the Y axis, through $(-2, 5)$
   20   Vertex at $(3, -2)$, axis parallel to the Y axis, through $(-3, -5)$

*Find the equation that is satisfied by the coordinates of each point in
the set that is equidistant from the given line and the given point in
each of Problems 21 to 24.*
   21   $x = 2$, $(6, 5)$ 22 $x = -1$, $(7, 1)$
   23   $y = -3$, $(0, 5)$ 24 $y = 4$, $(3, -4)$

   25   The width of a parabolic reflector is 12 in. and its depth is 4 in.
Locate the focus.

26   A parabola is 16 in. wide at a distance of 6 in. from the vertex. How wide is it at the focus?

27   If a ball is thrown vertically upward from the ground with an initial velocity of v_0 ft per sec, its distance above the ground after t sec is given approximately by $y = v_0 t - 16t^2$. Sketch the curve with $v_0 = 40$.

28   If a ball is thrown with an initial velocity of v_0 ft per sec at an angle of $45°$ with the horizontal, it traverses a path whose equation is approximately $y = x - 32x^2/v_0{}^2$. If v_0 is 96, find the horizontal distance traveled and the greatest height reached by the ball.

9.3   The ellipse

ellipse

foci

axis

We shall define and discuss another one of the conics. It is the ellipse and is characterized by the following statement: An *ellipse* is the set of all points in a plane such that the sum of the distances of each from two fixed points is a constant. The fixed points are called the *foci* and the line through them is the *axis*.

Our next task is that of finding the equation of the ellipse. In order to obtain an equation that has a relatively simple form, we shall choose the position of the coordinate axes so that the foci are at $F_1(h - c, k)$ and $F_2(h + c, k)$ as shown in Fig. 9.5. If we let $P(x, y)$ be any point on the ellipse and $2a$ be the sum of its distances from the foci, then the definition can be written symbolically as $|F_1P| + |F_2P| = 2a$. We shall now express this equation in terms of x, y, and appropriate constants by use of the distance formula. Thus,

$$\sqrt{(x - h + c)^2 + (y - k)^2} + \sqrt{(x - h - c)^2 + (y - k)^2} = 2a$$

If we think of $x - h$ as a single term, isolate the first radical, square, and solve for the only remaining radical, we get

$$\sqrt{[(x - h) - c]^2 + (y - k)^2} = a - \frac{c}{a}(x - h)$$

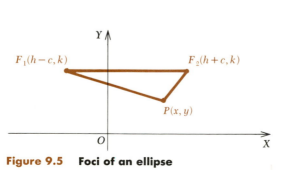

Figure 9.5   Foci of an ellipse

If we square, collect coefficients of like terms, and divide by $a^2 - c^2$, we obtain

$$\frac{(x-h)^2}{a^2} + \frac{(y-k)^2}{a^2 - c^2} = 1 \tag{1}$$

Since the distance between $F_1(h-c, k)$ and $F_2(h+c, k)$ is $2c$ and the sum of the distances F_1P and F_2P in Fig. 9.5 is $2a$ and since the sum of two sides of a triangle is greater than the third side, it follows that $a^2 - c^2$ is positive. We shall replace it by b^2 in (1) and have

$$\frac{(x-h)^2}{a^2} + \frac{(y-k)^2}{b^2} = 1 \tag{9.4}$$

in which $b = \sqrt{a^2 - c^2} < a$.

We have shown that $P(x, y)$ satisfies (9.4) if $F_1P + F_2P = 2a$ and can show that if $P(x, y)$ satisfies (9.4), then $F_1P + F_2P = 2a$ by reversing the order of the steps in the derivation. Hence, (9.4) is the equation of the ellipse.

eccentricity   The ratio c/a is called the *eccentricity* of the ellipse. The point (h, k)
center   that is midway between the foci called the *center*. The intersections
vertices   of an ellipse and its axis are known as the *vertices*. The lengths
semiaxes a and b are the *semimajor* and *semiminor* axes, respectively. Consequently,

$$\frac{(x-h)^2}{a^2} + \frac{(y-k)^2}{b^2} = 1 \tag{9.4}$$

standard   *is the equation of the ellipse with center at (h, k), major axis parallel*
equations   *to the X axis, and semiaxes a and b.*

It can be shown similarly that

$$\frac{(y-k)^2}{a^2} + \frac{(x-h)^2}{b^2} = 1 \tag{9.5}$$

is the equation of the ellipse with center at (h, k), major axis parallel to the Y axis, and semiaxes a and b.

Equations (9.4) and (9.5) are called *standard forms* of the equation of an ellipse.

focal chord   A *latus rectum*, or *focal chord*, of an ellipse is defined to be the segment of a line through a focus and perpendicular to the major axis intercepted by the ellipse. By solving the equation of the line along
length of   which it lies simultaneously with that of the ellipse and then using the
focal   distance formula, we find that *the length of the focal chord is $2b^2/a$.*
chord   A sketch of the curve determined by (9.4) is shown in Fig. 9.6.

Example   Find the equation of the ellipse with center at $(3, 4)$, a focus at $(6, 4)$, and corresponding vertex at $(8, 4)$. Sketch the curve after locating the other focus, other vertex, and ends of the focal chords.

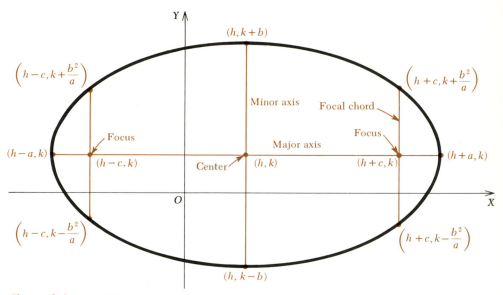

Figure 9.6   An ellipse

Solution: We can obtain the equation of an ellipse if we know the center, the semiaxes, and which of the standard forms to use. With the given data, we must use (9.4) since the center, focus, and vertex are on a line parallel to the X axis. Furthermore $a = 5$ and $c = 3$ since the distance between the center and a vertex is always a and a

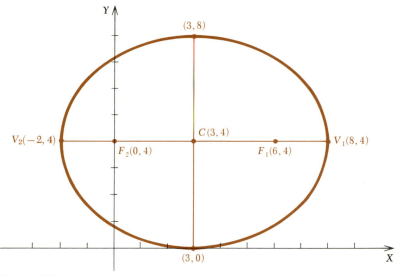

Figure 9.7

is $8 - 3 = 5$ in this problem and since the distance from center to focus is always c and c is $6 - 3 = 3$ from the given data. Hence, F_2 is at $(0, 4)$ and V_2 is at $(-2, 4)$. Now we shall determine the value of b. It can be found by use of the relation $b = \sqrt{a^2 - c^2}$; thus, $b = \sqrt{5^2 - 3^2} = 4$. Therefore, the desired equation is

$$\frac{(x - 3)^2}{5^2} + \frac{(y - 4)^2}{4^2} = 1$$

and the sketch of the ellipse is given in Fig. 9.7. The points determined by the ends of the focal chords as well as the ends of the semiaxis were used in drawing the curve.

EXERCISE 9.2

In each of Problems 1 to 4, find the coordinates of the center, foci, vertices, and ends of focal chords. Sketch each ellipse.

1 $\dfrac{(x - 1)^2}{5^2} + \dfrac{(y - 2)^2}{4^2} = 1$ 2 $\dfrac{(x + 3)^2}{13^2} + \dfrac{y^2}{5^2} = 1$

3 $\dfrac{(y - 2)^2}{5^2} + \dfrac{(x + 1)^2}{3^2} = 1$ 4 $\dfrac{(y + 4)^2}{17^2} + \dfrac{(x - 1)^2}{8^2} = 1$

Find the equation of each ellipse described in Problems 5 to 20. Sketch each one after finding the coordinates of the center, the vertices, the foci, and ends of focal chords.

5   $a = 5, b = 3$, major axis parallel to the X axis, center at $(2, 4)$
6   $a = 5, b = 4$, major axis parallel to the X axis, center at $(4, -1)$
7   $a = 13, b = 12$, major axis parallel to the Y axis, center at $(1, -3)$
8   $a = 13, b = 5$, major axis parallel to the Y axis, center at $(-2, 3)$
9   Center at $(-3, -1)$, a vertex at $(2, -1)$, a focus at $(1, -1)$
10   Center at $(0, 2)$, a vertex at $(0, 19)$, a focus at $(0, 10)$
11   Center at $(2, -2)$, a vertex at $(2, -15)$, a focus at $(2, 3)$
12   Center at $(6, -3)$, a vertex at $(1, -3)$, a focus $(10, -3)$
13   Vertices at $(-4, -1)$ and $(6, -1)$, a focus at $(1 + \sqrt{21}, -1)$
14   Vertices at $(2, -4)$ and $(2, 10)$, a focus at $(2, 3 - \sqrt{13})$
15   Foci at $(3, -4 - \sqrt{17})$ and $(3, -4 + \sqrt{17})$, a vertex at $(3, 5)$
16   Foci at $(-2 - \sqrt{11}, 3)$ and $(-2 + \sqrt{11}, 3)$, a vertex at $(4, 3)$
17   Ends of minor axis at $(-8, 1)$ and $(2, 1)$, a vertex at $(-3, 6)$
18   Ends of minor axis at $(4, -3)$ and $(0, -3)$, a vertex at $(2, 2)$
19   An end of the minor axis at $(4, 0)$, a vertex at $(0, 3)$, a focus at $(4 + \sqrt{7}, 3)$
20   An end of the minor axis at $(-5, -5)$, a vertex at $(2, 0)$, a focus at $(-5 + 2\sqrt{6}, 0)$

Find the equation that is satisfied by the coordinates of the points on the curve that is described in each of Problems 21 to 26.

21 The sum of the distances of each point from $(2, 5)$ and $(2, -1)$ is 10.

22 The sum of the distances of each point from $(3, 7)$ and $(3, -3)$ is 26.

23 The sum of the distances of each point from $(2, 3)$ and $(5, 8)$ is 7.

24 The sum of the distances of each point from $(-1, 4)$ and $(2, -3)$ is 9.

25 The distance of each point from $(0, -2)$ is equal to one-half its distance from $y = 4$.

26 The distance of each point from $(6, 0)$ is two-thirds its distance from the Y axis.

27 A rod PQ of length 24 units moves so that P is always on the Y axis and Q always on the X axis. A point M is on PQ two-thirds of the way from P to Q. Find the equation of the path traveled by M.

28 The orbit in which the earth travels about the sun is an ellipse with one focus at the sun. The semimajor axis of the ellipse is 92.9 million miles and its eccentricity is 0.0168. What are the greatest and least distances of the earth from the sun?

9.4 The hyperbola

hyperbola

foci

axis

We shall now define and discuss the hyperbola. It is characterized by the following statement: The *hyperbola* is the set of all points in a plane such that the difference of the distances of each from two fixed points is the same constant. The fixed points are called the *foci*, and the line through them is the *axis*.

Our next task is to find the equation of the hyperbola. In order to obtain an equation that is relatively simple in form, we shall choose the position of the coordinate axes so that the foci are at $F_1(h - c, k)$ and $F_2(h + c, k)$ as shown in Fig. 9.8. If we let $P(x, y)$ be any point on the hyperbola and $2a$ be the difference of its distances from the foci,

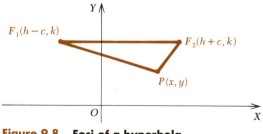

Figure 9.8 **Foci of a hyperbola**

then the definition can be written symbolically as $|F_1P| - |F_2P| = \pm 2a$. We shall now translate this equation in terms of x, y, and appropriate constants by use of the distance formula. Thus,

$$\sqrt{(x - h + c)^2 + (y - k)^2} - \sqrt{(x - h - c)^2 + (y - k)^2} = \pm 2a$$

If we think of $x - h$ as a single term, isolate the first radical, square, and solve for the only remaining radical, we get

$$\pm \sqrt{[(x - h) - c]^2 + (y - k)^2} = \frac{c}{a}(x - h) - a$$

If we square, collect coefficients of like terms, divide by $a^2 - c^2$, and write $-(c^2 - a^2)$ in place of $a^2 - c^2$, we obtain

$$\frac{(x - h)^2}{a^2} - \frac{(y - k)^2}{c^2 - a^2} = 1 \tag{1}$$

The inequality $a < c$ holds since the difference between two sides of a triangle is less than the third side. Therefore, $c^2 - a^2$ is positive; we shall replace it by b^2 and we have

$$\frac{(x - h)^2}{a^2} - \frac{(y - k)^2}{b^2} = 1 \tag{2}$$

in which $b = \sqrt{c^2 - a^2}$.

We have shown that $P(x, y)$ satisfies (2) if $|F_1P| - |F_2P| = \pm 2a$, and we can show, by reversing the order of steps, that if $P(x, y)$ satisfies (2), then $|F_1P| - |F_2P| = \pm 2a$. Hence, (2) is the equation of the hyperbola.

eccentricity   The ratio c/a is called the *eccentricity* of the hyperbola. The point
center (h, k) that is midway between the foci is called the *center*. The inter-
vertices   sections of a hyperbola and its axis are known as the *vertices*. The
numbers a and b are called the lengths of the *semitransverse* and
semiaxes   *semiconjugate* axes, respectively. Consequently,

$$\frac{(x - h)^2}{a^2} - \frac{(y - k)^2}{b^2} = 1 \tag{9.6}$$

is the equation of the hyperbola with center at (h, k), transverse axis parallel to the X axis, and semiaxes a and b.
It can be shown similarly that

$$\frac{(y - k)^2}{a^2} - \frac{(x - h)^2}{b^2} = 1 \tag{9.7}$$

is the equation of the hyperbola with center at (h, k), transverse axis parallel to the Y axis, and semiaxes a and b.

Equations (9.6) and (9.7) are called *standard forms* of the equations of the hyperbola.

*focal
chord*   A *latus rectum*, or *focal chord*, of a hyperbola is defined to be the segment of a line through a focus and perpendicular to the axis that is intercepted by the hyperbola. By solving the equation of the line along which it lies simultaneously with that of the hyperbola and then using the distance formula, we can find that *the length of the focal chord is* $2b^2/a$.

Example   Find the equation of the hyperbola with center at $C(3, 4)$, a focus at $F(8, 4)$, and corresponding vertex at $V(6, 4)$.

Solution: We can write the equation of a hyperbola if we know the center, the semiaxes, and which of the standard forms to use. With the given data, we must use (9.6) since the center, vertex, and focus are on a line parallel to the X axis. We find that the distance from the center to a vertex is $CV = a = 3$ and the distance from center to a focus is $CF = c = 5$. Now we need only determine b. This can be done by use of the relation $b = \sqrt{c^2 - a^2}$; thus, $b = \sqrt{5^2 - 3^2} = 4$. Therefore, the desired equation is

$$\frac{(x - 3)^2}{3^2} - \frac{(y - 4)^2}{4^2} = 1$$

9.5   Asymptotes

If we solve (9.6) for $y - k$, we get

$$y - k = \pm \frac{b}{a}\sqrt{(x - h)^2 - a^2} = \pm \frac{b}{a}(x - h)\sqrt{1 - \frac{a^2}{(x - h)^2}}$$

by multiplying and dividing by $x - h$. If the value of h is fixed, the value of $x - h$ becomes larger as x does. Hence, for a fixed, the value of $a^2/(x - h)^2$ approaches zero as x becomes larger. Consequently, the radicand $1 - a^2/(x - h)^2$ approaches 1, and we see that the hyperbola (9.6) approaches the lines

$$y - k = \pm \frac{b}{a}(x - h) \tag{9.8}$$

asymptotes   as x becomes larger. These lines are called the *asymptotes* of the hyperbola and are of considerable help in sketching the curve. The equations need not be remembered since they can be obtained readily by replacing the 1 of standard form by zero and solving for $y - k$.

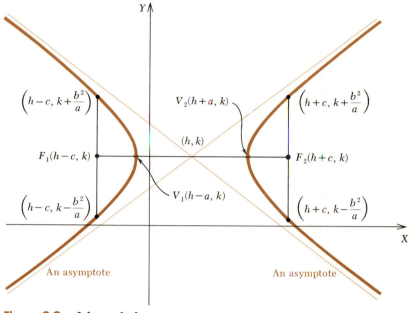

Figure 9.9 **A hyperbola**

The asymptotes of the hyperbola (9.7) can be obtained in a similar manner.

A sketch of the curve determined by (9.6) is shown in Fig. 9.9.

Example Sketch the graph of

$$\frac{(x-2)^2}{4^2} - \frac{(y+1)^2}{3^2} = 1$$

Solution: This equation is of the form (9.6) since $(x-h)^2 = (x-2)^2$ is preceded by a plus sign. Consequently, the transverse axis is parallel to the X axis. Furthermore, the center is at $(2, -1)$, $a = 4$, and $b = 3$. Therefore, the vertices are at $(2+4, -1) = (6, -1)$ and $(2-4, -1) = (-2, -1)$. The ends of the conjugate axes are 3 units above and below the center and at $(2, -4)$ and $(2, 2)$. The ends of the focal chords are $b^2/a = 9/4$ units above and below the foci; they can be located after we determine c. We can find c by use of the relation $c = \sqrt{a^2 + b^2}$; thus, $c = \sqrt{4^2 + 3^2} = 5$. Hence, the foci are at $(-3, -1)$ and $(7, -1)$. Consequently, the ends of the focal chord are at $(-3, -1 \pm 2.25)$ and $(7, -1 \pm 2.25)$. The equations of the asymptotes are $y + 1 = \pm 9/4 (x - 2)$. The center, vertices, foci, asymptotes, and the hyperbola are shown in Fig. 9.10.

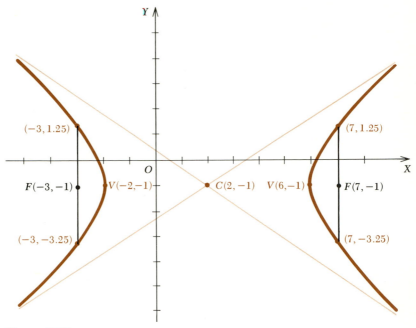

Figure 9.10

EXERCISE 9.3

In each of Problems 1 to 4, find the coordinates of the center, foci, vertices, and ends of the focal chords. Sketch each hyperbola and its asymptotes.

1 $\dfrac{(x-1)^2}{3^2} - \dfrac{(y-2)^2}{4^2} = 1$ 2 $\dfrac{(x+3)^2}{12^2} - \dfrac{y^2}{5^2} = 1$

3 $\dfrac{(y-2)^2}{4^2} - \dfrac{(x+1)^2}{3^2} = 1$ 4 $\dfrac{(y+4)^2}{8^2} - \dfrac{(x-1)^2}{15^2} = 1$

Find the equation of each hyperbola described in Problems 5 to 20. Sketch each one after finding the equations of the asymptotes and the coordinates of the center, vertices, foci, and ends of focal chords.

5 $a = 4$, $b = 3$, transverse axis parallel to the X axis, center at $(2, 4)$

6 $a = 3$, $b = 4$, transverse axis parallel to the X axis, center at $(4, -1)$

7 $a = 5$, $b = 12$, transverse axis parallel to the Y axis, center at $(1, -3)$

8 $a = 12$, $b = 5$, transverse axis parallel to the Y axis, center at $(-2, 3)$

9 Center at $(-3, -1)$, a vertex at $(1, -1)$, a focus at $(2, -1)$

10 Center at $(0, 2)$, a vertex at $(0, 10)$, a focus at $(0, 19)$

11 Center at $(2, -2)$, a vertex at $(2, 10)$, a focus at $(2, 11)$

12 Center at $(6, -3)$, a vertex at $(9, -3)$, a focus at $(1, -3)$

13 Vertices at $(1 \pm \sqrt{21}, -1)$, a focus at $(6, -1)$

14 Vertices at $(6, 3 \pm \sqrt{13})$, a focus at $(6, -4)$

15 Foci at $(3, -13)$, and $(3, 5)$, a vertex at $(3, -4 + \sqrt{17})$

16 Foci $(-8, 3)$ and $(4, 3)$, a vertex at $(3, 3)$

17 Ends of conjugate axis at $(-3 \pm 2\sqrt{6}, 1)$, a vertex at $(-3, 6)$

18 Ends of conjugate axis at $(0, -3)$ and $(4, -3)$, a vertex at $(2, -3 + \sqrt{21})$

19 An end of a conjugate axis at $(4, 6)$, a vertex at $(8, 3)$, a focus at $(-1, 3)$

20 An end of a conjugate axis at $(-5, 2\sqrt{6})$, a vertex $(0, 0)$, a focus at $(2, 0)$

Find the equation that is satisfied by the coordinates of the points in the set described in each of Problems 21 to 28.

21 The difference of the distances of each point from $(3, 4)$ and $(3, -4)$ is 6.

22 The difference of the distances of each point from $(1, 4)$ and $(1, -2)$ is 5.

23 The differences of the distances of each point from $(3, -2)$ and $(5, -5)$ is 4.

24 The differences of the distances of each point from $(4, 3)$ and $(-1, 0)$ is 6.

25 The distance of each point from $(2, -3)$ is twice its distance from $y = 1$.

26 The distance of each point from $(6, 0)$ is 1½ times its distance from the Y axis.

27 The distance of each point from $(4, 5)$ is three times its distance from $x = -2$.

28 The distance of each point from $(-3, 7)$ is twice its distance from $y = -1$.

9.6 Reduction to standard form

It is frequently desirable to convert the equation

$$Ax^2 + Cy^2 + Dx + Ey + F = 0$$

where A, C, D, E, and F are constants and where A and C are not both zero, to standard form since it is then a simple matter to determine pertinent facts concerning the conic. For this purpose, we begin by completing the square of each quadratic in the equation.

Example 1   Put $y^2 - 4y - 8x + 12 = 0$ in standard form.

Solution: We complete the square of the quadratic in y by adding $[\frac{1}{2}(-4)]^2 = 4$ to each member of the equation and thus have $y^2 - 4y + 4 - 8x + 12 = 4$. We now write $(y - 2)^2$ in place of $y^2 - 4y + 4$ and have $(y - 2)^2 - 8x + 12 = 4$. Next we add $8x - 12$ to each member so as to have the perfect square in a member by itself. Thus, we get $(y - 2)^2 = 4 + 8x - 12 = 8(x - 1)$. Therefore, $(y - 2)^2 = 4(2)(x - 1)$ is the standard form.

Example 2   Convert $16x^2 - 32x + 25y^2 - 100y = 284$ to standard form.

Solution: The first step in completing the squares is to factor the coefficient 16 of x^2 out of the terms in x^2 and x and factor the coefficient 25 of y^2 out of the terms in y^2 and y. Thus, we have $16(x^2 - 2x) + 25(y^2 - 4y) = 284$. We complete the squares by adding 1 in the first parentheses and 4 in the second; thereby, adding $16 + 100 = 116$ to the left member. This must be offset by adding 116 to the right member. When this is done, we have $16(x^2 - 2x + 1) + 25(y^2 - 4y + 4) = 284 + 116 = 400$. Now, dividing by 400 gives the standard form

$$\frac{(x - 1)^2}{5^2} + \frac{(y - 2)^2}{4^2} = 1$$

since $16/400 = 1/5^2$, $x^2 - 2x + 1 = (x - 1)^2$, $25/400 = 1/4^2$, and $y^2 - 4y + 4 = (y - 2)^2$. This is an ellipse with center at $(1, 2)$ and semiaxes 5 and 4.

9.7   Another definition of the conics

Instead of a separate definition for each type of conic, we can give *conic* one definition for all conics. We call a *conic* the set of points so *defined* located in a plane that the undirected distance of each point from a fixed point divided by its undirected distance from a fixed line is the *focus* same constant e. The fixed point is called the *focus*, the fixed line is *directrix* known as the *directrix* and the constant e is the *eccentricity*.

In deriving the equation we shall take the directrix along the Y axis, let the focus be at $F(p, 0)$, and let $P(x, y)$ be any point of the set that constitutes the conic as shown in Fig. 9.11. In terms of the symbols in the figure, the definition is $FP/DP = e$ or $FP = eDP$, and by use of the distance formula we get

$$\sqrt{(x - p)^2 + y^2} = ex$$

equation of If we square and collect coefficients of like terms, we find that
a conic

$$(1 - e^2)x^2 + y^2 - 2px + p^2 = 0 \qquad (9.9)$$

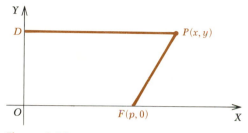

Figure 9.11

is the equation of the conic with eccentricity e, a focus at $(p, 0)$, and directrix along the Y axis.

It can be shown similarly that

$$(1 - e^2)y^2 + x^2 - 2py + p^2 = 0 \qquad (9.10)$$

is the equation of the conic with eccentricity e, a focus at $(0, p)$, and directrix along the X axis.

types of conics   The conic is an *ellipse,* a *parabola,* or a *hyperbola* according as e is less than, equal to, or greater than 1.

Example   Determine the type of conic represented by

$$3x^2 + 4y^2 - 32x + 64 = 0$$

by evaluating e. Locate a focus.

Solution: This equation contains x; hence, it must be compared with (9.9). Consequently, the coefficient of y^2 must be 1. Therefore, we must divide by the coefficient 4 of y^2. Thus, we get

$$\frac{3x^2}{4} + y^2 - 8x + 16 = 0$$

Consequently, $1 - e^2 = \frac{3}{4}$, $e^2 = \frac{1}{4}$, and $e = \frac{1}{2}$; hence, the conic is an ellipse. Finally, since $-2p = -8$, it follows that $p = 4$ and a focus is at $(4, 0)$.

EXERCISE 9.4

Convert the equation in each of Problems 1 to 12 to standard form.
1   $y^2 - 4y - 8y - 4 = 0$
2   $y^2 + 2y + 12x - 11 = 0$
3   $x^2 + 6x + 4y + 1 = 0$
4   $x^2 - 4x - 16y - 76 = 0$
5   $9x^2 + 25y^2 - 18x - 150y + 9 = 0$
6   $25x^2 + 169y^2 + 100x - 1352y - 69 = 0$
7   $9y^2 + 25x^2 - 54y + 50x - 119 = 0$

8    $16y^2 + 25x^2 + 128y - 150x + 81 = 0$
9    $9y^2 - 16x^2 + 36y + 32x - 124 = 0$
10   $25y^2 - 144x^2 + 50y + 288x - 3719 = 0$
11   $16x^2 - 9y^2 - 160x - 72y + 112 = 0$
12   $9x^2 - 16y^2 + 36x + 32y - 124 = 0$

Determine the value of e, type of conic, and location of a focus in each of Problems 13 to 24.

13   $y^2 - 4x + 4 = 0$
14   $x^2 + 2y^2 - 16x + 32 = 0$
15   $-x^2 + y^2 - 12x + 36 = 0$
16   $y^2 - 16x + 64 = 0$
17   $3x^2 + 4y^2 + 16x + 16 = 0$
18   $-2x^2 + 3y^2 - 18x + 27 = 0$
19   $x^2 - 2y + 1 = 0$
20   $2y^2 + 5x^2 + 50y + 125 = 0$
21   $-3y^2 + 4x^2 - 48y + 144 = 0$
22   $x^2 - 20y + 100 = 0$
23   $4y^2 + 9x^2 + 18y + 9 = 0$
24   $-5y^2 + 4x^2 - 16y + 16 = 0$

25   For what value or values of F does the graph of $x^2 + y^2 - 4x + 6y + F = 0$ become a point? This is a degenerate circle or point circle.

26   For what value or values of F does the graph $9x^2 + 18x - 4y^2 + 16y + F = 0$ become a pair of lines? This is a degenerate hyperbola.

27   For what value or values of F does $x^2 + 4x + 4y^2 - 8y + F = 0$ represent an imaginary ellipse. An ellipse is imaginary if $b^2(x - h)^2 + a^2(y - k)^2$ is equal to a negative number.

28   For what value or values of F does $x^2 - 2x + 4y^2 + 16y + F = 0$ represent a single point? This is a point ellipse.

9.8   Translation

We shall now investigate the effect on an equation of changing the position of the origin while leaving the axes parallel to and oriented as the original axes. If such a change is made, the axes and the origin are said to be *translated*. We shall now show that

translation formulas   *If the origin is translated to (x, k) and if (x, y) and (x′, y′) are the coordinates of a point P when referred to the original axes and the new axes, respectively, then*

$$x = x' + h \qquad and \qquad y = y' + k$$

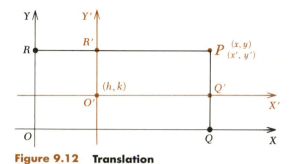

Figure 9.12 Translation

In order to prove this statement, we shall use the sketch shown in Fig. 9.12. It shows the original axes and the translated axes along with both pairs of coordinates of a point P. The origin is translated to the point (h, k). From the figure, we see that $RP = RR' + R'P$; consequently, $x = h + x'$ since $RP = x$, $RR' = h$, and $R'P = x'$. Furthermore, $QP = QQ' + Q'P$ and it follows that $y = k + y'$ since $QP = y$, $QQ' = k$, and $Q'P = y'$.

Example

Find the equation into which $x^2 - 2x + 4y^2 - 16y + 13 = 0$ is transformed if the origin is translated to $(1, 2)$.

Solution: In this problem, the origin is translated to $(1, 2)$. Hence, $h = 1$ and $k = 2$, and the translation equations become $x = x' + 1$ and $y = y' + 2$. Consequently, if we substitute these values in the given equation, we get

$$(x' + 1)^2 - 2(x' + 1) + 4(y' + 2)^2 - 16(y' + 2) + 13 = 0$$

$$x'^2 + 2x' + 1 - 2x' - 2 + 4y'^2 + 16y'$$
$$+ 16 - 16y' - 32 + 13 = 0 \qquad \textit{expanding,}$$
$$x'^2 + 4y'^2 - 4 = 0 \qquad \textit{collecting like terms}$$

9.9 Simplification by translation

The translation formulas are often used to eliminate the linear terms from an equation. If this is possible, it is done by replacing x and y, respectively, by $x' + h$ and $y' + k$, collecting coefficients of linear terms, setting them equal to zero, and solving for h and k. Thus, the values of h and k needed to eliminate the linear terms are determined, and the transformed equation is then readily obtained.

Example

Find the equation into which $4x^2 + y^2 - 8x + 4y + 4 = 0$ is transformed if the axes are translated so as to eliminate the linear terms.

Solution: If we replace x by $x' + h$ and y by $y' + k$ in the given equation, it becomes

$$4(x' + h)^2 + (y' + k)^2 - 8(x' + h) + 4(y' + k) + 4 = 0$$

$$4x'^2 + 8hx' + 4h^2 + y'^2 + 2ky' + k^2 - 8x' - 8h + 4y'$$
$$+ 4k + 4 = 0 \quad \text{\textit{performing the indicated operations}}$$

$$4x'^2 + y'^2 + (8h - 8)x' + (2k + 4)y' + 4h^2 + k^2 - 8h$$
$$+ 4k + 4 = 0 \quad \text{\textit{collecting coefficients of } x' \text{ and of } y'} \quad (1)$$

In order to eliminate the linear terms, we must have $8h - 8 = 0$ and $2k + 4 = 0$; hence, $h = 1$ and $k = -2$. If these values are substituted in (1), we find that

$$4x'^2 + y'^2 - (8 - 8)x' + (-4 + 4)y' + 4(1^2) + (-2)^2$$
$$-8(1) + 4(-2) + 4 = 0$$

and, combining similar terms, we have

$$4x'^2 + y'^2 - 4 = 0$$

as the transformed equation of the ellipse.

EXERCISE 9.5

Find the equation into which each equation in Problems 1 to 4 is transformed if the origin is translated to the given point.
1   $x^2 + y^2 - 2x - 4y - 4 = 0$, $(1, 2)$
2   $4x^2 + y^2 - 24x + 4y + 36 = 0$, $(3, -2)$
3   $y^2 - 8x - 8y - 8 = 0$, $(-3, 4)$
4   $-4x^2 + 9y^2 - 16x - 18y - 43 = 0$, $(-2, 1)$

Find the equation into which the equation in each of Problems 5 to 12 is transformed if the axes are translated so as to remove the linear terms if possible.
5   $4x^2 + 9y^2 - 8x - 36y + 4 = 0$
6   $x^2 + 4x + 4y - 4 = 0$
7   $4x^2 - y^2 + 24x + 4y + 36 = 0$
8   $x^2 + y^2 + 10x - 6y + 18 = 0$
9   $y^2 - 6y - 4x + 13 = 0$
10   $9x^2 - y^2 - 36x + 2y + 26 = 0$
11   $4x^2 + 9y^2 - 16x + 18y - 11 = 0$
12   $25x^2 + 4y^2 - 50x - 16y - 59 = 0$

13   Remove the linear terms from $(x - 2)^2 = 8(y - 1)$ by a translation, or tell why it cannot be done.
14   To what point must the origin be translated in order to remove the linear term in y and the constant terms from $y^2 + 2x - 2y - 7 = 0$?

15   Find the equation into which $ax + ay + c = 0$ is transformed if the origin is translated to $(-h, h)$.

16   Show that the distance between two points is not altered by a translation of axes.

17   Find the equation into which $ax + by + c = 0$ is transformed if the origin is translated to $(bh, -ah - c/b)$.

18   Find the equation into which $y = mx + b$ is transformed if the origin is translated to $(-b/m, 0)$.

19   What is $(x - h)^2 - (y - k)^2 = r^2$ transformed into if the origin is translated to (k, h)?

20   Show that the slope of a line is not altered by a translation of axes.

9.10   Rotation

We shall find the effect on an equation of leaving the origin fixed and turning both axes through the same angle about the origin. If this is done, we say that the axes have been *rotated*. We shall show that

If (x, y) designates a point before the axes are rotated and if (x', y') designates the same point after the axes are rotated through an angle θ, then

rotation
formulas
$$x = x' \cos \theta - y' \sin \theta \qquad and \qquad y = x' \sin \theta + y' \cos \theta$$

In order to prove this statement, we shall make use of Fig. 9.13. The first step in constructing it is to draw two pairs of coordinate axes with the same origin and making an angle of θ with one another. We then select any point P in the plane and let its coordinates be (x, y) relative to one pair of axes and (x', y') relative to the other pair. Now we drop perpendiculars from P to the X and X' axes, and we call their feet Q and R, respectively. We connect P and O and designate the angle ROP by the name ϕ. Then, $\cos (\phi + \theta) = OQ/OP$,

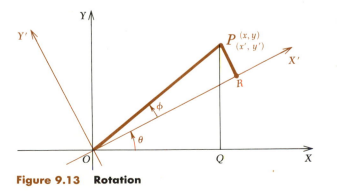

Figure 9.13   Rotation

$\cos \phi = OR/OP$, and $\sin \phi = RP/OP$. If we substitute these values in the identity $\cos (\phi + \theta) = \cos \phi \cos \theta - \sin \phi \sin \theta$ we get

$$\frac{OQ}{OP} = \frac{OR}{OP} \cos \theta - \frac{RP}{OP} \sin \theta$$

Now multiplying each member of the above equation by OP, we have

$$OQ = OR \cos \theta - RP \sin \theta$$

hence, $x = x' \cos \theta - y' \sin \theta$

since $OQ = x$, $OR = x'$, and $RP = y'$. This completes the derivation of the formula for x. The formula for y can be obtained similarly by use of the identity $\sin (\phi + \theta) = \sin \phi \cos \theta + \cos \phi \sin \theta$.

Example   Transform the equation $xy - 2 = 0$ by rotating the axes through $\pi/4$.

Solution: For $\theta = \pi/4$, the rotation formulas are $x = x' \cos (\pi/4) - y' \sin (\pi/4)$ and $y = x' \sin (\pi/4) + y' \cos (\pi/4)$. After substituting $1/\sqrt{2}$ for $\sin (\pi/4)$ and for $\cos (\pi/4)$, we have

$$x = \frac{x' - y'}{\sqrt{2}} \qquad \text{and} \qquad y = \frac{x' + y'}{\sqrt{2}}$$

Therefore, the given equation $xy - 2 = 0$ becomes

$$\frac{x' - y'}{\sqrt{2}} \frac{x' + y'}{\sqrt{2}} - 2 = 0$$

which can be expressed in the form

$$x'^2 - y'^2 - 4 = 0$$

9.11   Simplification by rotation

The most frequent use of rotation in connection with a quadratic equation in two variables is to eliminate the product term. We shall now find how to determine θ so that if the axes are rotated through that angle, the coefficient of the product term is zero. The first step in this procedure is to replace x by $x' \cos \theta - y' \sin \theta$ and y by $x' \sin \theta + y' \cos \theta$ in the given equation of form

$$Ax^2 + Bxy + Cy^2 + Dx + Ey + F = 0 \qquad B \neq 0 \qquad (1)$$

If this is done, the new equation is

$$A'x'^2 + B'x'y' + C'y'^2 + D'x' + E'y' + F' = 0 \qquad (2)$$

We find that

$$A' = A \cos^2 \theta + B \sin \theta \cos \theta + C \sin^2 \theta \tag{3}$$

$$B' = (C - A) \cdot 2 \sin \theta \cos \theta + B(\cos^2 \theta - \sin^2 \theta) \tag{4}$$

$$C' = A \sin^2 \theta - B \sin \theta \cos \theta + C \cos^2 \theta \tag{5}$$

$$D' = D \cos \theta + E \sin \theta$$

$$E' = E \cos \theta - D \sin \theta$$

Consequently, in order to eliminate the product term, we must have

$$(C - A) \cdot 2 \sin \theta \cos \theta + B(\cos^2 \theta - \sin^2 \theta) = 0 \tag{4}$$

By use of the identities for the sine and the cosine of twice an angle, Eq. (4) can be expressed in the form

$$(C - A) \sin 2\theta + B \cos 2\theta = 0$$

Hence, $$B \cos 2\theta = (A - C) \sin 2\theta$$

eliminating the product term Now dividing each member by $B \sin 2\theta \neq 0$, we see that the *product term in (1) is eliminated if the axes are rotated through an angle θ determined by*

$$\cot 2\theta = \frac{A - C}{B} \qquad B \neq 0 \tag{9.11}$$

There are infinitely many angles which satisfy (9.11), and the use of any one of them eliminates the product term. It is customary to choose the smallest possible positive angle. Hence, we ordinarily rotate through a positive acute angle θ, which is available since there is always an angle 2θ between 0 and π that satisfies (9.11).

Equation (9.11) gives the value of $\cot 2\theta$, but we must have the exact value of $\sin \theta$ and of $\cos \theta$ to apply the rotation formulas. These can be found by use of

$$\cot 2\theta = \frac{1 - \tan^2 \theta}{2 \tan \theta} \qquad \text{and} \qquad \tan \theta = \frac{\sin \theta}{\cos \theta} \tag{9.12}$$

since the first formula enables us to find $\tan \theta$ if $\cot 2\theta$ is known and the second can be used to find $\sin \theta$ and $\cos \theta$ after $\tan \theta$ is determined.

finding $\sin \theta$ and $\cos \theta$ In finding $\sin \theta$ and $\cos \theta$ when $\tan \theta$ is known, we must realize that, if $\tan \theta = a/b$, it does not follow that $\sin \theta = a$ and $\cos \theta = b$, but rather $\sin \theta = ka$ and $\cos \theta = kb$. Consequently, if we square each member of these two equations and add corresponding members of the new equations, we have $1 = k^2(a^2 + b^2)$ since $\sin^2 \theta + \cos^2 \theta = 1$. Therefore, it follows that $k = 1/\sqrt{a^2 + b^2}$. Consequently, *if* $\tan \theta = a/b$ then

$$\sin \theta = \frac{a}{\sqrt{a^2 + b^2}} \qquad \text{and} \qquad \cos \theta = \frac{b}{\sqrt{a^2 + b^2}} \tag{9.13}$$

Example   Find the sine and cosine of the angle θ through which the axes must be rotated so as to eliminate the product term from $9x^2 + 24xy + 2y^2 - 3 = 0$.

Solution: We must rotate the axes through an angle θ determined by

$$\cot 2\theta = \frac{A - C}{B} = \frac{9 - 2}{24} = \frac{7}{24}$$

Consequently, by the first of Eqs. (9.12), we have

$$\frac{1 - \tan^2 \theta}{2 \tan \theta} = \frac{7}{24}$$

$$12 - 12 \tan^2 \theta = 7 \tan \theta \qquad \textit{multiplying each member by 24 tan } \theta$$

adding $-7 \tan \theta$ to each member,

$$12 \tan^2 \theta + 7 \tan \theta - 12 = 0 \qquad \textit{rearranging terms}$$

$$(4 \tan \theta - 3)(3 \tan \theta + 4) = 0 \qquad \textit{factoring}$$

Therefore, $\tan \theta = \frac{3}{4}, -\frac{4}{3}$.

We shall use $\frac{3}{4}$ since we want to rotate through a positive acute angle. Substituting $a = 3$ and $b = 4$ in (9.13), we get $\sin \theta = \frac{3}{5}$ and $\cos \theta = \frac{4}{5}$.

If desired, the equation into which the given one is transformed can now be found by replacing $\sin \theta$ and $\cos \theta$ by the values just determined in the rotation formulas, substituting in the given equations, and then collecting coefficients of like terms.

9.12   The discriminant and the types of conics

discriminant   The number $B^2 - 4AC$ is called the *discriminant* of the equation

$$Ax^2 + Bxy + Cy^2 + Dx + Ey + F = 0 \qquad (1)$$

It can be shown that if (1) becomes

$$A'x^2 + B'x'y' + C'y'^2 + D'x' + E'y' + F' = 0 \qquad (2)$$

when the axes are rotated through any angle θ, then

$$B^2 - 4AC = B'^2 - 4A'C' \qquad (9.14)$$

In order to prove (9.14), we must make use of (3), (4), and (5) of Sec. 9.11 and several trigonometric identities. The details of the proof will be left as an exercise for the reader.

If θ is so chosen as to eliminate the product term from (1), then $B' = 0$ and (9.14) becomes $B^2 - 4AC = -4A'C'$. If we now compare (2) with $B' = 0$ and the standard forms of the conics, we see that

$$A'x'^2 + C'y'^2 + D'x' + E'y' + F' = 0$$

is the equation of:

1. An ellipse if $-4A'C' < 0$, since the coefficients of the second-degree terms have the same sign in the equation of an ellipse
2. A parabola if $-4A'C' = 0$, since one second-degree term is missing in the equation of a parabola
3. A hyperbola if $-4A'C' > 0$, since the coefficients of the second-degree terms have opposite signs in the equation of a hyperbola

types of   Consequently, we now know that (1) *is the equation of an ellipse,*
conics   *a parabola, or a hyperbola according as the discriminant* $B^2 - 4AC$
is negative, zero, or positive.

Example   What type of conic is represented by
   (*a*) $4x^2 + 6xy + 3y^2 - 5x + 7y - 8 = 0$?
   (*b*) $3x^2 + 6xy + 3y^2 - 5x + 7y - 8 = 0$?
   (*c*) $2x^2 + 6xy + 3y^2 - 5x + 7y - 8 = 0$?

Solution: The first equation represents an ellipse, since $B^2 - 4AC = 6^2 - 4(4)(3) = -12$ and $-12 < 0$. The second equation represents a parabola, since $B^2 - 4AC = 6^2 - 4(3)(3) = 0$. The third equation represents a hyperbola, since $B^2 - 4AC = 6^2 - 4(2)(3) = 12$ and $12 > 0$.

EXERCISE   9.6

Find the new equation if the axes are rotated through the indicated angle in each of Problems 1 to 4.
   1   $x^2 - 5xy + y^2 = 3$, $45°$
   2   $2x^2 - 3\sqrt{3}xy - y^2 = 5$, $60°$
   3   $11x^2 + 24xy + 4y^2 = 9$, $\theta = \sin^{-1} \tfrac{3}{5}$, θ acute
   4   $8x^2 + 3xy + 4y^2 = 3$, $\theta = \cos^{-1}(-1/\sqrt{10})$, θ obtuse

Identify the type of conic that is represented by the equation in each of Problems 5 to 12.
   5   $4x^2 - 3\sqrt{2}xy + y^2 + 3x = 9$
   6   $8x^2 + 8xy + 2y^2 - 5x - 7y = 4$
   7   $4x^2 + 7xy + 3y^2 = 24$
   8   $5x^2 + 7xy + 3y^2 = 8$
   9   $3x^2 - 2xy + 5y^2 - 2y = 9$
  10   $9x^2 - 6xy + y^2 + 4x - 7y = 0$
  11   $7x^2 + 11xy + 5y^2 = 48$

12 $6x^2 + 2xy - y^2 + 3x - y = 7$

Find the sine and cosine of the smallest positive angle through which the axes can be rotated so as to eliminate the product term in each of Problems 13 to 16.

13 $5x^2 + 2\sqrt{3}xy + 3y^2 = 3$
14 $3x^2 + xy + 3y^2 = 1$
15 $16x^2 + 7xy - 8y^2 = 17$
16 $7x^2 + 4xy + 4y^2 = 8$

17   Prove that $x^2 + y^2 = r^2$ is unaltered by a rotation.
18   Prove that both linear terms cannot be removed from $Ax^2 + Bxy + Cy^2 + Dx + Ey + F = 0$ by a rotation.
19   Show that the distance between two points is not altered by a rotation.
20   Find the equation into which $xy = a^2$ is transformed if the axes are rotated through $45°$.

Remove the product term from the equation in each of Problems 21 to 24; then, if possible, remove the linear terms.

21 $31x^2 + 10\sqrt{3}xy + 21y^2 - 144 = 0$
22 $23x^2 + 26\sqrt{3}xy - 3y^2 - 144 = 0$
23 $41x^2 + 24xy + 34y^2 + 25x + 50y - 25 = 0$
24 $3x^2 - 8xy - 3y^2 - 2\sqrt{5}x - 4\sqrt{5}y = 0$

Use $Ax^2 + Bxy + Cy^2 + Dx + Ey + F = 0$ in Problems 25 to 28.

25   Show that $B^2 - 4AC$ is unchanged by a rotation.
26   Show that $A + C + F$ is not changed by a rotation.
27   Show that $4(AC + CF + FA) - B^2 - D^2 - E^2$ is an invariant relative to rotation.
28   Show that $4ACF + BDE - (AE)^2 - (CD)^2 - (FB)^2$ is an invariant relative to translation.

10
Applications of the derivative

10.1   Angle of intersection of curves

If the two curves $y = f_1(x)$ and $y = f_2(x)$ intersect at a point P, their *angle of intersection* is defined to be the positive angle ϕ less than or equal to 90° between the tangents drawn to the curves at P as shown in Fig. 10.1.

angle of intersection

We can find this angle easily as follows:

1.   Determine the coordinates of a point P of intersection by solving the equations of the curves simultaneously.
2.   Find the slope of the tangent to each curve at P by use of the derivative, and call the slopes m_1 and m_2.
3.   Now make use of the fact that if θ_1 and θ_2 are the inclinations of the tangents, then $\phi = \theta_1 - \theta_2$ if $\theta_1 > \theta_2$; hence,

$$\tan \phi = \tan(\theta_1 - \theta_2) = \frac{\tan \theta_1 - \tan \theta_2}{1 + \tan \theta_1 \tan \theta_2} = \frac{m_1 - m_2}{1 + m_1 m_2}$$

and $\qquad \phi = \text{Arctan} \dfrac{m_1 - m_2}{1 + m_1 m_2} \qquad m_1 > m_2$

It can be shown similarly that if $\theta_2 > \theta_1$, then

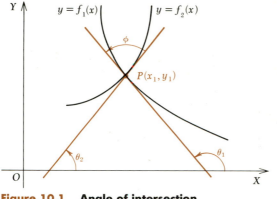

Figure 10.1   Angle of intersection

$$\phi = \text{Arctan} \frac{m_2 - m_1}{1 + m_2 m_1} \qquad m_2 > m_1$$

If there is more than one point of intersection, each one must be considered separately.

Example   At what angle do the parabolas $y = x^2/4$ and $y^2 = x/2$ intersect?

Solution:

1. If we solve the equations simultaneously, we find that the curves intersect at $P(2, 1)$ and at $O(0, 0)$ as shown in Fig. 10.2.
2. We next find the slope of each curve at P by use of their derivatives. Thus,

For $y = x^2/4$,

$$\frac{dy}{dx} = \frac{x}{2} \qquad m_1 = \frac{dy}{dx}\bigg|_{(2,1)} = 1$$

For $y^2 = x/2$,

$$2y\frac{dy}{dx} = \frac{1}{2} \qquad \frac{dy}{dx} = \frac{1}{4y} \qquad m_2 = \frac{dy}{dx}\bigg|_{(2,1)} = \frac{1}{4}$$

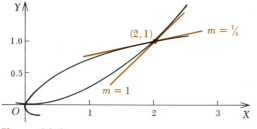

Figure 10.2

3.   The tangent of the required angle at $(2, 1)$ is then

$$\tan \phi = \frac{m_1 - m_2}{1 + m_1 m_2} = \frac{1 - \frac{1}{4}}{1 + 1(\frac{1}{4})} = \frac{3}{5}$$

at $(2, 1)$ and

$$\phi = \text{Arctan } \tfrac{3}{5} = 30°58'$$

Furthermore, $m_1\big|_{(0,0)} = 0$ and $m_2\big|_{(0,0)} = \frac{1}{6}$. Therefore, the angle between the curves at $(0, 0)$ is $90°$, since one of the curves is horizontal and the other vertical.

10.2   Equations of tangent and normal

It was proved in Sec. 2.8 that the equation of the line that passes through a given point (x_1, y_1) and has slope m is

$$y - y_1 = m(x - x_1)$$

equation of the tangent   Using this, we can obtain the equation of the line that is tangent to the curve $y = f(x)$ at a given point $P(x_1, y_1)$ on the curve, since the value of dy/dx at P is the slope of the required line.

equation of the normal   The line through P perpendicular to the tangent line is called the *normal line*. Since its slope is $-1/m$, its equation can also be found by using the above point-slope form of the equation of a line.

Example   Write the equation of the line that is tangent to the parabola $y^2 = x/2$ at $P(2, 1)$ and the equation of the normal through P.

Solution: Since $y^2 = x/2$, it follows that $2y \, dy/dx = \frac{1}{2}$ and

$$\frac{dy}{dx} = \frac{1}{4y} = \frac{1}{4} \qquad \text{at } (2, 1)$$

The equation of the line through $P(2, 1)$ with slope $\frac{1}{4}$ is

$$y - 1 = \frac{x - 2}{4}$$

or $$4y - x = 2$$

The slope of the normal at $(2, 1)$ is the negative reciprocal of $\frac{1}{4}$; hence it is -4. Therefore, the equation of the normal is

$$y - 1 = -4(x - 2)$$

or $$4x + y = 9$$

EXERCISE 10.1

In Problems 1 to 12 sketch each of the curves. Find the equation of the tangent and normal to each at the given point, and draw these lines.

1    $y = x^3 - 3$, $(2, 5)$
2    $y = x^3 - 3x^2 + 4x - 12$, $(2, -8)$
3    $y = x^3 - 3x + 4$, $(1, 2)$
4    $y = x^3 + 2x^2 - 5x - 10$, $(3, 20)$
5    $x^2 + 4y^2 = 25$, $(3, 2)$ 6 $x^2 + 9y^2 = 40$, $(-2, 2)$
7    $4x^2 + y^2 + 8x + 2y = 20$, $(-3, 2)$
8    $9x^2 + 4y^2 + 36x + 24y = 0$, $(0, 0)$
9    $xy = 6$, $(3, 2)$ 10 $xy - y = x^2$, $(2, 4)$
11    $x^2y - 4y - 4x = 0$, $(1, -\frac{4}{3})$
12    $4y^2 - xy^2 = x^3$, $(2, -2)$

In Problems 13 to 28 sketch each of the pairs of curves. Find the points of intersection and the tangent of each acute angle between the curves at the intersections.

13    $x^2 + y^2 = 25$, $x - 2y + 5 = 0$
14    $y = x^2 - 6x + 5$, $y = x - 1$
15    $x^2 + y^2 = 10y$, $x - y + 4 = 0$
16    $y = 2x^2 - 5x + 3$, $3x + y = 7$
17    $y^2 = 4x$, $y = 2x^2$
18    $x^2 + y^2 = 100$, $2y^2 = 9x$
19    $4x^2 + y^2 = 32$, $y^2 = 8x$
20    $x^2 + 8y^2 = 6x$, $4y = x^2$
21    $11x + xy = 48$, $11y - 4xy = -5$
22    $14x + xy = 60$, $28y - 5xy = 8$
23    $7xy - x - 11y = 7$, $xy - x + y = 7$
24    $8xy + 11x - 14y = 8$, $2xy - 11x - 20y = -20$
25    $x^2y = 1$, $y = 2x + 3$ 26 $x^2y + 4y = 8$, $2y^2 = x$
27    $xy^2 + 2x = 11$, $x^2 + y^2 = 10x$
28    $xy^2 + 3y + 1 = 0$, $x^2 + y^2 = -5y$

29   Find the equation of the line with slope 5 and tangent to $y = x^2 - 3x + 7$.
30   If a line is tangent to $y = x^2 - 4x - 5$ and has slope 6, what is its equation?
31   A line is parallel to $3x + y = 7$ and tangent to $y = x^2 + x - 6$. Find its equation.
32   Find the equation of a line that is parallel to $x + y = 3$ and tangent to $y = 2x^2 - 5x - 1$.
33   Prove that the equation of the tangent to the circle $x^2 + y^2 = r^2$ at (x_1, y_1) is $x_1x + y_1y = r^2$.

34   Prove that the equation of the tangent to the ellipse $x^2/a^2 + y^2/b^2 = 1$ at (x_1, y_1) is $x_1 x/a^2 + y_1 y/b^2 = 1$.

35   Prove that the equation of the normal to the hyperbola $x^2/a^2 - y^2/b^2 = 1$ at (x_1, y_1) is $b^2 x_1 y + a^2 y_1 x = (a^2 + b^2)x_1 y_1$.

36   Prove that the equation of the normal to the parabola $y^2 = 2px$ at (x_1, y_1) is $y_1 x + py = (p + x_1)y_1$.

37   Find the equations of the lines that are tangent to the ellipse $16x^2 + 25y^2 = 400$ and parallel to the line $3x - 5y + 15 = 0$.

38   A line has x intercept -4 and y intercept 3. Another line is perpendicular to the given line and tangent to the circle $x^2 + y^2 = 10x$ at a point in the first quadrant. At what point will this second line cross the Y axis?

39   Consider the parabola $y^2 = 4ax$, $a > 0$, and recall that its focus is at the point $F(a, 0)$. Join any point $A(x, y)$ on the parabola to F, and also draw a line through A parallel to the axis of the parabola. Prove that these two lines make equal angles with the tangent line to the parabola at A.

40   At what points on the hyperbola $9x^2 - 2y^2 = 14$ can a tangent of slope 4.5 be drawn?

41   Determine a, b, and c so that $y = ax^2 + bx + c$ has a horizontal tangent at $(1, 3)$ and passes through $(-1, 7)$.

42   Find the values of a, b, and c if $y = ax^2 + bx + c$ is parallel to $2x - y = 13$ at $(2, 1)$ and passes through $(1, 1)$.

43   Find the values of a, b, c, and d in $y = ax^3 + bx^2 + cx + d$ if the curve has a horizontal tangent at $(2, -13)$ and is tangent to $3x - y = 1$ at $(-1, -4)$.

44   Find the values of the coefficients in the equation $y = ax^3 + bx^2 + cx + d$ if the graph has a horizontal tangent line at $(0, 9)$ and is tangent to the line $8x - y = 7$ at $(2, 9)$.

10.3   Time rates

We have emphasized the fact that the derivative of a function $f(x)$ is a measure of the instantaneous rate of change of the value of the function relative to x. Suppose now that we have a relation

$$Q = f(t)$$

where t represents time. The value of dQ/dt, for any value of t, represents the instantaneous rate of change of the quantity Q *relative to the time*. If, for example, Q is in pounds and t is in minutes, then dQ/dt is the rate of change of Q in *pounds per minute*.

Example 1   If the radius of a circle increases at ¼ in. per min, at what rate is the area increasing at the instant when $r = 6$ in.?

Solution 1: We know the rate of change of r with respect to t, and we want to find the rate of change of A with respect to t. Hence, we must express dA/dt in terms of dr/dt. This can be done if we can express A as a differentiable function of r. We make use of the fact that $A = \pi r^2$. Differentiating with respect to t, we find that

$$\frac{dA}{dt} = 2\pi r \frac{dr}{dt}$$

If we now replace r and dr/dt by their values as given in the problem, we see that

$$\frac{dA}{dt} = 2\pi(6)\left(\frac{1}{4}\right) = 3\pi$$

in terms of square inches per minute.

Solution 2: We can solve this problem in a somewhat different manner by thinking of the radius as starting from zero when $t = 0$ and increasing at ¼ in. per min; hence, $r = t/4$. If we substitute this in $A = \pi r^2$, we see that

$$A = \pi\left(\frac{t}{4}\right)^2 = \frac{\pi t^2}{16}$$

Consequently, $\qquad \dfrac{dA}{dt} = \dfrac{2\pi t}{16} = \dfrac{\pi t}{8}$

We want the value of this for $r = 6$ in., hence, for $t = 24$ min since $r = t/4$. Therefore,

$$\frac{dA}{dt} = \frac{\pi(24)}{8} = 3\pi$$

for $r = 6$ in.

The above example illustrates the two procedures that we may employ in time-rate problems. We may express the quantity whose time rate is desired as a function of a second variable whose rate of change with respect to time is known, and then we use the fact that if

$$Q = f(x)$$

then $\qquad \dfrac{dQ}{dt} = \dfrac{dQ}{dx}\dfrac{dx}{dt}$

by use of the chain rule. The second procedure consists of expressing the quantity whose time rate of change is required as a function of t and then differentiating with respect to t.

In Examples 2 and 3 below, the first procedure is perhaps more convenient than the second.

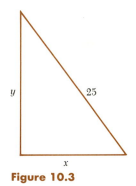

Figure 10.3

Example 2   A ladder 25 ft long rests against a wall and the lower end is being pulled out, directly away from the wall, at 6 ft per min. At what rate is the top of the ladder descending at the instant when the lower end is 7 ft from the wall?

Solution: We may let x denote the distance from the wall to the foot of the ladder as shown in Fig. 10.3. Then x is increasing at 6 ft per min and $dx/dt = 6$. We let y denote the distance from the floor to the top of the ladder. Then dy/dt is the time rate that we are to find. Now the relation between x and y is

$$x^2 + y^2 = 625$$

We may regard x and y as functions of t and differentiate with respect to t. Thus, we get

$$2x\frac{dx}{dt} + 2y\frac{dy}{dt} = 0$$

If we solve this for dy/dt we get

$$\frac{dy}{dt} = -\frac{x}{y}\frac{dx}{dt}$$

At the instant in question, $x = 7$, $y = \sqrt{625 - 49} = 24$, and $dx/dt = 6$. Hence,

$$\frac{dy}{dt} = -\frac{7}{24}(6) = -\frac{7}{4} \text{ ft per min}$$

The negative sign indicates that y is decreasing.

Example 3   A right circular cone is 10 in. in diameter and 10 in. deep. Water is poured into it at a rate of 4 cu in. per min. At what rate is the water level rising at the instant when the depth of water is 6 in.?

Solution: If we let V denote the volume of water in the cone, then

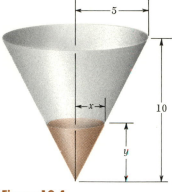

Figure 10.4

since V is increasing at 4 cu in. per min; it follows that $dV/dt = 4$. If we let y denote the depth of the water, then dy/dt is the rate that we are asked to find. Since the volume of a cone is one-third of the area of the base times the height, we have

$$V = \frac{\pi}{3} x^2 y$$

and from similar triangles, $x/y = \tfrac{5}{10}$ or $x = y/2$ as seen from Fig. 10.4. Thus we have

$$V = \frac{\pi}{12} y^3$$

and by differentiating with respect to t, we get

$$\frac{dV}{dt} = \frac{\pi}{12} \cdot 3y^2 \frac{dy}{dt} = \frac{\pi}{4} y^2 \frac{dy}{dt}$$

At the instant in question $y = 6$ and $dV/dt = 4$. Replacing y and dV/dt by these values and solving for dy/dt, we get

$$\frac{dy}{dt} = \frac{4}{9\pi} = 0.14 \text{ in. per min}$$

10.4 Velocity in rectilinear motion

As a special case of time rate, let us consider the situation that is illustrated by Fig. 10.5.

A particle moves along a straight line, and its directed distance from some fixed point A of this line at time t is s, where $s = f(t)$. If s and s_1 are the distances that correspond to times t and t_1, respectively, then the fraction

$$\frac{s - s_1}{t - t_1} = \frac{\Delta s}{\Delta t}$$

average   is called the *average velocity* of the particle over the time interval
velocity   from t to t_1. The *limit* of this fraction as t approaches t_1 is, by defini-
instantaneous   tion, the *instantaneous velocity at time t_1*. Thus,
velocity

$$v = \lim_{t \to t_1} \frac{s - s_1}{t - t_1} = \lim_{\Delta t \to 0} \frac{\Delta s}{\Delta t} = \frac{ds}{dt}$$

and the value of the derivative of s with respect to t, for any value
of t, is the instantaneous velocity of the particle. It may be either
positive or negative, depending on whether $s - s_1$ and $t - t_1$ have
speed   like or unlike signs for t very near t_1. The word *speed* is generally
used to denote the *absolute value of the velocity*, and speed is never
negative.

As an example, we shall consider the motion of a projectile that is
fired vertically upward from a point A with initial velocity $v_0 = 160$
ft per sec. If air resistance is neglected, the directed distance s from
A to the projectile at the end of t sec is given by the formula

$$s = v_0 t - \frac{gt^2}{2} \qquad g = 32 \text{ approximately}$$

In our case, $v_0 = 160$, and we have

$$s = 160t - 16t^2$$

The velocity of the projectile is $v = ds/dt = 160 - 32t$.

It follows that v is positive and the projectile is ascending until
$t = 160/32 = 5$ sec, at which time v becomes zero. For $t > 5$, v is
negative and the projectile is descending. We observe also that s is
positive if $0 < t < 10$ but negative if $t > 10$. Consequently, the projectile will be above A for the first 10 sec, 5 sec ascending and 5 sec
descending to A again. It will be below A for $t > 10$. If A is on the
ground, then the domain of t is $0 \le t \le 10$.

In this discussion we have arbitrarily taken upward as positive.
With this choice, our equation would be $s = -gt^2/2$ if the projectile
were dropped from A since then $v_0 = 0$. It would be $s = -160t -
gt^2/2$ if the projectile were thrown *downward* at a speed of 160 ft
per sec since then $v_0 = -160$.

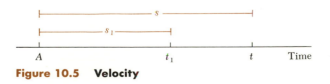

Figure 10.5   Velocity

EXERCISE 10.2

1   A lead ball is thrown vertically upward from a point A on the ground with initial velocity of 120 ft per sec. Find (a) its velocity at the end of 2 sec, at the end of 4 sec; (b) its greatest distance from the ground; and (c) the total time in the air.

2   Solve Problem 1 with the initial velocity of 160 ft per sec upward.

3   A small heavy object is dropped from the top of a cliff 600 ft high. With what velocity will it strike the ground below?

4   Solve Problem 3 if the object is thrown downward with initial velocity of 60 ft per sec; if it is thrown upward with this velocity.

5   A particle moves along a line such that its distance from the starting point is $s = t^2 - 3t$. Find the position and velocity of the particle for $t = 0, 1, 2, 3, 4, 5$.

6   A particle moves along the X axis such that its distance from the origin at the end of t min is given by $x = 6 + t^2/4 - 2t$. Find the position and velocity of the particle when $t = 0, 4, 6, 10$, and 12 min. Describe the motion.

7   A particle moves along the X axis such that its distance from the origin at the end of t min is given by $x = 3t - t^2/2 + 4$. Find its position and velocity when $t = 0, 2$, and $\sqrt{12}$ min. Describe the motion, discussing in particular the way in which the velocity varies.

8   A particle moves along the X axis such that its distance from the origin at the end of t sec is $x = \sqrt{t^2 + 1} - 2t$. Find its position and velocity for $t = 0, \sqrt{3}$, and $2\sqrt{2}$ sec. Describe the motion.

9   Two sides and the included angle of a triangle are x, y, and θ, respectively, as shown in Fig. 10.6. Show that if x is allowed to increase while y and θ are held constant, then the rate of increase of the area is given by the formula

$$\frac{dA}{dt} = \frac{y}{2} \sin \theta \frac{dx}{dt}$$

Check the result by observing that when x increases by 1 ft, the area added (shaded in figure) is $y/2 \sin \theta$ sq ft.

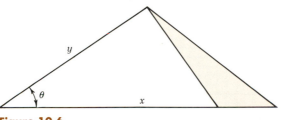

Figure 10.6

10 A man walking across a bridge at 6 ft per sec observes at a certain instant that a boat is passing directly under him. The bridge is 30 ft above the water; the boat is traveling at 12 ft per sec. At what rate is the distance s between the man and boat changing 2 sec later?

11 A boat B is 12 miles west of another boat A. B starts east at 8 mph and at the same time A starts north at 12 mph. At what rate is the distance between them changing at the end of ½ hr?

12 The legs a and b of a right triangle are 4 and 6 in., respectively. At the same instant, a starts increasing at 2 in. per min and b starts decreasing at 1 in. per min. Express the area of the triangle after t min as a function of t. Is the area increasing or decreasing at the end of 1 min and at what rate? What is the situation at the end of 4 min?

13 The side of a square is increasing at 2 in. per min. At what rate is the area increasing when each side is 8 in. long?

14 If the radius of a circle is increasing at the rate of 3 in. per min, find the rate of increase of the area when the radius is 6 in.

15 At what rate is the volume of a cube changing if the edge is 5 cm and is changing at a rate of 2 cm per sec?

16 A baseball player is running from second to third base at 25 ft per sec. At what rate is his distance from home plate changing when he is 20 ft from third base. A baseball diamond is 90 feet square.

17 At what rate does his shadow shorten as a 6-ft man walks at 6 mph on horizontal ground toward a street light that is 16 ft above the ground?

18 Water is being poured at 16 cu in. per min into a cone that is 10 in. in diameter and 6 in. deep. At what rate is the surface rising just as the cone is filled?

19 Sand is being poured into a conical pile in such a way that the height is always one-third of the radius. At what rate is sand being added to the pile when it is 4 ft high if the height is increasing at 2 in. per min?

20 A trough with triangular cross section is 20 ft long, 5 ft wide at the top, and 4 ft deep. At what constant rate is water entering the trough if the height of the water is 2 ft and is rising at 2.5 ft per min?

21 A 13-ft ladder is leaning against a vertical wall with its foot on the same horizontal plane as the base of the wall. If the lower end of the ladder is moving away from the wall horizontally at 4 ft per sec, how fast is the top of the ladder descending when the lower end is 5 ft from the wall?

22 A plane is directly over an air terminal at 10 A.M. and flying west at 300 mph. At 10:12 A.M., a second plane is over the terminal and flying N45°W at 400 mph. How fast are the planes separating at 10:24 A.M. if they are flying at the same altitude?

23   A ship is moving at 30 knots parallel to a straight shore and is 6 nautical miles from it. How fast is the ship approaching a lighthouse on the shore when it is 10 nautical miles from it?

24   A swimming pool whose floor is at a constant angle is 1.5 ft deep at one end, 11.5 ft deep at the other, 50 ft wide, and 120 ft long. How fast is the water level falling when the depth at the deep end is 4 ft if water is being discharged at 60 cu ft per min?

25   Oil is poured at 4 cu ft per min into a hemispherical container that is 18 ft in diameter. At what rate is the liquid surface rising when the depth is 6 ft? *Note:* The volume of a spherical segment of altitude h is $\pi h^2(3r - h)/3$ for a sphere of radius r.

26   The adiabatic law for the expansion of a gas is $PV^{1.4} = K$. If the volume is 20 cu ft and is decreasing at 1 cu ft per min, find the rate at which pressure is changing when it is 60 lb per sq ft.

27   A weight is attached to a rope that is 21 ft long. The rope runs through a pulley that is 11 ft from the ground. The weight is directly below the pulley. A man keeps the other end of the rope 3 ft above the ground and walks away from the weight at 10 ft per sec (Fig. 10.7). Find the rate at which the weight is rising when it is 7 ft from the ground. The figure shows weight and rope for $t = 0$ and $t = t_1$.

28   If p is the distance between a convex lens and an object and q is the distance between the lens and the image, then it is known that $1/q + 1/p = 1/f$ where f is the focal length of the lens. Find the rate of change of q for a lens of focal length 15 in. if $p = 8$ ft and is changing at 4.9 ft per sec.

29   A point moves along the parabola $y = x^2$, and its abscissa is increasing uniformly at 3 units per minute. At what rate is its distance from $(0, 0)$ increasing as it passes through $(3, 9)$?

30   When a given mass of air is allowed to expand adiabatically, its pressure p and volume v satisfy the relation $pv^k =$ constant, where k is a certain positive constant. Show that the ratio of the time rates of change of p and v is equal to $-k$ times the corresponding ratio of p to v, that is,

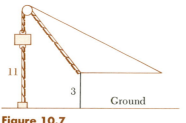

Figure 10.7

11

3

Ground

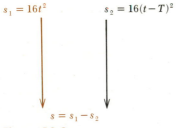

$$s_1 = 16t^2 \qquad\qquad s_2 = 16(t - T)^2$$

$$s = s_1 - s_2$$

Figure 10.8

$$\frac{dp/dt}{dv/dt} = (-k)\frac{p}{v}$$

What is the significance of the minus sign?

31   A ball is dropped from a hovering helicopter and T sec later a second ball is dropped. Prove that the difference in their velocities is a constant while both are falling. See Fig. 10.8.

32   From Problem 31 show that the difference of the velocities is a constant even if one is given an upward initial velocity and the other a downward initial velocity.

10.5 Maxima and minima

maximum value We say that $f(a)$ is a *maximum value* of the function f provided that $f(a)$ is greater than $f(x)$ for all numbers x that are sufficiently near but not equal to a. The graphical interpretation is shown in Fig. 10.9. The definition does not imply that $f(a)$ is the largest value that the function has for any x in the domain of f but rather that $f(a)$ is the largest function value corresponding to values of x in the "immediate neighborhood" of a. For this reason the term "relative maximum" is somewhat more appropriate, but we shall use the shorter term if no confusion is likely to result. The corresponding point on the graph is

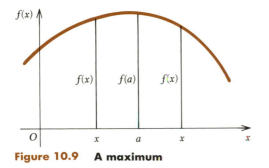

Figure 10.9   A maximum

maximum point   usually called a *maximum point*. The definitions of *minimum value* and *minimum point* are, of course, analogous.

We shall consider now the problem of determining the maximum and minimum values of a function.

Let the curve shown in Fig. 10.10 be the graph of the equation $y = f(x)$. The value of dy/dx at any point on the curve is the slope of the tangent to the curve at that point. By setting dy/dx equal at *zero* and solving for x, we can locate the points at which the tangent line to the graph is parallel to the X axis. These are called *critical points*.

critical point   In Fig. 10.10, we have

$$\frac{dy}{dx} = 0 \qquad \text{for } x = a, x = b, \text{ and } x = c$$

We found in Chap. 4 that a function is increasing if its derivative is positive and is decreasing if its derivative is negative. If, as x increases, the function value $f(x)$ is increasing for $x < x_1$, if $f'(x_1) = 0$, and if $f(x)$ is decreasing for $x > x_1$, then $f(x)$ is greater for $x = x_1$ than for any other value of x in the immediate neighborhood of $x = x_1$. Consequently, a relative maximum value of the function occurs for $x = x_1$. This discussion and conclusion can be summarized and made *test for relative maximum*   more useable by stating that if $f'(x_1 - \varepsilon) > 0$, $f'(x_1) = 0$, *and* $f'(x_1 + \varepsilon) < 0$, *then* $y = f(x)$ *has a relative maximum value* $f(x_1)$ *for* $x = x_1$ *where ε is any arbitrarily small positive number.*

This is the situation for $x = a$ in Fig. 10.10.

Example 1   Find the maximum value of $y = f(x) = -x^2 + 4x + 1$.

Solution: If we take the derivative of the given function, we find that

$$y' = f'(x) = -2x + 4$$

and that this is zero for $x = 2$. Consequently, if the given function has a maximum value, it occurs for $x = 2$. We shall now see if

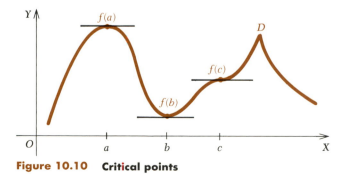

Figure 10.10   Critical points

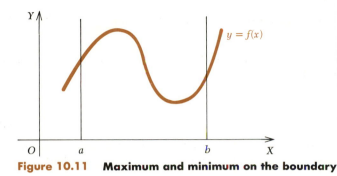

Figure 10.11   Maximum and minimum on the boundary

this possibility gives a maximum. In keeping with the italicized state-ment above, we must begin by evaluating $f'(2 - \varepsilon)$ and $f'(2 + \varepsilon)$. By substituting in $f'(x)$, we find that

$$f'(2 - \varepsilon) = -2(2 - \varepsilon) + 4 = -4 + 2\varepsilon + 4 = 2\varepsilon > 0$$

and $$f'(2 + \varepsilon) = -2(2 + \varepsilon) + 4 = -4 - 2\varepsilon + 4 = -2\varepsilon < 0$$

Consequently, we know that

$$f'(2 - \varepsilon) > 0 \qquad f'(2) = 0 \qquad f'(2 + \varepsilon) < 0$$

Therefore, $y = f(x) = -x^2 + 4x + 1$ has a relative maximum value for $x = 2$, and it is $f(2) = -(2)^2 + 4(2) + 1 = 5$.

A function which exists only over a limited interval or which is being considered only for a limited domain $a \le x \le b$ may have a maximum on either or both of the boundaries without the derivative of the function being equal to zero at the point. That this is the situation can be seen readily by considering the curve in Fig. 10.11. In it, the function is increasing as x approaches the right-hand boundary $x = b$; consequently, $f(b)$ is greater than the value of $f(x)$ for any x in the domain and sufficiently near b. We can now say that $f(x)$ has a relative maximum at the right-hand boundary of a function that is being considered for a limited domain if $f'(b) \ge 0$ and $f'(b - \varepsilon) > 0$. Similarly, $f(x)$ has a relative maximum at the left-hand boundary of a function that is being considered for a limited domain if $f'(a) \le 0$ and $f'(a + \varepsilon) < 0$.

It can be shown in a similar manner that

test for relative minimum   *If $f'(x_1 - \varepsilon) < 0, f'(x_1) = 0$, and $f'(x_1 + \varepsilon) > 0$, then $y = f(x)$ has a relative minimum value $f(x_1)$ for $x = x_1$ where ε is any arbitrarily small positive number; furthermore, if the domain of f is restricted to the interval $a \le x \le b$, then $f(x)$ has a minimum for $x = a$ if $f'(a) \ge 0$ and $f'(a + \varepsilon) > 0$ and a minimum for $x = b$ if $f'(b) \le 0$ and $f'(b - \varepsilon) < 0$.*

Example 2 Find the relative maxima and minima of $y = f(x) = 4x^3 + 3x^2 - 18x + 9$ for $x \leq 3$.

Solution: If we get the derivative and set it equal to zero, we find that

$$f'(x) = 12x^2 + 6x - 18 = 6(2x^2 + x - 3)$$
$$= 6(2x + 3)(x - 1) = 0 \qquad \text{if } x = -1.5, 1$$

We shall now test each of these critical values to see if it is a maximum or minimum. For this purpose, we find the value of $f'(x)$ for x arbitrarily near each critical value but less than it and greater than it. Thus,

$$f'(-1.5 - \varepsilon) = 6(-3 - 2\varepsilon + 3)(-1.5 - \varepsilon - 1) = 6(-2\varepsilon)(-2.5 - \varepsilon) > 0$$

and

$$f'(-1.5 + \varepsilon) = 6(-3 + 2\varepsilon + 3)(-1.5 + \varepsilon - 1) = 6(2\varepsilon)(-2.5 + \varepsilon) < 0$$

Consequently, $f(x)$ has a maximum value for $x = -1.5$, and that value is $f(-1.5) = 4(-1.5)^3 + 3(-1.5)^2 - 18(-1.5) + 9 = 29.25$. Furthermore,

$$f'(1 - \varepsilon) = 6(2 - 2\varepsilon + 3)(1 - \varepsilon - 1) = 6(5 - 2\varepsilon)(-\varepsilon) < 0$$

and $f'(1 + \varepsilon) = 6(2 + 2\varepsilon + 3)(1 + \varepsilon - 1) = 6(5 + 2\varepsilon)(\varepsilon) > 0$

Therefore, $f(x)$ has a minimum value for $x = 1$, and that value is $f(1) = 4(1)^3 + 3(1)^2 - 18(1) + 9 = -2$.

Finally, we must see whether $f(x)$ has either a maximum or a minimum for the right-hand boundary value $x = 3$. For this purpose we need only find the sign of $f'(x)$ for $x = 3$. Substituting 3 for x in $f'(x)$, we find that

$$f'(3) = 6(6 + 3)(3 - 1) > 0$$

Consequently, $f(x)$ is an increasing function for $x = 3$ and has a maximum for that value. It is $f(3) = 4(3^3) + 3(3^2) - 18(3) + 9 = 90$.

It is possible that there may be maximum or minimum points of a function that will not be discovered by the above procedure. This is *derivative* the case only if the function does not have a derivative at such a point. *at a* In other words, *if a function has a derivative at a maximum or* *minimum* *minimum point other than on a boundary, its value must be zero and any other maximum or minimum point must be one at which the derivative fails to exist.* The point D in Fig. 10.10 is an example. Another example is shown in Fig. 4.6. There, the equation of the curve is

$$y = 1 + \sqrt[3]{x^2}$$

Consequently,
$$\frac{dy}{dx} = \frac{2x^{-1/3}}{3} = \frac{2}{3\sqrt[3]{x}}$$

There is no value of x for which $dy/dx = 0$, but it is clear that this derivative is positive for all $x > 0$ and negative for all $x < 0$. At $x = 0$ the derivative fails to exist. We see immediately, however, that this is a minimum point. Thus, in the search for maximum and minimum values of a function, we should always investigate any points at which the derivative fails to exist.

Example 3   Find the maximum and minimum points on the curve whose equation is

$$y = \frac{x^3}{x^2 - 3}$$

and sketch the graph.

Solution: If we differentiate the given quotient, we find that

$$\frac{dy}{dx} = \frac{(x^2 - 3)(3x^2) - x^3(2x)}{(x^2 - 3)^2} = \frac{x^2(x^2 - 9)}{(x^2 - 3)^2}$$

Therefore, if $x = 0, 0, -3, 3$, the derivative is zero and the tangent to the curve is horizontal. Consequently, there are horizontal tangents at $(0, 0), (-3, -4.5)$, and $(3, 4.5)$ as shown in Fig. 10.12. We shall now test each of these points to see if it is a maximum, a minimum, or neither. For this purpose, we begin by finding that, for $\varepsilon > 0$ and arbitrarily near zero,

$$y'(0 - \varepsilon) = \frac{(-\varepsilon)^2(\varepsilon^2 - 9)}{(\varepsilon^2 - 3)^2} < 0 \quad \text{and} \quad y'(0 + \varepsilon) = \frac{\varepsilon^2(\varepsilon^2 - 9)}{(\varepsilon^2 - 3)^2} < 0$$

Therefore, there is neither a maximum nor a minimum for y at $x = 0$ since the derivative has the same sign on both sides of the point corresponding to $x = 0$. To test $x = -3$, we find that, for $\varepsilon > 0$ and arbitrarily, near zero,

$$y'(-3 - \varepsilon) = \frac{(-3 - \varepsilon)^2[(-3 - \varepsilon)^2 - 9]}{[(-3 - \varepsilon)^2 - 3]^2} > 0$$

and
$$y'(-3 + \varepsilon) = \frac{(-3 + \varepsilon)^2[(-3 + \varepsilon)^2 - 9]}{[(-3 + \varepsilon)^2 - 3]^2} < 0$$

Consequently, y has a maximum value for $x = -3$. Furthermore,

$$y'(3 - \varepsilon) = \frac{(3 - \varepsilon)^2[(3 - \varepsilon)^2 - 9]}{[(3 - \varepsilon)^2 - 3]^2} < 0$$

and
$$y'(3 + \varepsilon) = \frac{(3 + \varepsilon)^2[(3 + \varepsilon)^2 - 9]}{[(3 + \varepsilon)^2 - 3]^2} > 0$$

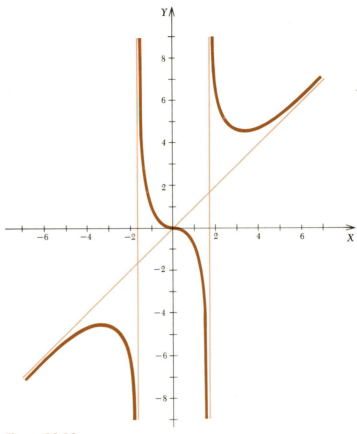

Figure 10.12

Hence, y has a minimum value for $x = 3$. The derivative fails to exist for $x = \pm\sqrt{3}$ but neither gives a maximum or minimum since the function is decreasing immediately to the left and to the right of each corresponding point.

In sketching the graph in Fig. 10.12, we must use the fundamental principles of analytic geometry in addition to the above information regarding maximum and minimum points. Thus, we note from the given equation that the graph is symmetric with respect to the origin, that its only intercept on either axis is at the origin, and that it has vertical asymptotes at $x = \pm\sqrt{3}$. Also, from the fact that $x^3/(x^2 - 3) \equiv x + 3x/(x^2 - 3)$ we may infer that the curve is asymptotic to the line $y = x$.

EXERCISE 10.3

Determine the values of x for which the curves whose equations are given in Problems 1 to 44 have maximum and minimum values.

1   $y = 6x - x^2$
2   $2y = x^2 + 4x - 3$
3   $y = ax^2 + bx + c$
4   $y = -x^2 + 4x - 1$
5   $y = x^3/4 - 3x^2/2$
6   $y = x^3 - 6x^2 + 9x - 2$
7   $y = x^3 + 3x^2 - 1$
8   $y = x^3 - 3x^2 + 4$
9   $y = 2x^3 - 9x^2 + 12x + 3$
10   $3y = x^3 - 3x^2 - 9x$
11   $y = x^3 - 6x^2 + 12x + 6$
12   $6y = 2x^3 + 3x^2 - 12x + 5$
13   $2y = x^4 - 4x^3 + 4$
14   $y = x^4 - 4x^3 + 16x - 13$
15   $y = 3x^4 + 4x^3 - 12x^2 - 4$
16   $y = 8x^2 - x^4$
17   $2y = x^2 + 16/x$
18   $4y = x^3 + 48/x$
19   $y = 1/x + 1/x^2$
20   $y = 4/x + 1/(1 - x)$
21   $y = 6/(x^2 + 1)$
22   $y = 18/(x^2 + 6)$
23   $y = 4/(x^2 - 2x + 2)$
24   $y = 3/(x^2 - 4x + 7)$
25   $y = (x - 2)/x^2$
26   $y = x^2/(x + 2)$
27   $y = 4x/(x^2 + 4)$
28   $y = 8x/(x^2 + 1)$
29   $y = (9 - x^2)/(x^2 - 4)$
30   $y = (8x + 4)/(x - 2)^2$
31   $y = 3x^2/(x^2 + 1)$
32   $y = (2x^2 + 5)/(x^2 - 3)$
33   $y = x^3/(3x + 6)$
34   $y = (x^2 + 2x + 10)/(x - 3)$
35   $y = (x^2 - 7x + 16)/(x - 4)$
36   $y = x^3/(x^2 - 12)$
37   $y = x\sqrt{16 - x^2}$
38   $y = x\sqrt{18 - x^2}$
39   $y = x^2\sqrt{24 - x^2}$
40   $y = x^2\sqrt{27 - x^2}$
41   $y = \sqrt[3]{(x - 4)^2}$
42   $y = 2 + \sqrt[3]{(2x - 1)^2}$
43   $y = 1 - \sqrt[5]{x^4}$
44   $y = 3 - \sqrt[5]{(9 - x)^2}$

Find the values of x in the domain of definition or limited interval for which the function has a maximum or minimum value.

45   $y = x^2 - 4x + 3,\ 1 \le x \le 4$
46   $y = x^2 - 6x + 1,\ 2 \le x \le 5$
47   $y = x^3 + 6x + 2,\ -1 \le x \le 3$
48   $y = x^3 + 3x^2 + 15x + 7,\ -2 \le x \le 4$

10.6   Applications of maxima and minima

Situations abound in everyday living in which it is important to achieve maximum results under a given set of conditions or to find methods for obtaining a given result in the most efficient way. A business man is surely interested in maximum profits and minimum costs, a manufacturer is constantly seeking the most efficient methods, and the operators of a packaging company would be interested in the shape of a container with a given volume that could be made with the least amount of material. The methods discussed in the previous sections of this chapter frequently yield the answers to such problems. Before discussing the general procedure, we shall illustrate the methods with a simple example.

A farmer has 100 ft of fencing with which to build a rectangular chicken pen. What should be the dimensions of the pen if it encloses the largest possible area?

In order to solve this problem we shall let

$$x = \text{length of pen}$$
$$50 - x = \text{width of pen}$$
$$A = x(50 - x) = 50x - x^2 = \text{area}$$

Now to get the replacement for x that will yield the maximum value of A, we set dA/dx equal to zero and solve for x, as indicated below.

$$\frac{dA}{dx} = 50 - 2x$$
$$50 - 2x = 0$$
$$x = 25$$

Hence the pen should be a 25 ft by 25 ft square.

In order to solve a particular problem, the student should first read the problem very carefully and make certain that he understands which of the quantities mentioned are fixed and which are variable. If necessary, he should draw a figure to illustrate the situation. After thus obtaining a clear understanding of just what the problem is, he may, at least in the simple cases, proceed as follows:

maximum and minimum in applications

1. *Select the variable quantity for which a maximum or minimum value is required;* it may be the volume of something, the cost of something, the time required to do something, etc. Denote this quantity by some letter. For the moment we shall call it Q.
2. *Select a variable x on which Q depends.* There may be several possible choices for this variable.
3. *Obtain the equation that expresses Q in terms of x, say $Q = f(x)$.* It may be simpler in some problems to express Q as a function of two variables, say, x and y, and then eliminate one of them by using a condition of the problem that relates y to x. Note any limitation that the physical conditions impose on x.
4. *Proceed to investigate the equation obtained in step 3* for the required maximum or minimum values by the methods of the preceding section.

Steps 1, 2, and 3 are to enable us to find an expression for the quantity that is to be a maximum or minimum. It is *essential* that this be done.

Example 1   A man in a rowboat is at a point C which is 5 miles off the straight shore line. He wishes to reach a point A that is on the shore line and 13 miles away. He can row 2 mph and walk 4 mph. At what point on the shore should he land in order to reach A in the least time?

Solution 1:   A sketch of the situation is shown in Fig. 10.13.

1.   *Quantity to be made a minimum:* Time required to go from C to A by way of B. Call this T.
2.   *Independent variable:* We assume that he may land at any point B between D and A and let $DB = x$ be the independent variable.
3.   He rows a distance $CB = \sqrt{x^2 + 25}$ at 2 mph and then walks a distance $BA = 12 - x$ at 4 mph. The total time T is then given in terms of x by

$$T = \frac{\sqrt{x^2 + 25}}{2} + \frac{12 - x}{4} \qquad 0 < x < 12$$

4. $$\frac{dT}{dx} = \frac{(\tfrac{1}{2})(x^2 + 25)^{-1/2}(2x)}{2} + \frac{-1}{4} = \frac{x}{2\sqrt{x^2 + 25}} - \frac{1}{4}$$

Setting $dT/dx = 0$ and solving for x, we have

$$\frac{x}{2\sqrt{x^2 + 25}} - \frac{1}{4} = 0 \qquad x = \frac{5}{\sqrt{3}} = 2.89 \text{ miles approx.}$$

The corresponding value of T is $(5\sqrt{3}/4 + 3)$ hr = 5.2 hr. Finally, we can show that the value of dT/dx is negative for x slightly less than, and positive for x slightly more than, $5/\sqrt{3}$ by replacing x by $5/\sqrt{3} - \varepsilon$ and $5/\sqrt{3} + \varepsilon$ in the expression for dT/dx, thus proving that we have found a relative minimum.

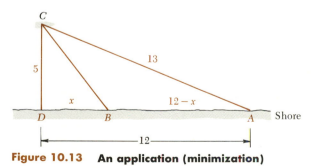

Figure 10.13   An application (minimization)

Solution 2:   In the above example, we could let $BA = y$ and $DB = x$ and then express T in terms of the two variables x and y. Thus,

$$T = \frac{\sqrt{x^2 + 25}}{2} + \frac{y}{4} \qquad \text{where } x + y = 12$$

What we actually did amounted to writing down these equations and then eliminating y in order to get an expression for T in terms of a single variable. However, it was not really necessary to eliminate y. Instead, we could use implicit differentiation, assuming that y is a function of x with a derivative dy/dx, and get

$$\frac{dT}{dx} = \frac{x}{2\sqrt{x^2 + 25}} + \frac{dy/dx}{4}$$

from the first equation and

$$1 + \frac{dy}{dx} = 0$$

from the second. If we set $dT/dx = 0$ and solve the resulting equation simultaneously with the equation $1 + dy/dx = 0$, by equating the expressions for dy/dx from the two equations, we get the result obtained above. In this particular case there is no advantage in this alternate procedure, but if the relation between x and y is more complicated, as in the next example, the alternate procedure may be preferable.

Example 2   A manufacturer packages his product in 1-gal cans that are right circular cylinders. If the material for the top costs five times as much per square inch as that used for the bottom and side, find the dimensions of the can if the cost of the material is to be a minimum.

Solution: We shall let the radius and height of the can be x and y, respectively, and the cost of material for side and bottom be k cents per sq in. Then the cost for the top is $5k$. Hence, if C is the total cost of material, we have

$$C = k\pi x^2 + 5k\pi x^2 + k \cdot 2\pi xy$$

where $\pi x^2 y = 231$ since 1 gal = 231 cu in. Therefore,

$$C = 6k\pi x^2 + 2k\pi xy \qquad \text{where } \pi x^2 y = 231 \tag{1}$$

If we wished to use the first procedure we would solve the equation $\pi x^2 y = 231$ for y in terms of x, substitute the resulting expression for y in the equation on the left, and thus obtain an expression for C in terms of x only. However, it may be simpler to use the second procedure, and we shall do that.

Differentiating the equations in (1), we have

$$\frac{dC}{dx} = 12k\pi x + 2k\pi x \frac{dy}{dx} + 2k\pi y \qquad \pi x^2 \frac{dy}{dx} + 2\pi xy = 0$$

If we set $dC/dx = 0$ and perform some obvious simplifications, we have

$$6x + x\frac{dy}{dx} + y = 0 \qquad x\frac{dy}{dx} + 2y = 0$$

These relations must hold between x, y, and dy/dx if $dC/dx = 0$ and x and y are related by the equation $\pi x^2 y = 231$. We shall eliminate dy/dx between them and get the relation that must hold between x and y for a minimum cost of material for the container. Thus, from the equation on the right, we have $x(dy/dx) = -2y$, and if we substitute this into the middle term of the equation on the left, we get

$$6x + (-2y) + y = 0 \qquad \text{or} \qquad y = 6x$$

This last result tells us that the height of the can must be six times the radius if the conditions imposed are to be satisfied. From the equations $y = 6x$ and $\pi x^2 y = 231$, we get

$$x = \sqrt[3]{\frac{38.5}{\pi}} = 2.31 \text{ in.} \qquad \text{and} \qquad y = 6(2.31) = 13.86 \text{ in.}$$

EXERCISE 10.4

1   Divide 25 into two parts whose product is a maximum.

2   Divide 25 into two parts such that the sum of their squares is a maximum.

3   Divide 25 into two parts such that the product of one of them and the square of the other is a maximum.

4   Divide 25 into two parts so that one of them plus the cube of the other is a maximum.

5   A rectangular area is to be enclosed by 440 ft of fence. Find the dimensions if the area is a maximum.

6   A rectangular area is bounded on one side by a wall and on the other three by 440 ft of fence. What should be the dimensions if the area is to be a maximum?

7   Find the volume of the largest box that can be made from a sheet of metal that is 40 cm long and 25 cm wide by cutting equal squares from the corners and turning up the edges.

8   A rectangular field is to be enclosed by a fence and then divided into two enclosures by a fence that is parallel to one of the sides. Find the dimensions of the largest such field if 600 ft of fence is available.

9   What is the smallest sum that can be obtained by adding a positive number and its reciprocal?

10   Find the relative dimensions of a right circular cylinder with top and bottom if the volume is fixed and the total area is to be a minimum.

11   Find the relative dimensions of a cylindrical can with an open top if the volume is fixed and the amount of material used is to be a minimum.

12   Find the area of the largest rectangle that can be inscribed in a right triangle with base 30 in. and altitude 18 in. if the base of the rectangle lies along the base of the triangle.

13   An isosceles triangle is inscribed in a given circle. What is the relation between the base and altitude if the perimeter of the triangle is a maximum?

14   Find the point on the hypotenuse of a right triangle from which perpendiculars can be dropped to the legs so as to form a rectangle of maximum area.

15   An isosceles trapezoid is inscribed in a semicircle of radius r. Find the length of the shorter base if the area of the trapezoid is to be a maximum. Assume that the longer base lies along the diameter.

16   Show that a square is the largest rectangle that can be inscribed in a circle.

17   Find the dimensions of the right circular cylinder of greatest volume that can be inscribed in a sphere of radius r.

18   Find the dimensions of the right circular cylinder of greatest volume that can be inscribed in a right circular cone of altitude h and radius r.

19   Determine the values of the constants a and b in $f(x) = x^3 + ax^2 + bx + c$ such that $f(-2)$ is a maximum and $f(\frac{2}{3})$ is a minimum.

20   Determine the values of the constants a and b in $f(x) = x^3 + ax^2 + bx + 4$ such that $f(-\frac{8}{3})$ is a maximum point and $f(-2)$ is a minimum.

21   Show that a cylindrical can with lid requires a minimum amount of material for a specified volume V if its diameter and height are equal.

22   A cylindrical tank with given volume V is to be made without a lid. The material for the bottom costs four times as much per unit area as that for the sides. What should be the ratio of height to radius for minimum cost?

23   A right circular cone of height y is circumscribed about a sphere whose radius is 4 in. Express the volume V of the cone as a function of y, and determine the value of y for which V is a minimum. Show that the volume of this smallest cone is twice that of the sphere.

24 A Norman window has the form of a rectangle surmounted by a semicircle. What relation between height of rectangle and radius of semicircle will give the greatest area for a given perimeter?

25 If the glass in the semicircular part of the window in Problem 24 is stained and admits only half as much light per unit area as the other portion, what should be the relation between the height of the rectangle and the radius of the semicircle to admit the most light for a given perimeter?

26 A figure consists of a rectangle surmounted by an equilateral triangle, one side of the triangle coinciding with the upper base of the rectangle. If the perimeter of the figure is a fixed number P, how long should the side of the triangle be for maximum area of the figure?

27 From a circle whose radius is less than 12 in. a sector with perimeter 24 in. is to be cut. What should be the radius of the circle if the area of this sector is to be a maximum?

28 A gutter with trapezoidal cross section is to be made from a long sheet of tin that is 15 in. wide by turning up one-third of the width on each side. What width across the top will give the greatest cross-sectional area?

29 A tank without a cover consists of a hemisphere surmounted by a right circular cylinder. Determine the proportions for minimum cost of material if the material for the hemisphere costs twice as much per unit area as the material for the cylinder.

30 A gutter with trapezoidal cross section is to be made from a long sheet of tin that is $4\sqrt{108}$ in. wide by turning up $\sqrt{108}$ in. on each side. What is the depth of the gutter for the greatest possible cross-sectional area?

31 Find the height of the right pyramid with square base and with largest volume which can be inscribed in a sphere of radius 9 in.

32 An oil can consists of a right circular cylinder surmounted by a cone. The can is to have a fixed volume V, and the height of the cone is to be 2.4 times its radius. Find the ratio of the height of the cylindrical part to the radius for minimum surface area.

33 A bus company will transport 100 passengers or less on an excursion trip for $12 each. If there are more than 100 passengers, the company agrees to reduce the price of every ticket 5 cents for each passenger in excess of 100 (e.g., if there are 102 passengers, each fare is $11.90). What number of passengers will produce the greatest gross revenue?

34 If each additional person beyond 100 in Problem 33 adds 1 cent to the cost of transporting each passenger, what number of passengers will produce the greatest profit? Assume that the profit for 100 passengers is $600.

35 A telephone company has 10,000 telephones in a certain city where the charge is $4 per month. The officials believe that if the charge is reduced, the number of telephones in use will increase at a rate of 200 additional telephones for each 5-cent reduction. What monthly charge would yield the greatest gross revenue on this basis?

36 What is the most favorable charge in Problem 35 if it must be assumed that each additional telephone increases operating expenses by 20 cents per month?

37 The cost per hour of fuel for operating a boat is proportional to the square of the speed and is $30 for a speed of 20 mph. Other costs are independent of the speed and are $120 per hour. What is the cost per mile in terms of speed? What speed gives the least cost per mile?

38 Find the equation of the ellipse of the smallest area that can be circumscribed about a rectangle that is 12 by 16 cm. If the semi-axes of an ellipse are a and b, its area is πab.

39 As shown in Fig. 10.14, a ray of light from A strikes a plane mirror at B and is reflected through C. Prove that angles α and β are equal if $AB + BC$ is a minimum.

40 If the sum of the areas of a cube and a sphere are a constant, find the ratio of the radius of the sphere to the edge of the cube if the sum of the volumes is a minimum.

41 Right circular cans of a given volume are to be made. Find the ratio of height to diameter of the most economical can if there is no wasted material in making the sides and if each end is cut from a square whose side is equal to the diameter of the can and whose corners are to be wasted.

42 Find the coordinates of the point on the upper half of the ellipse $4x^2 + 9y^2 = 900$ that is nearest the point $(5, 0)$, and the coordinates of the point that is farthest away. Sketch a graph showing how the distance S from $(5, 0)$ to a point (x, y) on this upper half varies with x, and explain why one cannot locate the point that is farthest away by setting $dS/dx = 0$ and solving for x.

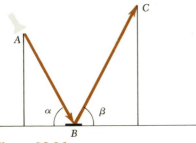

Figure 10.14

43 Find the shortest distance and the longest distance from the origin to a point on the ellipse $(x - 6)^2 + 4y^2 = 256$.

44 A point moves along the curve whose equation is $x^2 + (y - 1)^2 = 9$. Find its greatest and least distances from the origin.

45 A wire of length L is to be cut into two pieces. One of these pieces is to be bent into the form of a circle and the other into the form of a square. Find the length of each piece if the sum of the areas of the circle and square is to be as small as possible. Find also the lengths if the sum of the areas is to be as large as possible.

46 A 30-gal trash can is made in the form of a right circular cylinder with radius x and height y in. The bottom is stamped from a circular disk whose radius is 1 in. greater than that of the can. The extra inch is turned down to form a rim so that the bottom does not touch the ground. The side fits down over this rim. The top is stamped similarly from a disk whose radius is 2 in. greater than that of the can. Find the relation between y and x if the amount of material used is to be a minimum.

47 Assume that for the can described in Problem 46, a heavier material is used for the top and bottom and that it costs four-thirds as much per square inch as the material for the sides. Find the relation between y and x for minimum cost of material.

48 Sketch the curve whose equation is $y = 8x - x^2$, and write the equation of the chord that joins the points $(1, 7)$ and $(5, 15)$. Let $Q(x)$ denote the directed distance, measured parallel to the Y-axis, from a point (x, y) on this chord up to the curve. Write the equation that expresses $Q(x)$ as a function of x. Find the largest value of $Q(x)$. Show that the tangent to the curve at the point for which $Q(x)$ is a maximum is parallel to the chord.

11 Higher derivatives

11.1   Successive differentiation

We defined the derivative $f'(x)$ of $f(x)$ with respect to x as

$$\lim_{\Delta x \to 0} \frac{f(x + \Delta x) - f(x)}{\Delta x}$$

if that limit exists. We shall now continue our discussion of derivatives by saying that if

$$\lim_{\Delta x \to 0} \frac{f'(x + \Delta x) - f'(x)}{\Delta x}$$

second derivative exists, it is called the *second derivative* of $f(x)$ with respect to x. If we denote $f(x)$ by y and write $y = f(x)$, then we may indicate this second derivative by

$$\frac{d^2y}{dx^2}, \quad D_x^2 y, \quad f''(x), \quad \text{or} \quad y''$$

Just as the symbol d/dx or D_x denotes the operation of taking the derivative with respect to x of the function to which it is applied, the symbol d^2/dx^2 or D_x^2 denotes the operation of taking the second derivative. Thus $(d^2/dx^2)x^3$ or $D_x^2 x^3$ is $6x$ since $D_x x^3 = 3x^2$ and $D_x 3x^2 = 6x$. If the second derivative is a differentiable function of x, its derivative is called the *third derivative* of the original function, etc. The nth derivative may be indicated by

third derivative

$$\frac{d^n y}{dx^n}, \quad D_x^n y, \quad f^{(n)}(x), \quad \text{or} \quad y^{(n)}$$

Example   If

$$y = f(x) = x^2 + x^{-1} + \sqrt{x}$$

then

$$\frac{dy}{dx} = f'(x) = 2x - x^{-2} + \frac{x^{-1/2}}{2}$$

$$D_x^2 y = f''(x) = 2 + 2x^{-3} - \frac{x^{-3/2}}{4}$$

$$D_x^3 y = f'''(x) = -6x^{-4} + \frac{3x^{-5/2}}{8}$$

11.2   Successive differentiation of implicit functions

If y is determined in terms of x by an equation in x and y that is not solved for y, we can find the successive derivatives if we treat y and its derivatives as functions of x. The first derivative of y with respect to x can then be found in terms of x and y. The second derivative can be found in terms of x, y, and y' and then in terms of x and y alone by replacing y' in terms of x and y. At times, it may be desirable to find y'' without solving the equation in x, y, and y' for y'.

Example   If $x^2 + y^2 = r^2$, find y' and y''.

Solution: In differentiating, we must treat y as a function of x; hence,

$$2x + 2yy' = 0 \tag{1}$$

since $D_x(x^2 + y^2) = 2x + 2yy'$ and $D_x r^2 = 0$. Therefore,

$$y' = \frac{-x}{y} \qquad \textit{solving for } y' \tag{2}$$

Now differentiating Eq. (2) implicitly, we find that

$$y'' = \frac{dy'}{dx} = -\frac{d(x/y)}{dx} = -\frac{y \, D_x x - x \, D_x y}{y^2} = -\frac{y - xy'}{y^2}$$

This can be expressed in terms of x and y by replacing y' by its value as given in (2). Thus,

$$y'' = -\frac{y - x(-x/y)}{y^2}$$

$$= -\frac{y^2 + x^2}{y^3} \qquad \textcolor{brown}{\textit{multiplying by } y/y}$$

$$= -\frac{r^2}{y^3} \tag{3}$$

since $x^2 + y^2 = r^2$ is the given equation.

We found y'' by solving Eq. (1) for y' in terms of x and y and then differentiating. We shall now find y'' from (1) without solving for y'. Dividing each member of (1) by 2, we have $x + yy' = 0$. Now recalling that y and y' are functions of x and differentiating accordingly, we have

$$1 + yy'' + y'^2 = 0 \tag{4}$$

Therefore, $$y'' = -\frac{1 + y'^2}{y}$$

This does not appear to be the same as the expression we had earlier for y'', but it can be shown to be by replacing y' by $-x/y$ as found in (2).

We can use (4) or either form of the expression for y'' in order to find y'''. If we find it by differentiating each member of (4), we have

$$0 + yy''' + y''y' + 2y'y'' = 0$$

Consequently, $$y''' = -\frac{3y'y''}{y}$$

which can be expressed in terms of x and y by use of (2) and (3). Thus,

$$y''' = \frac{-3(-x/y)(-r^2/y^3)}{y} = -\frac{3r^2 x}{y^5}$$

EXERCISE 11.1

Find $D_x y$ and $D_x^2 y$ for the functions given in each of Problems 1 to 16.

1 $y = 2x^2 - 3x + 5$ 2 $y = -x^2 + x + 7$

3 $y = 4x^3 + 5x^2 - 7$ 4 $y = x^5 - 2x^3 + 3x^2 - 4$

5 $y = 2x - 1/x$ 6 $y = x^2 - 2/x^2$

7 $y = x^3 - 1/x^3$ 8 $y = x^2 + 1/x^2 - 1/x$

9 $y = x/(x + 1)$ 10 $y = (x - 3)/(2x + 1)$

11 $y = x/(x^2 - 1)$ 12 $y = (x - 2)/(x^2 - x)$

13 $y = \sqrt{x^2 + 1}$ 14 $y = \sqrt{2x - x^2}$

15 $y = \sqrt[3]{x^3 + 3x}$ 16 $y = \sqrt[5]{x^5 + 5x^4}$

Find y'' in terms of x and y in each of Problems 17 to 28 by use of implicit differentiation.

17 $y^2 = 2px$ 18 $x^2 + y^2 = r^2$

19 $x^2/a^2 + y^2/b^2 = 1$ 20 $x^2/a^2 - y^2/b^2 = 1$

21 $xy - x^2 - 8y = 0$ 22 $xy + y = x - 4$

23 $x^2 + (y + 4)^2 = 16$

24 $x^2 + 2y^2 - 2x - 12y + 15 = 0$

25 $x^{1/2} + y^{1/2} = a^{1/2}$ 26 $x^{2/3} + y^{2/3} = a^{2/3}$

27 $x^{1/2} + (y - 1)^{1/2} = a^{1/2}$

28 $(x + 2)^{2/3} + y^{2/3} = a^{2/3}$

Find the value of dy/dx and d^2y/dx^2 at the indicated points in Problems 29 to 32.

29 $y^2 = 8x$, $(2, 4)$ 30 $x^2 + y^2 = 169$, $(-5, 12)$

31 $xy + 2x = 3y + 6$, $(5, -2)$

32 $3xy - 5y = 4x - 14$, $(5, 1)$

In each of Problems 33 to 40, find the indicated derivative.

33 $x^2 + y^2 = r^2$, d^3y/dx^3 34 $x^2/a^2 + y^2/b^2 = 1$, d^3y/dx^3

35 $x^2/a^2 - y^2/b^2 = 1$, d^3y/dx^3

36 $x^2 = 2py$, d^3y/dx^3 37 $xy = 6$, d^ny/dx^n

38 $x^2y = 1$, d^ny/dx^n

39 $y = x^m$, m an integer, d^ny/dx^n

40 $y = x^m$, m rational but not an integer, d^ny/dx^n

41 Show, by differentiating $dy/dx = 1/(dx/dy)$ with respect to x, that

$$\frac{d^2y}{dx^2} = -\frac{d^2x/dy^2}{(dx/dy)^3}$$

42 Derive a formula for $D_x^2 uv$ under the assumption that the first and second derivatives of u and of v with respect to x exist.

43 Derive a formula for $D_x^3 uv$ under the assumptions of Problem 42 and the further assumption that $D_x^3 u$ and $D_x^3 v$ exist.

44 Verify the formula derived in Problem 43 for $u = x^4$ and $v = y$.

11.3   Concavity and the sign of $D_x^2 y$

Since $D_x^2 y$ is the derivative with respect to x of $D_x y$, its value at any point on $y = f(x)$ is the rate at which the value of the slope y' of the tangent is changing relative to x. Furthermore, if $y'' = dy'/dx > 0$ over an interval, then y' is an increasing function of x over this interval, and the tangent line at $P(x, y)$ turns in the counterclockwise direction as P moves to the right along the curve. If $y'' > 0$ over an interval, *concave upward* we say that the curve is *concave upward*.

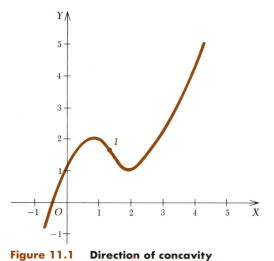

Figure 11.1   Direction of concavity

concave
downward

Similarly, if $y'' = dy'/dx < 0$ over an interval, then y' is a decreasing function of x over this interval. The tangent line at $P(x, y)$ turns in the clockwise direction as P moves to the right along the curve, and we say that the curve is *concave downward*.

Example

Find the interval in which the graph of

$$y = f(x) = x^3 - 4x^2 + 4x + 1$$

is concave downward and in which it is concave upward.

Solution: We can find by differentiating

$$y = f(x) = x^3 - 4x^2 + 4x + 1$$

that

$$y' = f'(x) = 3x^2 - 8x + 4$$

and

$$y'' = f''(x) = 6x - 8$$

Consequently, $f''(x) < 0$ for $x < \frac{4}{3}$ and $f''(x) > 0$ for $x > \frac{4}{3}$. Therefore, the graph of $y = f(x)$ is concave downward for the part of the curve in Fig. 11.1 to the left of $I(\frac{4}{3}, \frac{43}{27})$ and concave upward for the arc to the right of I. Furthermore, $y'' = f''(x)$ is zero for $x = \frac{4}{3}$, the value of x at I.

11.4   Point of inflection

We found in Sec. 11.3 that a curve $y = f(x)$ is concave upward over an interval if $y''(x) > 0$ for each value of x in the interval and concave downward in an interval if $y''(x) < 0$ for each value of x in the

inflection point

interval. A point at which the direction of concavity changes is called an *inflection point*. That is, an inflection point is a point at which d^2y/dx^2 changes sign. This can occur only at points where d^2y/dx^2 is zero or fails to exist.

Example 1

Find the maxima, minima, and inflection points of $y = f(x) = 2x^3 - 6x^2 + 6x + 9$.

Solution: We readily find that

$$f'(x) = 6x^2 - 12x + 6$$
$$= 6(x^2 - 2x + 1) = 6(x - 1)^2 = 0 \qquad \text{if } x = 1$$

and $f''(x) = 12(x - 1) = 0$ if $x = 1$

Therefore, $x = 1$ is the only possible value of x for a maximum or minimum value of y. It does not yield either, since $f'(x)$ does not change sign as x varies from slightly less than 1 to slightly greater than 1. Furthermore, 1 is the only value of x for which $f''(x) = 0$. We shall now see if $f''(x)$ changes sign as x increases through 1, so as to find whether the point on the graph whose abscissa is 1 is an inflection point. Thus, $f''(1 - \varepsilon) = 12(1 - \varepsilon - 1) = -12\varepsilon < 0$ and $f''(1 + \varepsilon) = 12(1 + \varepsilon - 1) = 12\varepsilon > 0$.   Consequently, $y = 2x^3 - 6x^2 + 6x + 9$ has a point of inflection for $x = 1$.

Example 2

Test $y = f(x) = x^{1/5}$ for inflection points.

Solution: Taking the first derivative, we find that

$$f'(x) = \frac{x^{-4/5}}{5} = \frac{1}{5}\frac{1}{x^{4/5}} > 0$$

for both positive and negative values of x. Hence, $y = f(x)$ is always rising and has a vertical tangent for $x = 0$.
   Taking the second derivative, we see that

$$f''(x) = \frac{1}{5}\left(-\frac{4}{5}\right)x^{-9/5} = -\frac{4}{25}\frac{1}{x^{9/5}}$$

is positive for $x < 0$ and negative for $x > 0$. Consequently, $f''(x)$ changes sign as x increases through zero, and $y = f(x)$ has an inflection point at $x = 0$. The graph is shown in Fig. 11.2.

inflection points of y as maximum or minimum of y'

   We have seen that in testing for points of inflection at which $f''(x) = 0$ that $f''(x_1 - \varepsilon) \gtrless 0, f''(x_1) = 0$, and $f''(x_1 + \varepsilon) \lessgtr 0$. If we make use of the fact that $f''(x) = D_x f'(x)$, we can now say that *inflection points at which $f''(x) = 0$ and maximum and minimum points of $f'(x)$ are the same*. Thus to test for inflection points at which $f''(x) = 0$, we need only test $y' = f'(x)$ for maxima and minima.

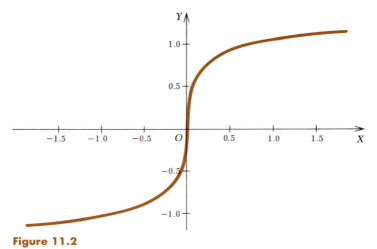

Figure 11.2

11.5   The second derivative test

We found in Sec. 10.5 that if $f'(x_1 - \varepsilon) > 0$, $f'(x_1) = 0$, and $f'(x_1 + \varepsilon) < 0$, then $y = f(x)$ has a maximum for $x = x_1$. If we notice the variation in $f'(x)$ as x increases through x_1, we see that $f'(x)$ is a decreasing function of x at a maximum of $f(x)$. Consequently, the *second derivative* $f''(x)$ of the decreasing function $f'(x)$ is negative for the *derivative* value of x for which $f(x)$ has a maximum. We can now say that *test for* if $f'(x_1) = 0$ *and* $f''(x_1) < 0$, *then* $f(x)$ *has a maximum for* $x = x_1$. *maximum* We can show similarly that *if* $f'(x_1) = 0$ *and* $f''(x_1) > 0$, *then* $f(x)$ *and has a minimum for* $x = x_1$.

minimum

Example   Find the maximum and minimum points on the curve whose equation is $y = f(x) = x^3 - 4x^2 + 4x + 1$.

Solution: See Fig. 11.1.

$$\frac{dy}{dx} = 3x^2 - 8x + 4 = (3x - 2)(x - 2) = 0 \qquad \text{if } x = \tfrac{2}{3}, 2$$

In order to test these critical values, we shall employ the second derivative. Thus,

$$\frac{d^2y}{dx^2} = f''(x) = 6x - 8$$

$f''(\tfrac{2}{3}) = -4$ is negative; hence, $y = f(x)$ has a maximum for $x = \tfrac{2}{3}$.

$f''(2) = 4$ is positive; hence, $y = f(x)$ has a minimum for $x = 2$.

The value of the second derivative is of no importance in this particular connection. We are interested only in whether it is positive or negative.

If $d^2y/dx^2 = 0$ at a point P where $dy/dx = 0$, then we have an inflection point with horizontal tangent at P if the sign of d^2y/dx^2 changes at this point. If the sign of d^2y/dx^2 does not change at P, then we have a maximum or minimum according as this second derivative remains negative or positive in the neighborhood of P. In this connection the student should examine the equations $y = x^3$ and $y = x^4$ for maxima and minima.

11.6   Acceleration in rectilinear motion

We have already found that if a point moves along a straight line such that its distance from a fixed point A on the path varies with the time according to the law

$$s = f(t)$$

the value of ds/dt at any instant is the velocity v of the moving point at that instant.

acceleration   The rate at which the velocity is changing with respect to time is called the *acceleration* of the moving point. This rate is given by the value of

$$\frac{dv}{dt} = \frac{d}{dt}\left(\frac{ds}{dt}\right) = \frac{d^2s}{dt^2}$$

If s is expressed in feet and t in seconds, the velocity is in feet per second. The acceleration is the rate of change of velocity with respect to time and may be expressed in feet per second per second which is often abbreviated as ft/sec^2. Acceleration may be expressed in terms of other units, for example, miles per hour per hour, miles per hour per second, and feet per second per minute.

EXERCISE 11.2

Find the value of x for which the graph of the equation in each of Problems 1 to 28 has a maximum, a minimum, or an inflection point. Sketch each graph.

1    $y = f(x) = x^3 - 12x + 3$
2    $y = f(x) = x^2 - 2x^3$
3    $y = f(x) = x^3 - 3x^2 - 9x + 2$
4    $y = f(x) = x^3 - 6x^2 + 9x - 5$
5    $y = f(x) = x^3 - 9x^2 + 24x - 20$

6    $y = f(x) = x^3 - 3x^2 + 3x + 2$
7    $y = f(x) = x^3 - 6x^2 + 9x + 5$
8    $y = f(x) = 2x^3 - 3x^2 - 12x + 6$
9    $y = f(x) = x^4 - 1$
10   $y = f(x) = 8x^2 - x^4$
11   $y = f(x) = x^4 - 4x^3 + 8x + 1$
12   $y = f(x) = x^4 - x^2 - 2x + 3$

13   $y = f(x) = \dfrac{3}{x} + \dfrac{2}{x^2}$ 14 $y = f(x) = \dfrac{x}{3} + \dfrac{3}{x}$

15   $y = f(x) = \dfrac{16}{x} + \dfrac{x^3}{3}$ 16 $y = f(x) = \dfrac{4}{x} - \dfrac{1}{x^2}$

17   $y = f(x) = \dfrac{2}{x^2 + 3}$ 18 $y = f(x) = \dfrac{8a^3}{x^2 + 4a^2}$

19   $y = f(x) = \dfrac{5}{x^3 - 54}$ 20 $y = f(x) = \dfrac{a}{x^3 + a^3}$

21   $y = f(x) = \dfrac{x^2 - 1}{x^2 - 4}$ 22 $y = f(x) = \dfrac{x^2 + 4}{x^2 - 1}$

23   $y = f(x) = \dfrac{x^2}{x^2 + 24}$ 24 $y = f(x) = \dfrac{x^2}{x^2 - 5}$

25   $y = f(x) = \dfrac{2x + 1}{(x - 2)^2}$ 26 $y = f(x) = \dfrac{3x - 1}{(x + 1)^2}$

27   $y = f(x) = \dfrac{3x}{x^2 + 3}$ 28 $y = f(x) = \dfrac{x}{x^2 + 1}$

29   Show that the cubic curve $y = ax^3 + bx^2 + cx + d$ always has an inflection point and that the abscissa of this point is $-b/3a$.

30   Show that the equation of Problem 29 takes the form $y = ax^3 + Cx$ when the axes are translated such that the new origin is at the inflection point. Hence, infer that the cubic curve is always symmetric with respect to its inflection point.

31   Under what condition is the inflection point of $y = ax^3 + bx^2 + cx + d$ on the Y axis?

32   Show that if P is on the graph of $y = ax^3 + bx^2 + cx + d$ and if I is the inflection point of the graph, then P' is also on the curve provided that PIP' is a straight line and $PI = IP'$. Such a curve is said to be *centrosymmetric* about I. Compare with Problem 30.

33   Determine the coefficients a, b, c, and d in $y = ax^3 + bx^2 + cx + d$ such that the curve has a maximum at $(-1, 9)$ and an inflection point at $(\tfrac{1}{3}, \tfrac{7}{27})$.

34   Show that for r a rational number, the function $f(x) = (1 + x)^r - (1 + rx)$, $x \geq -1$, has a minimum for $x = 0$.

35   The function T defined by $T(x) = (ax + b)/(cx + d)$ is called a *linear fractional transformation.* Show that $T'(x)$ is never zero.

36   Find the values of the coefficients in $y = ax^3 + bx^2 + cx + d$ such that it has a minimum at $(3, -8)$ and an inflection point at $(1, 8)$.

37   A particle moves along a straight line in such a way that its distance S in feet from a fixed point A on the line at the end of t min is given by $S = t^2 - 8t + 18$. Find the position, velocity, and acceleration of the particle at the end of 2 min.

38   Solve Problem 37 for the equation $S = 9t^2 - t^3$.

39   A particle moves along a straight line in such a way that its distance S in feet from a fixed point A on the line at the end of t min is $S = t(t - 10)^2$. Find the position, velocity, and acceleration of the particle at the end of 2 min. During what time interval is it moving toward A?

40   Solve Problem 39 for the equation $S = t^2(t - 6)^2$.

41   A particle moves along the Y axis in such a way that its distance in inches from the origin at the end of t sec is given by $y = \sqrt{3t + 5}$. Find its velocity and acceleration at the end of $20\frac{2}{3}$ sec.

42   A particle moves along the Y axis; its distance in feet from the origin after t sec is given by $y = \sqrt{2t + 4}$. Find its velocity and acceleration at the end of 16 sec.

43   Explain the following statement: The condition that $f'(x_1) = 0$ and $f''(x_1) = k < 0$ is sufficient, but not necessary, for a maximum value of $f(x)$ at $x = x_1$.

44   Explain the following statement: The condition that $f''(x_1) = 0$ is neither necessary nor sufficient for an inflection point on the graph of $f(x)$ at $x = x_1$.

12

Differentiation of the trigonometric functions

12.1 Definitions and simple identities

The main purpose of this chapter is to present a discussion of differentiation of the trigonometric and inverse trigonometric functions, but before doing that we shall give a résumé of the part of trigonometry that we shall need. We begin by pointing out that if the vertex of an angle is at the origin and the initial side of the angle is along the positive X axis as in Fig. 12.1, the angle is said to be in *standard position*. Both θ and α are in standard position. The angle θ is measured in a counterclockwise direction and is said to be a positive angle; whereas α is a negative angle since it is measured in a clockwise direction.

If $P(x, y)$ is a point on the terminal side of an angle θ in standard position and if $P(x, y)$ is a distance r from the origin, then we define the trigonometric functions of the angle as follows:

standard position

definitions of the trigonometric functions

$$\sin \theta = \frac{y}{r} \qquad \csc \theta = \frac{r}{y}$$

$$\cos \theta = \frac{x}{r} \qquad \sec \theta = \frac{r}{x}$$

$$\tan \theta = \frac{y}{x} \qquad \cot \theta = \frac{x}{y}$$

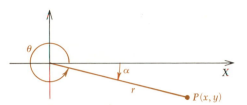

Figure 12.1 Positive and negative angles

We can readily determine the signs of the trigonometric functions of an angle in standard position by use of the definitions and the usual conventions for the signs of x, y, and r. Thus, if θ is a fourth-quadrant angle as in Fig. 12.1, $\sin \theta$ is the quotient of a negative and a positive number; hence $\sin \theta$ is negative. Similarly, $\cos \theta$ is positive.

We can find the values of the functions of $30°$, $45°$, and $60°$ by use of Fig. 12.2 and the definitions of the functions. Thus, $\sin 30° = \frac{1}{2}$, $\cos 45° = 1/\sqrt{2}$, and $\cot 60° = 1/\sqrt{3}$.

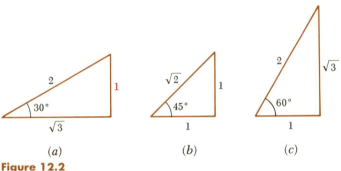

(a)   (b)   (c)

Figure 12.2

If the value of one function of an angle and the quadrant in which the angle lies are known, we can find the values of the other functions by making use of the relation $x^2 + y^2 = r^2$ between the coordinates and radius vector of a point. Thus, if $\sin \theta = \frac{3}{5}$ and $90° < \theta < 180°$, then $x^2 + y^2 = r^2$ becomes $x^2 + 3^2 = 5^2$ with x negative; hence, $x = -4$. Therefore, $\cos \theta = -\frac{4}{5}$, $\tan \theta = 3/(-4)$, $\cot \theta = -\frac{4}{3}$, $\sec \theta = 5/(-4)$, and $\csc \theta = \frac{5}{3}$.

There are eight fundamental relations between the functions of an angle. They are known as fundamental identities and can be proved by means of the definition of the functions of an angle. The identities are

identities

$$\sin^2 \theta + \cos^2 \theta = 1$$
$$1 + \tan^2 \theta = \sec^2 \theta$$
$$1 + \cot^2 \theta = \csc^2 \theta$$

Pythagorean identities

$$\left. \begin{array}{l} \tan \theta = \dfrac{\sin \theta}{\cos \theta} \\[2ex] \cot \theta = \dfrac{\cos \theta}{\sin \theta} \end{array} \right\} \text{\textit{ratio identities}}$$

$$\left. \begin{array}{l} \sin \theta \csc \theta = 1 \\[1ex] \cos \theta \sec \theta = 1 \\[1ex] \tan \theta \cot \theta = 1 \end{array} \right\} \text{\textit{reciprocal identities}}$$

These identities are used to change the form of an expression that involves trigonometric functions.

Since zero divided by any nonzero number is zero and division by zero is not a defined operation, we see that two of the functions of any integral multiple of 90° are zero and two are undefined. Furthermore, since one of the coordinates of a point on a coordinate axis is numerically equal to the radius vector of the point, it follows that two of the trigonometric functions of an integral multiple of 90° are 1 or −1. An angle of 180° is shown in standard position in Fig. 12.3. The values of the functions of 180° are:

$$\sin 180° = \frac{0}{r} = 0 \qquad\qquad \cos 180° = \frac{-r}{r} = -1$$

$$\tan 180° = \frac{0}{-r} = 0 \qquad\qquad \cot 180° \text{ is undefined}$$

$$\sec 180° = \frac{r}{-r} = -1 \qquad\qquad \csc 180° \text{ is undefined}$$

The values of the functions of other integral multiples of 90° can be found in a similar manner.

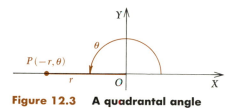

Figure 12.3   A quadrantal angle

12.2   Functions of a composite angle

By use of the distance formula and the definition of the trigonometric functions of an angle, we can prove, for any two angles A and B, that

$$\cos (A + B) = \cos A \cos B - \sin A \sin B \tag{1}$$

If we make use of the fact that $\sin(-B) = -\sin B$ and $\cos(-B) = +\cos B$, we can use (1) to show that

$$\cos(A - B) = \cos A \cos B + \sin A \sin B \tag{2}$$

By replacing B by A in (1), we find that

$$\cos 2A = \cos^2 A - \sin^2 A \tag{3a}$$

$$= 2\cos^2 A - 1 \qquad \text{since } \sin^2 A = 1 - \cos^2 A \tag{3b}$$

$$= 1 - 2\sin^2 A \qquad \text{since } \cos^2 A = 1 - \sin^2 A \tag{3c}$$

Now, solving (3b) for $\cos A$, we have

$$\cos A = \pm\sqrt{\frac{1 + \cos 2A}{2}} \tag{4}$$

where the sign is determined by the quadrant in which A lies.

If we make use of the fact that $\sin(\pi/2 - \alpha) = \cos \alpha$ and $\cos(\pi/2 - \alpha) = \sin \alpha$, we can obtain

$$\sin(A + B) = \sin A \cos B + \cos A \sin B \tag{5}$$

from (1). Furthermore, we can obtain

$$\sin(A - B) = \sin A \cos B - \cos A \sin B \tag{6}$$

and

$$\sin 2A = 2\sin A \cos A \tag{7}$$

from (5) and

$$\sin A = \pm\sqrt{\frac{1 - \cos 2A}{2}} \tag{8}$$

from (3c).

Another group of four identities can be obtained from (1) to (8) by making use of $\tan \alpha = \sin \alpha/\cos \alpha$. They are

$$\tan(A + B) = \frac{\tan A + \tan B}{1 - \tan A \tan B} \tag{9}$$

$$\tan(A - B) = \frac{\tan A - \tan B}{1 + \tan A \tan B} \tag{10}$$

$$\tan 2A = \frac{2\tan A}{1 - \tan^2 A} \tag{11}$$

$$\tan A = \pm\sqrt{\frac{1 - \cos 2A}{1 + \cos 2A}} \tag{12a}$$

$$= \frac{1 - \cos 2A}{\sin 2A} \tag{12b}$$

$$= \frac{\sin 2A}{1 + \cos 2A} \tag{12c}$$

By combining (5) and (6) by addition and then subtraction, we find that

$$2 \sin A \cos B = \sin (A + B) + \sin (A - B) \qquad (13)$$

and
$$2 \cos A \sin B = \sin (A + B) - \sin (A - B) \qquad (14)$$

Similarly by equating the sums and also the differences of the corresponding members of (1) and (2), we obtain

$$2 \cos A \cos B = \cos (A + B) + \cos (A - B) \qquad (15)$$

and
$$2 \sin A \sin B = -\cos (A + B) + \cos (A - B) \qquad (16)$$

12.3   Law of sines and of cosines

Relations between sides and angles of a triangle are given by the law of sines, law of cosines, law of tangents, half-angle formulas for triangles, and Molleweide's equations. These relations are discussed in most trigonometry books.[*] We shall state two of them.

Law of Sines   In any triangle the three ratios obtained by dividing a side by the sine of the angle opposite it are equal. Thus,

$$\frac{a}{\sin A} = \frac{b}{\sin B} = \frac{c}{\sin C}$$

Law of Cosines   The square of any side of a triangle is equal to the sum of the squares of the other sides decreased by twice their product times the cosine of the included angle. Thus,

$$a^2 = b^2 + c^2 - 2bc \cos A$$

12.4   The graphs of the trigonometric functions

The functions $y = \sin x$, $y = \cos x$, and $y = \tan x$ are all periodic and continuous, except that tangent is not continuous for odd multiples of $\pi/2$. They are all periodic with 2π as the period of the first two and π the period of $y = \tan x$. Their graphs for $0 \leq x \leq 2\pi$ are shown in Figs. 12.4 to 12.6.

The graphs of the other three functions will not be given, but they can be obtained from the graphs given above by making use of the reciprocal relations.

[*] See Sparks and Rees, "Plane Trigonometry," 5th ed., Prentice-Hall, Englewood Cliffs, N.J., 1965.

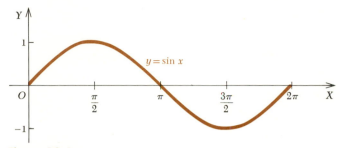

Figure 12.4 $y = \sin x$

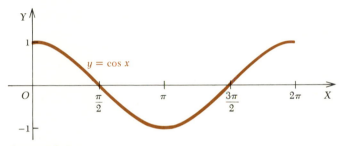

Figure 12.5 $y = \cos x$

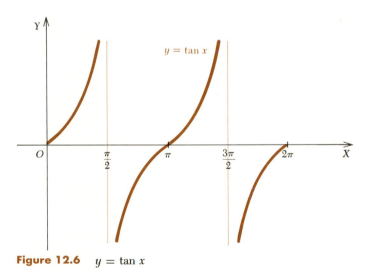

Figure 12.6 $y = \tan x$

EXERCISE 12.1

Prove that the equation in each of Problems 1 to 8 is an identity.

1   $\sin A \sec A = \tan A$

2   $\sin^2 A + \tan^2 A + \cos^2 A = \sec^2 A$

3 $\sec^2 A + \csc^2 A = \sec^2 A \csc^2 A$

4 $\cos A + \sin A \tan A = \sec A$

5 $\dfrac{\sec^2 A}{1 + \cot^2 A} = \tan^2 A$

6 $\dfrac{\cot A}{\cos A} + \dfrac{\sec A}{\cot A} = \sec^2 A \csc A$

7 $\dfrac{\sin x}{\csc x - 1} + \dfrac{\sin x}{\csc x + 1} = 2 \tan^2 x$

8 $\cot x + \tan x = \sec x \csc x$

Find the value of the combination of functions in each of Problems 9 to 12.

9 $\sin 30° \cos 120° + \sin 240° \cos 300°$

10 $\dfrac{\sin 300°}{1 - \cos 240°} - \tan 210°$

11 $\tan 225° \sec 315° + \cos 135° \sin 45°$

12 $\sin 120° \tan 135° + 2 \sin 300° \cos 150°$

Derive the identity in each of Problems 13 to 16 in the indicated manner.

13 $\cos (x + y) = \cos x \cos y - \sin x \sin y$ by use of a figure

14 $\sin (x - y) = \sin x \cos y - \cos x \sin y$ from the formula for $\sin (x + y)$

15 $\tan (x + y) = \dfrac{\tan x + \tan y}{1 - \tan x \tan y}$ from the formulas for $\cos (x + y)$
and $\sin (x + y)$

16 $\tan A = (1 - \cos 2A)/\sin 2A$ from (12a) of Sec. 12.2

Find the other functions of the angle in each of Problems 17 to 20.

17 $\sin A = \frac{5}{13}$, A in the first quadrant

18 $\cos A = \frac{3}{5}$, A in the fourth quadrant

19 $\tan A = -\frac{4}{3}$, A in the second quadrant

20 $\sec A = -\frac{17}{8}$, A in the third quadrant

In each of Problems 21 to 24, find $\sin (A + B)$, $\cos (A - B)$, $\sin 2A$,
and $\tan A/2$.

21 $\sin A = \frac{3}{5}$, A in quadrant II; $\cos B = -\frac{3}{5}$, B in quadrant III

22 $\cos A = \frac{5}{13}$, A in quadrant I; $\tan B = \frac{3}{4}$, B in quadrant III

23 $\tan A = -\frac{8}{15}$, A in quadrant IV; $\sin B = \frac{8}{17}$, B in quadrant II

24 $\cot A = \frac{4}{3}$, A in quadrant III; $\sec B = \frac{13}{5}$, B in quadrant IV

Prove each identity in Problems 25 to 32.

25 $\cos 5A \cos A + \sin 5A \sin A = 1 - 4 \sin^2 A \cos^2 A - \sin^2 2A$

26 $(1 - \sin 2\theta)(1 + 2 \sin 2\theta) = \sin 2\theta + \cos 4\theta$

27    $\dfrac{\csc A - 2 \sin A}{2 \cos A} = \cot 2A$

28    $\dfrac{2 \tan A}{1 + \tan^2 A} = \sin 2A$

29    $\dfrac{2 \sin \theta - \sin 2\theta}{2 \sin \theta + \sin 2\theta} = \tan^2 \dfrac{\theta}{2}$

30    $\dfrac{\cos 2\theta - \cos 4\theta}{\sin 2\theta + \sin 4\theta} = \tan \theta$

31    $\dfrac{1 + 2 \sin 2\theta + \cos 2\theta}{2 + \sin 2\theta - 2 \cos 2\theta} = \cot \theta$

32    $\dfrac{\{1 + [\sin \theta/(1 - \cos \theta)]^2\}^{3/2}}{1/(1 - \cos \theta)^2} = 4 \sqrt{\dfrac{1 - \cos \theta}{2}}$

Solve the equation in each of Problems 33 to 36 for values of the angle in $[0, 360°)$.

33    $2 \sin \theta \cos \theta - \sin \theta - 2 \cos \theta + 1 = 0$

34    $2\sqrt{3} \cos \theta \tan 2\theta - 2 \cos \theta - 3 \tan 2\theta + \sqrt{3} = 0$

35    $\cos^2 \theta - \cos 2\theta = 0$

36    $3 \cos \theta + 2 \sin^2 \theta - 3 = 0$

12.5   The limit of $(\sin x)/x$ as x approaches zero

This is a fundamental limit since its value is used in obtaining the derivative of the sine of an angle, which in turn is used in finding the derivatives of the other trigonometric and the inverse trigonometric functions. In order to evaluate the limit, we shall use Fig. 12.7. In it, the radius OP of the circle is taken as our unit of linear measure, and x is the radian measure of a central angle with $0 \leq x \leq \pi/2$. We have

$$\text{Area of } ORQ < \text{area } OPQ < \text{area } OPS$$

$$\frac{(OR)(RQ)}{2} < \frac{(OP)^2 x}{2} < \frac{(OP)(PS)}{2} \qquad x \text{ in radians} \qquad (1)$$

Since $\cos x = OR$, $\sin x = RQ$, $OP = 1$, $\tan x = PS$, this can be put in the form

$$\frac{\cos x \sin x}{2} < \frac{x}{2} < \frac{\tan x}{2}$$

Now, dividing each member of the inequality by the positive number $\frac{1}{2} \sin x$, we have

$$\cos x < \frac{x}{\sin x} < \frac{1}{\cos x}$$

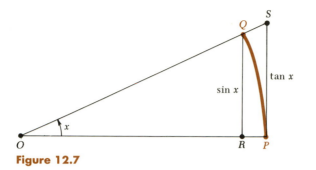

Figure 12.7

Consequently, taking the reciprocal of each member and reversing the order of each inequality sign, we find that

$$\frac{1}{\cos x} > \frac{\sin x}{x} > \cos x$$

Thus, $(\sin x)/x$ lies between $\cos x$ and $1/\cos x = \sec x$, and the limit of each as x approaches zero is 1. Therefore, it follows that

$$\lim_{x \to 0} \frac{\sin x}{x} = 1 \tag{12.1}$$

We have considered only the case in which $x > 0$, but the same limit is obtained if $x < 0$ since $[\sin(-x)]/(-x) = (\sin x)/x$.

Since the area of the sector in (1) is as given if and only if x is in radians, it follows that (12.1) is true if and only if the angle is measured in radians.

Example 1   Evaluate $\lim_{x \to 0} [(\sin^2 x)/x]$.

Solution: If we make use of (12.1) and the fact that the limit of a product is the product of the limits, we find that

$$\lim_{x \to 0} \frac{\sin^2 x}{x} = \lim_{x \to 0} \frac{\sin x}{x} \frac{\sin x}{1}$$

$$= \lim_{x \to 0} \frac{\sin x}{x} \lim_{x \to 0} \frac{\sin x}{1} = 1 \cdot 0 = 0$$

Example 2   Evaluate $\lim_{x \to \pi} \dfrac{\tan x}{2(x - \pi)}$.

Solution: In order to evaluate this limit, we shall make use of the fact $\tan(\pi - x) = -\tan x$ and have

$$\lim_{x \to \pi} \frac{\tan x}{2(x - \pi)} = \lim_{x \to \pi} \frac{-\tan(\pi - x)}{2(x - \pi)} = \lim_{x \to \pi} \frac{\tan(\pi - x)}{2(\pi - x)}$$

$$= \lim_{x \to \pi} \frac{\sin(\pi - x)}{2(\pi - x)\cos(\pi - x)}$$

$$= \lim_{x \to \pi} \frac{\sin(\pi - x)}{\pi - x} \lim_{x \to \pi} \frac{1}{2\cos(\pi - x)} = (1)\left(\frac{1}{2}\right) = \frac{1}{2}$$

by use of (12.1) and since $\cos(\pi - x) \to 1$ as $x \to \pi$.

12.6   The derivative of sin *u*

As a first step in finding $D_x \sin u$, we shall find $D_x \sin x$ by use of the fundamental limit process. If

$$y = \sin x$$

then $$y + \Delta y = \sin(x + \Delta x)$$

and $$\frac{\Delta y}{\Delta x} = \frac{\sin(x + \Delta x) - \sin x}{\Delta x}$$

$$= \frac{2\cos(x + \tfrac{1}{2}\Delta x)\sin \tfrac{1}{2}\Delta x}{\Delta x} \qquad \textit{by use of (14) of Sec. 12.2}$$

$$= \cos(x + \tfrac{1}{2}\Delta x)\frac{\sin \tfrac{1}{2}\Delta x}{\tfrac{1}{2}\Delta x}$$

Now, taking the limit as Δx approaches zero, we have

$$\lim_{\Delta x \to 0}\frac{\Delta y}{\Delta x} = \lim_{\Delta x \to 0}\cos(x + \tfrac{1}{2}\Delta x)\frac{\sin \tfrac{1}{2}\Delta x}{\tfrac{1}{2}\Delta x}$$

$$= \lim_{\Delta x \to 0}\cos(x + \tfrac{1}{2}\Delta x)\lim_{\tfrac{1}{2}\Delta x \to 0}\frac{\sin \tfrac{1}{2}\Delta x}{\tfrac{1}{2}\Delta x}$$

since $\tfrac{1}{2}\Delta x \to 0$ as $\Delta x \to 0$,

$$\lim_{\Delta x \to 0}\frac{\Delta y}{\Delta x} = \cos x \qquad \textit{by use of (12.1)}$$

Consequently, we have shown that

$$D_x \sin x = \cos x \qquad x \text{ in radians} \qquad (12.2a)$$

If *u* is a differentiable function of *x*, then

$$D_x \sin u = \cos u \, D_x u \qquad (12.2b)$$

by use of the chain rule.

Example   Find the derivative with respect to *x* of $y = \sin 3x$.

Solution: If

$$y = \sin 3x$$

then, by use of (12.2*b*), we have

$$D_x y = D_x \sin 3x = \cos 3x \; D_x 3x = 3 \cos 3x$$

12.7   The derivative of $\cos u$

In order to find the derivative of $\cos u$, we shall make use of the identity $\cos u = \sin (\pi/2 - u)$. If this is done, we have

$$D_x \cos u = D_x \sin \left(\frac{\pi}{2} - u\right) = \cos \left(\frac{\pi}{2} - u\right) D_x \left(\frac{\pi}{2} - u\right)$$

$$= \sin u \; D_x \left(\frac{\pi}{2} - u\right)$$

$$= -\sin u \; D_x u$$

Consequently,

$$D_x \cos u = -\sin u \; D_x u \tag{12.3}$$

Example 1

$$D_x \cos (2x - 1) = -\sin (2x - 1) \; D_x(2x - 1)$$

$$= -[\sin (2x - 1)](2) = -2 \sin (2x - 1)$$

Example 2   Find $D_x \sin 5x \cos 3x$.

Solution: Since the given function is a product, we have

$$D_x \sin 5x \cos 3x = \sin 5x \; D_x \cos 3x + \cos 3x \; D_x \sin 5x$$

$$= \sin 5x(-3 \sin 3x) + \cos 3x(5 \cos 5x)$$

$$= -3 \sin 5x \sin 3x + 5 \cos 3x \cos 5x$$

The form of this answer, like any function of trigonometric functions, can be expressed in a variety of forms. Thus,

$$5 \cos 3x \cos 5x - 3 \sin 5x \sin 3x = 4 \cos 3x \cos 5x$$
$$- 4 \sin 3x \sin 5x$$
$$+ \cos 3x \cos 5x$$
$$+ \sin 3x \sin 5x$$
$$= 4 \cos 8x + \cos 2x$$

by use of (1) and (2) of Sec. 12.2. Furthermore,

$$\sin 5x \cos 3x = \tfrac{1}{2} \sin 8x + \tfrac{1}{2} \sin 2x \qquad \text{\textit{Eq. (13) of Sec. 12.2}}$$

Therefore, $D_x \sin 5x \cos 3x = D_x(\tfrac{1}{2} \sin 8x + \tfrac{1}{2} \sin 2x)$

$$= \tfrac{1}{2}(\cos 8x)(8) + \tfrac{1}{2}(\cos 2x)(2)$$

$$= 4 \cos 8x + \cos 2x$$

This is the form into which the expression for $D_x \sin 5x \cos 3x$ was transformed, and it illustrates the fact that the work of differentiating a function of trigonometric functions can be simplified at times by changing the form of the given function before differentiating.

EXERCISE 12.2

In Problems 1 to 16 evaluate the limits if they exist.

1 $\displaystyle\lim_{x\to 0} \frac{\tan x}{x}$

2 $\displaystyle\lim_{x\to \pi} \frac{\sin 2x}{\sin x}$

3 $\displaystyle\lim_{x\to \pi} \frac{\cos 2x}{\cos x}$

4 $\displaystyle\lim_{x\to \pi/2} \frac{\cos^2 x}{1 - \sin x}$

5 $\displaystyle\lim_{x\to 0} \frac{2 - 2\sec^2 x}{\sin^2 x}$

6 $\displaystyle\lim_{x\to 0} \frac{1 - \cos x}{x \sin x}$

7 $\displaystyle\lim_{x\to 0} \frac{4 - 4\cos x}{x^2}$

8 $\displaystyle\lim_{x\to 3} \frac{\cos (\pi/x)}{x - 3}$

9 $\displaystyle\lim_{x\to 0} \frac{\sin^2 x}{x}$

10 $\displaystyle\lim_{x\to 0} \frac{\tan 3x}{5x}$

11 $\displaystyle\lim_{x\to \pi/2} \frac{\cos x}{\pi - 2x}$

12 $\displaystyle\lim_{x\to 0} \frac{\sin 5x}{\sin 3x}$

13 $\displaystyle\lim_{x\to 0} \frac{\tan 4x \csc 2x}{2x}$

14 $\displaystyle\lim_{x\to \pi} \frac{\sin x}{\pi - x}$

15 $\displaystyle\lim_{x\to 0} \frac{x^2 + 3x}{\sin 3x}$

16 $\displaystyle\lim_{h\to 0} \frac{\cos (a + h) - \cos h}{h}$

In Problems 17 to 32 differentiate each function.

17 $y = 3 \cos 4x$

18 $y = 2 \sin 5x$

19 $y = \sin x - \cos 2x$

20 $y = 3 \cos 5x - 5 \sin 3x$

21 $y = 2 \sin^3 \pi x$

22 $y = \cos^2 2x$

23 $y = \sin^2 3x + \cos^2 3x$

24 $y = (\sin x + \cos x)^2$

25 $y = 4x \cos 2x$

26 $y = 3x \sin \pi x$

27 $y = x^2 \cos^2 x$

28 $y = x^3 \sin^2 x$

29 $y = \dfrac{1 + \cos 2x}{\sin 2x}$

30 $y = \dfrac{1 - \sin x}{1 + \sin x}$

31 $y = \dfrac{1 - \cos x}{\sin x}$

32 $y = \dfrac{\sin x + \cos x}{\sin x - \cos x}$

Find d^2y/dx^2 for the function in each of Problems 33 to 40.

33 $y = 5 \sin^2 kx$

34 $y = 2 \cos^2 ax$

35 $y = 2 \sin x + \cos 2x$

36 $y = \sin 2x + \cos^2 x$

37 $y = x \sin x$

38 $y = \dfrac{\cos x}{x}$

39 $y = \dfrac{\sin x}{1 - \cos x}$

40 $y = a \sin^3 x \cos x$

In each of Problems 41 to 44, sketch both curves for $x = 0$ to $x = 2\pi$, and find their angle or angles of intersection.

41 $y = \sin x, \ y = \cos x$
42 $y = \sin x, \ y = \cos 2x$
43 $y = 2 \sin^2 x, \ y = \cos 2x$
44 $y = 4 \cos^2 x, \ y = \sin 2x$

In each of Problems 45 to 52, find the values of x in $[0, 2\pi)$ for which the function has a maximum, a minimum, or an inflection point.

45 $y = 2 \sin^2 x$

46 $y = \cos^2 x$

47 $y = \sin^3 x$

48 $y = \cos^4(x/2)$

49 $y = \sin x + \cos x$

50 $y = \cos x + \sin x \cos x$

51 $y = 2 \cos x - \sin 2x$

52 $y = x + \sin x$

53 Find the range of $f(t) = 4 \sin t \cos^2 t$.

54 Show $n + \frac{1}{2}$ is the abscissa of a maximum or of a minimum point on $y = \pi x - \tan (\pi x/2)$ according as n is an even or an odd integer.

55 The percentage error due to a small error in reading an electric current by use of a tangent galvanometer is proportional to $\sin x/\cos x + \cos x/\sin x$. For what value of x is this percentage error least?

56 The force

$$F = \dfrac{\mu w}{\cos \theta + \mu \sin \theta}$$

if directed at an angle θ above the horizontal, will move a weight w along the horizontal where μ is a constant and is called the coefficient of friction. Show that the value of θ for which F is smallest is given by $\tan \theta = \mu$.

57 A cylinder is inscribed in a sphere of radius a. Find the maximum volume.

58 A right circular cone is inscribed in a sphere of radius a. Find the height of the cone if its lateral surface is a maximum. Make use of the fact that the slant height is $2a \cos \theta$, the height is $2a \cos^2 \theta$, and the radius of the base is $2a \sin \theta \cos \theta$.

59 Solve Problem 58 if lateral surface is maximized, rather than volume.

60 A right circular cone is circumscribed about a sphere of radius a. Show that the volume of the cone is a minimum if $\sin \theta = \frac{1}{3}$ where the angle at the vertex is 2θ.

12.8   The derivatives of $\tan u$, $\cot u$, $\sec u$, $\csc u$

In order to obtain the derivative of $\tan u$, we shall make use of $\tan u$ in terms $\sin u$ and $\cos u$ and then apply the formula for the derivative of a quotient. If this is done, we have

$$D_x \tan u = D_x \frac{\sin u}{\cos u} \qquad\qquad \textit{since } \tan u = \sin u/\cos u$$

$$= \frac{\cos u \, D_x \sin u - \sin u \, D_x \cos u}{\cos^2 u}$$

$$= \frac{\cos^2 u + \sin^2 u}{\cos^2 u} \, D_x u$$

$$= \frac{1}{\cos^2 u} \, D_x u \qquad\qquad \textit{since } \cos^2 u + \sin^2 u = 1$$

$$= \sec^2 u \, D_x u \qquad\qquad \textit{since } 1/\cos^2 u = \sec^2 u$$

Consequently, we know that

$$D_x \tan u = \sec^2 u \, D_x u \qquad\qquad (12.4)$$

We can show in a similar manner that

$$D_x \cot u = -\csc^2 u \, D_x u \qquad\qquad (12.5)$$

Example   If
$$y = \sqrt{\tan x}$$

then $D_x y = D_x \, (\tan x)^{1/2} = \dfrac{(\tan x)^{-1/2}}{2} \, D_x \tan x = \dfrac{1}{2\sqrt{\tan x}} \sec^2 x$

We shall continue our study of the derivatives of the trigonometric functions by finding the derivative of $\sec u$. Thus,

$$D_x \sec u = D_x(\cos u)^{-1} \qquad\qquad \textit{since } \sec u \cos u = 1$$

$$= -1(\cos u)^{-2} \, D_x \cos u$$

$$= -1(\cos u)^{-2}(-\sin u) \, D_x u$$

$$= \frac{\sin u}{\cos^2 u} \, D_x u$$

$$= \frac{1}{\cos u} \frac{\sin u}{\cos u} \, D_x u$$

$$= \sec u \tan u \, D_x u$$

Therefore, if u is a differentiable function, then

$$D_x \sec u = \sec u \tan u \, D_x u \qquad\qquad (12.6)$$

We can show in a similar manner that

$$D_x \csc u = -\csc u \cot u \, D_x u \qquad (12.7)$$

EXERCISE 12.3

Find the derivative of the function in each of Problems 1 to 24.

1	$y = \tan 3x$	**2**	$y = \cot 5x$
3	$y = \sec 2x$	**4**	$y = \csc 6x$
5	$y = \csc(x^3 - 1)$	**6**	$y = \tan(x^2 - 3x)$
7	$y = \cot(x^2 - x)$	**8**	$y = \sec(x - x^3)$
9	$y = \sec^3(x^2 - 3)$	**10**	$y = \csc^2\sqrt{x^2 + 4x}$
11	$y = \tan^2\sqrt{x^2 - 2x}$	**12**	$y = \cot^3\sqrt[3]{x^3 - 3x}$
13	$y = \cot^2\sqrt{x^4 - x} + 2x$		
14	$y = \sec^3(x^3 + 3) + x$	**15**	$y = \csc^2 3x + 5x$
16	$y = 2x + \tan^3(x + 1)$	**17**	$y = x\tan^2(x^2 - x)$
18	$y = x^2\cot^3(2x - 5)$	**19**	$y = x^2\sec^2\sqrt{x^2 + 6x}$
20	$y = x\sec^4\sqrt{x^2 + 4}$		

21 $\quad y = \dfrac{\cot x}{1 - \cot^2 x}$ $\qquad$ **22** $\quad y = \dfrac{\sec x + \tan x}{\sec x - \tan x}$

23 $\quad y = \dfrac{2\tan x - 1}{4\tan x + 1}$ $\qquad$ **24** $\quad y = \dfrac{\csc x - \cot x}{\csc x + \cot x}$

Find d^2y/dx^2 for the function in each of Problems 25 to 32.

25	$y = 2\tan(x/2)$	**26**	$y = 4\tan^2(x/2)$
27	$y = \cot 2x$	**28**	$y = \cot^2 3x$
29	$y = \sec^2 x$	**30**	$y = \sec 2x$
31	$y = \csc^3 2x$	**32**	$y = \csc 3x$

In Problems 33 to 36, sketch both curves in $[0, 2\pi)$, and find the angle between them at each intersection.

33 $\quad y = \sin x, \ y = \frac{1}{2}\tan x$

34 $\quad y = \tan 2x, \ y = \cot x$

35 $\quad y = \sqrt{2}\cos x, \ y = \tan x$

36 $\quad y = \sqrt{3}\sec x, \ y = 4\sin x$

37 Show that $y = \tan x$ is a solution of $y'' - y'\tan x - y\sec^2 x = 0$.

38 Show that $y = \sec x$ is a solution of $y'' - y'\tan x - y\sec^2 x = 0$.

39 Show that $y = \cot x$ is a solution of $y''\sin^2 x - 2y = 0$.

40 Show that $y = \csc x$ is a solution of $2y''\sin^2 x + y'\sin 2x - 2y = 0$.

41 Show that the derivative of tan x with respect to sin x is $\sec^3 x$.
42 Show that the derivative of sin x with respect to tan x is $\cos^3 x$.
43 Show that the derivative of sec x with respect to cos x is $-\sec^2 x$.
44 Show that the derivative of sec x with respect to csc x is $-\tan^3 x$.

12.9 The inverse trigonometric functions

We discussed inverse functions in Sec. 1.8 and shall now relate our discussion to those inverse functions that arise in connection with the trigonometric functions. If a value is given in $[-1, 1]$ for x in $x = \sin y$, there is at least one number between zero and 2π for y. Furthermore, any number that differs from this value by an integral multiple of 2π also satisfies the equation $x = \sin y$. Consequently, the value of y is not uniquely determined if x is given; thus, we do not have a function.

It is customary to indicate that $x = \sin y$ has been solved for y by writing $y = \arcsin x$ or $y = \sin^{-1} x$ and reading "y is an angle whose sine is x." However, we must have one and only one value of y for a given x if y is to be a function of x. We indicate that a value has been selected by writing Arcsin x instead of arcsin x and select the value to be used from among the possibilities by requiring that:

Arcsin x

$$\text{If} \quad y = \text{Arcsin } x \quad \text{then} \quad \frac{-\pi}{2} \le y \le \frac{\pi}{2}$$

as indicated by the darker part of the curve in Fig. 12.8. In keeping with this, $y = \text{Arcsin } \frac{1}{2} = \pi/6$ since $\pi/6$ is the number between $-\pi/2$ and $\pi/2$ whose sine is $\frac{1}{2}$.

In order to find the derivative of $y = \text{Arcsin } u$, we make use of the fact that

$$y = \text{Arcsin } u \quad \text{and} \quad u = \sin y$$

give the same relation between u and y for $-\pi/2 \le y \le \pi/2$. If $u = \sin y$, then

$$\frac{du}{dy} = \cos y$$

Consequently,

$$\frac{dy}{du} = \frac{1}{\cos y} = \frac{1}{\sqrt{1 - u^2}}$$

since $\cos y > 0$ for $-\pi/2 < y < \pi/2$.

Hence, if u is differentiable, we have

$$\frac{dy}{dx} = \frac{dy}{du}\frac{du}{dx} = \frac{1}{\sqrt{1 - u^2}}\frac{du}{dx}$$

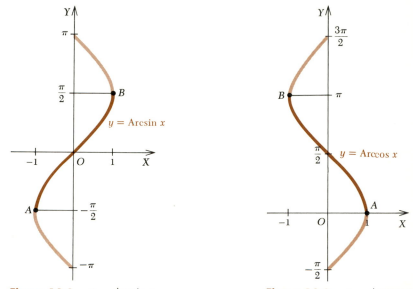

Figure 12.8 $y = \text{Arcsin } x$ **Figure 12.9** $y = \text{Arccos } x$

We now know that if u is a differentiable function of x, then

$$D_x \text{ Arcsin } u = \frac{1}{\sqrt{1 - u^2}} D_x u \qquad (12.8)$$

If we solve the equation $x = \cos y$ for y, we get $y = \arccos x = \cos^{-1} x$. We select one value from among the possibilities by means of the statement:

Arccos x **If** $y = \text{Arccos } x$ **then** $0 \leq y \leq \pi$

as indicated in Fig. 12.9. For example, $y = \text{Arccos } 0.5 = \pi/3$ since that is the number between 0 and π whose cosine is 0.5.

The derivative of $y = \text{Arccos } u$ can be obtained in several ways. We shall make use of the fact that $\text{Arccos } u = \pi/2 - \text{Arcsin } u$. Consequently,

$$D_x \text{ Arccos } u = D_x \left(\frac{\pi}{2} - \text{Arcsin } u \right) = \frac{-1}{\sqrt{1 - u^2}} D_x u$$

Thus, if u is a differentiable function of x, then

$$D_x \text{ Arccos } u = - \frac{1}{\sqrt{1 - u^2}} D_x u \qquad (12.9)$$

Example 1 $$D_x \text{ Arccos } 2x = \frac{-1}{\sqrt{1 - (2x)^2}} D_x 2x = \frac{-2}{\sqrt{1 - 4x^2}}$$

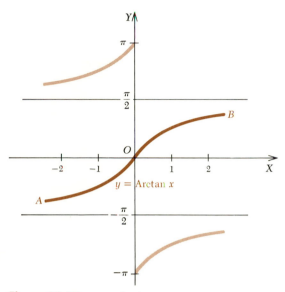

Figure 12.10 $y = \text{Arctan } x$

If we solve the equation $x = \tan y$ for y, we get $y = \text{arctan}$ $x = \tan^{-1} x$. We must select one value from among the possibilities so as to have a single value of y for each value of x. The value is chosen such that

$$\text{If} \quad y = \text{Arctan } u \quad \text{then} \quad \frac{-\pi}{2} < y < \frac{\pi}{2}$$

as shown in the darker part of the curve in Fig. 12.10. Thus, $y = \text{Arctan}$ $\sqrt{3} = \pi/3$ since that is the value in $(-\pi/2, \pi/2)$.

We shall now find the derivative of Arctan u by making use of the fact that $y = \text{Arctan } u$ and $u = \tan y$ are equivalent for the range of y given above. If $u = \tan y$, then

$$D_x u = D_x \tan y = \sec^2 y \, D_x y$$

Therefore,

$$D_x y = \frac{1}{\sec^2 y} D_x u = \frac{1}{1 + \tan^2 y} D_x u = \frac{1}{1 + u^2} D_x u$$

Thus, if u is a differentiable function of x, then

$$D_x \text{ Arctan } u = \frac{1}{1 + u^2} D_x u \tag{12.10}$$

If we make use of the fact that Arccot $u = \pi/2 - \text{Arctan } u$ and differentiate, we find that

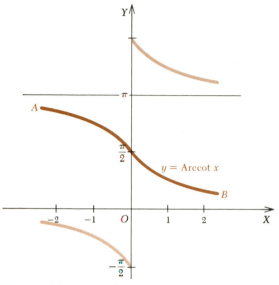

Figure 12.11 $y = \text{Arccot } x$

$$D_x \text{ Arccot } u = \frac{-1}{1 + u^2} D_x u \tag{12.11}$$

The graph of $y = \text{Arccot } x$ is shown in Fig. 12.11.

Example 2   If $y = \text{Arctan } (x^2 - 1)$, find $D_x y$.

Solution: If we apply (12.10), we find that

$$D_x \text{ Arctan } (x^2 - 1) = \frac{1}{1 + (x^2 - 1)^2} D_x (x^2 - 1)$$

$$= \frac{2x}{x^4 - 2x^2 + 2}$$

Example 3   Find $D_x y$ if $y = x^2 \text{ Arccot } 2x$.

Solution: If we use (12.11) along with the derivative of a product, we get

$$D_x (x^2 \text{ Arccot } 2x) = x^2 D_x \text{ Arccot } 2x + \text{Arccot } 2x \, D_x x^2$$

$$= x^2 \left(\frac{-1}{1 + 4x^2} \right) (2) + (\text{Arccot } 2x)(2x)$$

$$= \frac{-2x^2}{1 + 4x^2} + 2x \text{ Arccot } 2x$$

We shall define the Arcsec u and Arccsc u by means of

$$\text{Arcsec } u = \text{Arccos } \frac{1}{u}$$

and

$$\text{Arccsc } u = \text{Arcsin } \frac{1}{u}$$

They are functions with domain $\{u \le -1\} \cup \{u \ge 1\}$ and ranges as given by

$$0 \le \text{Arcsec } u \le \pi \qquad \text{except } \frac{\pi}{2}$$

and

$$\frac{-\pi}{2} \le \text{Arccsc } u \le \frac{\pi}{2} \qquad \text{except } 0$$

domain and range of Arcsec u and Arccsc u

We shall now find the derivative of $y = \text{Arcsec } u$ by making use of the fact that $y = \text{Arcsec } u$ and $u = \sec y$ are equivalent for $0 \le y \le \pi$. If

$$u = \sec y \qquad 0 \le y \le \pi$$

then

$$D_x u = D_x \sec y = \sec y \tan y \, D_x y$$

Hence,

$$D_x y = \frac{1}{\sec y \tan y} D_x u$$

Now, if we make use of the fact that $\sec y = u$ and, consequently, that

$$\tan y = \pm \sqrt{\sec^2 y - 1}$$
$$= \pm \sqrt{u^2 - 1}$$

we have

$$D_x \text{ Arcsec } u = \frac{1}{u(\pm \sqrt{u^2 - 1})} D_x u$$

We shall investigate the ambiguous sign. This sign is determined by the sign of $\tan y$; hence, it is plus for $0 < y < \pi/2$ and minus for $-\pi/2 < y < 0$. Furthermore, $u > 0$ for $0 < y < \pi/2$ and $u < 0$ for $\pi/2 < y < \pi$. Therefore,

$$D_x \text{ Arcsec } u = \begin{cases} \dfrac{1}{u \sqrt{u^2 - 1}} D_x u & \text{if } u > 0 \\[3mm] \dfrac{1}{-u \sqrt{u^2 - 1}} D_x u & \text{if } u < 0 \end{cases}$$

Consequently,

$$D_x \text{ Arcsec } u = \frac{1}{|u| \sqrt{u^2 - 1}} D_x u \qquad (12.12)$$

This result could have been obtained by finding $D_x \text{ Arccos } (1/u)$.

By use of the relation $\text{Arccsc } u = \pi/2 - \text{Arcsec } u$ and differentiation, we find that

$$D_x \text{ Arccsc } u = \frac{-1}{|u| \sqrt{u^2 - 1}} D_x u \qquad (12.13)$$

Example 4   Find $D_x y$ if $y = \text{Arcsec } (2x - 1)$.

Solution: If we make use of (12.12), we get

$$D_x \text{ Arcsec } (2x - 1) = \frac{1}{|2x - 1| \sqrt{(2x - 1)^2 - 1}} D_x(2x - 1)$$

$$= \frac{2}{|2x - 1| \sqrt{4x^2 - 4x}}$$

$$= \frac{1}{|2x - 1| \sqrt{x^2 - x}}$$

EXERCISE 12.4

Evaluate the trigonometric function of the angle given in each of Problems 1 to 8.

1   $\sin \frac{1}{2}(\text{Arccos } \frac{3}{5})$
2   $\cos 2(\text{Arctan } \frac{1}{3})$
3   $\tan 2[\text{Arcsin } (-\frac{3}{5})]$
4   $\cos \frac{1}{2}[\text{Arccos } (-\frac{1}{2})]$
5   $\cos (\text{Arcsin } \frac{2}{5} + \text{Arccos } \frac{2}{5})$
6   $\tan (\text{Arcsin } \frac{3}{5} + \text{Arctan } \frac{1}{3})$
7   $\tan (\text{Arctan } \frac{1}{2} + \text{Arctan } \frac{1}{3})$
8   $\tan [2(\text{Arctan } \frac{1}{3}) + \text{Arctan } \frac{1}{7}]$

9   Solve $\text{Arctan } \frac{1}{3} + \text{Arctan } x = \text{Arctan } \frac{4}{7}$ for x.
10   Solve $\text{Arctan } 2x + \text{Arctan } 3x = \pi/4$ for x.
11   Solve $\text{Arccos } 2x + \text{Arccos } x = \pi/6$ for x.
12   Solve $\tan (\pi/4 + \text{Arctan } x) = \text{Arctan } 5$ for x.

Find dy/dx in each of Problems 13 to 36.

13   $y = \text{Arccos } (x/2)$ 14 $y = \text{Arcsin } 3x$
15   $y = 2 \text{ Arctan } (x/2)$ 16 $y = \text{Arccot } 2x$
17   $y = \text{Arcsec } x^2$ 18 $y = \text{Arccsc } (1/x)$
19   $y = \text{Arcsin } (x/a)$ 20 $y = \text{Arcsin } \sqrt{x}$

21   $y = \text{Arccos } \dfrac{1 - x}{x}$ 22 $y = \text{Arccos } \dfrac{1}{\sqrt{1 + x^2}}$

23   $y = \text{Arctan } \dfrac{x}{\sqrt{1 - x^2}}$ 24 $y = \text{Arcsin } \dfrac{1 - x}{1 + x}$

25   $y = \frac{1}{4} \text{Arctan } (x/4)$ 26 $y = \text{Arccot } \dfrac{x + 4}{1 - 4x}$

27   $y = \text{Arcsec } \dfrac{x + 1}{x}$ 28 $y = \text{Arccsc } \dfrac{x}{x^2 - 1}$

29   $y = \sqrt{a^2 - x^2} + a \text{ Arcsin } (x/a), \ a > 0$

30   $y = \text{Arcsin } \dfrac{x}{2} + \dfrac{\sqrt{4 - x^2}}{x}$

31   $y = x\sqrt{1 - x^2} + \text{Arcsin } x$

32   $y = x \text{ Arcsin } x + \sqrt{1 - x^2}$

33   $y = x^2 \text{ Arctan } x$

34   $y = \sqrt{x^2 - 4} + 2 \text{ Arccot } \frac{1}{2}\sqrt{x^2 - 4}$

35   $y = 4x^2 \text{ Arccos } x$

36   $y = x \text{ Arcsec } (1/x)$

Find d^2y/dx^2 in each of Problems 37 to 44.

37   $y = \text{Arcsin } (x/2)$

38   $y = 3 \text{ Arccos } 5x$

39   $y = \text{Arctan } 2x$

40   $y = \text{Arccot } (x/3)$

41   $y = \text{Arcsin } \dfrac{1}{\sqrt{1 + x^2}} , x < 0$

42   $y = \text{Arctan } \dfrac{x}{1 - x^2}$

43   $y = \text{Arcsec } x^2$

44   $y = \text{Arccsc } x^4$

45   Find the inflection point of $y = \text{Arctan } x$.

46   Show that $y = \text{Arcsin } x$ is concave upward for $x > 0$ and concave downward for $x < 0$.

47   At what angle does $y = \text{Arcsin } x$ intersect the X axis?

48   If u and v are differentiable functions of x and $y = \text{Arctan } (u/v)$, show that

$$\frac{dy}{dx} = \frac{v \, D_x u - u \, D_x v}{u^2 + v^2}$$

49   At what rate is the vertex angle of a right circular cone changing when the radius is 3 ft and the altitude is 4 ft if the altitude is increasing at 6 in. per min and the radius is constant?

50   An airplane is moving horizontally at 150 mph over an observer at a height of 4,000 ft. How fast is the angle between the vertical line and the line of sight from the observer to the plane changing?

51   A picture that is 5 ft tall is hanging on a wall so that the lower edge is 2 ft above the observer's eye. How far from the wall should he stand so that the angle subtended by the picture is a maximum?

52   Find the angle of intersection of $y = $ Arctan x and $y = $ Arctan $(2x + 1)$.

13

The exponential and logarithmic functions

13.1   The laws of exponents

The student will recall that in algebra we first define the symbol a^n for n a positive integer. That definition is

$$a^n = a \cdot a \cdot a \cdots \text{ to } n \text{ factors}$$

We call this the *nth power* of a, and we call a the *base* and n the *exponent* of the power.

From this definition one easily proves that the following laws apply to positive integral exponents:

$$a^m a^n = a^{m+n} \tag{1}$$

$$\frac{a^m}{a^n} = \begin{cases} a^{m-n} & \text{if } m > n \text{ and } a \neq 0 \\ \dfrac{1}{a^{n-m}} & \text{if } m < n \text{ and } a \neq 0 \end{cases} \tag{2}$$

$$(a^m)^n = a^{mn} \tag{3}$$

$$(ab)^n = a^n b^n \tag{4}$$

$$\left(\frac{a}{b}\right)^n = \frac{a^n}{b^n} \tag{5}$$

When we extend our definition of a^n so as to give meaning to symbols as 8^0, $3^{1/2}$, and 6^{-3}, we require that the above laws hold for all exponents in deciding upon the needed definitions. For example, we desire that

$$\frac{2^5}{2^5} = 2^{5-5} = 2^0 \qquad \text{and} \qquad 2^0 \cdot 2^6 = 2^{0+6} = 2^6$$

These statements are true if and only if $2^0 = 1$. If we replace 2 by any real number a and examine the situation carefully, we see that the above laws hold if and only if $a^0 = 1$ *for all values of a except zero.* Hence, we make the definition:

$$a^0 = 1 \qquad a \neq 0 \tag{6}$$

We do not define the symbol 0^0.

Next, if we require that

$$2^3 \cdot 2^{-3} = 2^{3-3} = 2^0 = 1 \qquad \text{and} \qquad \frac{2^5}{2^{-3}} = 2^{5-(-3)} = 2^8$$

we must define 2^{-3} to mean $1/2^3$. More generally, *we define*

$$a^{-n} = \frac{1}{a^n} \qquad a \neq 0 \tag{7}$$

If we want

$$a^{1/2} \cdot a^{1/2} = a^{1/2+1/2} = a^1 = a$$

we must define $a^{1/2}$ to denote a number whose square is a; that is, a square root of a. To avoid ambiguity, we define it to stand for the *positive* square root if a is any positive number. More generally, we define the symbol $a^{1/n}$, where n is a positive integer, to denote the *principal* nth root of a. This is the positive root if a is positive, and the negative one if a is negative and n is odd. Thus $16^{1/2} = 4$, and $(-8)^{1/3} = -2$.

Finally, we consider the symbol $a^{m/n}$, where m and n are relatively prime positive integers and a is positive if n is even. If (3) is to hold, we must have

$$(a^{1/n})^m = (a^m)^{1/n} = a^{m/n} \tag{8}$$

Consequently, we define $a^{m/n}$ to denote the mth power of the principal nth root of a, or the principal nth root of a^m. These are equal under the conditions specified.

If the exponent x is irrational, the base a in the symbol a^x is restricted to positive numbers. We shall not discuss the situation in detail, but shall assume that there is a unique real number that is properly denoted by the symbol $5^{\sqrt{2}}$ and that this number is between $5^{1.41}$ and $5^{1.42}$, between $5^{1.414}$ and $5^{1.415}$, etc.

13.2 The exponential function

We shall assume that for any given positive number a, different from 1, the symbol a^x denotes a unique number for every real value of x. We can then draw an approximation to the graph of the function by the usual procedure of making a table of corresponding values of x and a^x, plotting the points, and drawing a smooth curve through them. Thus, for the equation $y = 2^x$, we have the graph shown in Fig. 13.1.

The function a^x, for a positive, is called an *exponential function*. Its value is positive for all real values of x, and it increases as x increases if $a > 1$. The graph has the general shape shown in Fig. 13.1. If $0 < a < 1$, the value of a^x decreases as x increases. It turns out that the function is differentiable for all values of x. We shall be concerned with the problem of finding its derivative later in this chapter and can then verify the statements made above relative to increasing and decreasing.

13.3 The logarithmic function

If $a^y = x$, then y is called the *logarithm of x to the base a* and is abbreviated by writing $y = \log_a x$. For example,

$$\text{If} \qquad 10^{2.64} = 437 \qquad \text{then} \qquad \log_{10} 437 = 2.64$$

logarithmic function The function $\log_a x$ is called the *logarithmic function* and is defined

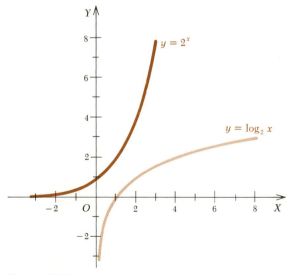

Figure 13.1

for all $x > 0$. It increases as x increases if $a > 1$, and it has a derivative for every positive value of x. We shall soon attack the problem of computing this derivative. We should note that the equation

$$y = \log_a x \qquad \text{is equivalent to} \qquad x = a^y$$

It follows that the graph of $y = \log_a x$ is identical with that of $x = a^y$. It is therefore simply the graph of $y = a^x$ with the axes interchanged. Consequently, $y = \log_a x$ is the inverse of $y = a^x$. The graphs of $y = \log_2 x$ and $y = 2^x$ are shown in Fig. 13.1.

The student will recall from his study of algebra that the logarithmic function has the following properties:

$$\log_a (MN) = \log_a M + \log_a N$$

$$\log_a \left(\frac{M}{N}\right) = \log_a M - \log_a N$$

$$\log_a (N^q) = q \log_a N$$

It follows from the first of these properties that

$$\log_a (10N) = \log_a N + \log_a 10$$

If we use 10 as the base, we have

$$\log_{10} (10N) = \log_{10} N + \log_{10} 10 = \log_{10} N + 1$$

Thus, multiplying a number by 10 increases the logarithm to the base 10 by one.

Example   If $\log_{10} 4.87 = 0.6875$, then $\log_{10} 48.7 = 1.6875$ and $\log_{10} 487 = 2.6875$. Similarly, $\log_{10} 0.487 = 0.6875 - 1$ or $9.6875 - 10$.

The logarithms with which a student is primarily concerned in his work in algebra and trigonometry are those to the base 10. These are *common* called *common logarithms*. It turns out, however, that the derivatives *logarithms* of the exponential and logarithmic functions are simpler if the base used is a certain irrational number that is denoted by the letter e and is 2.718 to three decimal places. We shall discuss it briefly in the next section.

13.4   The number e

We shall consider the function

$$f(h) = (1 + h)^{1/h}$$

and think of h as becoming smaller in absolute value, in fact, approaching zero. The following table gives a partial answer to what happens to $f(h)$.

h	-0.5	-0.1	-0.01	-0.001	$\cdots$	0.001	0.01	0.1	0.5
$f(h)$	4.000	2.868	2.732	2.720	$\cdots$	2.717	2.705	2.594	2.250

It appears probable from this table that as $h \to 0$, the value of $(1 + h)^{1/h}$ approaches a number in the neighborhood of 2.7 as a limit. This is true, but we shall not give a proof here.

We shall assume that the limit does exist and approximate it by letting h approach zero by taking successively the values

$$1, \tfrac{1}{2}, \tfrac{1}{3}, \tfrac{1}{4}, \tfrac{1}{5}, \ldots$$

in which the nth term is $1/n$. For each of these values of h, the exponent $1/h$ is an integer. By use of the binomial theorem and for any such value of h, we have

$$(1 + h)^{1/h} = 1 + \frac{1}{h}h + \frac{(1/h)(1/h - 1)}{2!}h^2$$

$$+ \frac{(1/h)(1/h - 1)(1/h - 2)}{3!}h^3 + \cdots + h^{1/h}$$

$$= 1 + 1 + \frac{1 - h}{2!} + \frac{(1 - h)(1 - 2h)}{3!} + \cdots + h^{1/h}$$

If h is allowed to approach zero through the sequence of values indicated above, this expansion is valid for each value of h, and the number of terms in the expansion increases indefinitely as $h \to 0$. This suggests that the limit we seek *may* be approximated to any desired degree of accuracy by letting $h \to 0$ in the expansion on the right and taking a sufficiently large number of terms. This conclusion is valid, but no rigorous justification of it can be given at this point. If we denote the limit by e, we have

$$e = \lim_{h \to 0} (1 + h)^{1/h}$$

$$= 1 + 1 + \frac{1}{2!} + \frac{1}{3!} + \frac{1}{4!} + \frac{1}{5!} + \cdots$$

$$= 2.718 \cdots$$

The corresponding functions e^x and $\log_e x$ are special cases of the functions a^x and $\log_a x$, respectively.

natural
logarithms
Logarithms to the base e are called *natural logarithms*. Since we shall use them almost exclusively in our work, we shall agree to write *ln x* to represent the natural logarithm of x. If the base is any number other than e, we shall always write it. Thus ln 64 will mean $\log_e 64$. The logarithm of 64 to the base 10 will be written $\log_{10} 64$.

The general formula°

$$\log_b N = \log_a N \log_b a \tag{1}$$

change of base for change in base enables us to obtain a relation between the natural logarithm of x and the common logarithm of x. The relation is

$$\ln x = \frac{\log_{10} x}{\log_{10} e} \tag{2}$$

as obtained by using $N = x$, $a = e$, and $b = 10$ and then dividing by $\log_{10} e$. We can write this in the form

$$\ln x = 2.3026 \log_{10} x$$

by making use of the fact that $1/\log_{10} e = 2.3026$. This relation enables one to compute the natural logarithm of a number from its common logarithm. There are, however, tables that give the natural logarithms directly. Table IV in this book at the end of the text is one such table.

Since $b = e^{\ln b}$, any positive number b can be expressed in the form e^k; consequently, the function b^{rx} is equivalent to $(e^k)^{rx} = e^{krx}$. Therefore, we can always use e as the base for an exponential function if we so wish. We usually do prefer to do this because it turns out that the expression for the derivative of the exponential function is simpler if e is the base.

Example   Express the function $5^{2.64x}$ in the form e^{kx}.

Solution: Since $5 = e^{\ln 5}$, it follows that

$$5^{2.64x} = (e^{\ln 5})^{2.64x}$$

$$= (e^{1.609})^{2.64x} \qquad \text{\textit{by use of Table IV}}$$

$$= e^{4.25x} \qquad \text{\textit{multiplying exponents}}$$

EXERCISE 13.1

Express the function in each of Problems 1 to 12 in a different form.

1 $e^{\ln x}$ 2 $e^{\ln (1/x)}$ 3 $e^{-\ln x}$

4 $e^{\ln x^2 - \ln x}$ 5 $e^{\ln x^2 + \ln 2}$ 6 $e^{2 \ln x^2}$

7 $e^{x + \ln x}$ 8 $e^{1/x - \ln x}$ 9 $\ln e^x$

10 $\ln e^{x^2 - x}$ 11 $\ln x e^{x^2}$ 12 $\ln (x/e^x)$

If m and n are positive integers, prove the statements in Problems 13 to 16.

° See Rees and Sparks, "College Algebra," 5th ed., p. 348, McGraw-Hill Book Company, New York, 1967.

13 $a^m a^n = a^{m+n}$ **14** $a^m/a^n = a^{m-n}, m > n$
15 $a^m/a^n = 1/(a^{n-m}), n > m$ **16** $(a^m)^n = a^{mn}$

Evaluate the number in each of Problems 17 to 20.
17 $5(64)^{-2/3} + 3(8)^0 + (-8)^{-2/3}$
18 $16^{-3/4} + 9^{3/2} + 7^0$
19 $12/4^{3/2} - 4/8^{2/3} + (32)^{-0.4}/2$
20 $1/64^{2/3} + 1/64^{5/6} - 3(8^{-5/3})$

Solve for x in each of Problems 21 to 24.
21 $(x/2)^{3/2} = \frac{1}{64}$ **22** $(x/3)^{2/3} = \frac{4}{9}$
23 $16x^{-2/3} = 9$ **24** $(2/x)^{-3/4} = (\frac{2}{3})^3$

25 Prove that $\log_a (MN) = \log_a M + \log_a N$.
26 Prove that $\log_a M/N = \log_a M - \log_a N$.
27 Prove that $\log_a N^q = q \log_a N$.
28 Prove that $\log_b N = \log_a N \log_b a$.
29 The equation $y = 10^{2.17x}$ is equivalent to $y = e^{kx}$ for a value of k. Find that value.
30 The equation $y = 4(10)^{1.54x}$ is equivalent to $y = 4e^{kx}$ for a properly chosen value of k. Find that value.
31 The equation $y = 1.8(2)^{1.14x}$ is equivalent to $y = 1.8e^{kx}$ for a properly chosen value of k. Find the value.
32 The equation $y = 8(2)^{3x-4}$ is equivalent to $y = Ae^{kx}$ for properly chosen values of A and k. Find a pair of these numbers.

Express the number in each of Problems 33 to 36 in the form Ae^{kt}.
33 $5.4(3)^{1.6t-2}$ **34** $0.6(\frac{1}{8})^{0.25t+2/3}$
35 $3.6(\frac{1}{27})^{0.5t+2/3}$ **36** $3.2(5)^{t/3-2}$

Sketch the graph of the equation in each of Problems 37 to 44.
37 $y = 2^{-0.5x}$ **38** $y = 2^{-x^2/2}$
39 $y = 3^{-0.4x}$ **40** $y = 4e^{-0.2x^2}$
41 $y = \ln (x - 2)$ **42** $y = \log_2 (x + 1)$
43 $y = 10 \ln \sqrt{x}$ **44** $y = \ln x^2$

In each of Problems 45 to 48, find the natural logarithms of the numbers.
45 37.4, 2870   **46** 0.197, 0.00243
47 8.62, 5640   **48** 22.6, 1150

In each of Problems 49 to 52, find to three digits the numbers whose natural logarithms are given.
49 $\ln A = 2.2247, \ln B = 7.2854 - 10$
50 $\ln A = 1.9206, \ln B = 8.7485 - 10$
51 $\ln A = 4.3184, \ln B = 6.2175 - 10$
52 $\ln A = 5.4172, \ln B = 9.4267 - 10$

13.5   The derivative of $\log_a u$

We should recall that the derivative of a function f is defined by the statement that

$$f'(x) = \lim_{x \to x_1} \frac{f(x) - f(x_1)}{x - x_1} \quad \text{or} \quad \lim_{\Delta x \to 0} \frac{f(x + \Delta x) - f(x)}{\Delta x}$$

We shall use the second form in finding the derivative of $\log_a x$. If the limit exists,

$$\frac{d}{dx} \log_a x = \lim_{\Delta x \to 0} \frac{\log_a (x + \Delta x) - \log_a x}{\Delta x} = \lim_{\Delta x \to 0} \frac{1}{\Delta x} \log_a \frac{x + \Delta x}{x}$$

$$= \lim_{\Delta x \to 0} \frac{1}{\Delta x} \log_a \left(1 + \frac{\Delta x}{x}\right)$$

In order to evaluate this limit, we first multiply and divide by x and then make use of the fact that $q \log_a N = \log_a N^q$:

$$\frac{d}{dx} \log_a x = \lim_{\Delta x \to 0} \frac{1}{x} \frac{x}{\Delta x} \log_a \left(1 + \frac{\Delta x}{x}\right)$$

$$= \lim_{\Delta x \to 0} \left[\frac{1}{x} \log_a \left(1 + \frac{\Delta x}{x}\right)^{x/\Delta x}\right]$$

$$= \frac{1}{x} \lim_{\Delta x \to 0} \log_a \left(1 + \frac{\Delta x}{x}\right)^{x/\Delta x}$$

$$= \frac{1}{x} \log_a \left[\lim_{\Delta x \to 0} \left(1 + \frac{\Delta x}{x}\right)^{x/\Delta x}\right]$$

if we make use of the fact that $y = \log_a x$ is continuous.

Now, as $\Delta x \to 0$, the quantity $(1 + \Delta x/x)^{x/\Delta x}$ approaches the number e as a limit since it is of the form $(1 + h)^{1/h}$ with h approaching zero. We thus have

$$D_x \log_a x = \frac{1}{x} \log_a e \qquad (13.1a)$$

We can now find by use of the chain rule that

$D_x \log_a u$

$$D_x \log_a u = \frac{1}{u} (\log_a e) D_x u \qquad (13.1)$$

If the base is the number e, the factor $\log_a e$ becomes $\log_e e = 1$ and (13.1) reduces to

$D_x \ln u$

$$D_x \ln u = \frac{1}{u} D_x u \qquad (13.1b)$$

furthermore, (13.1a) becomes

$D_x \ln x$

$$D_x \ln x = \frac{1}{x} \qquad (13.1c)$$

Formulas (13.1a), (13.1b), and (13.1c) are all special cases of (13.1) and are obtained from it; hence, they need not be memorized.

Example 1 $$\frac{d}{dx}\log_{10}(x^3 + 4) = \frac{1}{x^3 + 4}\log_{10}e \quad \frac{d}{dx}(x^3 + 4) = \frac{3x^2}{x^3 + 4}\log_{10}e$$

Example 2 $$\frac{d}{dx}\ln(x^3 + 4) = \frac{1}{x^3 + 4}\frac{d}{dx}(x^3 + 4) = \frac{3x^2}{x^3 + 4}$$

It is in order to avoid the factor $\log_{10}e = 0.4343$ that we prefer to use natural logarithms instead of common logarithms in the calculus.

13.6 The derivative of a^u

We shall first develop a formula for the derivative of e^u and then show how it can be used to obtain the derivative of a^u. In finding $D_x e^u$, we shall make use of the fact that if $y = e^u$, then $u = \ln y$. Consequently

$$D_x u = \frac{1}{y}D_x y \quad \text{by (13.1}b)$$

Hence, $$D_x y = y\, D_x u$$
$$= e^u\, D_x u \quad \text{since } y = e^u$$

Thus, we see that

$D_x e^u$
$$D_x e^u = e^u\, D_x u \tag{13.2}$$

provided that u is a differentiable function of x.

We shall now make use of the fact that $a = e^{\ln a}$ as pointed out in Sec. 13.4 and as evident from the definition of $\ln a$. By use of this equation, we see that

$$a^u = (e^{\ln a})^u = e^{u\ln a}$$

Consequently,

$$D_x a^u = D_x e^{u\ln a} = e^{u\ln a}\, D_x u\ln a = a^u\ln a\, D_x u$$

since $e^{u\ln a} = a^u$ and $D_x u\ln a = \ln a\, D_x u$.

Therefore,

$D_x a^u$
$$D_x a^u = a^u\ln a\, D_x u \tag{13.2a}$$

Example 1 $$D_x 4^{x^2+5} = D_x e^{(\ln 4)(x^2+5)} \qquad \text{since } 4 = e^{\ln 4}$$
$$= e^{(\ln 4)(x^2+5)}\, D_x(\ln 4)(x^2 + 5) \qquad \text{by use of (13.2)}$$
$$= 4^{x^2+5}(\ln 4)(2x) \qquad \text{since } e^{(\ln 4)(x^2+5)} = 4^{x^2+5}$$

Example 2 If
$$Q = 4.84e^{-0.2t}$$

then
$$\frac{dQ}{dt} = (4.84)(e^{-0.2t})\frac{d}{dt}(-0.2t)$$

$$= (4.84)(e^{-0.2t})(-0.2) = -0.968e^{-0.2t}$$

Example 3 If
$$y = e^{-x/2}\ln x$$

then
$$D_x y = e^{-x/2}D_x\ln x + \ln x\, D_x e^{-x/2}$$

$$= e^{-x/2}\frac{1}{x} + (\ln x)(e^{-x/2})D_x\left(-\frac{x}{2}\right)$$

$$= e^{-x/2}\frac{1}{x} + (\ln x)(e^{-x/2})\left(-\frac{1}{2}\right)$$

$$= e^{-x/2}\left(\frac{1}{x} - \frac{1}{2}\ln x\right) = e^{-x/2}\left(\frac{1}{x} - \ln\sqrt{x}\right)$$

13.7 The derivative of u^n

We have proved that $D_x x^n = nx^{n-1}$ holds for n an integer, and we shall now prove it for n non-integral rational and irrational values of x. To do this, we shall let

$$y = x^n$$

then
$$\ln y = n\ln x$$

Differentiating with respect to x, we have

$D_x x^n$, n
a constant
$$\frac{1}{y}\frac{dy}{dx} = n\frac{1}{x}$$

hence,
$$\frac{dy}{dx} = \frac{ny}{x} = \frac{nx^n}{x} = nx^{n-1}$$

We now can use the chain rule to find that

$$D_x u^n = nu^{n-1}\, D_x u \tag{13.3}$$

Example If $y = x^{\sqrt{2}}$, then $dy/dx = \sqrt{2}x^{\sqrt{2}-1}$ for $x > 0$.

13.8 Logarithmic differentiation

In the two preceding sections we have found it convenient to use the fact that if $y = f(x)$, then $\ln y = \ln f(x)$ in order to compute dy/dx from a given relation between y and x. We shall now continue our study of the use of this device.

Example 1   If $y = x\sqrt{(x^2 + 1)/(x^2 + 4)}$, find dy/dx.

Solution: We may simplify the work by noting that ln y must be equal to the logarithm of the expression on the right. Thus,

$$\ln y = \ln x \sqrt{\frac{x^2 + 1}{x^2 + 4}} = \ln x + \frac{1}{2}\ln (x^2 + 1) - \frac{1}{2}\ln (x^2 + 4)$$

Differentiation with respect to x, we get

$$\frac{1}{y}\frac{dy}{dx} = \frac{1}{x} + \frac{1}{2}\left(\frac{2x}{x^2 + 1}\right) - \frac{1}{2}\left(\frac{2x}{x^2 + 4}\right)$$

Therefore, multiplying by y and removing common factors, we have

$$\frac{dy}{dx} = y\left(\frac{1}{x} + \frac{x}{x^2 + 1} - \frac{x}{x^2 + 4}\right)$$

If we want an expression for dy/dx in terms of x alone, we may now replace y by its value in terms of x. For most purposes this is not necessary.

It sometimes happens that we need to find $D_x y$ from a relation of the form

$$y = u^v$$

a variable power of a variable   where u and v are functions of x. This can be accomplished by using the fact that if $y = u^v$, then ln $y = v \ln u$.

Example 2   In order to find $D_x y$ from the relation $y = x^x$, we first observe that this relation is equivalent to

$$\ln y = x \ln x \qquad x > 0$$

Now, differentiating with respect to x, we get

$$\frac{1}{y}\frac{dy}{dx} = x\left(\frac{1}{x}\right) + (\ln x)(1)$$

Therefore, $$\frac{dy}{dx} = y(1 + \ln x) = x^x(1 + \ln x)$$

EXERCISE 13.2

Find the derivative of the function in each of Problems 1 to 56.

1   $y = \ln x^2$
2   $y = \ln x^3$
3   $y = \ln \sqrt{x}$
4   $y = \ln x^{2/3}$
5   $y = \ln \sqrt{x^2 + 2x + 3}$
6   $y = \ln \sqrt{x^2 - 4x}$

7 $y = \ln \sqrt[3]{3x^2 - 9x + 1}$ 8 $y = 3 \ln \sqrt[3]{x^2 - x + 1}$

9 $y = \log_3 (x^2 - x + 1)$ 10 $y = \log_5 (x^2 + x - 1)$

11 $y = \log_{10} (3x - 4)$ 12 $y = \log_{10} (2x + 7)$

13 $y = \ln \sin x$ 14 $y = 2 \ln \sec x$

15 $y = \ln \tan x$ 16 $y = \ln \cos x$

17 $y = x \ln x^3$ 18 $y = x^3 \ln x$

19 $y = x^2 \ln x^2$ 20 $y = x \ln (x + 1)$

21 $y = x \ln \sec x$ 22 $y = x^2 \ln \cot x$

23 $y = x \ln \sin^2 x$ 24 $y = x^2 \ln \tan^2 x$

25 $y = \ln (x + \sqrt{x^2 + a^2})$ 26 $y = \ln x \sqrt{x^2 + a^2}$

27 $y = \ln x \sqrt{6x + 5}$ 28 $y = \ln x^2 \sqrt{x^2 + 1}$

29 $y = \ln \sqrt{\dfrac{4 + x^2}{4 - x^2}}$ 30 $y = \ln \sqrt{\dfrac{1 - \cos x}{1 + \cos x}}$

31 $y = 2 \ln \sqrt{\dfrac{1 + x^3}{1 - x^3}}$ 32 $y = 2 \ln \sqrt{\dfrac{1 + \sin x}{1 - \sin x}}$

33 $y = 4e^{-3x}$ 34 $2e^{-x^2/2}$

35 $y = e^{\sin x}$ 36 $y = e^{\tan x}$

37 $y = 2^{x^2-1}$ 38 $y = 5^{5x+4}$

39 $y = 3^{x^2+2}$ 40 $y = 7^{x^2+2x}$

41 $y = x^2 e^{-x/2}$ 42 $y = 4xe^{-x/4}$

43 $y = xe^{-x}$ 44 $y = (x - 2)e^{x^2-4x}$

45 $y = e^{-x} \ln x$ 46 $y = -(e^{-x}/2) \cos 2x$

47 $y = e^{-2x} \sin (x/2)$ 48 $y = e^{x^2} \ln x^2$

49 $y = e^x/\ln x$ 50 $y = e^{\ln x}/x$

51 $y = e^{x^2-2}e^{\ln(x^2-2)}$ 52 $y = (x^2 - 4)/e^{\ln(x-2)}$

53 $y = \sin x \, e^{\cos x}$ 54 $y = e^{\tan x} \ln \sec x$

55 $y = e^{\sin x} \ln (\sec x + \tan x)$

56 $y = e^{\cos x} \ln \cos x$

57 Show that $D_x y = 2 \sec^3 x$ if $y = \sec x \tan x + \ln (\sec x + \tan x)$.

58 Show that $D_x y = -\csc x$ if $y = \ln \sqrt{(1 + \cos x)/(1 - \cos x)}$.

59 Show that $y = x \ln x$ is concave upward for all $x > 0$. Show that y is a minimum for $x = 1/e$.

60 Show that $D_x y = \csc x$ if $y = \ln [\sin x/(1 + \cos x)]$.

61 Show that $y'' + 2y' + 2y = 0$ if $y = e^{-x} \sin x$.

62 Show that $y'' - 2y' + 5y = 0$ if $y = e^x \cos 2x$.

63 Show that $y'' - 6y' + 13y = 0$ if $y = e^{3x} \sin 2x$.

64 Show that $y'' - y' - y = 3e^{2x} \cos x$ if $y = e^{2x} \sin x$.

65 Show that $y'' - 4y' + 4y = 0$ if $y = Ae^{2x} + Bxe^{2x}$ and A and B are arbitrarily constants.

66 For what value or values of m does $y = Ae^{mx}$ satisfy $y'' + y' = 6y$?

67   For what value or values of m is $y = Ae^{mx}$ a solution of $y'' - 3y' + 2y = 0$?

68   Determine m so that $y = Ae^{mx}$ is a solution of $y''' - 6y'' + 11y' - 6y = 0$.

69   Show that $\ln (x^2 + 4)$ increases at a rate of 0.3 unit per unit of increase in x when $x = 6$.

70   A quantity E varies with the time t in keeping with

$$E = \ln \frac{8t}{4t^2 + 5}$$

where t is in minutes. Find the rate of change of E with respect to t for $t = 5$.

71   A quantity Q varies with the time t in keeping with $Q = 4.8e^{0.3t}$. Evaluate $D_t Q$ for $t = 2$.

72   A quantity Q varies with time in keeping with $Q = Ae^{kt}$ when A and k are constants. Show that the rate at which Q changes at any instant is proportional to the value of Q at that instant. Does this statement hold in the more general case in which $Q = Ab^{kt}$ where b is any constant?

Find dy/dx in Problems 73 to 80.

73 $y = x^{\sin x}$

74 $y = x^{\tan x}$

75 $y = (\sin x)^x$

76 $y = (\tan x)^x$

77 $y = x^{\sqrt{x}}$

78 $y = (\ln x)^x$

79 $y = (\sec x)^{\cos x}$

80 $y = (\tan x)^{\sin x}$

13.9   The hyperbolic functions

The exponential functions

$$\frac{e^x - e^{-x}}{2} \quad \text{and} \quad \frac{e^x + e^{-x}}{2}$$

occur frequently in applied mathematics; in fact, so frequently that it is desirable to assign names to them and to tabulate their values. The names assigned are the *hyperbolic sine of x* and the *hyperbolic cosine of x*. They are abbreviated as $\sinh x$ and $\cosh x$ and defined by

$\sinh x$

$$\sinh x = \frac{e^x - e^{-x}}{2} \tag{13.4}$$

$\cosh x$ and

$$\cosh x = \frac{e^x + e^{-x}}{2} \tag{13.5}$$

The hyperbolic tangent of x is abbreviated as tanh x and defined by

tanh x

$$\tanh x = \frac{e^x - e^{-x}}{e^x + e^{-x}} \tag{13.6}$$

It can be shown to be sinh $x/\cosh x$. As might be suspected, there are relations between the hyperbolic functions that are similar to, but not necessarily the same as, the corresponding relations for the trigonometric functions.

The hyperbolic cotangent, hyperbolic secant, and hyperbolic cosecant are defined as the reciprocals of the hyperbolic tangent, hyperbolic cosine, and hyperbolic sine, respectively. The values of e^x, e^{-x}, sinh x, cosh x, and tanh x are given in Table V.

Many of the properties of the hyperbolic functions are analogous to those of the circular, or trigonometric, functions. These properties can be derived from the definitions.

Example 1   Show that sinh $2x = 2$ sinh x cosh x.

Solution:

$$\sinh 2x = \frac{e^{2x} - e^{-2x}}{2} \qquad \textit{by definition}$$

$$= 2 \frac{e^x - e^{-x}}{2} \frac{e^x + e^{-x}}{2} \qquad \textit{factoring and introducing ½}$$

$$= 2 \sinh x \cosh x$$

Example 2   Find the values of the other hyperbolic functions if sinh $x = -\tfrac{3}{4}$.

Solution: We shall make use of the identities between hyperbolic functions in order to evaluate the remaining five functions. If we substitute the given value of sinh x in the indentity given in Problem 1 of Exercise 13.3, we find that

$$\cosh^2 x - (-\tfrac{3}{4})^2 = 1$$

Therefore,

$$\cosh^2 x = 1 + \tfrac{9}{16} = \tfrac{25}{16}$$

Consequently, as a purely algebraic operation, we find that cosh $x = \pm\tfrac{5}{4}$. However, cosh $x = (e^x + e^{-x})/2$ is never negative; thus we have only cosh $x = \tfrac{5}{4}$. Now

$$\tanh x = \frac{\sinh x}{\cosh x} = \frac{-\tfrac{3}{4}}{\tfrac{5}{4}} = -\frac{3}{5}$$

and by use of the reciprocal relations, we get coth $x = -\tfrac{5}{3}$, sech $x = \tfrac{4}{5}$, and csch $x = -\tfrac{4}{3}$.

EXERCISE 13.3

Prove the identities given in Problems 1 to 16.

1   $\cosh^2 x - \sinh^2 x = 1$ 2 $\tanh^2 x + \operatorname{sech}^2 x = 1$
3   $\coth^2 x - \operatorname{csch}^2 x = 1$ 4 $\cosh x + \sinh x = e^x$
5   $\cosh x - \sinh x = e^{-x}$ 6 $\sinh (-x) = -\sinh x$
7   $\cosh (-x) = \cosh x$ 8 $\tanh (-x) = -\tanh x$
9   $\sinh 2x = 2 \sinh x \cosh x$
10   $\cosh 2x = \cosh^2 x + \sinh^2 x$
11   $\cosh 2x = 1 + 2 \sinh^2 x$
12   $\cosh 2x = 2 \cosh^2 x - 1$
13   $\sinh (x \pm y) = \sinh x \cosh y \pm \cosh x \sinh y$
14   $\cosh (x \pm y) = \cosh x \cosh y \pm \sinh x \sinh y$

15   $\tanh (x \pm y) = \dfrac{\tanh x \pm \tanh y}{1 \pm \tanh x \tanh y}$

16   $\cosh x \cosh y = \cosh (x + y) + \cosh (x - y)$

17   Sketch the graph of $y = \cosh x$.
18   Sketch the graph of $y = \sinh x$.
19   Sketch the graph of $y = \tanh x$.
20   Sketch the graph of $y = a(e^{x/a} + e^{-x/a})/2$. This curve is called a
catenary. It is the curve in which a homogeneous flexible chain hangs
if suspended from two points and acted on only by its own weight.

*Find the value or values of the other hyperbolic functions if one of
them has the value given in each of Problems 21 to 28.*

21   $\sinh x = \frac{8}{15}$ 22 $\cosh x = \frac{5}{4}$
23   $\tanh x = -\frac{3}{5}$ 24 $\coth x = -\frac{13}{12}$
25   $\operatorname{sech} x = \frac{5}{13}$ 26 $\operatorname{csch} x = \frac{4}{3}$
27   $\cosh x = \frac{25}{7}$ 28 $\sinh x = \frac{12}{5}$

29   Prove that $(\cosh x + \sinh x)^n = \cosh nx + \sinh nx$.
30   Show that $P_1(x_1, y_1)$ with $x_1 = \cosh u$ and $y_1 = \sinh u$ is on the
curve represented by $x^2 - y^2 = 1$.
31   Show that the tangent to $x^2 - y^2 = 1$ at $P_1(\cosh u, \sinh u)$
intersects the coordinate axes at $(\operatorname{sech} u, 0)$ and $(0, -\operatorname{csch} u)$.
32   Show that if $\sinh x = \cot \theta$, θ in $(0, \pi/2)$, then $\cosh x = \csc \theta$,
$\tanh x = \cos \theta$, $\coth x = \sec \theta$, $\operatorname{sech} x = \sin \theta$, and $\operatorname{csch} x = \tan \theta$.

13.10   The derivatives of the hyperbolic functions

We can find the derivatives of the hyperbolic functions readily since,
for u a differentiable function of x, we know the derivative of e^u and
the derivative of a sum, a difference, and a quotient. Thus,

$$D_x \sinh u = \frac{D_x(e^u - e^{-u})}{2} = \frac{e^u + e^{-u}}{2} D_x u = \cosh u \, D_x u$$

Consequently, for u a differentiable function of x,

$D_x \sinh u$
$$D_x \sinh u = \cosh u \, D_x u \qquad (13.7)$$

and we can show similarly that

$D_x \cosh u$
$$D_x \cosh u = \sinh u \, D_x u \qquad (13.8)$$

We shall now differentiate the hyperbolic tangent. By use of its definition, we have

$$D_x \tanh u = D_x \frac{\sinh u}{\cosh u}$$

$$= \frac{\cosh u \, D_x \sinh u - \sinh u \, D_x \cosh u}{\cosh^2 u}$$

$$= \frac{\cosh^2 u - \sinh^2 u}{\cosh^2 u} D_x u$$

$$= \frac{1}{\cosh^2 u} D_x u \qquad \textit{since } \cosh^2 u - \sinh^2 u = 1$$

$$= \operatorname{sech}^2 u \, D_x u \qquad \textit{since } \cosh u \operatorname{sech} u = 1$$

Therefore, for u a differentiable function of x, we have

$D_x \tanh u$
$$D_x \tanh u = \operatorname{sech}^2 u \, D_x u \qquad (13.9)$$

Similarly,

$D_x \coth u$
$$D_x \coth u = -\operatorname{csch}^2 u \, D_x u \qquad (13.10)$$

If we make use of the reciprocal relation between $\cosh u$ and $\operatorname{sech} u$, we have

$$D_x \operatorname{sech} u = D_x(\cosh u)^{-1}$$

$$= -1(\cosh u)^{-2} D_x \cosh u$$

$$= -\frac{\sinh u}{\cosh^2 u} D_x u$$

$$= -\operatorname{sech} u \tanh u \, D_x u \qquad \textit{since } \sinh u/\cosh u = \tanh u$$

Hence,

$D_x \operatorname{sech} u$
$$D_x \operatorname{sech} u = -\operatorname{sech} u \tanh u \, D_x u \qquad (13.11)$$

and we can show similarly that

$D_x \operatorname{csch} u$
$$D_x \operatorname{csch} u = -\operatorname{csch} u \coth u \, D_x u \qquad (13.12)$$

Example   Find the derivative of $\sinh^3 2x$.

Solution: We are to differentiate a power of a function; hence, we must make use of $D_x u^n = n u^{n-1} D_x u$. If this is done, we have

$$D_x \sinh^3 2x = 3 \sinh^2 2x\, D_x \sinh 2x = 3 \sinh^2 2x \cosh 2x\, D_x 2x$$

$$= 6 \sinh^2 2x \cosh 2x$$

13.11   The inverse hyperbolic functions

We indicate that the equation $x = \sinh y$ has been solved for y in terms of x by writing $y = \sinh^{-1} x$ and reading "y is the inverse *definition of* hyperbolic sine of x." We define $\sinh^{-1} x$ by saying that *sinh⁻¹ x*

$$y = \sinh^{-1} x \qquad \textit{is equivalent to} \qquad x = \sinh y$$

The domain and range of the inverse hyperbolic sine are both $-\infty$ *to* ∞.

If $x = \cosh y$, there are two values of y for each value of x. This is clear from the definition of $\cosh y = (e^y + e^{-y})/2$ since replacing y *definition of* by $-y$ leaves the function value unchanged. We now define the *cosh⁻¹ x* inverse hyperbolic cosine by saying that

$$y = \cosh^{-1} x \qquad \textit{is equivalent to} \qquad x = \cosh y$$

Consequently, there are two values of $y = \cosh^{-1} x$ for each value of x, but there must be only one if we are to say that the inverse hyperbolic cosine is a function. This situation is taken care of by agreeing that *the domain of* $y = \cosh^{-1} x$ *is* $[1, \infty)$ *and the range is* $[0, \infty)$.

Example 1   Find $w = \cosh^{-1} 1.337$.

Solution: As a first step in the solution, we change the given equation to the equivalent form $\cosh w = 1.337$. Now, Table V shows that $w = 0.8$ since that is the value of the variable across from the entry 1.337 in the hyperbolic cosine column.

definition of   We now define the inverse hyperbolic tangent by saying that *tanh⁻¹ x*

$$y = \tanh^{-1} x \qquad \textit{is equivalent to} \qquad x = \tanh y$$

There is only one value of y for each value of x. *The domain of* $y = \tanh^{-1} x$ *is* $(-1, 1)$ *and the range is* $(-\infty, \infty)$.

The other three inverse hyperbolic functions are seldom used but *other inverse* are defined by means of the reciprocal relations. Thus, $\coth^{-1} u =$ *hyperbolic* $\tanh^{-1}(1/u)$, $\operatorname{sech}^{-1} u = \cosh^{-1}(1/u)$, $\operatorname{csch}^{-1} u = \sinh^{-1}(1/u)$. *functions* We shall now express $y = \sinh^{-1} x$ in another form. If $y = \sinh^{-1} x$, then $x = \sinh y$; hence,

$$x = \frac{(e^y - e^{-y})}{2}$$

$$2xe^y = e^{2y} - 1 \qquad \textcolor{brown}{\textit{multiplying by } 2e^y}$$

$$e^{2y} - 2xe^y - 1 = 0$$

This is a quadratic in e^y. Solving it as such, we have

$$e^y = \frac{2x \pm \sqrt{(2x)^2 - 4(1)(-1)}}{2} = x \pm \sqrt{x^2 + 1}$$

but we cannot use $x - \sqrt{x^2 + 1}$ since e^y is never negative. Therefore,

$$e^y = x + \sqrt{x^2 + 1}$$

Consequently, $y = \ln(x + \sqrt{x^2 + 1})$ $\textcolor{brown}{\ln e^y = y}$

and we see that

formula for sinh⁻¹ *x* $\sinh^{-1} x = \ln(x + \sqrt{x^2 + 1}) \qquad -\infty < x < \infty \qquad (1)$

If we make use of the fact that $y = \cosh^{-1} x$ and $x = \cosh y$ are equivalent equations, we can express $\cosh^{-1} x$ in logarithmic form. Thus, if $x = \cosh y$, then

$$x = \frac{(e^y + e^{-y})}{2}$$

Hence, multiplying by $2e^y$ and collecting all terms in one member, we have

$$e^{2y} - 2xe^y + 1 = 0$$

Therefore, the quadratic formula yields

$$e^y = \frac{2x \pm \sqrt{4x^2 - 4}}{2} = x \pm \sqrt{x^2 - 1}$$

and we must choose the sign of the radical so that $e^y \geq 1$ since $y \geq 0$. Since the domain of $\cosh^{-1} x$ is $[1, \infty)$, we must discard the negative sign before the radical because, if $x > 1$, then

$$x - \sqrt{x^2 - 1} < 1$$

To prove this, we start with $\sqrt{x - 1} < \sqrt{x + 1}$, multiply each member by $\sqrt{x - 1}$ and get $x - 1 < \sqrt{x^2 - 1}$, and then add $1 - \sqrt{x^2 - 1}$ to each member and get $x - \sqrt{x^2 - 1} < 1$. If $x > 1$, then $x + \sqrt{x^2 - 1}$ is the sum of a number larger than 1 and a positive number; hence, it is greater than 1. Consequently,

$$e^y = x + \sqrt{x^2 - 1} \qquad \text{and} \qquad y = \ln(x + \sqrt{x^2 - 1})$$

hence,

formulas for
cosh⁻¹ x
and
tanh⁻¹ x

$$\cosh^{-1} x = \ln\left(x + \sqrt{x^2 - 1}\right) \qquad x \geq 1 \tag{2}$$

It can be shown by a procedure similar to that used in obtaining (1) and (2) that

$$\tanh^{-1} x = \frac{1}{2} \ln \frac{1 + x}{1 - x} \qquad -1 < x < 1 \tag{3}$$

We shall make use of (1) to (3) in finding the derivatives of the inverse hyperbolic functions. Thus,

$$D_x \sinh^{-1} u = D_x \ln\left(u + \sqrt{u^2 + 1}\right) = \frac{D_x\left(u + \sqrt{u^2 + 1}\right)}{u + \sqrt{u^2 + 1}}$$

$$= \frac{1}{u + \sqrt{u^2 + 1}} \left(1 + \frac{u}{\sqrt{u^2 + 1}}\right) D_x u = \frac{D_x u}{\sqrt{u^2 + 1}}$$

Consequently, for u a differentiable function of x,

$D_x \sinh^{-1} u$

$$D_x \sinh^{-1} u = \frac{D_x u}{\sqrt{u^2 + 1}} \tag{13.13}$$

We can show similarly that

$D_x \cosh^{-1} u$

$$D_x \cosh^{-1} u = \frac{D_x u}{\sqrt{u^2 - 1}} \tag{13.14}$$

$D_x \tanh^{-1} u$ and

$$D_x \tanh^{-1} u = \frac{D_x u}{1 - u^2} \tag{13.15}$$

Example 2   Find the derivative of $y = \tanh^{-1}(\sin 2x)$.

Solution 1:   If we make use of (13.15), we have

$$D_x \tanh^{-1}(\sin 2x) = \frac{D_x \sin 2x}{1 - \sin^2 2x} = \frac{2 \cos 2x}{\cos^2 2x} = 2 \sec 2x$$

Solution 2:   If we use (3), we have

$$D_x \tanh^{-1}(\sin 2x) = D_x \frac{1}{2} \ln \frac{1 + \sin 2x}{1 - \sin 2x}$$

$$= \frac{1}{2} \frac{1 - \sin 2x}{1 + \sin 2x} \frac{(1 - \sin 2x)(2 \cos 2x) - (1 + \sin 2x)(-2 \cos 2x)}{(1 - \sin 2x)^2}$$

$$= \frac{1}{2} \frac{1 - \sin 2x}{1 + \sin 2x} \frac{4 \cos 2x}{(1 - \sin 2x)^2}$$

$$= \frac{2 \cos 2x}{(1 + \sin 2x)(1 - \sin 2x)} = \frac{2 \cos 2x}{1 - \sin^2 2x}$$

$$= \frac{2 \cos 2x}{\cos^2 2x} = \frac{2}{\cos 2x} = 2 \sec 2x$$

EXERCISE 13.4

By use of Table V, evaluate each inverse hyperbolic function given in Problems 1 to 12. Find the first eight to the nearest entry in the table and interpolate in Problems 9 to 12 so as to obtain one more significant figure than can be read directly from the table.

1 $\sinh^{-1} 1.027$ 2 $\cosh^{-1} 1.004$
3 $\tanh^{-1} 0.987$ 4 $\sinh^{-1} 36.84$
5 $\cosh^{-1} 4.368$ 6 $\tanh^{-1} 0.917$
7 $\sinh^{-1} 3.001$ 8 $\cosh^{-1} 3.333$
9 $\tanh^{-1} 0.926$ 10 $\sinh^{-1} 9.437$
11 $\cosh^{-1} 5.396$ 12 $\tanh^{-1} 0.088$

Find dy/dx in each of Problems 13 to 36.

13 $y = \sinh 2x$ 14 $y = \cosh 3x$
15 $y = \tanh (x/2)$ 16 $y = \sinh (x/2)$
17 $y = \tanh^2 x$ 18 $y = \sinh^3 x$
19 $y = \cosh^2 2x$ 20 $y = \sqrt{\cosh x}$
21 $y = \ln \cosh x$ 22 $y = \ln \tanh x$
23 $y = \ln \sinh x$ 24 $y = \ln \operatorname{sech} x$
25 $y = e^x \sinh 2x$ 26 $y = e^x \cosh 2x$
27 $y = e^{-2x} \operatorname{sech} x$ 28 $y = e^{2x} \tanh x$
29 $y = \sinh^{-1} 2x$ 30 $y = \cosh^{-1} 2x$
31 $y = \tanh^{-1} 3x$ 32 $y = \sinh^{-1} (1/x)$
33 $y = \cosh^{-1} (\sec^2 x),\ 0 < x < \pi/2$
34 $y = \tanh^{-1} (\cos x)$
35 $y = \sinh^{-1} (\tan x),\ 0 \le x < \pi/2$
36 $y = \cosh^{-1} (\csc x),\ 0 < x \le \pi/2$
37 Derive Eq. (3).
38 Derive Eq. (13.14).
39 Derive Eq. (13.15).
40 Prove that $D_x \operatorname{sech}^{-1} u = (D_x u)/u\sqrt{1 - u^2}$.

14

Rolle's theorem, mean value theorem, indeterminate forms

14.1 Rolle's theorem

This chapter will cover several theorems that are important in calculus and in other areas of mathematics. They will be used from time to time in the later chapters of this book. The proofs of these theorems depend on a fact that we used in Chap. 7 when dealing with maxima and minima. It is: *If a maximum or minimum of a function $y = f(x)$ in an interval is $f(c)$ and occurs at an interior point of the interval and if $f(x)$ has a derivative for $x = c$, then $f'(c) = 0$.* That this is true can be seen from the fact that if $f'(c)$ is different from zero, then $f(x)$ is either increasing or decreasing for $x = c$ and hence cannot have a maximum or a minimum at the corresponding interior point. In proving the following theorem, we shall also make use of the fact that if $y = f(x)$ is continuous in a closed interval $a \leq x \leq b$, then $f(x)$ has a greatest value M and a least value m in the interval.*

Rolle's theorem   If $f(x)$ *is continuous in a closed interval* $a \leq x \leq b$, *if* $f(a) = f(b) = 0$, *and if* $f'(x)$ *exists for* $a < x < b$, *then there is at least one value c of x in the open interval between a and b such that* $f'(c) = 0$.

---

* A proof can be found in most any book on advanced calculus; for example, see Widder, "Advanced Calculus," p. 147, Prentice-Hall, Englewood Cliffs, N.J., 1947.

This statement is known as Rolle's theorem and we shall now prove it. We know that $f(x)$ has maximum and minimum values since it is continuous in a closed interval. If $f(x) \equiv 0$, then $f'(c) = 0$ for c any value in the interval. If $f(x)$ is positive anywhere, it has a maximum M as indicated in Fig. 14.1, and if it is negative anywhere, it has a minimum. Therefore, in each case there is a value c of x such that $f'(c) = 0$.

14.2   The law of the mean

We shall now give the second of our series of theorems. It is called the *law of the mean,* or *the mean value theorem,* and is

*mean value
theorem*   *If $f(x)$ is continuous in the interval $a \leq x \leq b$ and if $f'(x)$ exists for every interior point of the interval, then there is at least one value c of x between a and b such that*

$$f'(c) = \frac{f(b) - f(a)}{b - a} \tag{14.1}$$

*geometric
interpretation*   Before proving this theorem, we shall point out, as indicated in Fig. 14.2, the geometric interpretation of the theorem: that there is a value c of x between a and b such that the tangent at the corresponding point on the curve is parallel to the secant that connects A and B. From (14.1) the geometric interpretation is clearly true since $f'(c)$ is the slope of the tangent for $x = c$ and the right-hand member of (14.1) is the slope of the secant through A and B.

In order to prove the mean value theorem, we shall use the function

$$F(x) = f(x) - f(a) - \frac{f(b) - f(a)}{b - a}(x - a)$$

where $f(x)$ is the function of the theorem. We shall now see that $F(x)$ satisfies the hypotheses of Rolle's theorem and that, consequently, the

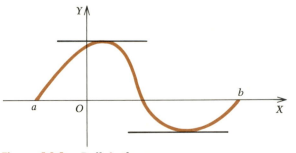

Figure 14.1   Rolle's theorem

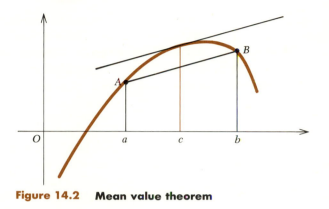

Figure 14.2 Mean value theorem

conclusion of that theorem is justified for $F(x)$. We shall consider the hypotheses in the order in which they are given in Rolle's theorem. Since $f(x)$ and $x - a$ are continuous for $a \leq x \leq b$, it follows that $F(x)$ is continuous in that interval. Furthermore, $F(a) = F(b) = 0$ since

$$F(a) = f(a) - f(a) - \frac{f(b) - f(a)}{b - a}(a - a) = 0$$

and

$$F(b) = f(b) - f(a) - \frac{f(b) - f(a)}{b - a}(b - a)$$

$$= f(b) - f(a) - f(b) + f(a) = 0$$

Finally, $F'(x)$ exists for $a < x < b$ since the difference of two differentiable functions, $f(x)$ and $\frac{f(b) - f(a)}{b - a}(x - a)$, is differentiable. Consequently, there is at least one value c of x in $a < x < b$ such that $F'(c) = 0$. By differentiating, we see that

$$F'(x) = f'(x) - \frac{f(b) - f(a)}{b - a}$$

hence,

$$F'(c) = f'(c) - \frac{f(b) - f(a)}{b - a} = 0$$

Now, solving for $f'(c)$, we have $f'(c) = \dfrac{f(b) - f(a)}{b - a}$.

Example

Find the points on $f(x) = x^3$ between $A(-2, -8)$ and $B(3, 27)$ at which the tangent is parallel to the secant through A and B.

Solution: In this problem, $a = -2$, $f(a) = -8$, $b = 3$, and $f(b) = 27$. Therefore,

$$\frac{f(b) - f(a)}{b - a} = \frac{27 - (-8)}{3 - (-2)} = 7$$

Consequently, we must determine c so that $f'(c) = 7$. Since $f(x) = x^3$, it follows that $f'(x) = 3x^2$; hence, we must solve $3c^2 = 7$ for c. Thus $c = \pm\sqrt{7}/\sqrt{3} = \pm\sqrt{21}/3$. Therefore, the tangents at $(\sqrt{21}/3, 7\sqrt{21}/9)$ and $(-\sqrt{21}/3, -7\sqrt{21}/9)$ are parallel to the secant through $(-2, -8)$ and $(3, 27)$ as shown in Fig. 14.3.

In applying the law of the mean, we must be certain that both hypotheses are satisfied by the function for the interval under consideration. For example, if $f(x) = x^{1/3}$, $a = -2$, and $b = 3$, then $f'(x) = x^{-2/3}/3$ does not exist for all values of x in the interval between $a = -2$ and $b = 3$. Specifically, $f'(x)$ fails to exist for $x = 0$. Hence, the law of the mean is not valid, even though $f(x) = x^{1/3}$ is continuous in $[-2, 3]$. There is no value c of x in $(-2, 3)$ for which (14.1) holds.

functions with the same derivative We shall now use the law of the mean to prove that *if the two functions $F(x)$ and $G(x)$ have the same derivative for all x in the interval $a < x < b$, then $F(x)$ and $G(x)$ differ at most by a constant; that is $G(x) = F(x) + C$.*

For this purpose we consider the function

$$\phi(x) = G(x) - F(x)$$

Hence,

$$\phi'(x) = G'(x) - F'(x) = 0$$

Now for c any value of x in $a < x < b$, the law of the mean tells us that

$$\phi(x) - \phi(c) = (x - c)\phi'(x_1)$$

where x_1 is a value of x between x and c. Since $\phi'(x_1) = 0$, it follows that $\phi(x) - \phi(c) = 0$. Therefore, since $\phi(x) = G(x) - F(x)$ and $\phi(c)$ is a constant, say C, we see that $G(x) = F(x) + C$ as was to be shown.

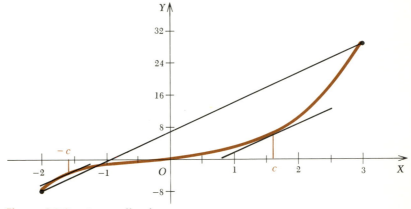

Figure 14.3   An application

Consequently, if $F(x)$ is an integral of $f(x)$ for $a < x < b$, then all integrals are given by

$$\int f(x)\, dx = F(x) + c \tag{14.2}$$

EXERCISE 14.1

Verify the fact that the hypotheses of Rolle's theorem are satisfied by the functions and indicated intervals in Problems 1 to 8. Find each value of c.

1   $f(x) = x^2 - 3x,\ a = 0,\ b = 3$
2   $f(x) = x^2 - 2x - 8,\ a = -2,\ b = 4$
3   $f(x) = 6x^2 - x^3,\ a = 0,\ b = 6$
4   $f(x) = x^3 - 7x + 6,\ a = -3,\ b = 1$
5   $f(x) = x^3 + 5x^2 - x - 5,\ a = -5,\ b = 1$
6   $f(x) = x^3 + 3x^2 - 6x - 8,\ a = -4,\ b = 2$
7   $f(x) = x^4 - 5x^3 + 9x^2 - 7x + 2,\ a = 1,\ b = 2$
8   $f(x) = x^4 - 4x^3,\ a = 0,\ b = 4$

Verify the fact that the hypotheses of the mean value theorem are satisfied by the functions and intervals in Problems 9 to 16. Find each value of c.

9   $f(x) = x^2 - 3x,\ 1 \le x \le 5$
10   $f(x) = 8x - x^2,\ 2 \le x \le 6$
11   $f(x) = 2x^3 - 3x^2 - 12x + 5,\ -4 \le x \le 3$
12   $f(x) = x^3 - 12x + 3,\ -3 \le x \le 4$
13   $f(x) = x^3 - 6x^2 + 9x + 1,\ [0, 3]$
14   $f(x) = x^3 + x^2 - x + 3,\ [-2, 1]$
15   $f(x) = (x - 2)/(x + 1),\ [0, 2]$
16   $f(x) = (2x - 3)/(x + 2),\ [1, 5]$

In each of Problems 17 to 24, either verify that the mean value theorem holds and find each c, or tell why it is not applicable.

17   $f(x) = (x + 2)/(x - 1),\ [0, 2]$
18   $f(x) = (2x - 3)/(3x - 2),\ [0.5, 0.75]$
19   $f(x) = \left\{ \begin{array}{l} 2 + x^2,\ (-\infty, 3] \\ x^2 + x - 1,\ (3, \infty) \end{array} \right\}\quad [1, 4]$
20   $f(x) = \left\{ \begin{array}{l} x^3 - 2x^2 + 1,\ (-\infty, 2) \\ x^3 - x^2 - 3,\ [2, \infty) \end{array} \right\}\quad [1, 3]$
21   $f(x) = x^{2/3},\ -3 \le x \le 1$
22   $f(x) = (x + 3)^{1/5},\ -5 \le x \le 0$
23   $f(x) = (x^2 - 4)^{3/5},\ -3 \le x \le 3$
24   $f(x) = (x^2 - 3)^{5/7},\ -2 \le x \le 2$

The conclusion of the mean value theorem is that c exists in the open interval from a to b such that

$$f(b) = f(a) + (b - a)f'(c)$$

If b is near a, it follows that c is also near a and the approximation $f'(c) \doteq f'(a)$ leads to the approximation

$$f(b) \doteq f(a) + (b - a)f'(a)$$

Use this approximation to calculate the numbers called for in Problems 25 to 28.

25    $\sqrt{26}$ by using $f(x) = \sqrt{x}$, $a = 25$, $b = 26$

26    $\sqrt[3]{28}$ by using $f(x) = \sqrt[3]{x}$, $a = 27$, $b = 28$

27    $\frac{1}{101}$ by using $f(x) = 1/x$, $a = 100$, $b = 101$

28    $1/\sqrt{17}$ by using $f(x) = 1/\sqrt{x}$, $a = 16$, $b = 17$

14.3   The indeterminate forms $0/0$ and ∞/∞, l'Hospital's rule

If the denominator of a fraction is zero and the numerator is not zero, there is no number to represent the fraction. If the numerator and denominator of the fraction $g(x)/h(x)$ are both zero for $x = a$, the

indeterminate form

fraction takes the form $0/0$ and is called an *indeterminate form*. This is not a new idea for us since the ratio $\Delta y/\Delta x$, which is the foundation of differential calculus, is an indeterminate form $0/0$ for $\Delta x = 0$. We shall now investigate the limit, as x approaches a, of $g(x)/h(x)$.

l'Hospital's rule

In fact, we shall prove that *if $g(x)$ and $h(x)$ are two continuous functions that are both differentiable with continuous derivatives for every number other than a in some interval and if $\lim\limits_{x \to a} g(x) = \lim\limits_{x \to a} h(x) = 0$ and $g'(x) \neq 0$ for $x \neq a$, then*

$$\lim_{x \to a} \frac{g(x)}{h(x)} = \lim_{x \to a} \frac{g'(x)}{h'(x)}$$

if the latter limit exists. This theorem is *l'Hospital's rule.*

We shall now prove it for the case in which $\lim\limits_{x \to a} h'(x) \neq 0$. Since $\lim\limits_{x \to a} g(x) = \lim\limits_{x \to a} h(x) = 0$ it follows that

$$\lim_{x \to a} g(x) \equiv \lim_{x \to a} [g(x) - g(a)] \qquad \text{and} \qquad \lim_{x \to a} h(x) \equiv \lim_{x \to a} [h(x) - h(a)]$$

Consequently, by use of the mean value theorem,

$$g(x) - g(a) = (x - a)g'(x_1) \qquad \text{for some } x_1 \text{ in } (a, x)$$

and

$$h(x) - h(a) = (x - a)h'(x_2) \qquad \text{for some } x_2 \text{ in } (a, x)$$

Hence, $$\frac{g(x)}{h(x)} = \frac{g(x) - g(a)}{h(x) - h(a)} = \frac{(x-a)g'(x_1)}{(x-a)h'(x_2)} = \frac{g'(x_1)}{h'(x_2)}$$

As x approaches a, both x_1 and x_2 approach a since both are in (a, x). Therefore,

$$\lim_{x\to a}\frac{g(x)}{h(x)} = \lim_{x\to a}\frac{g'(x_1)}{h'(x_2)} = \lim_{x\to a}\frac{g'(x)}{h'(x)} \qquad h'(a) \neq 0$$

If $h'(a) = 0$ and $g'(a) \neq 0$, then it is customary to say that $\lim\limits_{x\to a}$ $g(x)/h(x) = \infty$. Furthermore, if $g'(a) = h'(a) = 0$, then $g'(x)$ and $h'(x)$ satisfy the conditions of the mean value theorem and

$$\lim_{x\to a}\frac{g(x)}{h(x)} = \lim_{x\to a}\frac{g'(x)}{h'(x)} = \lim_{x\to a}\frac{g''(x)}{h''(x)} = \frac{g''(a)}{h''(a)} \qquad \text{if } h''(a) \neq 0$$

This procedure can be continued as long as the hypotheses are satisfied.

Example   Evaluate $$\lim_{x\to 2}\frac{x^2 - 2x}{x^2 - 6x + 8}$$

Solution: This fraction approaches the form $0/0$ as x approaches 2; hence, it is a function to which we may apply l'Hospital's rule. Thus,

$$\lim_{x\to 2}\frac{x^2 - 2x}{x^2 - 6x + 8} = \lim_{x\to 2}\frac{2x - 2}{2x - 6} = \frac{4 - 2}{4 - 6} = -1$$

This result can also be obtained by taking the common factor $x - 2$ from each member of the fraction. Thus,

$$\lim_{x\to 2}\frac{x^2 - 2x}{x^2 - 6x + 8} = \lim_{x\to 2}\frac{x(x - 2)}{(x - 2)(x - 4)} = \lim_{x\to 2}\frac{x}{x - 4} = -1$$

more indeterminate forms   L'Hospital's rule is also applicable if the fraction $g(x)/h(x)$ takes the form ∞/∞ as x approaches a and if it takes the form $0/0$ or ∞/∞ as x approaches infinity. These facts will be used but the proofs will not be given.

EXERCISE 14.2

Explain why l'Hospital's rule does not apply in each of Problems 1 to 4, and find the limit.

1 $\lim\limits_{x\to 3}\dfrac{x^2 - 7}{x + 3}$

2 $\lim\limits_{x\to 2}\dfrac{x^2 - 4}{x - 1}$

3 $\lim\limits_{x\to 1^+}\dfrac{x^2 + 2}{x - 1}$

4 $\lim\limits_{x\to 3}\dfrac{x^2 - 3x + 1}{x^2 - 6x + 8}$

Evaluate each of the following limits.

5. $\lim\limits_{x \to 2} \dfrac{2x^3 + 2x^2 + 2x + 6}{x^2 + 4x + 3}$

6. $\lim\limits_{x \to 0} \dfrac{x^3}{x^2 - x}$

7. $\lim\limits_{x \to 0} \dfrac{3x^2 + 5x}{4x^2 + 3x}$

8. $\lim\limits_{x \to 1} \dfrac{x^2 + x - 2}{2x^2 - x - 1}$

9. $\lim\limits_{x \to 0} \dfrac{2 - \sqrt{x + 4}}{x}$

10. $\lim\limits_{x \to 0} \dfrac{\sqrt{1 + x} - \sqrt{1 - x}}{x}$

11. $\lim\limits_{x \to 0} \dfrac{\sqrt{9 + x} - 3}{x}$

12. $\lim\limits_{x \to \infty} \dfrac{P(x)}{e^{mx}}$, $P(x)$ a polynomial and $m > 0$

13. $\lim\limits_{x \to 0} \dfrac{\sin x}{x}$

14. $\lim\limits_{x \to 0} \dfrac{1 - \cos x}{x^2}$

15. $\lim\limits_{x \to 0} \dfrac{\sec x - 1}{x}$

16. $\lim\limits_{x \to 0} \dfrac{\tan x}{x}$

17. $\lim\limits_{x \to 0} \dfrac{1 - \cos^4 x}{x^2}$

18. $\lim\limits_{x \to 0} \dfrac{x^2 \cos x}{\sin^2 \frac{1}{2} x}$

19. $\lim\limits_{x \to 0} \dfrac{\sin 4x}{\tan 2x}$

20. $\lim\limits_{x \to 0} \dfrac{\sin x - x}{\tan x - x}$

21. $\lim\limits_{x \to 0} \dfrac{\ln \sec x}{x^2}$

22. $\lim\limits_{x \to \infty} \dfrac{\ln (x + 1)^3}{6x}$

23. $\lim\limits_{x \to 0} \dfrac{8^x - 2^x}{x}$

24. $\lim\limits_{x \to 0} \dfrac{e^x + e^{-x} - 2}{1 - \cos x}$

25. $\lim\limits_{x \to \infty} \dfrac{4x + 5}{5x^2 + 3x}$

26. $\lim\limits_{x \to \infty} \dfrac{6x^3 + 9x}{2x^4 + 5}$

27. $\lim\limits_{x \to \infty} \dfrac{7x^2}{2x^2 - 3}$

28. $\lim\limits_{x \to \infty} \dfrac{e^x}{x^4}$

29. $\lim\limits_{x \to 0} \dfrac{\cot 3x}{\cot x}$

30. $\lim\limits_{x \to \pi/2} \dfrac{\tan 4x}{\tan 2x}$

31. $\lim\limits_{x \to \pi/2} \dfrac{\sec^2 x}{\sec^2 3x}$

32. $\lim\limits_{x = \pi/4} \dfrac{\tan 2x}{\cot (x - \pi/4)}$

33. $\lim\limits_{x \to \infty} \dfrac{\ln x}{x}$

34. $\lim\limits_{x \to \infty} \dfrac{x^n}{e^x}$, n a positive integer

35. $\lim\limits_{x \to \infty} \dfrac{\ln x}{x^a}$, $a > 0$

36. $\lim\limits_{x \to 2} \dfrac{\ln (x - 2)}{\cot \pi x}$

14.4   The indeterminate forms $0 \cdot \infty$ and $\infty - \infty$

At times a function is not in a form to which l'Hospital's rule is applicable but can be changed to such a form by use of an algebraic transformation.

If $g(x)h(x)$ takes the form $0 \cdot \infty$ or $\infty \cdot 0$ as x approaches a, we can write

$$\lim_{x \to a} g(x)h(x) = \lim_{x \to a} \frac{g(x)}{1/h(x)}$$

$$= \frac{0}{0} \text{ or } \frac{\infty}{\infty}$$

according as $g(x)$ approaches zero and $h(x)$ approaches ∞, or $g(x)$ approaches ∞ and $h(x)$ approaches zero. In either case, l'Hospital's rule is applicable to the transformed function.

If $\lim_{x \to a} f(x) = \infty$ and $\lim_{x \to a} h(x) = \infty$, then $g(x) - h(x)$ takes on the

$\infty - \infty$ form $\infty - \infty$ as x approaches a. We can then divide numerator and denominator by $h(x)g(x)$ and have

$$g(x) - h(x) = \frac{[g(x) - h(x)]/g(x)h(x)}{1/g(x)h(x)}$$

$$= \frac{1/h(x) - 1/g(x)}{1/h(x)g(x)}$$

Now as x approaches a, this form of $g(x) - h(x)$ becomes $0/0$. Consequently l'Hospital's rule is applicable. At times it is possible to get $g - h$ into a form to which l'Hospital's rule is applicable by dividing by a simpler expression than gh.

Example   Evaluate $\lim_{x \to 1} (x - 1) \tan (\pi x/2)$.

Solution: Since the given limit is of the form $0 \cdot \infty$, we must change its form before we can apply l'Hospital's rule. If we make use of the reciprocal relation between the tangent and cotangent of an angle, we have

$$\lim_{x \to 1} (x - 1) \tan \frac{\pi x}{2} = \lim_{x \to 1} \frac{x - 1}{\cot (\pi x/2)} = \frac{0}{0}$$

$$= \lim_{x \to 1} \frac{1}{-(\pi/2) \csc^2 (\pi x/2)}$$

$$= -\frac{2}{\pi}$$

since $\csc (\pi x/2) = 1$ and 1 divided by $- \pi/2$ is $-2/\pi$.

14.5   The indeterminate forms 0^0, 1^∞, ∞^0

Each of the indeterminate forms listed in the section heading is of the form $N = f(x)^{g(x)}$ and can be changed to one of the forms already considered by use of logarithms. If $N = f(x)^{g(x)}$, then $\ln N = \ln f(x)^{g(x)} = g(x) \ln f(x) = 0 \cdot -\infty$, $\infty \cdot 0$, or $0 \cdot \infty$ which can be expressed in a form to which l'Hospital's rule is applicable by dividing $\ln f(x)$ by $1/g(x)$ instead of multiplying by $g(x)$.

use of logarithms with f^g

Example 1   Evaluate $N = \lim_{x \to 0} x^x$.

Solution: If we take the logarithm of each member, we have

$$\ln N = \lim_{x \to 0} \ln x^x = \lim_{x \to 0} x \ln x = 0 \cdot -\infty$$

$$= \lim_{x \to 0} \frac{\ln x}{x^{-1}} = \frac{0}{0}$$

Now, applying l'Hospital's rule, we get

$$\ln N = \lim_{x \to 0} \frac{1/x}{-1/x^2} = \lim_{x \to 0} (-x) = 0$$

Therefore, $$N = \lim_{x \to 0} x^x = e^0 = 1$$

Example 2   Evaluate $N = \lim_{y \to 0} (\csc y)^{\sin y}$.

Solution: Since the limit takes the form ∞^0, it is an exponential indeterminate form, and we begin by taking the logarithm of N. Thus,

$$\ln N = \lim_{y \to 0} \ln (\csc y)^{\sin y}$$

$$= \lim_{y \to 0} \sin y \ln \csc y = 0 \cdot \infty$$

$$= \lim_{y \to 0} \frac{\ln \csc y}{\csc y} = \frac{\infty}{\infty} \qquad \textit{since } \sin y \csc y = 1$$

This is a form to which l'Hospital's rule is applicable. Therefore,

$$\ln N = \lim_{y \to 0} \frac{-\csc y \cot y/\csc y}{-\csc y \cot y} = \lim_{y \to 0} \frac{1}{\csc y} = 0$$

Consequently, $$N = \lim_{y \to 0} (\csc y)^{\sin y} = e^0 = 1$$

Example 3   Evaluate $N = \lim_{x \to 0^+} (1 + ax)^{1/x}$.

Solution: The use of 0^+ in the limit indicates that x approaches zero through values greater than zero. The limit takes the exponential form 1^∞; hence, we begin by taking the logarithm of N. Thus,

$$\ln N = \lim_{x \to 0^+} \ln (1 + ax)^{1/x} = \lim_{x \to 0^+} \frac{1}{x} \ln (1 + ax) = \infty \cdot 0$$

$$= \lim_{x \to 0^+} \frac{\ln (1 + ax)}{x} = \frac{0}{0}$$

Now we use l'Hospital's rule and get

$$\ln N = \lim_{x \to 0^+} \frac{a/(1 + ax)}{1} = a$$

Consequently, $$N = \lim_{x \to 0^+} (1 + ax)^{1/x} = e^a$$

EXERCISE 14.3

Evaluate each of the following limits.

1   $\lim_{x \to \infty} (x^2 - x)$

2   $\lim_{x \to \infty} (x - \sqrt{x^2 + 1})$

3   $\lim_{x \to 2^+} \left(\dfrac{1}{x - 2} - \dfrac{1}{\sqrt{x - 2}} \right)$

4   $\lim_{x \to 0} \left(\dfrac{1}{x^2} - \dfrac{1}{x} \right)$

5   $\lim_{x \to \pi/2} (\sec x - \tan x)$

6   $\lim_{x \to 0} \left(\dfrac{1}{\sin^2 x} - \dfrac{1}{x^2} \right)$

7   $\lim_{x \to 0} \left(\dfrac{1}{x} - \dfrac{1}{\text{Arctan } x} \right)$

8   $\lim_{x \to \pi/2^+} \left(\dfrac{1}{\cos x} - \dfrac{1}{x - \pi/2} \right)$

9   $\lim_{x \to \infty} x \sin (1/x)$

10   $\lim_{x \to 1} (x - 1) \tan (\pi x/2)$

11   $\lim_{x \to 0^+} x \ln x$

12   $\lim_{x \to \pi/2} \sec x \cos 3x$

13   $\lim_{x \to 0^+} (1 + 1/x)^x$

14   $\lim_{x \to \infty} x^{1/x}$

15   $\lim_{x \to \infty} (x + e^x)^{2/x}$

16   $\lim_{x \to 0^+} (\cot x)^{\sin x}$

17   $\lim_{x \to 0^+} x^{\sin x}$

18   $\lim_{x \to 0^+} (\sin x)^x$

19   $\lim_{x \to 2} (x^2 - 4)^{x-2}$

20   $\lim_{x \to 0^+} (x + \sin x)^{\tan x}$

21   $\lim_{x \to \infty} (1 + 1/x)^x$

22   $\lim_{x \to \pi/2} (1 + \cos x)^{2 \sec x}$

23   $\lim_{x \to 0} (1 - 3x)^{5/x}$

24   $\lim_{x \to 1^+} (x - 1)^{1/(2x-2)}$

15

Polar coordinates and parametric equations

15.1   The polar-coordinate system

pole

polar axis

vectorial angle

radius vector

The frame of reference for the polar-coordinate system is a ray such as *OS* in Fig. 15.1 that originates at a point *O* and usually extends to the right. The point *O* is called the *pole* and the ray is the *polar axis*. If we choose a point *P* in the plane and connect the point to the pole with the straight line segment *OP*, we have the angle *SOP*. We call this angle the *vectorial angle* and designate it by the Greek letter θ; we shall require that the polar axis be the initial side.

The angle θ is positive if the direction of rotation from the polar axis to *OP* is counterclockwise and negative if the direction is clockwise. For example, if in Fig. 15.1 the magnitudes of the angles *SOP* and *SOT* are equal, then angle *SOP* = θ is positive and angle *SOT* = $-\theta$ is negative. The line segment *OP* is called the *radius vector* of *P*. If the length of the line segments *OP* and *OT* is *r*, then we say that the polar coordinates of *P* are (r, θ) and those of *T* are $(r, -\theta)$. Consequently, the position of a point is fully determined if the radius vector *r* and the vectorial angle θ are given. Furthermore, any point in the plane determines (1) a unique distance *r* from the pole to the point and (2) an angle θ from the polar axis to the line joining the pole to the point.

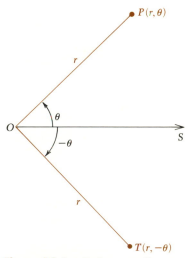

Figure 15.1   Polar coordinates

In order to plot the point whose polar coordinates are $(10, 55°)$, we first construct the ray OR in Fig. 15.2 making the positive angle $55°$ with the polar axis OS, and then, starting at O, we lay off 10 units on OR and thus arrive at the point P. In Fig. 15.2 we show two other points together with polar coordinates of each.

In the above discussion we assumed that the radius vector r was positive, and we measured the distance r on the terminal side of the vectorial angle. If r stands for a negative number, we extend the terminal side of the vectorial angle through the pole and then lay off

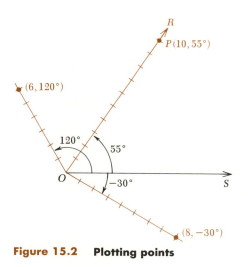

Figure 15.2   Plotting points

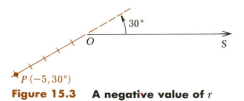

Figure 15.3   A negative value of r

the distance $|r|$ on the ray thus obtained. For example, the point P with the polar coordinates $(-5, 30°)$ is shown in Fig. 15.3.

We now call attention to the fact that the ordered pair (r, θ) determines a unique point P in Fig. 15.4. In contrast to this, however, a point may have an unlimited number of pairs of polar coordinates. For example, four pairs of polar coordinates of the point P in Fig. 15.4 are $(7, 30°)$, $(7, -330°)$, $(-7, 210°)$, and $(-7, -150°)$. In each of these pairs the absolute value of the vectorial angle is less than $360°$. If the vectorial angles in $(7, 30°)$ and $(-7, 210°)$ are increased by $360°$ or 2π radians and the angle in $(7, -330°)$ and $(-7, 150°)$ are decreased by $360° = 2\pi$ radians, four additional pairs of polar coordinates of P are obtained. Consequently, P has an unlimited number of pairs of polar coordinates.

Example   Plot the points whose polar coordinates are (*a*) $(10, 30°)$, (*b*) $(8, -60°)$, (*c*) $(-5, 70°)$, and (*d*) $(-7, -120°)$.

Solution: We first construct the vectorial angle. Next, if the radius vector is positive, we lay off the given distance on the terminal side of the angle. If the radius vector is negative, we produce the terminal side of the angle through the pole and lay off the given distance on the ray thus obtained. The four points are shown in Fig. 15.5.

15.2   The graph from a polar equation

The graph of the equation $f(r, \theta) = 0$ is the set of points whose polar coordinates satisfy the equation. As in the case of cartesian coordinates,

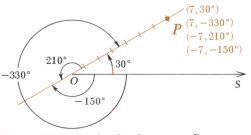

Figure 15.4   Lack of unique coordinates

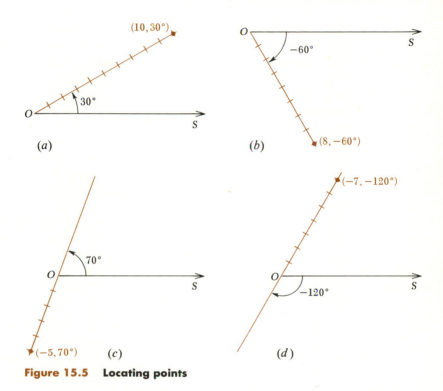

Figure 15.5 Locating points

we employ the following procedure for obtaining the graph of an equation in polar coordinates.

finding the graph

1. Assign a set of values to one of the variables in the equation, and obtain each corresponding value of the other variable. Usually it is more convenient to assign values to θ.
2. Plot the point determined by each pair of corresponding values obtained in step 1.
3. Draw a smooth curve through the set of points obtained in step 2.

We shall illustrate the procedure with the following examples.

Example 1 Plot the graph of the equation $r = 4 \cos \theta$.

Solution: We shall assign values to θ at intervals of $15°$, starting with $0°$ and ending with $180°$, compute each corresponding value of r, and tabulate the results below.

θ	0°	15°	30°	45°	60°	75°	90°	105°	120°	135°	150°	165°	180°
$\cos \theta$	1	0.97	0.87	0.71	0.50	0.26	0	−0.26	−0.50	−0.71	−0.87	−0.97	−1
r	4	3.9	3.5	2.8	2.0	1.0	0	−1.0	−2.0	−2.8	−3.5	−3.9	−4

Now we plot the points determined by the corresponding pairs of values of r and θ, draw a smooth curve through them, and thus obtain the graph in Fig. 15.6. It should be noticed that if $0 \leq \theta \leq 90°$, then $r \geq 0$; hence, the distance r is measured on the terminal side of the angle. Therefore, the points (r, θ), $0 \leq \theta \leq 90°$ are either on or above the polar axis. If, however, $90° < \theta \leq 180°$, then r is negative; hence, the distance $|r|$ is measured on the extension through the pole of the terminal side of θ. Therefore, the points (r, θ), $90° < \theta < 180°$ are below the polar axis. If $\theta = 180°$, $r = -4$, and the point $(-4, 180°)$ coincides with the point $(4, 0°)$. Furthermore, if we assign values greater than $180°$ to θ, we obtain no additional points. For example, if $\theta = 225°$, $\cos \theta = \cos(180° + 45°) = -\cos 45° = -0.71$; hence, $r = -2.8$. It is readily verified that the point $(-2.8, 225°)$ coincides with the point $(2.8, 45°)$. The curve in Fig. 15.6 appears to be a circle, and in fact, it is as we shall prove in a later section.

Example 2   Construct the graph of $r = 1 - \cos \theta$ for θ in the interval $0° \leq \theta \leq 180°$.

Solution: We shall assign values at intervals of $15°$ from 0 to $180°$ to θ, and calculate each corresponding value of r. For example, if $\theta = 30°$, $\cos \theta = 0.87$, and $r = 1 - 0.87 = 0.13$. Furthermore, if $\theta = 135°$, $\cos \theta = -0.71$ and $r = 1 - (-0.71) = 1.71$. The corresponding values of θ and r are tabulated.

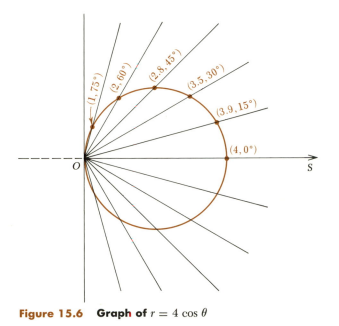

Figure 15.6   Graph of $r = 4 \cos \theta$

θ	0°	15°	30°	45°	60°	75°	90°	105°	120°	135°	150°	165°	180°
r	0	0.03	0.13	0.29	0.5	0.74	1	1.26	1.5	1.71	1.87	1.97	2

We next plot the points determined by the corresponding ordered number pairs in the table, connect them with a smooth curve, and get the graph in Fig. 15.7.

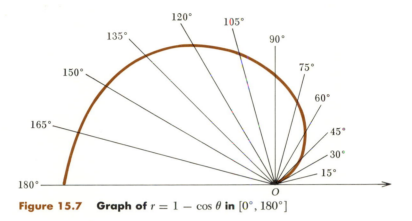

Figure 15.7   **Graph of $r = 1 - \cos \theta$ in $[0°, 180°]$**

EXERCISE 15.1

Plot the points whose coordinates are given in Problems 1 to 4.

1   $(5, 45°), (-5, 45°), (5, -45°), (-5, -45°)$
2   $(3, 60°), (-3, 60°), (3, -60°), (-3, -60°)$
3   $(4, 120°), (-4, 120°), (4, -120°), (-4, -120°)$
4   $(6, 135°), (-6, 135°), (6, -135°), (-6, -135°)$

In each of Problems 5 to 8 find three other pairs of polar coordinates with $|\theta| < 360°$ that represent the same point.

5   $(5, 40°)$ 6 $(-4, 60°)$ 7 $(3, -30°)$ 8 $(-4, -120°)$

In Problems 9 to 12 the polar axis, or its extension through the pole O, is the perpendicular bisector of the line segment BA at the point C, and the point A is above B. In each problem, find the polar coordinates of A and B.

9   The point C is to the right of O, $BA = 2$, and $OC = 1$.
10   The point C is to the right of O, $BA = 2\sqrt{3}$, and $OC = 1$.
11   The point C is to the left of O, $OB = 2$, and $OC = 1$.
12   The point C is to the left of O, $BA = 2$, and $OC = \sqrt{3}$.

Plot the graphs of the following polar equations. Assign integral multiples of 30° to θ, unless smaller intervals are needed to determine the nature of the curve.

13   $r = 6$ 14 $r = 4$ 15 $r + 8 = 0$
16   $r + 3 = 0$ 17 $\theta = 45°$ 18 $\theta = 60°$
19   $\theta = -30°$ 20 $\theta = -90°$ 21 $r = \sin \theta$
22   $r = -\cos \theta$ 23 $r = \sin(180° + \theta)$
24   $r = \sin(90° - \theta)$ 25 $r = 1 + \cos \theta$
26   $r = 1 + \sin \theta$ 27 $r = 1 - \sin \theta$
28   $r = \cos \theta - 1$ 29 $r \sin \theta = 1$
30   $r \cos \theta = 1$
31   $r \sin(30° + \theta) = 4,\ -60° \le \theta \le 120°$
32   $r \cos(30° - \theta),\ -30° \le \theta \le 90°$
33   $r = 6/(3 + 2 \cos \theta),\ 0 \le \theta \le 180°$
34   $r = 6/(3 - 2 \cos \theta),\ 0 \le \theta \le 180°$
35   $r = 6/(3 + 2 \sin \theta),\ -90° \le \theta \le 90°$
36   $r = 6/(3 - 2 \cos \theta),\ -90° \le \theta \le 90°$

15.3   Intercepts and symmetry

In this chapter we shall discuss methods that often enable us to sketch the graph of a polar equation without recourse to point-by-point plotting. We shall find it helpful to extend the polar axis through the pole and to construct a line through the pole perpendicular to the *normal axis* polar axis as in Fig. 15.8. We shall call the latter line the *normal axis.*

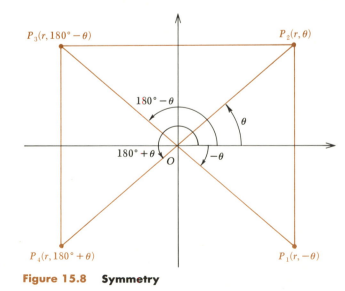

Figure 15.8   Symmetry

The first step in obtaining the sketch of the graph of a polar equation is to determine the intersections of the curve and the coordi- *intercepts* nate axes. The radius vector of each such point is an *intercept*. To obtain the intercepts of a given equation, we assign successively the integral multiples of 90° to θ and solve the resulting equation for r. Frequently, it is not necessary to assign values greater than 360° to θ since the values of r corresponding to such angles are often the same as those already obtained. We shall illustrate the procedure.

Example 1   Find the intercepts of the graph defined by $r = 1 + \sin \theta$.

Solution: In this problem it is not necessary to assign values greater than 360° to θ since $\sin \theta$ is periodic with 360° as a period. Consequently, we obtain the intercepts by assigning 0, 180, 90, and 270° to θ. Thus we get $r = 1 + \sin 0 = 1 + 0 = 1$, $r = 1 + \sin 180° = 1 + 0 = 1$, $r = 1 + \sin 90° = 1 + 1 = 2$, and $r = 1 + \sin 270° = 1 + (-1) = 0$. Consequently the intercepts on the polar axis and on its extension are 1. The intercepts on the normal axis are 2 and 0.

The next step in the procedure for sketching a curve is to determine the axes of symmetry. It should be recalled that a curve is symmetric with respect to a line if the line bisects every chord of the curve that is perpendicular to it. Furthermore, a curve is symmetric with respect to a point if the point bisects every chord of the curve that passes through it. There are many tests that enable us to determine the axes, or points, of symmetry of a curve. The following three, however, will suffice for our purposes.

tests for
symmetry

1.   If the equation $r = f(\theta)$ is unchanged when θ is replaced by $-\theta$, the curve is symmetric with respect to the polar axis or its extension.
2.   If $r = f(\theta)$ is not changed when θ is replaced by $180° - \theta$, the curve is symmetric to the normal axis.
3.   If $r = f(\theta)$ is not changed when θ is replaced by $180° + \theta$, the curve is symmetric with respect to the pole.

We shall refer to Fig. 15.8 in proving these statements.

Proof of 1:   Since $r = f(\theta) = f(-\theta)$, the points $P_1(r, -\theta)$ and $P_2(r, \theta)$ are the extremities of a chord of the curve for each value of θ. Furthermore, the triangle P_1OP_2 is isosceles, and the polar axis bisects the angle at O. Therefore, the polar axis is the perpendicular bisector of the chord P_1P_2. Consequently, the curve is symmetric with respect to the polar axis.

Proof of 2:   Here we have $r = f(\theta) = f(180° - \theta)$, and it follows that the triangle P_2OP_3 is isosceles with the angle at O bisected by the normal axis. Consequently, the normal axis is the perpendicular bisector of the chord P_3P_2, and the curve is symmetric with respect to the normal axis.

Proof of 3: Since $r = f(\theta) = f(180° + \theta)$, the points $P_4(r, 180° + \theta)$, O, and $P_2(r, \theta)$ are on a straight line segment with O as the midpoint. Hence the pole O bisects the chord P_2P_4 and the curve is symmetric with respect to the origin.

As previously stated, these tests do not exhaust the possibilities.

Example 2   Apply the tests for symmetry to the equation $r = 1 + \sin \theta$.

Solution: If we replace θ by $-\theta$ in the given equation, we get

$$r = 1 + \sin (-\theta)$$
$$= 1 - \sin \theta \qquad \textit{since} \sin (-\theta) = -\sin \theta$$

Hence, since the equation is changed by this replacement, this test fails to reveal symmetry with respect to the polar axis or to its extension.

If θ is replaced by $180° - \theta$, we have

$$r = 1 + \sin (180° - \theta)$$
$$= 1 + \sin \theta \qquad \textit{since} \sin (180° - \theta) = \sin \theta$$

Consequently, the equation is not changed by this replacement, and the graph is symmetric with respect to the normal axis.

The replacement of θ by $180° + \theta$ yields

$$r = 1 + \sin (180° + \theta)$$
$$= 1 - \sin \theta \qquad \textit{since} \sin (180° + \theta) = -\sin \theta$$

Therefore this test fails to reveal symmetry with respect to the pole.

We conclude this discussion by stating the following four steps in the procedure for sketching the graph of a polar equation:

procedure to obtain the graph

1. Determine the intercepts of the graphs and plot the point corresponding to each.
2. Apply the three tests given for symmetry.
3. If necessary, determine the coordinates of a few additional points on the graph, and locate the points.
4. Sketch the curve.

Example 3   Sketch the graph of $r = 1 + \sin \theta$.

Solution: In Example 1 we found that the intercepts of the graph of this equation on the polar axis and its extension are each 1 and that

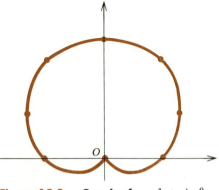

Figure 15.9   Graph of $r = 1 + \sin \theta$

the normal intercepts are 0 and 2. We plot these points as indicated in Fig. 15.9. In Example 2 we found that the graph is symmetric with respect to the normal axis. We now assign -60, -30, 30, and 60° to θ, compute each corresponding value of r, and obtain the following table:

θ	$-60°$	$-30°$	$30°$	$60°$
r	0.1	0.5	1.5	1.9

After plotting these points and drawing a smooth curve through them and those determined by the polar and normal intercepts, we obtain the curve to the right of the normal axis in Fig. 15.9. Since the normal axis is an axis of symmetry, the left half of the curve is a reflection of the right half with respect to the normal axis, as indicated in Fig. 15.9.

Frequently we can obtain an approximate sketch of a curve by considering the variation of r as θ increases through certain intervals. We shall illustrate the procedure in Example 4.

Example 4   Sketch the graph of the equation $r = \cos 2\theta$.

Solution: The tests for symmetry reveal the fact that the curve is symmetric with respect to both axes and the pole. We shall discover the intercepts as we proceed with the following argument.

As θ varies from 0 to 45°, 2θ varies from 0 to 90°, and $r = \cos 2\theta$ decreases from 1 to 0. Hence the portion of the curve yielded by corresponding values of r and θ in this interval is numbered 1 in Fig. 15.10. Similarly, the information tabulated below enables us to sketch the portions of the curve indicated.

Vectorial angle	Radius vector	No. of portion of curve in Fig. 15.10
$45° \leq \theta \leq 90°$	r decreases from 0 to -1	2
$90° \leq \theta \leq 135°$	r increases from -1 to 0	3
$135° \leq \theta \leq 180°$	r increases from 0 to 1	4

Similarly, by considering the variation of r as θ passes through the intervals $180° \leq \theta \leq 225°$, $225° \leq \theta \leq 270°$, $270° \leq \theta \leq 315°$, and $315° \leq \theta \leq 360°$, we obtain the portions of the graph numbered 5, 6, 7, and 8 in the figure.

In the above discussion we made no use of the knowledge that we have about the symmetry of the curve. If we use this information, we need only consider the intervals $0 \leq \theta \leq 45°$ and $45° \leq \theta \leq 90°$ and obtain the curves numbered 1 and 2. Then since the graph is symmetric with respect to the polar axis, we can sketch curves 7 and 8, and finally the curve numbered 4, 5, 6, and 3 by the use of symmetry with respect to the normal axis.

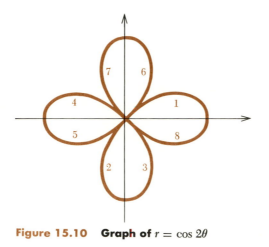

Figure 15.10   Graph of $r = \cos 2\theta$

EXERCISE 15.2

Find the intercepts of the curves whose equations are given in Problems 1 to 32, apply the tests for symmetry to each, and sketch the curve.

1 $r = \sin \theta$	2 $r = -\sin \theta$	3 $r = \cos \theta$
4 $r = -4 \cos \theta$	5 $r = \sin 2\theta$	6 $r = -\cos 2\theta$
7 $r = \sin 3\theta$	8 $r = \cos 3\theta$	9 $r = 1 + 2 \sin \theta$

10   $r = 1 + 2 \cos \theta$ 11 $r = 1 - 2 \sin \theta$
12   $r = 1 - 2 \cos \theta$ 13 $r = \sin^2 \theta$
14   $r = \cos^2 \theta$ 15 $r^2 = 4 \sin 2\theta$
16   $r^2 = 4 \cos 2\theta$ 17 $r^2 = 9 \sin^2 \theta + 16 \cos^2 \theta$
18   $r^2 = \cos^2 \theta - 9 \sin^2 \theta$
19   $r = 4 \cot \theta / \sin \theta$ 20 $r = 6 \tan \theta / \cos \theta$
21   $r = 6/(2 + \cos \theta)$ 22 $r = 12/(3 + \cos \theta)$
23   $r = 10/(3 - 2 \sin \theta)$ 24 $r = 30/(4 - \sin \theta)$
25   $r = 6/(1 - \cos \theta)$ 26 $r = 4/(1 + \cos \theta)$
27   $r = 4/(1 - \sin \theta)$ 28 $r = 6/(1 + \sin \theta)$
29   $r = 6/(1 + 2 \cos \theta)$ 30 $r = 6/(1 + 2 \sin \theta)$
31   $r = 7/(3 - 4 \cos \theta)$ 32 $r = 7/(3 + 4 \sin \theta)$

15.4   Relations between polar and rectangular coordinates

Frequently it is desirable to transform an equation from one system of coordinates to an equation in the other system. We accomplish this by use of one of the relations that we shall next develop.

As indicated in Fig. 15.11 we shall superimpose the polar plane on the cartesian plane so that the pole in the former coincides with the origin in the latter and the polar axis coincides with the positive part of the X axis. Next we choose a point P in the plane with polar coordinates (r, θ) and cartesian coordinates (x, y). Now we see from the figure that

rectangular to polar coordinates

$$x = r \cos \theta \qquad \text{and} \qquad y = r \sin \theta \qquad (15.1)$$

We employ Eqs. (15.1) to transform an equation from cartesian coordinates into polar coordinates, as illustrated in the following examples.

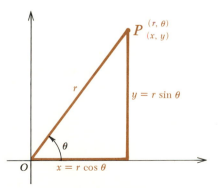

Figure 15.11   Rectangular and polar coordinates

Example 1   Transform the equation $x^2 + y^2 - x + 3y = 3$ into polar coordinates.

Solution: We replace x by $r \cos \theta$ and y by $r \sin \theta$ and get

$$r^2 \cos^2 \theta + r^2 \sin^2 \theta - r \cos \theta + 3r \sin \theta = 3$$

We now simplify this equation by the following steps:

$$r^2(\cos^2 \theta + \sin^2 \theta) - r(\cos \theta - 3 \sin \theta) = 3 \qquad \textit{by the distributive axiom}$$

$$r^2 - r(\cos \theta - 3 \sin \theta) = 3 \qquad \textit{since } \sin^2 \theta + \cos^2 \theta = 1$$

Example 2   Prove that a polar form of $y^2 - 3x^2 + 12x = 9$ is $r = 3/(1 + 2 \cos \theta)$.

Solution: We replace x and y by $r \cos \theta$ and $r \sin \theta$, respectively, and get

$$r^2 \sin^2 \theta - 3r^2 \cos^2 \theta + 12r \cos \theta = 9$$

$$r^2 - r^2 \cos^2 \theta - 3r^2 \cos^2 \theta + 12r \cos \theta - 9 = 0 \qquad \textit{replacing } \sin^2 \theta \textit{ by } 1 - \cos^2 \theta$$

$$r^2 - 4r^2 \cos^2 \theta + 12r \cos \theta - 9 = 0 \qquad \textit{combining terms}$$

$$r^2 - (4r^2 \cos^2 \theta - 12r \cos \theta + 9) = 0 \qquad \textit{by the distributive axiom}$$

We now factor the left member of the latter equation as indicated below.

$$r^2 - (2r \cos \theta - 3)^2 = 0$$

$$(r + 2r \cos \theta - 3)(r - 2r \cos \theta + 3) = 0 \qquad (1)$$

Consequently, any pair of numbers that satisfies either of the following equations also satisfies (1).

$$r + 2r \cos \theta - 3 = 0 \qquad (2)$$

$$r - 2r \cos \theta + 3 = 0 \qquad (3)$$

Therefore the graph of (1) is the curve composed of the graphs of (2) and (3). We shall now show that the graph of (3) is the same as the graph of (2).

We first note that (r, θ) and $(-r, 180° + \theta)$ are coordinates of the same point. Hence if $P(r', \theta')$ is on the graph of (2), we have

$$r' + 2r' \cos \theta' - 3 = 0 \qquad (4)$$

Furthermore, P is also a point on the graph of (3) since $(-r', 180° + \theta')$ satisfy (3), as we shall show next. If we replace (r, θ) by $(-r', 180° + \theta')$ in the left member of (3), we have

$$-r' + 2r' \cos (180° + \theta') + 3$$

$$= -r' - 2r' \cos \theta' + 3 \qquad \textit{since } \cos (180° + \theta') = -\cos \theta'$$

$$= -(r' + 2r' \cos \theta' - 3)$$

$$= -(0) \qquad \textit{by (4)}$$

$$= 0$$

Consequently any point on the graph of (2) is also a point on the graph of (3). By a similar argument we can prove that any point on the graph of (3) is also a point on the graph of (2). Therefore, the graphs of (2) and (3) coincide and form one curve; furthermore, this curve is the graph of (1). Hence, in the remainder of this discussion we shall consider only Eq. (2).

If $\theta \neq 2\pi/3$, we can solve (2) for r and get

$$r = \frac{3}{1 + 2 \cos \theta} \tag{5}$$

and this is the polar form required.

We next note that if $\theta = 2\pi/3$, no value of r exists. If, however, θ approaches $2\pi/3$, $|1 + 2 \cos \theta|$ approaches zero. Hence $|r|$ increases indefinitely. It follows that the graph of $\theta = 2\pi/3$ is an asymptote of the graph of (5).

In order to transform an equation that is in polar form to one in cartesian form, we employ one or more of the following relations. Each is evident from Fig. 15.11.

polar to rectangular coordinates

$$\sin \theta = \frac{y}{r} \qquad \cos \theta = \frac{x}{r} \qquad r = \sqrt{x^2 + y^2} \tag{15.2}$$

$$\theta = \arctan \frac{y}{x} \qquad r = \sqrt{x^2 + y^2} \tag{15.3}$$

If the equation involves sines, cosines, secants, or cosecants, we use (15.2). It is usually advisable to replace $\sin \theta$ and $\cos \theta$ by y/r and x/r, respectively, as a first step, then to replace r by $\sqrt{x^2 + y^2}$ and simplify the result. If the resulting equation involves radicals of the second order, it should be rationalized if possible.

Example 3 Transform the equation $r = 4/(2 - 3 \sin \theta)$ to an equation in cartesian coordinates.

Solution: We proceed as follows:

$$r = \frac{4}{2 - 3y/r} \qquad \text{\textit{replacing} $\sin \theta$ \textit{by} y/r}$$

$$= \frac{4r}{2r - 3y} \qquad \text{\textit{simplifying the complex fraction}}$$

$$2r - 3y = 4 \qquad \text{\textit{multiplying each member by} $\frac{2r - 3y}{r}$}$$

$$2\sqrt{x^2 + y^2} = 4 + 3y \qquad \text{\textit{replacing} r \textit{by} $\sqrt{x^2 + y^2}$ \textit{and adding} $3y$ \textit{to each member}}$$

$$4x^2 + 4y^2 = 9y^2 + 24y + 16 \qquad \text{\textit{equating the squares of the members}}$$

$$4x^2 - 5y^2 - 24y = 16 \qquad \text{\textit{adding} $-9y^2 - 24y$ \textit{to each member}}$$

If the polar equation involves tangents and cotangents, it is advisable to use Eqs. (15.3).

Example 4 Change the equation $r = \tan \theta + \cot \theta$ into an equation in cartesian coordinates.

Solution: Since $\tan \theta \neq -\cot \theta$, it follows that $r \neq 0$. Hence, the graph of $r = \tan \theta + \cot \theta$ does not pass through the pole. Therefore in the transformation of this equation to cartesian coordinates, we must remember that $(x, y) \neq (0, 0)$. By (15.3), $\arctan (y/x) = \theta$, and it follows that $\tan \theta = y/x$ and $\cot \theta = x/y$. Therefore,

$$\sqrt{x^2 + y^2} = \frac{y}{x} + \frac{x}{y}$$ *replacing r by $\sqrt{x^2 + y^2}$, $\tan \theta$ by y/x, and $\cot \theta$ by x/y*

$$xy\sqrt{x^2 + y^2} = x^2 + y^2$$ *multiplying each member by xy*

$$x^2y^2 = x^2 + y^2$$ *equating the squares of the members and dividing by $x^2 + y^2$*

EXERCISE 15.3

Transform the equation in each of Problems 1 to 20 to an equation in polar coordinates.

1 $x + 4y = 2$	2 $3x - 4y = 1$	3 $2x = 1 + 5y$
4 $ax + by = c$	5 $x^2 = 4y$	6 $y^2 = 6x$
7 $(x - 1)^2 = 4y + 1$	8 $(y - 2)^2 = -2x + 4$	
9 $x^2 + y^2 - 2x = 0$	10 $x^2 + y^2 - 4y = 0$	
11 $9(x - 1)^2 + y^2 = 9$	12 $x^2 + 4(y - 1)^2 = 4$	
13 $x^2 + y^2 = 9$	14 $x^2 - y^2 = 16$	
15 $xy = 8$	16 $x^2 - 2xy - y^2 = 1$	
17 $y^2 + 6x = 9$	18 $x^2 - 2y = 1$	
19 $3x^2 + 4y^2 + 6x - 9 = 0$	20 $x^2 - 3y^2 + 8y = 4$	

Transform the equation in each of Problems 21 to 40 to an equation in cartesian coordinates.

21 $r = 2 \sec \theta$	22 $r = 4 \csc \theta$
23 $r = 1/(\sin \theta + 2 \cos \theta)$	24 $r = 3/(2 \sin \theta - 5 \cos \theta)$
25 $r = 4 \cos \theta$	26 $r = -3 \sin \theta$
27 $r = \tan \theta$	28 $r = -\cot \theta$
29 $r = \sin \theta + \cos \theta$	30 $r = \sec \theta + \csc \theta$
31 $r = \tan \theta + \cot \theta$	32 $r = 2 \cos \theta - \sin \theta$
33 $r^2 = 2/\sin 2\theta$	34 $r^2 = 1/\cos 2\theta$
35 $r^2 = \tan 2\theta$	36 $r^2 = 1/(\sin 2\theta + \cos 2\theta)$
37 $r = 1/(1 + \cos \theta)$	38 $r = 1/(1 - \sin \theta)$
39 $r = 2/(1 - 3 \cos \theta)$	40 $r = 6/(2 + 3 \sin \theta)$

15.5   Lines and circles in polar coordinates

In order to find the equation of a line in polar coordinates, we shall consider any line L as in Fig. 15.12, let the length of the normal ON from the pole to the line be p, let the angle from the polar axis to the normal be ω, and let $P(r, \theta)$ be any point on the given line. Therefore, angle NOP is $\theta - \omega$, and we have

$$\cos(\theta - \omega) = \frac{p}{r}$$

from the right triangle ONP. Consequently,

equation of a line

$$r\cos(\theta - \omega) = p \qquad (15.4)$$

is the polar equation of a line that is a distance p from the pole and whose normal makes an angle ω with the polar axis.

The rectangular equation of a line is often more convenient than the polar equation (15.4), but there are four special cases of (15.4) that are worthy of consideration.

If $\omega = 0$ or π, the line is parallel to the normal axis and has the equation

$$r\cos\theta = p \qquad \text{for } \omega = 0 \qquad (15.4a)$$

special cases   and

$$r\cos\theta = -p \qquad \text{for } \omega = \pi \qquad (15.4b)$$

If $\omega = \pi/2$ or $3\pi/2$, the line is parallel to the polar axis and has the equation

$$r\sin\theta = p \qquad \text{for } \omega = \pi/2 \qquad (15.4c)$$

and

$$r\sin\theta = -p \qquad \text{for } \omega = 3\pi/2 \qquad (15.4d)$$

The last four equations are $x = p$, $x = -p$, $y = p$, and $y = -p$, respectively, in rectangular form.

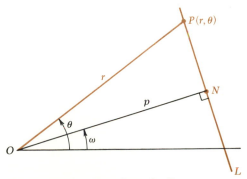

Figure 15.12   Equation of a line

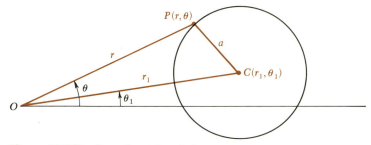

Figure 15.13   Equation of a circle

We shall now derive the equation of a circle in polar form. We shall consider a circle of radius a and center at $C(r_1, \theta_1)$. If we let $P(r, \theta)$ be any point on the circle and use the law of cosines in connection with triangle OPC in Fig. 15.13, we see that

$$a^2 = r^2 + r_1{}^2 - 2r_1 r \cos(\theta - \theta_1) \tag{15.5}$$

equation of a circle   *is the polar form of the equation of a circle of radius a and with center at (r_1, θ_1).*

Several special cases of this equation are simple and useful.

If the center is at the pole, then $r_1 = \theta_1 = 0$ and the equation is

$$r = a \tag{15.5a}$$

If the center is at $(a, 0°)$ or (a, π), the equation is

special cases

$$r = 2a \cos \theta \qquad \text{or} \qquad r = -2a \cos \theta \tag{15.5b}$$

If the center is at $(a, \pi/2)$ or $(a, 3\pi/2)$, the equation is

$$r = 2a \sin \theta \qquad \text{or} \qquad r = -2a \sin \theta \tag{15.5c}$$

Example   Describe the graphs of the following equations: (a) $r \cos \theta = 6$, (b) $r \sin \theta = -8$, (c) $r = 12 \cos \theta$, (d) $r = -10 \sin \theta$.

Solution: The graphs are: (a) a straight line perpendicular to the polar axis and 6 units to the right of the pole; (b) a straight line parallel to the polar axis and 8 units below the pole; (c) a circle of radius 6 with the center at $(6, 0°)$; and (d) a circle of radius 5 with the center at $(5, -90°)$.

15.6   Polar equations of the conics

In order to obtain the equation of a noncircular conic in polar form, we shall make use of the definition in terms of eccentricity. It states that the set of points $P(r, \theta)$ in a plane such that the ratio of the

conic distance of each from a fixed point to its distance from a fixed line is
directrix the same constant is called a *conic*. The fixed point and fixed line are
called the *focus* and *directrix*, respectively, and the constant is called
eccentricity the *eccentricity* and is represented by e. We choose the pole as a focus
and a line perpendicular to the polar axis and p units to the left of
the pole as the directrix as shown in Fig. 15.14. The definition of the
conic may now be written in the form

$$FP = eDP$$
$$r = e(DQ + QP) = e(p + r \cos \theta) \tag{1}$$

since $DQ = p$ and $\cos \theta = \cos FPQ = QP/r$. Now, solving (1) for r,
we see that

equation of
a conic
$$r = \frac{ep}{1 - e \cos \theta} \tag{15.6}$$

is a polar form of the equation of the noncircular conic of eccentricity e
with a focus at the pole and the corresponding directrix perpendicular
to the polar axis and p units to the left of the pole.

We can show in a similar manner that

$$r = \frac{ep}{1 + e \cos \theta} \tag{15.7}$$

is an equation of the noncircular conic of eccentricity e with a focus
at the pole and the corresponding directrix perpendicular to the polar
axis and p units to the right of the pole.

Furthermore, we can show that

$$r = \frac{ep}{1 - e \sin \theta} \tag{15.8}$$

and
$$r = \frac{ep}{1 + e \sin \theta} \tag{15.9}$$

are equations of the noncircular conics of eccentricity e with a focus
at the pole and the corresponding directrix perpendicular to the
normal axis and p units below the pole and above the pole, respectively.

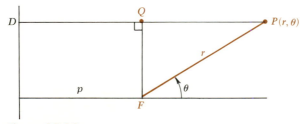

Figure 15.14

The conic is a parabola if $e = 1$, an ellipse if $e < 1$ and a hyperbola if $e > 1$.

Example 1   Describe the conic defined by $r = 12/(1 + 2 \cos \theta)$.

Solution: If we compare this equation with (15.7) we see that $e = 2$; hence, the conic is a hyperbola. Furthermore, a focus is at the pole, $ep = 12$, and $p = 6$. Hence, a directrix is 6 units to the right of the pole and is perpendicular to the polar axis.

Example 2   Describe the conic defined by $r = 8/(4 - 3 \sin \theta)$.

Solution: Since the first term in the denominator of the standard forms (15.6) to (15.9) is 1, we divide the numerator and denominator of the right member by 4 and get

$$r = \frac{2}{1 - \frac{3}{4} \sin \theta}$$

This equation is in the form (15.8) with $e = \frac{3}{4}$ and $p = 2 \div \frac{3}{4} = \frac{8}{3}$. Consequently, the conic is an ellipse with a focus at the pole and a directrix $\frac{8}{3}$ units below it.

Example 3   Describe the graph of the equation $r = 12/(3 - 3 \cos \theta)$.

Solution: Dividing the numerator and denominator of the right member by 3, we get

$$r = \frac{4}{1 - \cos \theta}$$

The equation is in the form (15.6) with $e = 1$ and $p = 4$. Hence, the graph is a parabola with a focus at the pole and a directrix 4 units to the left of the pole and perpendicular to the extension of the polar axis. Therefore the parabola opens to the right.

The following examples illustrate the method for obtaining the equation of a conic if we know the value of e and the position of the directrix.

Example 4   Find the equation of the conic if a focus is at the pole, the corresponding directrix is vertical and passes through $(6, 0°)$, and $e = \frac{2}{3}$.

Solution: Since a focus is at the pole and $e = \frac{2}{3}$, the conic is an ellipse; furthermore, a directrix is $p = 6$ units to the right of the pole. Hence, we must use Eq. (15.7). If we substitute the above values for e and p, we get the equation

$$r = \frac{4}{1 + \frac{2}{3} \cos \theta}$$

Example 5   Find the equation of the conic if a directrix is perpendicular to the normal axis and passes through $(6, -\pi/2)$, $e = \frac{5}{3}$, and a focus is at the pole.

Solution: In this problem a directrix is below the focus, $p = 6$, and $ep = (\frac{5}{3})(6) = 10$. Hence, we use equation (15.8) and get

$$r = \frac{10}{1 - \frac{5}{3} \sin \theta}$$

EXERCISE 15.4

Describe the graph of the equation in each of Problems 1 to 20.

1	$\theta = 70°$	2	$\theta = -3\pi/4$	3	$r = 6$
4	$r = 12$	5	$r = 18 \cos \theta$	6	$r = 20 \sin \theta$
7	$r = -8 \cos \theta$	8	$r = -16 \sin \theta$	9	$r = 6/(1 + \cos \theta)$
10	$r = 4/(1 + \sin \theta)$		11	$r = 7/(1 - \cos \theta)$	
12	$r = 5/(1 - \sin \theta)$		13	$r = 6/(1 + \frac{1}{2} \cos \theta)$	
14	$r = 12/(1 - \frac{2}{3} \sin \theta)$		15	$r = 9/(4 + 3 \sin \theta)$	
16	$r = 12/(6 - 4 \sin \theta)$		17	$r = 8/(1 + 2 \cos \theta)$	
18	$r = 24/(1 - 4 \cos \theta)$		19	$r = 10/(2 + 5 \sin \theta)$	
20	$r = 15/(3 - 8 \sin \theta)$				

Find the polar equation of the line or circle in each of Problems 21 to 24.

21   The line with $\omega = 30°$ and $r = 5$.
22   The line with $\omega = 315°$ and $r = 2$.
23   The radius of the circle is 7 and the center is at the pole.
24   The radius of the circle is 15 and the center is at the pole.

Find the equation of the circle in each of Problems 25 to 28.

25   The radius is 4 and the center is at $(4, 0)$.
26   The radius is 9 and the center is at $(9, \pi/2)$.
27   The radius is 7 and the center is at $(7, \pi)$.
28   The radius is 8 and the center is at $(8, -\pi/2)$.

Find the polar equation of the conic determined by the data in each of Problems 29 to 40. In each case the pole is at a focus and the directrix is vertical or horizontal.

29   Directrix through $(5, 0)$, $e = 1$
30   Directrix through $(8, \pi)$, $e = 1$
31   Directrix through $(6, \pi/2)$, $e = 1$
32   Directrix through $(10, -\pi/2)$, $e = 1$
33   Directrix through $(7, 0)$, $e = \frac{3}{5}$
34   Directrix through $(15, \pi/2)$, $e = \frac{3}{4}$
35   Directrix through $(6, \pi)$, $e = \frac{2}{3}$

36   Directrix through $(4, -\pi/2)$, $e = \frac{5}{8}$
37   Directrix through $(9, -\pi/2)$, $e = 2$
38   Directrix through $(12, \pi)$, $e = \frac{5}{3}$
39   Directrix through $(8, \pi/2)$, $e = \frac{3}{2}$
40   Directrix through $(9, 0)$, $e = 2$

15.7   Intersections of polar curves

If we solve a pair of equations in cartesian coordinates simultaneously, we find the coordinates of those points that have equal values of x and equal values of y on both curves. Since a point has only one pair of coordinates in the cartesian system, we thus obtain all of the points of intersection.

If we solve a pair of equations in polar coordinates simultaneously, we find the coordinates of those points that have equal values of r and equal values of θ. Since a point may have more than one pair of polar coordinates, it may have one pair on one curve and another *not all* pair on the other curve. Hence, we do not necessarily get all points of *intersections* intersection by solving the two equations simultaneously.

In order to solve two polar equations simultaneously and thereby obtain the points that have equal values of r and equal values of θ on the two curves, we solve each equation for r in terms of θ, equate these expressions in θ, solve the resulting equations for θ, and thus find each corresponding value of r.

Example   Solve

$$r = -\sin \theta \tag{1}$$

and

$$r = 2 - 3 \sin \theta \tag{2}$$

simultaneously.

Solution: The equations are already solved for r. By equating the two expressions in θ, we have

$$-\sin \theta = 2 - 3 \sin \theta \tag{3}$$

Now solving for $\sin \theta$, we find $\sin \theta = 1$. Hence, $\theta = 90°$ is the only solution in $[0°, 360°)$ of (3). The corresponding value of r as obtained from either equation is -1. Hence, the only point that has the same pair of coordinates on both curves is $(-1, 90°)$. The pole is an intersection of the curves even though it is not a simultaneous solution of the pair of equations since $(0, 0)$ satisfies (1) and $(0, \arcsin \frac{2}{3})$ is a point on the graph of (2).

WILDE LAKE HIGH SCHOOL
DEPARTMENT OF MATHEMATICS
REQUEST FOR TEST

NAME_____DATE_____

COURSE_____

UNIT_____PART_____

FORM_____

TEACHER_____

EXERCISE 15.5

Solve the following pairs of equations simultaneously after sketching both curves about the same axes.

1    $r = 2, r = 2 \sin \theta$ 2 $r = 2 \cos \theta, r = \sqrt{3}$
3    $r = 3, r \sin \theta = 6$ 4 $r \cos \theta = 2, r = 4$
5    $r = \sin 2\theta, r = \sqrt{3} \cos 2\theta$ 6 $r = \sin^2 \theta, r = \cos^2 \theta$
7    $r = 3 \cos \theta, r = 6 \cos \theta$ 8 $r = 2 \sin \theta, r = -2 \cos \theta$
9    $r = a \cos \theta, r = a \sec \theta$ 10 $r = a \sin \theta, r = a \csc \theta$
11    $r = 2 \tan \theta, r = 2 \cot \theta$ 12 $r = 2 \sin \theta, r = \sin 2\theta$
13    $r = 3 \sin \theta, r^2 = 9 \sin \theta$ 14 $r = 2 \cos \theta, r^2 = 4 \cos \theta$
15    $r = \sin \theta, r^2 = \sin 2\theta$
16    $r = \cos \theta, r^2 = \cos 2\theta$
17    $r = 4 \sin \theta, r = 1/(1 - \sin \theta)$
18    $r = -\cos \theta, r = 1/(3 + 2 \cos \theta)$
19    $r = -2 \cos \theta, r = 4/(1 - \cos \theta)$
20    $r = 3 \sin \theta, r = 12/(3 + \sin \theta)$
21    $r = 1 - \cos \theta, r = 3 \cos (\theta/2)$
22    $r = 1 + \cos \theta, r = \cos (\theta/2)$
23    $r = 1 - \sin \theta, r = 1/(1 + \sin \theta)$
24    $r = 4.5 - 2 \sin \theta, r = 2 + 3 \sin \theta$

15.8   Introduction to parametric equations

So far we have studied two methods of representing equations, the one employing rectangular coordinates and the other polar coordinates. We shall now consider another method of representation by introducing a third variable for use in connection with the rectangular coordinates of a point. If, in the cartesian equation of a curve, x and y are expressed in terms of a third variable in such a way that the given relation between x and y holds, the third variable is called a *parameter*, and the equations defining x and y in terms of the third variable are said to be a *parametric representation of the curve*.

parameter

Example

Give a parametric representation of the hyperbola $4x^2 - 9y^2 = 36$.

Solution: If we let $x = 3 \sec t$ and $y = 2 \tan t$, then

$$4x^2 - 9y^2 = 4(9 \sec^2 t) - 9(4 \tan^2 t) = 36(\sec^2 t - \tan^2 t) = 36$$

Consequently, $x = 3 \sec t, y = 2 \tan t$ is a parametric representation of the hyperbola.

15.9   Finding a parametric representation

The procedure for plotting the curves corresponding to some equations in either rectangular or polar form involves very tedious computation. Often it is possible to introduce parametric equations that render the curve much easier to plot. For example, to plot the curve represented by the equation

$$x^{2/3} + y^{2/3} = 1 \tag{1}$$

we would solve the equation for y and get

$$y = \pm\sqrt{(1 - x^{2/3})^3} \tag{2}$$

When we assign a number to x, however, we find that the computation required to obtain the corresponding value of y comprises several steps. If, instead, as suggested by the radicand in (2), we let $x = \cos^3 \theta$, we have $y = \pm\sqrt{(1 - \cos^2 \theta)^3} = \pm\sin^3 \theta$, which is much easier to plot. Hence, if we want to find a parametric form of (1), one such form is

$$x = \cos^3 \theta \qquad y = \sin^3 \theta \tag{3}$$

In Eq. (3), if a value is assigned to θ, only one computation is needed to obtain x or y.

The above example illustrates a procedure for obtaining a parametric representation of a given equation in two variables. The steps consist of: choosing one of the variables and setting it equal to an arbitrary function of the parameter, substituting this function for the chosen variable in the given equation, and, if possible, solving the resulting equation for the other variable in terms of the parameter.

Unless the function that is selected to replace the first variable is properly chosen, it may be difficult or even impossible to solve the equation obtained in the second step. Furthermore, unless each of the variables in the given equation can be and is expressed as a reasonably simple function of the parameter, little advantage is gained by use of a parametric form.

Frequently a parametric form of an equation in x and y can be obtained by replacing y by tx in the equation, solving the resulting equation for x in terms of t, and replacing x in $y = tx$ by this solution. We shall illustrate this method.

Example 1   Obtain a parametric representation of

$$x^3 + 3x^2 - y^2 = 0 \tag{4}$$

Solution: We shall let $y = tx$ and proceed as below:

$$x^3 + 3x^2 - t^2x^2 = 0$$

$$x^2(x + 3 - t^2) = 0 \qquad \textit{factoring the left member}$$

Since the first factor does not involve t, we set the second factor equal to zero, solve for x in terms of t and get

$$x = t^2 - 3$$

Hence, since $y = tx$, we have $y = t(t^2 - 3) = t^3 - 3t$. Therefore, the desired parametric equations are

$$x = t^2 - 3 \qquad y = t^3 - 3t$$

In some cases a parametric representation of an equation in two variables defines only part of the graph, and another parametric representation may be required for the remaining part. We shall now illustrate such a situation.

Example 2   (a) Obtain a parametric representation for the equation

$$y^2 = 4(x + 1) \tag{5}$$

by setting $y = 2 \sin \theta$ as the first step. (b) Determine the portion of the graph of (5) that corresponds to the parametric equations obtained. (c) Obtain a parametric representation for the remainder of the graph of (5).

Solution of (a)

$y^2 = 4(x + 1)$	*given equation*
$y^2 = 4x + 4$	*by the distributive axiom*
$y^2 - 4 = 4x$	*adding -4 to each member*
$4 \sin^2 \theta - 4 = 4x$	*replacing y by $2 \sin \theta$*
$\sin^2 \theta - 1 = x$	*dividing each member by 4*
$-\cos^2 \theta = x$	*since $\sin^2 \theta + \cos^2 \theta = 1$*

Hence, the desired parametric equations are

$$y = 2 \sin \theta \qquad x = -\cos^2 \theta \tag{6}$$

Solution of (b):   Since $|\sin \theta| \leq 1$ and $|\cos \theta| \leq 1$, the above equations yield values of x and y such that $-2 \leq y \leq 2$ and $-1 \leq x \leq 0$. Therefore, Eqs. (6) determine only the portion of the curve that is to the left of the Y axis. Since the graph of (5) is a parabola with the vertex at $(-1, 0)$ and opening to the right, it extends to the right of the Y axis and only part of the curve is determined by the parametric representation.

Solution of (c):   If we let $y = 2 \sec \theta$, substitute in (5), and solve for x, we obtain $x = \tan^2 \theta$. Hence, a second parametric form of (5) is

$$x = \tan^2 \theta \qquad y = 2 \sec \theta$$

In this case, $x \geq 0$ and $|y| \geq 2$. Therefore, these equations determine the portion of the graph that is to the right of the Y axis.

polar to
parametric
form

It is often a simple matter to change an equation from polar form to a parametric form. If the polar form is, or can be, expressed as $r = f(\theta)$, then we can use the relation between rectangular and polar coordinates to express the equation in parametric form. Thus, if $r = f(\theta)$, then $x = r \cos \theta$ and $y = r \sin \theta$ take the parametric form

$$x = f(\theta) \cos \theta \qquad \text{and} \qquad y = f(\theta) \sin \theta$$

Example 3   Express the equation $r = 4/(1 - \cos \theta)$ of a parabola in parametric form.

Solution: If we put the expression given for r in the equation of the parabola into $x = r \cos \theta$ and $y = r \sin \theta$, we see that

$$x = \frac{4 \cos \theta}{1 - \cos \theta} \qquad \text{and} \qquad y = \frac{4 \sin \theta}{1 - \cos \theta}$$

are parametric equations of the given parabola.

15.10   Parametric equations for the circle and central conics

In this section we shall derive parametric equations of the circle, the ellipse, and the hyperbola with the center in each case at the origin.

The rectangular equation of a circle of radius a with the center at the origin is

$$x^2 + y^2 = a^2 \tag{1}$$

To obtain a parametric representation of this equation, we seek two functions of a parameter such that the sum of their squares is a^2. The two functions $a \cos \theta$ and $a \sin \theta$ satisfy this condition since $a^2 \cos^2 \theta + a^2 \sin^2 \theta = a^2$. Hence a parametric representation of Eq. (1) is

parametric
equations of
a circle

$$x = a \cos \theta \qquad y = a \sin \theta \tag{2}$$

The geometric interpretation of these equations is shown in Fig. 15.15. From the figure it is evident that as θ increases from 0 to $360°$, the point $P(a \cos \theta, a \sin \theta)$ traces the entire circle.

If the center of an ellipse is at the origin, the major axis is $2a$, the minor axis is $2b$, and the foci are on the X axis, then the equation of the ellipse is

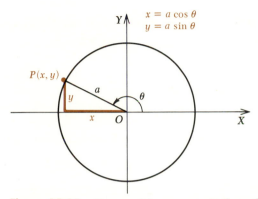

$$x = a \cos \theta$$
$$y = a \sin \theta$$

Figure 15.15   Parametric representation of a circle

$$\frac{x^2}{a^2} + \frac{y^2}{b^2} = 1 \tag{3}$$

To find a parametric representation of (3) we seek two functions $f(\theta)$ and $g(\theta)$ such that

parametric equations of an ellipse

$$\left[\frac{f(\theta)}{a}\right]^2 + \left[\frac{g(\theta)}{b}\right]^2 = 1$$

The functions $a \cos \theta$ and $b \sin \theta$ satisfy this condition since $(a \cos \theta/a)^2 + (b \sin \theta/b)^2 = \cos^2 \theta + \sin^2 \theta = 1$. Hence a parametric representation of (3) is

$$x = a \cos \theta \qquad y = b \sin \theta \tag{4}$$

Similarly, it can be shown that for the ellipse

$$\frac{x^2}{b^2} + \frac{y^2}{a^2} = 1$$

whose foci are on the Y axis, parametric equations of the ellipse are

$$x = b \cos \theta \qquad y = a \sin \theta \tag{5}$$

The equation of a hyperbola with the center at the origin and the foci on the X axis is

$$\frac{x^2}{a^2} - \frac{y^2}{b^2} = 1 \tag{6}$$

parametric equation of a hyperbola

provided that the transverse axis is of length $2a$ and the conjugate axis is of length $2b$. To obtain a parametric representation of Eq. (6), we seek two functions $f(\theta)$ and $g(\theta)$ such that $[f(\theta)/a]^2 - [g(\theta)/b]^2 = 1$. The functions $a \sec \theta$ and $b \tan \theta$ satisfy this condition. Hence we have the following parametric equations of the hyperbola

$$x = a \sec \theta \qquad y = b \tan \theta \tag{7}$$

If the foci of the hyperbola are on the Y axis, the equation is

$$\frac{y^2}{a^2} - \frac{x^2}{b^2} = 1 \tag{8}$$

and parametric equations are

$$x = b \tan \theta \qquad y = a \sec \theta \tag{9}$$

If the center of a circle of radius a is at (h, k), the equation is

$$(x - h)^2 + (y - k)^2 = a^2 \tag{10}$$

more *general* *equations*
In this case we let $x - h = a \cos \theta$ and $y - k = a \sin \theta$ and obtain the parametric equations

$$x = h + a \cos \theta \qquad y = k + a \sin \theta$$

Parametric equations of the ellipse or hyperbola with the center at (h, k) can be obtained in a similar manner.

EXERCISE 15.6

Express the equations in Problems 1 to 20 in parametric form, making use of the given relation between x or y and the parameter or between x and y and the parameter.

1. $x^2 = 9y + 3$, $y = t^2 - \frac{1}{3}$
2. $3x - xy + 2 = 0$, $y = 3 + t$
3. $xy + 3x = 3y + 6$, $y = t - 3$
4. $y^2 - x - 6y + 8 = 0$, $y = t + 3$
5. $x^2 + y^2 = 25$, $x = 5 \cos \theta$
6. $x^2 - y^2 = 49$, $y = 7 \tan \theta$
7. $4x^2 + 9y^2 = 36$, $x = 3 \cos \theta$
8. $9x^2 - 16y^2 = 144$, $x = 4 \sec \theta$
9. $(x - 1)^2 + y^2 = 9$, $y = 3 \cos \theta$
10. $x^2 - y^2 + 6y = 25$, $y = 3 + 4 \tan \theta$
11. $(x - 2)^2/9 + (y + 1)^2/25 = 1$, $x = 2 + 3 \cos \theta$
12. $(x + 3)^2/16 - (y + 1)^2/25 = 1$, $x = -3 + 4 \sec \theta$
13. $x^2 = 2y + 4$, $x = 2 \sec \theta$
14. $y^2 - 4y + x + 3 = 0$, $y = 2 + \sec \theta$
15. $x^2 - 4x - 5y - 5 = 0$, $y = (9 \tan^2 \theta)/5$
16. $y^2 + 2y + x - 1 = 0$, $x = -2 \tan^2 \theta$
17. $x^3 - y^2 - 2xy = 0$, $y = tx$
18. $y^2(x - a) = x^3$, $y = tx$
19. $y(y^2 - x^2) = x^4$, $y = tx$
20. $xy + x^2 = y - x$, $y = tx$

Obtain a parametric representation of the equation in each of Problems 21 to 32.

21	$r = 6 \sin \theta$	22	$r = 1 + 3 \cos \theta$
23	$r = 2/(1 + 3 \sin \theta)$	24	$r \cos \theta = 2$
25	$xy = y - x$	26	$x^2 + xy - y = 0$
27	$x^3 - y^3 + 2xy = 0$	28	$x^2y - 2xy = x^2 + y^2$
29	$(x - 1)^2 + (y + 2)^2 = 9$		
30	$(x + 2)^2 + 4y^2 = 4$	31	$4x^2 - (y - 3)^2 = 36$
32	$4(x - 3)^2 - (y + 2)^2 = 16$		

33   A point is moving in a plane with horizontal and vertical velocities of 4 ft per sec and 7 ft per sec, respectively. Find the parametric equations of the curve traced by the point if it starts from the origin. (Use $t = $ time in seconds as the parameter.)

34   A bullet is fired horizontally from a gun at the rate of v ft per sec; at the end of t sec the bullet has fallen $gt^2/2$ ft. Find the parametric equations of the curve traced by the bullet. (Take the origin at the muzzle of the gun and the X axis along the barrel.)

35   A line segment 16 units in length moves with its extremities in contact with the coordinate axes. Find the parametric equations of the curve traced by the point that is 4 units from the end of the line that touches the X axis. Use the acute angle that the line makes with the X axis as the parameter.

36   A point starts at the origin and moves along a line with the slope $5/12$ at the rate of 3 ft per sec. At the same time the point of intersection of the line and the X axis moves to the right at the rate of 4 ft per sec. Find the equation of the locus of the point. (Use $t = $ time in seconds as the parameter.)

15.11   Elimination of the parameter

If the equation of a curve is given in parametric form, it may be possible to obtain the rectangular equation or the polar equation by eliminating the parameter from the given equations. The procedure for eliminating the parameter depends upon the equations, and at times no procedure exists. We shall illustrate three methods in the following examples.

Example 1   Eliminate the parameter from

$$x = t - 1 \qquad y = t^2 + 1$$

Solution: We shall solve the first equation for t in terms of x, substitute the result for t in the second, and simplify. Thus we obtain

$$t = x + 1 \qquad \text{\textit{solving the first equation for t}}$$

$$y = (x + 1)^2 + 1 \qquad \text{\textit{substituting x + 1 for t in the second equation}}$$

$$y - 1 = (x + 1)^2 \qquad \text{\textit{adding} -1 \textit{to each member}}$$

The last equation is the equation of a parabola that opens upward with the vertex at $(-1, 1)$.

Example 2   Eliminate the parameter from

$$x = \frac{1}{\sqrt{1 + t^2}} \tag{1}$$

$$y = \frac{t}{\sqrt{1 + t^2}} \tag{2}$$

Solution: Since $\sqrt{1 + t^2}$ is positive for t any real number, it follows that $x > 0$ and the graph of Eqs. (1) and (2) in the cartesian plane passes through no point whose abscissa is zero or negative. Furthermore, $y = t(1/\sqrt{1 + t^2}) = tx$; hence, $t = y/x$. Consequently, we can eliminate t from (1) and (2) by replacing t by y/x in either equation. We shall choose (1) and proceed as follows:

$$x = \frac{1}{\sqrt{1 + y^2/x^2}} \qquad \text{\textit{replacing t by y/x in (1)}}$$

$$x = \frac{x}{\sqrt{x^2 + y^2}} \qquad \text{\textit{multiplying the numerator and denominator of the right member by} $x > 0$}$$

$$\sqrt{x^2 + y^2} = 1 \qquad \text{\textit{multiplying each member by} $\sqrt{x^2 + y^2}/x$}$$

The graph of Eqs. (1) and (2) is the portion of the circle determined by $x^2 + y^2 = 1$ that is to the right of the Y axis since

$$x = 1/\sqrt{1 + t^2} > 0$$

for all real values of t.

Example 3   Eliminate the parameter from

$$x = 2 + 3 \cos \theta \tag{1}$$

$$y = 4 \sin \theta \tag{2}$$

Solution: We shall add -2 to each member of (1), then divide each member of the resulting equation by 3 and get

$$\frac{x - 2}{3} = \cos \theta \tag{3}$$

Next, we shall divide each member of (2) by 4 and have

$$\frac{y}{4} = \sin \theta \tag{4}$$

Now we add the squares of the corresponding members of (3) and (4) and get

$$\left(\frac{x-2}{3}\right)^2 + \left(\frac{y}{4}\right)^2 = \cos^2\theta + \sin^2\theta = 1$$

Consequently, we have

$$\frac{(x-2)^2}{9} + \frac{y^2}{16} = 1 \tag{5}$$

Equation (5) is the standard form of the ellipse with the center at (2, 0), the major axis 8, the minor axis 6, and the foci on a line parallel to the Y axis.

15.12   Plotting a curve determined by parametric equations

In order to plot the curve determined by parametric equations, we assign several values to the parameter, calculate the corresponding values of x and y, plot the points determined by the resulting ordered number pairs (x, y), and draw a smooth curve through them. We shall illustrate the procedure.

Example 1   Construct the graph defined by the equations

$$x = \cos^3\theta \qquad y = \sin^3\theta$$

Solution: We assign the values in the first line of the following table to θ and then calculate the corresponding values of x and y shown in the fourth and fifth lines. Finally, we plot the points determined by each pair of corresponding values (x, y), draw a smooth curve through them, and obtain the curve above the X axis in Fig. 15.16.

θ	0	30°	45°	60°	90°	120°	135°	150°	180°
$\cos\theta$	1	0.87	0.71	0.5	0	−0.5	−0.71	−0.87	−1
$\sin\theta$	0	0.5	0.71	0.87	1	0.87	0.71	0.5	0
$x = \cos^3\theta$	1	0.66	0.36	0.13	0	−0.13	−0.36	−0.66	−1
$y = \sin^3\theta$	0	0.13	0.36	0.66	1	0.66	0.36	0.13	0

Since $\cos(-\theta) = \cos\theta$ and $\sin(-\theta) = -\sin\theta$, it follows that if we assign to θ the negatives of the values occurring in the first line of the above table, the corresponding values of $x = \cos^3\theta$ will be the same as those appearing in the table and the values of $y = \sin^3\theta$ will be the negatives of the numbers appearing in the last line of the table.

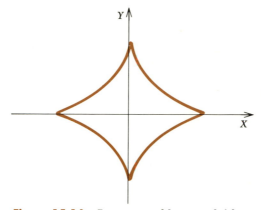

Figure 15.16   Four cusped hypocycloid

Consequently, the curve is symmetric with respect to the X axis, and the portion below the X axis can be sketched by use of this symmetry. The curve is called the *hypocycloid of four cusps*.

Example 2   Construct the graph of

$$x = \frac{2t}{1 + t^2} \qquad y = \frac{2t^2}{1 + t^2}$$

Solution: We assign each number in the first line of the following table to t, calculate each corresponding value of x and y, and thus obtain the values of x shown in the second line and the corresponding values of y shown in the third line. Finally, we plot the points determined by the ordered pairs (x, y), draw a smooth curve through them, and obtain the right half of the curve shown in Fig. 15.17.

t	⅕	⅓	½	⅗	1	6/5	2	3	6
x	0.38	0.6	0.8	0.88	1	0.98	0.8	0.6	0.32
y	0.08	0.2	0.4	0.53	1	1.2	1.6	1.8	1.9

If we assign the negative of each number in the first line of the table to t, each corresponding value of x will be the negative of the number in the second line of the table, and each value of y will be the number in the third line, since changing the sign of t changes the sign of x but does not change the sign of y. The ordered pairs of values (x, y) found by use of the negatives of the values of t in the table determine the left half of the curve in the figure.

We can find, by eliminating the parameter from the given equations, that the rectangular equation is $x^2 + y^2 - 2y = 0$. This equation can be converted to $x^2 + (y - 1)^2 = 1$, which is the standard form of the circle of radius 1 with center at $(0, 1)$.

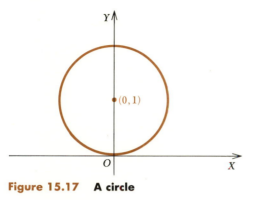

Figure 15.17 A circle

Example 3   Construct the graph of the parametric equations

$$x = \cos^2 \theta \qquad y = \sin^2 \theta$$

Then eliminate the parameter, and finally compare the graph of the rectangular equation thus obtained with the graph of the parametric equations.

Solution: We first note that neither x nor y can be negative since each is the square of a real number. Now, we assign the numbers in the first line of the following table to θ, calculate the values of $\cos^2 \theta$ and $\sin^2 \theta$, and thus obtain the values of x and y shown in the second and third lines of the table.

θ	0°	30°	45°	60°	90°
x	1	¾	½	¼	0
y	0	¼	½	¾	1

Next we plot the points determined by the corresponding values of x and y in the table and draw a smooth curve through them. Thus, we obtain the line segment in the first quadrant of Fig. 15.18. Because of the periodicity of $\sin \theta$ and $\cos \theta$, there is no need to assign values to θ less than 0° or greater than 90°, since such assignments would yield no new points.

We eliminate the parameter by equating the sums of the corresponding members of the parametric equations and get

$$x + y = 1 \qquad \textit{since } \sin^2 \theta + \cos^2 \theta = 1$$

The graph of the cartesian equation is the complete line in Fig. 15.18 obtained by extending the graph of the parametric equations in both directions.

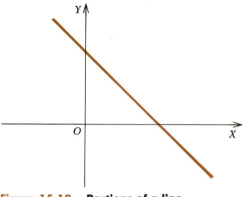

Figure 15.18   Portions of a line

Parametric equations of the portion of the line $x + y = 1$ below the
X axis are

$$x = \sec^2 \theta \qquad y = -\tan^2 \theta$$

since $y \le 0$, $x \ge 1$, and $\sec^2 \theta - \tan^2 \theta = 1$. Similarly, the parametric equations of the ray to the left of the Y axis are

$$x = -\tan^2 \theta \qquad y = \sec^2 \theta$$

EXERCISE 15.7

*Sketch the curve determined by the pair of parametric equations in
each of the following problems. Obtain the rectangular form by
eliminating the parameter, and determine if the entire graph of the
rectangular equation is defined by the parametric form.*

1	$x = 2t,\ y = t + 1$	2	$x = 1 + 3t,\ y = 6t$
3	$x = t - 2,\ y = t + 4$	4	$x = 2t - 1,\ y = t + 2$
5	$x = 4t^2,\ y = 2t^2 - 3$	6	$x = t^2 - 1,\ y = t^2 + 3$
7	$x = 3t^2 + 2,\ y = 2t^2 - 1$	8	$x = 6t^3 + 3,\ y = 9t^3 + 1$
9	$x = t^2,\ y = t - 1$	10	$x = 2t^2 - 1,\ y = t + 1$
11	$x = 3t^2 - 2,\ y = 2t + 1$	12	$x = 1 - 6t^2,\ y = 2 + 4t^2$
13	$x = 4\cos^2 \theta,\ y = 2\sin^2 \theta$	14	$x = 3\cos^2 \theta,\ y = -2\sin^2 \theta$
15	$x = 3\sec^2 \theta,\ y = \tan^2 \theta$	16	$x = 9\sec^2 \theta,\ y = -6\tan^2 \theta$
17	$x = 2\cos \theta,\ y = 2\sin \theta$	18	$x = 3\sec \theta,\ y = 3\tan \theta$
19	$x = 3\cos \theta,\ y = 2\sin \theta$	20	$x = 4\sec \theta,\ y = 3\tan \theta$

21 $\quad x = 2 + \cos \theta,\ y = 1 - \sin \theta$
22 $\quad x = 3 - \sec \theta,\ y = 2 + \tan \theta$
23 $\quad x = 1 + 3\sin \theta,\ y = 2 + \cos \theta$
24 $\quad x = 4 + 3\sec \theta,\ y = 2 - 5\tan \theta$
25 $\quad x = \cos \theta,\ y = \sec \theta$

26    $x = 3 - \sin \theta$, $y = \csc \theta$

27    $x = 5 - \tan \theta$, $y = 2 + \cot \theta$

28    $x = 4 + 2 \cot \theta$, $y = 2 + 3 \tan \theta$

29    $x = t^2/(t - 1)$, $y = t/(t - 1)$

30    $x = 2t/(t + 1)$, $y = 4t^2/(t + 1)$

31    $x = t/(t^2 + 1)$, $y = t^2/(t^2 + 1)$

32    $x = 2t^2/(t^2 - 1)$, $y = 2t/(t^2 - 1)$

33    $x = 8 \cos^3 \theta$, $y = 8 \sin^3 \theta$

34    $x = \cos^4 \theta$, $y = \sin^4 \theta$

35    $x = 4 \cos \theta + 2 \sin \theta$, $y = -2 \cos \theta + 4 \sin \theta$. *Hint:* To eliminate θ, equate the sums of the squares of the corresponding members.

36    $x = 5 \sec \theta + 3 \tan \theta$, $y = 3 \sec \theta + 5 \tan \theta$

16

Derivatives from parametric equations, curvature, vectors

16.1 The derivatives dx/dt, dy/dt, and dy/dx

We shall consider the equations

$$x = g(t) \qquad y = h(t) \qquad a \leq t \leq b$$

The derivative dx/dt is the rate of change of x with respect to t, and dy/dt is the rate of change of y with respect to t. The rate of change of y relative to x for any value of t is the quotient of these two rates. The formal proof is as follows: Consider a value t_1 of the parameter for which x and y have the values $g(t_1)$ and $h(t_1)$, respectively. Then think of a neighboring value t of the parameter for which x and y have the values $g(t)$ and $h(t)$. Then $\Delta x = g(t) - g(t_1)$ and $\Delta y = h(t) - h(t_1)$. By definition, the derivative of y with respect to x is

$$\lim_{t \to t_1} \frac{\Delta y}{\Delta x} = \lim_{t \to t_1} \frac{h(t) - h(t_1)}{g(t) - g(t_1)}$$

If we divide numerator and denominator by $t - t_1$, we have

$$\frac{dy}{dx} = \lim_{t \to t_1} \frac{[h(t) - h(t_1)]/(t - t_1)}{[g(t) - g(t_1)]/(t - t_1)}$$

The limit of the numerator, if it exists, is $h'(t) = dy/dt$, and that of the denominator is $g'(t) = dx/dt$. We may therefore conclude that *if $x = g(t)$, $y = h(t)$ is a parametric representation of a curve, then*

dy/dx from parametric equation

$$\frac{dy}{dx} = \frac{dy/dt}{dx/dt} = \frac{h'(t)}{g'(t)} \qquad g'(t) \neq 0$$

Example

If the relation between y and x is given by the parametric equations

$$y = t^3(3 - 2t) \qquad \text{and} \qquad x = t^2(3 - 2t)$$

then

$$\frac{dy}{dt} = t^2(9 - 8t) \qquad \text{and} \qquad \frac{dx}{dt} = 6t(1 - t)$$

The first of these measures the rate of change of y with respect to t, and the second gives the rate of change of x with respect to t. If we want the rate of change of y relative to x for any value of t, we have

$$\frac{dy}{dx} = \frac{dy/dt}{dx/dt} = \frac{t^2(9 - 8t)}{6t(1 - t)} = \frac{t(9 - 8t)}{6(1 - t)} \qquad \text{if } t \neq 1 \qquad (1)$$

This derivative, of course, gives the slope of the tangent to the curve in terms of t. Thus, for $t = 1.5$, $dy/dx = \frac{3}{2}$. This is the slope at the origin since that is the point on the graph corresponding to $t = \frac{3}{2}$. The origin also corresponds to $t = 0$ and the slope of the corresponding part of the curve is zero. If $t = 1$, then $x = y = 1$ and $dx/dt = 0$. Furthermore, x increases for $0 < t < 1$ and x decreases for $t > 1$ since $dx/dt > 0$ for $0 < t < 1$ and $dx/dt < 0$ for $t > 1$. The tangent line at this point is vertical as seen from (1). Similarly, $dy/dt = 0$ for $t = \frac{9}{8}$; y increases with t for $t < \frac{9}{8}$ and y decreases with t for $t > \frac{9}{8}$. The tangent line at the point corresponding to $t = \frac{9}{8}$ is horizontal.

16.2   The second derivative from parametric equations

If a curve is defined by use of parametric equations and if we want the second derivative of y with respect to x, we must not lose sight of the fact that it is the *derivative with respect to x of dy/dx* that we seek and that dy/dx is in terms of the parameter t. Consequently, *if x, y, and dy/dx are in terms of the parameter t, then*, by use of the chain rule,

second derivative from parametric form

$$\frac{d^2y}{dx^2} = \frac{d}{dx}\left(\frac{dy}{dx}\right) = \frac{d}{dt}\left(\frac{dy}{dx}\right)\frac{dt}{dx}$$

Example 1

Find the second derivative of y with respect to x if $x = 2t - 3$ and $y = t^2 - 2t$.

Solution: If we regard t as the independent variable, we have $dx/dt = 2$ and $dy/dt = 2t - 2$. Consequently,

$$\frac{dy}{dx} = \frac{dy/dt}{dx/dt} = \frac{2(t-1)}{2} = t - 1$$

Therefore, $\dfrac{d^2y}{dx^2} = \dfrac{d}{dx}\left(\dfrac{dy}{dx}\right)$ *definition*

$$= \frac{d}{dx}(t-1) \qquad\qquad dy/dx = t - 1$$

$$= \frac{d}{dt}(t-1)\frac{dt}{dx}$$

$$= \frac{1}{2}$$

since $d(t-1)/dt = 1$ and $dt/dx = 1 \div dx/dt = \frac{1}{2}$.

Example 2   Sketch the graph of $r = 2a \cos\theta$, $a > 0$, after expressing it in parametric form with θ as the parameter. Find the maximum and minimum points on the curve.

Solution: If we replace r by $2a \cos\theta$ in the equations $x = r \cos\theta$ and $y = r \sin\theta$ which give the relations between rectangular and polar coordinates, we have $x = 2a \cos\theta \cos\theta = 2a \cos^2\theta$ and

$$y = 2a \cos\theta \sin\theta = a \sin 2\theta$$

The graph is shown in Fig. 16.1. It can be obtained by assigning values to θ, evaluating each corresponding number pair (x, y), and drawing a smooth curve through the points thus determined or by recognizing the given equation as that of a circle of radius a with center at $(a, 0)$.

We shall now find the maximum and minimum points. These are the points at which $dy/dx = 0$; thus,

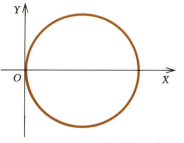

Figure 16.1   **Graph of $r = 2a \cos\theta$**

$$\frac{dy}{dx} = \frac{dy/d\theta}{dx/d\theta} = \frac{2a \cos 2\theta}{-4a \cos \theta \sin \theta} = \frac{2a \cos 2\theta}{-2a \sin 2\theta} = -\cot 2\theta$$

$$= 0 \qquad \text{for } 2\theta = \pi/2, 3\pi/2$$

Therefore, if there is a maximum or minimum, it occurs for $\theta = \pi/4$ or $3\pi/4$. We continue the investigation by use of the second derivative of y with respect to x. Thus,

$$\frac{d^2y}{dx^2} = \frac{d}{dx}\left(\frac{dy}{dx}\right) = \frac{d}{dx}(-\cot 2\theta)$$

$$= \frac{d(-\cot 2\theta)}{d\theta} \frac{d\theta}{dx}$$

$$= \frac{2 \csc^2 2\theta}{-2a \sin 2\theta} \qquad \text{since } d\theta/dx = -1/2a \sin 2\theta$$

$$= -\frac{\csc^3 2\theta}{a}$$

Hence
$$\left.\frac{d^2y}{dx^2}\right|_{\theta=\pi/4} = -\frac{1}{a} < 0 \qquad \text{for } a > 0$$

and
$$\left.\frac{d^2y}{dx^2}\right|_{\theta=3\pi/4} = \frac{1}{a} > 0 \qquad \text{for } a > 0$$

Consequently, the maximum and minimum points on the curve occur for $\theta = \pi/4$ and $\theta = 3\pi/4$, respectively. The coordinates of each can be found by replacing θ by the appropriate value in the parametric equations. Thus, the maximum is at (a, a) and the minimum is at $(a, -a)$ since $\cos 45° = 1/\sqrt{2}$, $\cos 135° = -1/\sqrt{2}$, $\sin 90° = 1$, and $\sin 270° = -1$.

EXERCISE 16.1

In each of Problems 1 to 16, find $D_x y$ and $D_x^2 y$ in terms of the parameter.

1   $x = t/2, y = 4/t$ 2 $x = t^2 - 2t, y = 2t^3 - 6t$

3   $x = t^2 + 1, y = t^3 + 3t$ 4 $x = 4t, y = 3\sqrt{1 - t^2}$

5   $x = 4 \cos t, y = 2 \sin t$

6   $x = 3 \cos t + 4, y = 5 \sin t - 2$

7   $x = 6 \cos t, y = 3 \sin^2 t$

8   $x = 4 \cos t, y = \cos 2t$

9   $x = 8 - 8 \sin t, y = 4 \sin t$

10   $x = \cos^3 t, y = 2 \sin^3 t$

11   $x = 3 \sin^2 t, y = 3 \cot t$

12 $x = 4e^{-t}, y = 2e^t$
13 $r = 6 \sin \theta$ 14 $r = 4 \cos \theta$
15 $r = 3(1 + \cos \theta)$ 16 $r = 2 \sin \theta + 1$

Find the equations of the tangent and normal to the curve for the given value of t or θ in each of Problems 17 to 20.
17 $x = t^2 - 1, y = t^2 - 4, t = 2$
18 $x = t + e^t, y = \ln(t + e^t), t = 0$
19 $r = 1 - \sin \theta, \theta = \pi/2$
20 $r = 2 - 3 \cos \theta, \theta = \pi/2$

Find the maximum, minimum, and inflection points of the curve in each of Problems 21 to 24.
21 $x = t^2 - 1, y = t^2 - 4$ 22 $x = \cos t, y = \sin^2 t$
23 $x = \sin \theta, y = \cos 2\theta$ 24 $x = 2 \tan t, y = 2 \sec t$

25 Show $y'' = 0$ for every point on $x = 3 \sin^2 t, y = 6 \cos^2 t$.
26 Show that $y'' = y'$ for all points on $x = \ln t, y = t + 4$.
27 Compute y'' from $x = a \sec t, y = b \tan t$, and then compute it again from the rectangular form of the equation. Thus show they are equal.
28 If $x = f(t)$ and $y = g(t)$, show that

$$y'' = \frac{\begin{vmatrix} dx/dt & dy/dt \\ d^2x/dt^2 & d^2y/dt^2 \end{vmatrix}}{(dx/dt)^3}$$

16.3   Differential of arc length

If a curve C, as in Fig. 16.2, is given by the parametric equations $x = f(t)$ and $y = g(t)$ where $f(t)$ and $g(t)$ have continuous derivatives which are not both zero for the same value of t, then for a fixed point

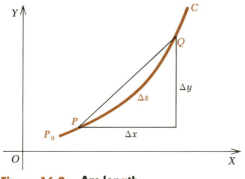

Figure 16.2   Arc length

P_0 and a variable point P, an arc P_0P of length s is determined. Furthermore s is a function of t since it depends on the coordinates of P. In order to find an expression for ds/dt, we shall let t take on an increment Δt and designate the corresponding increments of x, y, and s by Δx, Δy, and Δs, respectively. Then

$$\frac{\Delta s}{\Delta t} = \frac{\Delta s}{PQ}\frac{PQ}{\Delta t} = \frac{\Delta s}{PQ}\frac{\sqrt{(\Delta x)^2 + (\Delta y)^2}}{\Delta t} = \frac{\Delta s}{PQ}\sqrt{\left(\frac{\Delta x}{\Delta t}\right)^2 + \left(\frac{\Delta y}{\Delta t}\right)^2}$$

It can be shown[*] that $\Delta s/PQ$ approaches 1 as Δt approaches zero; consequently,

$$\frac{ds}{dt} = \sqrt{\left(\frac{dx}{dt}\right)^2 + \left(\frac{dy}{dt}\right)^2}$$

Now multiplying by dt, we see that

$$ds = \sqrt{(dx)^2 + (dy)^2} \tag{16.1}$$

This can be expressed in a variety of forms including

$$ds = \sqrt{1 + \left(\frac{dy}{dx}\right)^2}\,dx \qquad \text{\textit{multiplying and dividing by dx}} \quad (16.1a)$$

$$ds = \sqrt{\left(\frac{dx}{dy}\right)^2 + 1}\,dy \qquad \text{\textit{multiplying and dividing by dy}} \quad (16.1b)$$

$$ds = \sqrt{\left(\frac{dx}{dt}\right)^2 + \left(\frac{dy}{dt}\right)^2}\,dt \qquad \text{\textit{multiplying and dividing by dt}} \quad (16.1c)$$

We shall express ds in terms of polar coordinates as a final form. If we make use of the relations

$$x = r\cos\theta \qquad \text{and} \qquad y = r\sin\theta$$

we obtain

$$dx = -r\sin\theta\,d\theta + \cos\theta\,dr$$

$$dy = r\cos\theta\,d\theta + \sin\theta\,dr$$

Therefore,

$$(dx)^2 + (dy)^2 = (-r\sin\theta\,d\theta + \cos\theta\,dr)^2 + (r\cos\theta\,d\theta + \sin\theta\,dr)^2$$
$$= r^2(d\theta)^2 + (dr)^2$$

as obtained by expanding and collecting terms. Therefore, (16.1) becomes

$$ds = \sqrt{r^2(d\theta)^2 + (dr)^2}$$

[*] See most any advanced calculus, for example, Goursat-Hedrick, "A Course in Mathematical Analysis," vol. I, page 163, Ginn, Boston, 1916.

Now, multiplying and dividing by $d\theta$, we have

$$ds = \sqrt{r^2 + \left(\frac{dr}{d\theta}\right)^2}\, d\theta \qquad\qquad (16.1d)$$

It is not necessary to remember the variations of (16.1) since, as demonstrated above, each is readily obtained from (16.1).

Example 1   If $x = t^2 - 3t$ and $y = t^3 + t$, find ds.

Solution: Since x and y are given in terms of a parameter t, we use (16.1c). Consequently, we must find dx/dt and dy/dt. They are

$$\frac{dx}{dt} = 2t - 3 \qquad \text{and} \qquad \frac{dy}{dt} = 3t^2 + 1$$

Hence, (16.1c) becomes

$$ds = \sqrt{(2t - 3)^2 + (3t^2 + 1)^2}\, dt = \sqrt{9t^4 + 10t^2 - 12t + 10}\, dt$$

Example 2   If $y = x^2 + 3x + 7$, find ds for $x = 0$.

Solution: Since the equation is given in rectangular form, we use (16.1a). By differentiating with respect to x, we have $y' = 2x + 3$. Hence

$$1 + y'^2 = 1 + (2x + 3)^2 = 1 + (4x^2 + 12x + 9) = 2(2x^2 + 6x + 5)$$

Therefore,

$$ds = \sqrt{2}\, \sqrt{2x^2 + 6x + 5}\, dx = \sqrt{10}\, dx \qquad \text{for } x = 0$$

EXERCISE 16.2

Find the value of ds/dx for the specified value of x in each of Problems 1 to 24.

1   $y = x^2$, $x = \frac{1}{2}$

2   $y = x^2 - 2x$, $x = 1$

3   $y = x^3$, $x = 0$

4   $y = x^4 - 2x^3$, $x = 1$

5   $y = 1/(x^2 + 1)$, $x = 1$

6   $y = 4x/(x^2 + 1)$, $x = 1$

7   $y = 2x/(x^2 - 2)$, $x = 2$

8   $y = 3x/(2x^2 - 3)$, $x = 1$

9   $x^2 + 4y^2 = 8$, $x = 2$

10   $x^2 + y^2 = 10x$, $x = 2$

11   $9x^2 - y^2 = 9$, $x = -2$

12   $(x - 1)^2 - (y - 2)^2 = 4$, $x = 5$

13   $y = \sin x$, $x = x_1$

14   $y = \cos x$, $x = x_1$

15   $y = \sec x$, $x = \pi/4$

16   $y = \tan x$, $x = \pi/3$

17   $y = e^x$, $x = 0$

18   $y = e^{-x}$, $x = -1$

19   $y = \sinh x$, $x = x_1$

20   $y = \cosh x$, $x = x_1$

21   $y = \ln x$, $x = x_1$

22   $y = \ln \sqrt{x^2 + 1}$, $x = 0$

23   $y = \sinh^{-1} x$, $x = 2\sqrt{2}$

24   $y = \cosh^{-1} x$, $x = 3$

Find the value of ds/du or ds/dθ for the specified value of u or θ in each of Problems 25 to 36.

25 $x = 4 - u$, $y = 4u - u^2$, $u = 2$

26 $x = u^2$, $y = u + 1$, $u = \sqrt{2}$

27 $x = (u - 1)/(u + 1)$, $y = (u + 1)/(u - 1)$, $u = 0$

28 $x = (2u - 1)/u$, $y = (u + 2)/(u - 1)$, $u = 2$

29 $x = 2 \cos u$, $y = 2 \sin u$, $u = u_1$

30 $x = \sin u$, $y = \cos^2 u - 1$, $u = \pi/6$

31 $x = \tan u - \cot u$, $y = \tan u + \cot u$, $u = \pi/4$

32 $x = a(u - \sin u)$, $y = a(1 - \cos u)$, $u = \pi/3$

33 $r = 4 - \sin \theta$, $\theta = \pi/2$

34 $r = \cos \theta$, $\theta = \theta_1$ 35 $r = 4/(1 + \cos \theta)$, $\theta = \pi/2$

36 $r = 6/(2 - \sin \theta)$, $\theta = \pi$

37 Show that if, in moving along a curve, a point passes through a maximum or minimum at which $dy/dx = 0$, then the arc length is changing at the same rate as x.

38 Show that if θ is the inclination of the tangent line to a curve at any point $P(x, y)$ on it, then $dx/ds = |\cos \theta|$ and $dy/ds = |\sin \theta|$.

39 For what points on a curve is $ds/dx = ds/dy$?

40 Show $ds/dx = y$ for any point on $y = \cosh x$.

16.4 Curvature

We shall now investigate the rate of change of direction of a curve. More specifically, we shall find the rate of change of the angle of inclination of the tangent to the curve with respect to the arc length of the curve. Symbolically, we want to find an expression for $d\theta/ds$. In order to do this, we shall use the graph of $y = f(x)$ and shall assume that y' and y'' exist for all points on it that are under consideration. We then let P in Fig. 16.3 be any point on the curve and Q any other point such that arc PQ is Δs. We represent the angles of inclination of the tangents at P and Q by θ and $\theta + \Delta\theta$, respectively. Consequently, $\Delta\theta$ is the angle through which the tangent line turns as a point on the curve moves from P to Q. Therefore, $\Delta\theta/\Delta s$ is the average rate of change of inclination with respect to arc

curvature defined length. We now represent *curvature* at P by K and define it by the statement that

$$K = \left| \lim_{\Delta s \to 0} \frac{\Delta\theta}{\Delta s} \right| = \left| \frac{d\theta}{ds} \right|$$

In order to obtain an expression for curvature in terms of $dy/dx = y'$ and $d^2y/dx^2 = y''$, we use the fact that the value of y' at P is $\tan \theta$; hence, $\theta = \arctan y'$. Now differentiating with respect to s, we obtain

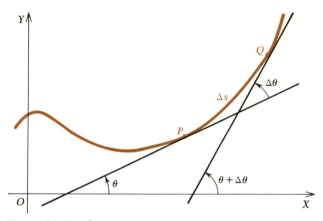

Figure 16.3 **Curvature**

$$\frac{d\theta}{ds} = \frac{d\theta}{dx}\frac{dx}{ds} = \frac{d}{dx}(\arctan y')\frac{dx}{ds} = \frac{dy'/dx}{1+y'^2}\frac{dx}{ds}$$

$$= \frac{y''}{1+y'^2}\frac{dx}{ds} = \frac{y''}{1+y'^2}\frac{1}{\sqrt{1+y'^2}}$$

formula
for
curvature
since $dx/ds = 1 \div (ds/dx) = 1/\sqrt{1+y'^2}$. Consequently, *the curva-*
ture K is given by

$$K = \left|\frac{d\theta}{ds}\right| = \frac{|y''|}{(1+y'^2)^{3/2}} \tag{16.2}$$

The absolute-value sign is not needed on $(1 + y'^2)^{3/2}$ since it is positive.

This expression for curvature is given in terms of rectangular coordinates, but its form can be changed for use with polar or parametric equations by means of the usual transformations.

Example 1   Find the curvature of $y = x^3$ at $(1, 1)$.

Solution: Differentiating, we see that

$$\frac{dy}{dx} = 3x^2 = 3 \qquad \text{at } (1, 1)$$

$$\frac{d^2y}{dx^2} = 6x = 6 \qquad \text{at } (1, 1)$$

and $$K = \frac{|y''|}{(1+y'^2)^{3/2}} = \frac{6}{(1+3^2)^{3/2}} = \frac{6\sqrt{10}}{100}$$

Example 2   Find the curvature of $x = t^2 - 2t$, $y = 1 - 4t$ for $t = 2$.

Solution: In order to use (16.2), we must have $D_x y$ and $D_x^2 y$; we shall now find them.

$$D_x y = \frac{D_t y}{D_t x} = \frac{-4}{2t - 2} = \frac{2}{1 - t} = -2 \qquad \text{for } t = 2$$

since $D_t y = -4$ and $D_t x = 2(t - 1)$.

$$D_x^2 y = D_x \left(\frac{2}{1 - t} \right) = D_t \left(\frac{2}{1 - t} \right) D_x t$$

$$= \frac{2}{(1 - t)^2} \frac{1}{2(t - 1)} = 1 \qquad \text{for } t = 2$$

Therefore, $$K = \frac{1}{[1 + (-2)^2]^{3/2}} = \frac{1}{5\sqrt{5}}$$

16.5   Curvature of a circle

If P and Q are any two points on a circle of radius r, if Δs is the arc length between them, and if $\Delta\theta$ is the angle between the tangents drawn at P and Q as in Fig. 16.4, then

$$\frac{\Delta\theta}{\Delta s} = \frac{\Delta\theta}{r\Delta\theta} \qquad since \ \Delta s = r \ \Delta\theta$$

curvature Consequently, $$K = \left| \lim_{\Delta s \to 0} \frac{\Delta\theta}{\Delta s} \right| = \frac{1}{r}$$
of a circle

It follows that the curvature of a circle is the reciprocal of the radius of the circle.

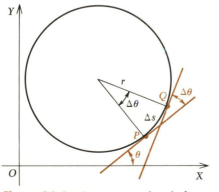

Figure 16.4   Curvature of a circle

16.6   Radius and circle of curvature

radius of curvature   We found in the last section that the radius of a circle is equal to the reciprocal of the curvature. By analogy, we define the *radius of curvature* of a curve at any point P on it to be the reciprocal of the curvature at that point. Thus, the radius of curvature is

$$R = \frac{[1 + (dy/dx)^2]^{3/2}}{|d^2y/dx^2|} \qquad (16.3)$$

circle of curvature   Furthermore, the circle that is tangent to the curve at P has its center on the normal to the curve at P, and has a radius equal to R as given by (16.3) is called the *circle of curvature*.

We shall now derive formulas for the coordinates (h, k) of the center of the circle of curvature for the case in which the given curve is concave upward at $P(x_1, y_1)$ and the slope at P is positive. From Fig. 16.5, we see that

$$h = x_1 - QP = x_1 - R \sin \theta_1$$

and

$$k = y_1 + QC = y_1 + R \cos \theta_1$$

The forms of these equations can be changed by making use of the value of R as given by (16.3) and the fact that

$$\tan \theta_1 = y_1' = \frac{y_1'}{1}$$

$$= \frac{y_1'/\sqrt{1 + y_1'^2}}{1/\sqrt{1 + y_1'^2}} \qquad \textit{dividing each member by } \sqrt{1 + y'^2}$$

Consequently, by use of (9.13),

$$\sin \theta_1 = \frac{y_1'}{\sqrt{1 + y_1'^2}} \qquad \text{and} \qquad \cos \theta_1 = \frac{1}{\sqrt{1 + y_1'^2}}$$

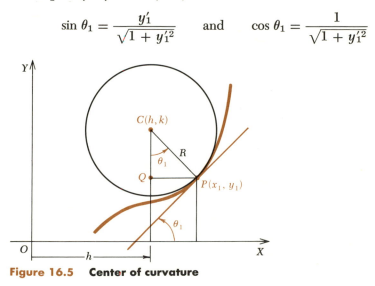

Figure 16.5   Center of curvature

Therefore,

center of
curvature
$$h = x_1 - \frac{(1 + y_1'^2)^{3/2}}{y_1''} \frac{y_1'}{\sqrt{1 + y_1'^2}} = x_1 - \frac{y_1'(1 + y_1'^2)}{y_1''}$$

$$k = y_1 + \frac{(1 + y_1'^2)^{3/2}}{y_1''} \frac{1}{\sqrt{1 + y_1'^2}} = y_1 + \frac{1 + y_1'^2}{y_1''}$$

It can be shown that these expressions for h and k hold even though the curve is not concave upward and the slope is not positive at P.

Example   Find the equation of the circle of curvature of $y = x^3$ at $(1, 1)$.

Solution: We can write the equation of the circle if we know the radius and the coordinates of the center. We shall now find these numbers. We need $y_1' = y'(1)$ and $y_1'' = y''(1)$ in determining h, k, and R. They are $y' = 3x^2 = 3$ for $x = 1$ and $y'' = 6x = 6$ for $x = 1$. Hence,

$$h = 1 - \frac{3(1 + 3^2)}{6} = 1 - 5 = -4$$

$$k = 1 + \frac{1 + 3^2}{6} = \frac{8}{3}$$

$$R = \frac{(1 + 3^2)^{3/2}}{6} = \frac{5\sqrt{10}}{3}$$

Consequently, the circle of curvature of $y = x^3$ at $(1, 1)$ is

$$(x + 4)^2 + (y - \tfrac{8}{3})^2 = \left(\frac{5\sqrt{10}}{3}\right)^2$$

EXERCISE 16.3

Find the curvature and radius of curvature of each curve in Problems 1 to 28 for the specified value.

1   $y = x^2 - 1$, $x = 1$ 2 $y = x^2 - 4x - 5$, $x = 2$
3   $y = 3x^2 - x^3$, $x = 1$ 4 $y = x^2(x^2 - 4)$, $x = 0$
5   $y = x^3 - 5x^2 + 12$, $x = 3$ 6 $y = x^3 + 3x - 1$, $x = 1$
7   $y = x^4 - 2x^2$, $x = 1$ 8 $y = x^5 - x^3 + 1$, $x = 1$
9   $9x^2 + 4y^2 = 36x$, $x = 2$ 10 $x^2 + y^2 + 4x + 6y = 0$, $x = 0$
11   $4x^2 - y^2 = 20$, $x = 3$ 12 $x^4 + y^4 = 32$, $x = 2$
13   $y = 2 \cos x$, $x = 0$ 14 $y = \sin^2 x$, $x = \pi/4$
15   $y = \sin x + \cos x$, $x = \pi/4$
16   $y = \sec x$, $x = \pi/3$ 17 $y = \ln x$, $x = \frac{1}{2}$
18   $y = \ln \sec x$, $x = \pi/3$
19   $y = e^{-x^2}$, $x = 0$ 20 $y = xe^{-x}$, $x = 1$
21   $x = t^2 + 1$, $y = t^3 + t$, $t = 1$

22 $x = 4t,\ y = 3\sqrt{1 - t^2},\ t = 0$
23 $x = 4\cos t,\ y = \cos 2t,\ t = \pi$
24 $x = a(t - \sin t),\ y = a(1 - \cos t),\ t = \pi$
25 $r = 4 - \sin\theta,\ \theta = \pi/2$
26 $r = \cos\theta,\ \theta = \pi/6$
27 $r = 4/(1 + \cos\theta),\ \theta = \pi/2$
28 $r = 6/(2 - \sin\theta),\ \theta = 0$

In each of Problems 29 to 36, locate the minimum point, and evaluate the curvature at that point.

29 $3y = x^3 - 6x^2$ 30 $y = e^{x/2} + e^{-x/2}$
31 $y = x\ln x$ 32 $y = x^2 e^{-x}$
33 $y = \sin^3 x,\ x$ in $(0, 2\pi)$
34 $y = \cos x + \tfrac{1}{2}\sin 2x,\ x$ in $[0, \pi]$
35 $y = \tan x,\ \pi/4 \leq x < \pi/3$
36 $y = \cos x,\ 0 \leq x \leq \pi/3$

Locate the point of maximum curvature and evaluate the curvature thereat for the curve in each of Problems 37 to 40.

37 $y = e^x$ 38 $xy = 1$
39 $y = \ln\sec x$ 40 $y = x^2$

Find the equation of the circle of curvature at the specified point in each of Problems 41 to 44.

41 $y = x^2,\ (1, 1)$ 42 $y = x^3 - x,\ (1, 0)$
43 $y = e^x,\ (0, 1)$ 44 $y = \ln x,\ (1, 0)$

45 Prove that if the center of curvature of $x^2/a^2 + y^2/b^2 = 1$ at one end of the minor axis lies at the other end of it, then the ellipse has $e = \sqrt{2}/2$.

46 Show that the curvature of $y = f(x)$ at a point $P(x, y)$ is $|y'' \cos^3\theta|$ where θ is the inclination of the tangent line to the curve at P.

47 Prove that $(1 + y'^2)y''' = 3y'y''^2$ at a point of maximum or minimum curvature if all three derivatives are continuous at the point.

48 Derive the equation

$$R = \frac{[1 + (dx/dy)^2]^{3/2}}{d^2x/dy^2}$$

from (16.3) by expressing dy/dx and d^2y/dx^2 in terms of dx/dy and d^2x/dy^2.

16.7 Vectors

vector quantity A quantity that is determined by a magnitude and a direction is called a *vector quantity*. Thus, a force in a specified direction is a vector quantity. Such a quantity can be and often is represented by

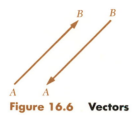

Figure 16.6   Vectors

designating the initial and terminal points. Thus, *AB* represents the vector from *A* to *B*, and *BA* represents the vector from *B* to *A* as shown in Fig. 16.6. They are equal in magnitude but measured in opposite directions; hence, we write *AB* = −*BA*, *BA* = −*AB*, or *AB* +

addition of vectors

BA = 0. We indicate geometrically that the vector *BC* is added to *AB* by drawing *BC* from the end of *AB*; furthermore, *AC* is the sum of *AB* and *BC*. If a vector *AC* can be given the same direction as *AB* by rotating *AC* through an angle *θ*, we say that *θ* is an angle between them as in Fig. 16.7. Since the sum of two vectors is a vector, we can add three vectors by adding the third to the sum of the first two.

associative law

It can be proved and we shall assume that the associative law holds for vectors; hence, the order of adding is immaterial. We subtract one vector from another by adding its negative. We say that two vectors

equal vectors

are *equal* if they have the same magnitude and same direction. They may be coincident but need not be. A vector is sometimes represented by a single boldface letter such as **A** or **v**.

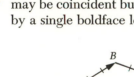

Figure 16.7   Angle between vectors

scalar

A number that specifies magnitude only is called a *scalar* and is represented by a numeral or a letter in italics as *m*. The product of the scalar *m* and the vector **A** is denoted by *m***A** or **A***m* and represents a vector with a magnitude |*m*| times that of **A** and with the same direction as, or the opposite direction to, **A** according as *m* is positive or negative. If *m* = 0, we get a zero or point vector.

If, as in Fig. 16.8, *OP* = **A** and *PQ* = **B**, then *OQ* = **A** + **B**.

product of a scalar and a vector

Furthermore, if *OR* = *m***A**, and *RS* = *m***B** then *OS* = *m*(**A** + **B**), and we see that

$$m(\mathbf{A} + \mathbf{B}) = m\mathbf{A} + m\mathbf{B}$$

Thus, the distributive law holds for multiplication of vectors by scalars.

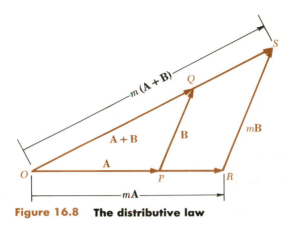

Figure 16.8   The distributive law

16.8   Components of a vector

components

If, by adding two or more vectors, we obtain a vector **v**, then the vectors added are called *components* of **v**. We shall be primarily interested in finding two components of a vector that are parallel to the coordinate axis. For this purpose we choose a vector one unit in length, called a *unit vector,* along each coordinate axis and then express the components in terms of these units. The unit vector along the X axis is designated by **i** and the one along the Y axis by **j**. Consequently, if $AB = a\mathbf{i} + b\mathbf{j}$, it is a vector originating at A, terminating at the point B, and with components of a units parallel to the X axis and b units parallel to the Y axis as shown in Fig. 16.9. The scalars a and b are called the coefficients of the vector.

unit
vector

coefficients of
a vector

In Sec. 16.7, we stated that two vectors are equal if they have the same magnitude and same direction. This condition can be expressed algebraically by saying that $v_1 = a_1\mathbf{i} + b_1\mathbf{j}$ and $v_2 = a_2\mathbf{i} + b_2\mathbf{j}$ are equal if and only if $a_1 = a_2$ and $b_1 = b_2$. Furthermore,

operation
on vectors

$$v_1 + v_2 = (a_1 + a_2)\mathbf{i} + (b_1 + b_2)\mathbf{j}$$

$$v_1 - v_2 = (a_1 - a_2)\mathbf{i} + (b_1 - b_2)\mathbf{j}$$

$$c(a\mathbf{i} + b\mathbf{j}) = ca\mathbf{i} + cb\mathbf{j}$$

where c is a scalar. If $c > 0$, then $c\mathbf{v}$ has horizontal and vertical components directed as those of **v**, and if $c < 0$, the horizontal and vertical components of $c\mathbf{v}$ are directed oppositely to those of **v**.

The direction of $\mathbf{v} = a\mathbf{i} + b\mathbf{j}$ is given by the angle **v** makes with the horizontal. The tangent of this angle is b/a and is the *slope* of the vector.

The length, or magnitude, of $\mathbf{v} = a\mathbf{i} + b\mathbf{j}$ is denoted by $|\mathbf{v}|$ or $|a\mathbf{i} + b\mathbf{j}|$ and, by use of the Pythagorean theorem and Fig. 16.9,

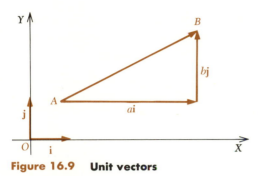

Figure 16.9   Unit vectors

we can see that

$$|ai + bj| = \sqrt{a^2 + b^2}$$

absolute value   This is often referred to as the *absolute value* of $ai + bj$. If the absolute value of a vector is equal to the unit of length used along *unit vector* the coordinate axes, we say that the vector is a *unit vector* as stated above. If **u** is a unit vector that makes an angle of θ with the positive X axis as in Fig. 16.10, then its horizontal and vertical components are

$$\mathbf{u}_x = \mathbf{i} \cos \theta \qquad \text{and} \qquad \mathbf{u}_y = \mathbf{j} \sin \theta$$

respectively. Therefore the vector can be written in the form

$$\mathbf{u} = \mathbf{i} \cos \theta + \mathbf{j} \sin \theta$$

This may be thought of as the unit vector obtained by rotating the vector **i** counterclockwise through an angle θ. Thus

$$\mathbf{u} = \frac{\sqrt{3}}{2}\mathbf{i} + \frac{1}{2}\mathbf{j}$$

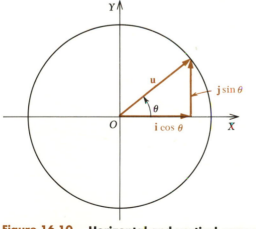

Figure 16.10   Horizontal and vertical components

is a unit vector since its absolute value is 1; furthermore, it can be thought of as being obtained by rotating **i** through 30° since $\cos 30° = \sqrt{3}/2$ and $\sin 30° = \frac{1}{2}$.

Example   Find a unit vector parallel to and directed as $\mathbf{v} = 5\mathbf{i} + 12\mathbf{j}$.

Solution: Since $|\mathbf{v}| = \sqrt{5^2 + 12^2} = 13$, we must divide by 13 in order to have a unit vector. Thus,

$$\mathbf{u} = \tfrac{5}{13}\,\mathbf{i} + \tfrac{12}{13}\,\mathbf{j}$$

is a unit vector parallel to and directed as **v** since **u** is 1 in absolute value and was obtained from **v** by multiplying by a positive scalar $\frac{1}{13}$.

16.9   The dot product of two vectors

dot product defined   If **A** and **B** are two vectors, then their *dot product,* indicated by **A** · **B**, is defined by

$$\mathbf{A} \cdot \mathbf{B} = |\mathbf{A}|\,|\mathbf{B}|\,\cos \theta$$

where θ is an angle between **A** and **B**. This product can be expressed in terms of the components of **A** and **B** by use of the angles the vectors **A** and **B** make with the X axis as we shall now show. If, as in Fig. 16.11,

$$\mathbf{A} = a_1\mathbf{i} + b_1\mathbf{j} \qquad \text{and} \qquad \mathbf{B} = a_2\mathbf{i} + b_2\mathbf{j}$$

make angles of θ_1 and θ_2 with the positive X axis, then

$$\sin \theta_1 = \frac{b_1}{|\mathbf{A}|} = \frac{b_1}{\sqrt{a_1{}^2 + b_1{}^2}} \qquad \cos \theta_1 = \frac{a_1}{\sqrt{a_1{}^2 + b_1{}^2}}$$

$$\sin \theta_2 = \frac{b_2}{|\mathbf{B}|} = \frac{b_2}{\sqrt{a_2{}^2 + b_2{}^2}} \qquad \cos \theta_2 = \frac{a_2}{\sqrt{a_2{}^2 + b_2{}^2}}$$

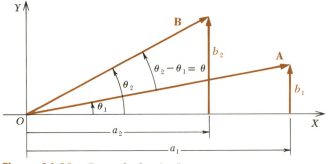

Figure 16.11   Formula for A · B

Therefore,

formula for
dot product

$$\mathbf{A} \cdot \mathbf{B} = |\mathbf{A}||\mathbf{B}| \cos \theta = \sqrt{a_1{}^2 + b_1{}^2} \sqrt{a_2{}^2 + b_2{}^2} \cos (\theta_2 - \theta_1)$$

$$= \sqrt{a_1{}^2 + b_1{}^2} \sqrt{a_2{}^2 + b_2{}^2} (\cos \theta_2 \cos \theta_1 + \sin \theta_2 \sin \theta_1)$$

$$= \sqrt{a_1{}^2 + b_1{}^2} \sqrt{a_2{}^2 + b_2{}^2} \frac{a_1 a_2 + b_1 b_2}{\sqrt{a_1{}^2 + b_1{}^2} \sqrt{a_2{}^2 + b_2{}^2}}$$

$$= a_1 a_2 + b_1 b_2$$

EXERCISE 16.4

Write the vector from the first point to the second in the form $a\mathbf{i} + b\mathbf{j}$, and find its absolute value.

1   $(2, 5)$, $(6, 8)$ 2 $(-1, -8)$, $(4, 4)$
3   $(-11, 0)$, $(13, 7)$ 4 $(1, -6)$, $(-7, 9)$

Find the sum of the two vectors both algebraically and geometrically.

5   $2\mathbf{i} + 3\mathbf{j}$, $5\mathbf{i} - \mathbf{j}$ 6 $-5\mathbf{i} + 11\mathbf{j}$, $3\mathbf{i} - 4\mathbf{j}$
7   $3\mathbf{i} + 4\mathbf{j}$, $8\mathbf{i} - 8\mathbf{j}$ 8 $17\mathbf{i} - 16\mathbf{j}$, $10\mathbf{i} + 8\mathbf{j}$

Find the unit vector that makes the specified angle with the X axis.

9   $30°$ 10 $150°$ 11 $225°$ 12 $300°$

Find the absolute value of the vector and the angle the vector makes with the X axis.

13   $\sqrt{3}\mathbf{i} + \mathbf{j}$ 14 $\mathbf{i} - \mathbf{j}$
15   $3\mathbf{i} + 4\mathbf{j}$ 16 $-5\mathbf{i} - 12\mathbf{j}$

Find the unit vector parallel to and directed as the given vector.

17   $3\mathbf{i} - 4\mathbf{j}$ 18 $-12\mathbf{i} + 5\mathbf{j}$
19   $4\mathbf{i} + 5\mathbf{j}$ 20 $-2\mathbf{i} - 7\mathbf{j}$

Find the unit vector that is parallel to the tangent to the given curve at the given point.

21   $y = x^3$, $(1, 1)$ 22 $y = x^2 - x$, $(3, 6)$
23   $y = \sin x$, $(\pi/3, \sqrt{3}/2)$ 24 $y = e^x$, $(0, 1)$

Find the dot product of the two vectors.

25   $2\mathbf{i} + 3\mathbf{j}$, $4\mathbf{i} - 5\mathbf{j}$ 26 $3\mathbf{i} - 2\mathbf{j}$, $-5\mathbf{i} + 3\mathbf{j}$
27   $7\mathbf{i} + 6\mathbf{j}$, $5\mathbf{i} - 4\mathbf{j}$ 28 $-4\mathbf{i} + 7\mathbf{j}$, $7\mathbf{i} - 4\mathbf{j}$

Find an angle between the given vectors.

29   $4\mathbf{i} - \mathbf{j}$, $3\mathbf{i} + 5\mathbf{j}$ 30 $-2\mathbf{i} + 7\mathbf{j}$, $3\mathbf{i} - 4\mathbf{j}$
31   $5\mathbf{i} - 3\mathbf{j}$, $-3\mathbf{i} + 5\mathbf{j}$ 32 $7\mathbf{i} + 5\mathbf{j}$, $-2\mathbf{i} + 3\mathbf{j}$

33   Prove that $\mathbf{A} \cdot (\mathbf{B} + \mathbf{C}) = \mathbf{A} \cdot \mathbf{B} + \mathbf{A} \cdot \mathbf{C}$.
34   Prove that $\mathbf{A}$ and $\mathbf{B}$ are perpendicular if and only if $\mathbf{A} \cdot \mathbf{B} = 0$.

35 Prove that **A** and **B** are parallel if and only if $\mathbf{A} \cdot \mathbf{B} = \pm|\mathbf{A}||\mathbf{B}|$.

36 Show that $\mathbf{i} \cdot \mathbf{i} = \mathbf{j} \cdot \mathbf{j} = 1$ and $\mathbf{i} \cdot \mathbf{j} = \mathbf{j} \cdot \mathbf{i} = 0$.

Use the results of Problems 34 and 35 to determine which pairs of vectors are perpendicular and which are parallel.

37 $3\mathbf{i} - 4\mathbf{j}, -6\mathbf{i} + 8\mathbf{j}$ **38** $5\mathbf{i} + 7\mathbf{j}, -2.5\mathbf{i} - 3.5\mathbf{j}$

39 $2\mathbf{i} + 5\mathbf{j}, 15\mathbf{i} - 6\mathbf{j}$ **40** $3\mathbf{i} - 6\mathbf{j}, 2\mathbf{i} + \mathbf{j}$

16.10   Differentiating vectors

In this section, we shall let x and y be functions of the parameter t that have first and second derivatives and then shall find the first and *position vector* second derivatives of the *position vector*

$$\mathbf{R} = \mathbf{i}x + \mathbf{j}y \tag{16.4a}$$

with respect to t. The vector $\mathbf{R}$ is called a position vector since it is determined if the scalars x and y are given.

If $x = f(t)$ and $y = g(t)$, then $\mathbf{R} = \mathbf{i}f(t) + \mathbf{j}g(t)$; furthermore, if t takes on an increment Δt and if we represent the corresponding increments in x, y, and $\mathbf{R}$ by Δx, Δy, and $\Delta \mathbf{R}$, respectively, (16.4a) becomes

$$\mathbf{R} + \Delta \mathbf{R} = \mathbf{i}(x + \Delta x) + \mathbf{j}(y + \Delta y) \tag{16.4b}$$

Consequently, subtracting each member of (16.4a) from the corresponding member of (16.4b), we get

$$\Delta \mathbf{R} = \mathbf{i}\,\Delta x + \mathbf{j}\,\Delta y$$

Therefore, dividing by Δt and taking the limit as Δt approaches zero, we find that

$$\frac{d\mathbf{R}}{dt} = \mathbf{i}\frac{dx}{dt} + \mathbf{j}\frac{dy}{dt} \tag{16.5a}$$

By use of the equations $x = f(t)$ and $y = g(t)$, this can be put in the form

$$\frac{d\mathbf{R}}{dt} = \mathbf{i}\frac{df(t)}{dt} + \mathbf{j}\frac{dg(t)}{dt} \tag{16.5b}$$

We shall use Fig. 16.12 in investigating the magnitude and direction of $d\mathbf{R}/dt$. Its magnitude is

$$\left|\frac{d\mathbf{R}}{dt}\right| = \left|\mathbf{i}\frac{dx}{dt} + \mathbf{j}\frac{dy}{dt}\right| = \sqrt{\left(\frac{dx}{dt}\right)^2 + \left(\frac{dy}{dt}\right)^2} = \frac{ds}{dt}$$

Since $d\mathbf{R}/dt$ rises dy/dt units as it moves horizontally dx/dt units as seen from (16.5a), it follows that the inclination of $d\mathbf{R}/dt$ is $(dy/dt) \div (dx/dt) = dy/dx$. Furthermore, dy/dx is the slope of the

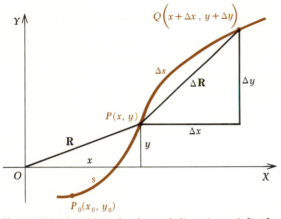

Figure 16.12   Magnitude and direction of $d\mathbf{R}/dt$

tangent at P to the curve traced by the terminus of $\mathbf{R}$. Consequently, $d\mathbf{R}/dt$ is parallel at P to the tangent to the curve traced by the position vector. Therefore, if t represents time, the vector $d\mathbf{R}/dt$ drawn from P represents the velocity of $\mathbf{R}$ along the path traced by P; *velocity vector* hence, $d\mathbf{R}/dt$ is the *velocity vector*.

If we take the derivative with respect to t of the velocity vector, we find that

$$\frac{d^2\mathbf{R}}{dt^2} = \mathbf{i}\frac{d^2x}{dt^2} + \mathbf{j}\frac{d^2y}{dt^2} \tag{16.6a}$$

acceleration This is the acceleration since, by definition, acceleration is the rate of change of velocity with respect to time. In terms of t, (16.6a) becomes

$$\frac{d^2\mathbf{R}}{dt^2} = \mathbf{i}\frac{d^2f(t)}{dt^2} + \mathbf{j}\frac{d^2g(t)}{dt^2} \tag{16.6b}$$

Example   If a particle moves along the parabola

$$x = \frac{t^2}{4} \qquad y = t + 1$$

find $\mathbf{R}$, $\mathbf{v}$, and $\mathbf{a}$.

Solution: The position vector is given by (16.4a) and is

$$\mathbf{R} = \mathbf{i}\frac{t^2}{4} + \mathbf{j}(t + 1)$$

Therefore,

$$\mathbf{v} = \mathbf{i}\frac{t}{2} + \mathbf{j}$$

and

$$\mathbf{a} = \frac{\mathbf{i}}{2}$$

16.11   Tangential and normal vectors

In Sec. 16.10, we thought of the position vector $\mathbf{R} = \mathbf{i}x + \mathbf{j}y$ of a particle P moving on a curve as depending on the coordinates of P. We shall now think of $\mathbf{R}$ as determined by the arc length s from some arbitrarily chosen point P_0 of reference on the curve and investigate $d\mathbf{R}/ds$.

If arc P_0P is of length s and arc PQ of length Δs, then

$$\frac{\Delta \mathbf{R}}{\Delta s} = \mathbf{i}\frac{\Delta x}{\Delta s} + \mathbf{j}\frac{\Delta y}{\Delta s}$$

Furthermore, $\Delta \mathbf{R}/\Delta s$ is a vector whose magnitude is chord PQ divided by arc PQ. Consequently,

$$\frac{d\mathbf{R}}{ds} = \lim_{\Delta s \to 0} \frac{\Delta \mathbf{R}}{\Delta s}$$

is a *unit vector* since

$$\lim_{\Delta s \to 0} \left| \frac{\Delta \mathbf{R}}{\Delta s} \right| = \lim_{\Delta s \to 0} \frac{\text{chord } PQ}{\text{arc } PQ}$$

and the ratio of the chord and arc approaches unity as Δs approaches zero. We shall now investigate the direction of this unit vector. Since Δs is a scalar, it follows that $\Delta \mathbf{R}/\Delta s$ has the same direction as $\Delta \mathbf{R} = PQ$ if Δs is positive and the opposite direction if Δs is negative. Figure 16.13 shows the situation for $\Delta s > 0$. In both cases, $\Delta \mathbf{R}/\Delta s$ points in the direction of increasing s and is along the chord PQ. Hence, as Δs approaches zero and Q approaches P, the vector $\Delta \mathbf{R}/\Delta s$

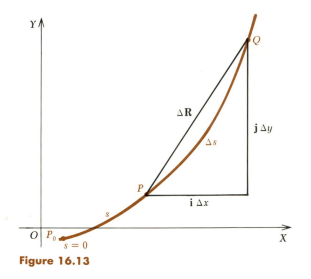

Figure 16.13

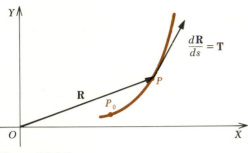

Figure 16.14

approaches the direction of the tangent to the curve at P. Conse-
direction of   quently, *the direction of $d\mathbf{R}/ds$ is along the tangent at P and away
dR/ds   from or toward P_0 according as Δs is positive or negative. Thus, we
may write

$$\frac{d\mathbf{R}}{ds} = \mathbf{T} \tag{16.7}$$

charcteri-   where $\mathbf{T}$ *is a unit vector that is tangent to the curve at P* as shown in
zation of   Fig. 16.14. It can be evaluated by means of
dR/ds

$$\frac{d\mathbf{R}}{ds} = \mathbf{i}\frac{dx}{ds} + \mathbf{j}\frac{dy}{ds}$$

if the equation of the curve is known.

Example   Find the unit vector $\mathbf{T}$ that is tangent to the circle

$$x = a \tanh t \qquad y = a \operatorname{sech} t$$

at any point $P(x, y)$ on it.

Solution: If we obtain the differentials, we find that

$$dx = a \operatorname{sech}^2 t\, dt \qquad \text{and} \qquad dy = -a \operatorname{sech} t \tanh t\, dt$$

Therefore,

$$(ds)^2 = (dx)^2 + (dy)^2 = a^2 \operatorname{sech}^4 t\, (dt)^2 + a^2 \operatorname{sech}^2 t \tanh^2 t\, (dt)^2$$
$$= a^2 \operatorname{sech}^2 t(\operatorname{sech}^2 t + \tanh^2 t)\, (dt)^2 = a^2 \operatorname{sech}^2 t\, (dt)^2$$

Hence, $ds = \pm a \operatorname{sech} t\, dt$

Consequently,

$$\mathbf{T} = \frac{d\mathbf{R}}{ds} = \mathbf{i}\frac{dx}{ds} + \mathbf{j}\frac{dy}{ds} = \mathbf{i}\frac{dx/dt}{ds/dt} + \mathbf{j}\frac{dy/dt}{ds/dt}$$

$$= \mathbf{i}\frac{a \operatorname{sech}^2 t}{\pm a \operatorname{sech} t} - \mathbf{j}\frac{a \operatorname{sech} t \tanh t}{\pm a \operatorname{sech} t}$$

$$= \pm\mathbf{i} \operatorname{sech} t \mp \mathbf{j} \tanh t$$

If s is measured counterclockwise from $(0, a)$ with $s = 0$ at $(0, a)$, then s is an increasing function of t and the plus sign should be used with ds.

If, as shown in Fig. 16.15, θ is the positive angle from the positive X axis to the unit tangential vector $\mathbf{T}$, then we may write

$$\mathbf{T} = \mathbf{i} \cos \theta + \mathbf{j} \sin \theta$$

because this is a unit vector with slope $\tan \theta$. Consequently,

$$\frac{d\mathbf{T}}{d\theta} = -\mathbf{i} \sin \theta + \mathbf{j} \cos \theta$$

is of magnitude

$$\left| \frac{d\mathbf{T}}{d\theta} \right| = \sqrt{(-\sin \theta)^2 + (\cos \theta)^2} = 1$$

and makes the angle ϕ with the X axis, where

$$\tan \phi = \frac{\cos \theta}{-\sin \theta} = -\cot \theta$$

We note that $d\mathbf{T}/d\theta$ is perpendicular to $\mathbf{T}$ because the product of their slopes is -1. Therefore,

characteri-
zation of
$d\mathbf{T}/d\theta$

$$\mathbf{N} = \frac{d\mathbf{T}}{d\theta} \qquad (16.8)$$

is a unit vector that is perpendicular to the unit vector $\mathbf{T}$. It is obtained by rotating $\mathbf{T}$ through $90°$ in a counterclockwise direction and is called the *unit normal* vector.

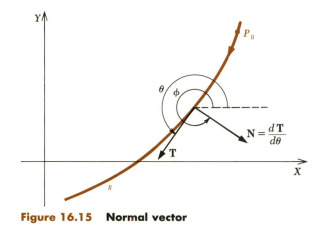

Figure 16.15   Normal vector

EXERCISE 16.5

The vector from the origin to the variable point P(x, y) in each of Problems 1 to 12 is given by R = ix + jy for any value of t. Find the velocity and acceleration in terms of t, and evaluate each for the given value of t.

1. $x = t^2, y = t^3, t = 0$
2. $x = t, y = t^2, t = 1$
3. $x = \cos^3 t, y = \sin^3 t, t = \pi/4$
4. $x = \sin 2t, y = \sin t, t = \pi/2$
5. $x = \tan t, y = \sec t, t = 0$
6. $x = \ln (t + 2), y = t + 1, t = -1$
7. $x = \cosh t, y = \sinh t, t = 0$
8. $x = e^{-t}, y = e^{2t}, t = 1$
9. $x = a \sin t, y = a \csc t, t = 3\pi/2$
10. $x = h + a \sec t, y = k + b \tan t, t = \pi/3$
11. $x = 3 \sin t, y = 3 \cos 2t, t = \pi$
12. $x = 2 - a \sin t, y = 1 + b \cos t, t = \pi/6$

*If **R** = ix + jy is the vector from the origin to the variable point P(x, y) in each of Problems 13 to 24, find **T** and **N** in each case in terms of t or for the given value of t.*

13. $x = 2 + t, y = 1 + 2t$
14. $x = t^2, y = 1 + t$
15. $x = t + t^{-1}, y = t - t^{-1}$
16. $x = (2t - 1)/t, y = t/(2t - 1)$
17. $x = e^t - e^{-t}, y = e^t + e^{-t}, t = 0$
18. $x = e^t, y = e^{3t}, t = 0$
19. $x = \text{sech } t, y = \tanh t, t = 0$
20. $x = \cosh t, y = \sinh t, t = 0$
21. $x = 2 \cot t, y = 2 \tan t, t = \pi/4$
22. $x = \sin t, y = 1 + \cos^2 t, t = \pi/6$
23. $x = \ln t, y = t + 1, t = \frac{1}{3}$
24. $x = a(t - \sin t), y = a(1 - \cos t), t = \pi/3$

25. By use of (16.8) and the definition of curvature K, show that

$$\frac{d\mathbf{T}}{ds} = NK$$

26. Show that if x and y are thought of as functions of arc length s and if $dx/ds = \cos \theta$ and $dy/ds = \sin \theta$ for θ the inclination of the tangent line, then

$$\mathbf{T} = \mathbf{i}\frac{dx}{ds} + \mathbf{j}\frac{dy}{ds} \quad \text{and} \quad \mathbf{N} = -\mathbf{i}\frac{dy}{ds} + \mathbf{j}\frac{dx}{ds}$$

27 Show that if $\mathbf{R} = \mathbf{i}e^t \cos t + \mathbf{j}e^t \sin t$, then the acceleration vector is perpendicular to it for all t.

28 Show that if $\mathbf{R} = A\mathbf{i}e^t + B\mathbf{j}e^{2t}$, $A \neq 0$, $B \neq 0$, then the acceleration vector is never parallel to it.

16.12 Tangential and normal components of velocity and acceleration

The instantaneous speed ds/dt, the acceleration d^2s/dt^2, and the path and curvature of the path are used in discussing the motion of a particle P. These quantities could be expressed in terms of their vertical and horizontal components, but we shall use their tangential and normal components.

The velocity of the position vector OP is given by

$$\mathbf{v} = \frac{d\mathbf{R}}{dt}$$

in $(16.5a)$ and can be put in the form

$$\mathbf{v} = \frac{d\mathbf{R}}{dt}\frac{ds}{ds}$$

$$= \frac{d\mathbf{R}}{ds}\frac{ds}{dt} \qquad \textit{by the chain rule}$$

$$= \mathbf{T}\frac{ds}{dt} \qquad \textit{by use of (16.7)}$$

where $\mathbf{T}$ is a unit tangent to the curve at P. In order to emphasize it, we shall rewrite

velocity formula
$$\mathbf{v} = \mathbf{T}\frac{ds}{dt} \qquad (16.9)$$

We shall now make use of the fact that acceleration is the rate of change of velocity with respect to time. We differentiate each member of (16.9) with respect to t and have

$$\mathbf{a} = \mathbf{T}\frac{d^2s}{dt^2} + \frac{ds}{dt}\frac{d\mathbf{T}}{dt}$$

$$= \mathbf{T}\frac{d^2s}{dt^2} + \frac{ds}{dt}\frac{d\mathbf{T}}{ds}\frac{ds}{dt}$$

$$= \mathbf{T}\frac{d^2s}{dt^2} + \frac{ds}{dt}\,\mathbf{N}K\,\frac{ds}{dt} \qquad \textit{Problem 25 of Exercise 16.5}$$

This equation can be put in the form

acceleration formula

$$\mathbf{a} = \mathbf{T}\frac{d^2s}{dt^2} + \mathbf{NK}\left(\frac{ds}{dt}\right)^2 \tag{16.10}$$

It gives the acceleration in terms of its tangential and normal components since $\mathbf{T}$ and $\mathbf{N}$ are tangential and normal unit vectors and their coefficients are scalars.

If we represent the horizontal, vertical, tangential, and normal components of acceleration by a_x, a_y, a_T, and a_N, respectively, we can write

$$|\mathbf{a}|^2 = a_x{}^2 + a_y{}^2 = a_T{}^2 + a_N{}^2$$

by use of the Pythagorean theorem. Now solving for a_N, we have a_N in terms $\mathbf{a}$ and a_T, as given by

$$a_N = \sqrt{|\mathbf{a}|^2 - a_T{}^2} \tag{16.11}$$

If $\mathbf{R} = \mathbf{i}x + \mathbf{j}y$, where x and y are in terms of t, then as found earlier

$$\mathbf{v} = \mathbf{i}\frac{dx}{dt} + \mathbf{j}\frac{dy}{dt}$$

collection of formulas

$$\mathbf{a} = \frac{d\mathbf{v}}{dt}$$

$$v_T = \frac{ds}{dt} = |\mathbf{v}| = \text{tangential component of velocity}$$

$$a_T = \frac{d^2s}{dt^2} = \text{tangential component of acceleration}$$

Example

If $\mathbf{R} = \mathbf{i}e^t \sin t + \mathbf{j}e^t \cos t$, find a_N.

Solution: In order to be able to find a_N, we must first find $\mathbf{v}$, $\mathbf{a}$, v_T, and a_T. By use of the equations collected just before this example, we find that

$$\mathbf{v} = \mathbf{i}e^t(\cos t + \sin t) + \mathbf{j}e^t(\cos t - \sin t)$$

$$\mathbf{a} = \frac{d\mathbf{v}}{dt} = \mathbf{i}2e^t \cos t - \mathbf{j}2e^t \sin t$$

$$v_T = \frac{ds}{dt} = |\mathbf{v}| = e^t\sqrt{2}$$

$$a_T = \frac{d}{dt}\left(\frac{ds}{dt}\right) = \frac{d}{dt}e^t\sqrt{2} = e^t\sqrt{2}$$

Therefore, $a_N = \sqrt{|\mathbf{a}|^2 - a_T{}^2}$

becomes $a_N = \sqrt{(2e^t)^2 - (\sqrt{2}e^t)^2} = e^t\sqrt{2}$

EXERCISE 16.6

In each of Problems 1 to 12, find a_T and a_N in terms of t and for the given value of t.

1 $R = i(2t^2 - 4) + j(t^4 - 8t^2)$, $t = 2$

2 $R = i(t^2/2 - 2) + j(t^4/4 - 4t^2)$, $t = 2\sqrt{2}$

3 $R = i(4 - t) + jt(t - 4)^2$, $t = 4$

4 $R = i(2t + 1) + jt^3$, $t = 0$

5 $R = i4 \cos t + j4 \sin t$, $t = \pi/6$

6 $R = i(2 + t \sin t) + j(5 \cos t - 3)$, $t = \pi$

7 $R = i(5 + 5 \sin t) + j(3 \cos t - 3)$, $t = \pi/4$

8 $R = i(1 - \sin t) + j4 \cos t$, $t = \pi/3$

9 $R = i \ln(t^2 + 1) - j(t + 2 \operatorname{arccot} t)$, $t = 3\pi/2$

10 $R = i \ln t + j(\ln t)/t$, $t = e$

11 $R = i \cosh t + j \sinh t$, $t = \pi$

12 $R = i \tanh t + j \operatorname{sech} t$, $t = 0$

13 If $u_r = i \cos \theta + j \sin \theta$ and $u_\theta = -i \sin \theta + j \cos \theta$, prove that $du_r/d\theta = u_\theta$ and $du_\theta/d\theta = -u_r$.

14 If $R = OP$, show that $R = ru_r$ provided that the coordinates of P are (r, θ).

15 Show that $du_r/dt = u_\theta \, d\theta/dt$, $du_\theta/dt = -u_r \, d\theta/dt$, and

$$v = \frac{dR}{dt} = u_r \frac{dr}{dt} + u_\theta r \frac{d\theta}{dt}$$

16 Show by use of $a = dv/dt$ that

$$a = u_r \left[\frac{d^2 r}{dt^2} - r \left(\frac{d\theta}{dt} \right)^2 \right] + u_\theta \left(r \frac{d^2\theta}{dt^2} + 2 \frac{dr}{dt} \frac{d\theta}{dt} \right)$$

Make use of the expressions given for v and a in Problems 15 and 16 to find v and a in Problems 17 to 20.

17 $r = 2 \cos \theta$, $d\theta/dt = 2$

18 $r = 4 \sin 2\theta$, $d\theta/dt = 2t$

19 $r = 2(1 - \cos \theta)$, $\theta = 3t$

20 $r = 4(1 + \sin \theta)$, $\theta = 2t$

17

Standard integration formulas

17.1 Introduction

We discussed the concept of the indefinite integral in Chap. 7 and explained the method of evaluating $\int u^n \, du$ where $n \neq -1$. The definite integral was presented in Chap. 8 and some of the applications of it were studied. In this chapter we shall employ the definition of the indefinite integral along with the derivatives studied in Chaps. 12 and 13 to obtain several additional integration formulas.

17.2 Logarithmic and exponential forms

By the definition of the indefinite integral, $\int f(u) \, du = F(u) + C$ if and only if the differential of $F(u)$ is $f(u) \, du$. Consequently, since, if $u > 0$,

$$d \ln u = \frac{d}{du} \ln u \, du = \frac{1}{u} \, du = \frac{du}{u}$$

it follows that

$$\int \frac{du}{u} = \ln u + C$$

Furthermore, if $u < 0$, then $-u > 0$, and $d \ln (-u) = -du/-u = du/u$. Therefore, in this case, we have

$$\int \frac{du}{u} = \int \frac{-du}{-u} = \int \frac{d(-u)}{-u} = \ln (-u) + C$$

Now since $|u| = u$ if $u > 0$, and $|u| = -u$ if $u < 0$, we have in each case,

$$\int \frac{du}{u} = \ln |u| + C \qquad\qquad (17.1)$$

Example 1   Evaluate

$$\int \frac{(2x - 1)\, dx}{x^2 - x}$$

Solution: Since the numerator of the integrand is the differential of the denominator, we have

$$\int \frac{(2x - 1)\, dx}{x^2 - x} = \ln |x^2 - x| + C$$

Example 2   Evaluate

$$\int \frac{x^2\, dx}{x^3 + 1}$$

Solution: In this example the differential of the denominator of the integrand is $3x^2\, dx$. Hence, we multiply the integrand by 3 and the integral by ⅓ and get

$$\int \frac{x^2\, dx}{x^3 + 1} = \frac{1}{3} \int \frac{3x^2\, dx}{x^3 + 1} = \frac{1}{3} \ln |x^3 + 1| + C$$

$\int e^u\, du$ Since the differential of e^u is $e^u\, du$, we have, by the definition of the indefinite integral

$$\int e^u\, du = e^u + C \qquad\qquad (17.2)$$

Example 3   Evaluate $\int e^{\sin x} \cos x\, dx$.

Solution: Since $d \sin x = \cos x\, dx$, we have

$$\int e^{\sin x} \cos x\, dx = e^{\sin x} + C$$

We shall now employ (17.2) to evaluate $\int a^v\, dv$. By the definition of a natural logarithm

$$L = \ln a \qquad\qquad (1)$$

if and only if

$$a = e^L \qquad\qquad (2)$$

Now, replacing L in (2) by its value from (1), we have $a = e^{\ln a}$. Consequently,

$$a^v = (e^{\ln a})^v = e^{v(\ln a)}$$

Therefore, $\int a^v \, dv = \int e^{v(\ln a)} \, dv$

Furthermore, if in (17.2), we let $u = v(\ln a)$, then $du = \ln a \, dv$ and

$$\int e^{v(\ln a)} \, dv = \frac{1}{\ln a} \int e^{v(\ln a)} \ln a \, dv$$

$$= \frac{1}{\ln a} e^{v(\ln a)} + C = \frac{1}{\ln a} a^v + C$$

$\int a^v \, dv$ since $\int e^{v(\ln a)} \ln a \, dv$ is in the form $\int e^u \, du$, and $e^{v(\ln a)} = a^v$. Hence,

$$\int a^v \, dv = \frac{1}{\ln a} a^v + C \qquad (17.2a)$$

Example 4   Evaluate $\int 5^x \, dx$.

Solution: The integral in this example is of the form $\int a^v \, dv$ with $a = 5$ and $v = x$. Hence, by (17.2a), we have

$$\int 5^x \, dx = \frac{5^x}{\ln 5} + C$$

Example 5   Evaluate $\int \dfrac{3^{\arctan\,(x/2)} \, dx}{x^2 + 4}$

Solution: The numerator of the integrand suggests the use of (17.2a) with $a = 3$ and $v = \arctan(x/2)$. However in this case

$$dv = \frac{dx/2}{x^2/4 + 1} = \frac{2 \, dx}{x^2 + 4}$$

Consequently, we multiply the integrand by 2 and the integral by ½ and complete the evaluation as follows:

$$\int \frac{3^{\arctan\,(x/2)} \, dx}{x^2 + 4} = \frac{1}{2} \int 3^{\arctan\,(x/2)} \frac{2 \, dx}{x^2 + 4} = \frac{3^{\arctan\,(x/2)}}{2 \ln 3} + C$$

by (17.2a), since the integrand on the right is in the form $a^v \, dv$.

EXERCISE 17.1

Perform the indicated integrations by use of the integration formulas of this and earlier chapters.

1   $\displaystyle\int \frac{2 \, dx}{2x - 1}$

2   $\displaystyle\int \frac{(2x - 3) \, dx}{x^2 - 3x}$

3   $\displaystyle\int_2^3 \frac{3(x^2 - 1) \, dx}{x^3 - 3x}$

4   $\displaystyle\int \frac{(4x^3 - 6x^2 - 3) \, dx}{x^4 - 2x^3 - 3x}$

5   $\int \dfrac{(x-1)\,dx}{x^2 - 2x}$

6   $\int_2^5 \dfrac{dx}{3x + 2}$

7   $\int \dfrac{(x+2)\,dx}{x^2 + 4x}$

8   $\int \dfrac{x^2\,dx}{2x^3 + 1}$

9   $\int_0^{\pi/4} \dfrac{\sec^2 x\,dx}{\tan x + 2}$

10   $\int \dfrac{\sin x\,dx}{2 - \cos x}$

11   $\int \cot x\,dx = \int \dfrac{\cos x\,dx}{\sin x}$

12   $\int_0^{\pi/3} \tan x\,dx = \int_0^{\pi/3} \dfrac{\sin x\,dx}{\cos x}$

13   $\int \dfrac{e^x}{e^x + 3}\,dx$

14   $\int \dfrac{3^x \ln 3}{3^x + 2}\,dx$

15   $\int \dfrac{xe^{x^2}\,dx}{e^{x^2} + 1}$

16   $\int \dfrac{2^x\,dx}{2^x + 5}$

17   $\int e^{2x}\,dx$

18   $\int (e^x + e^{-x})\,dx$

19   $\int x^2 e^{x^3}\,dx$

20   $\int \dfrac{e^{\ln x}}{x}\,dx$

21   $\int e^{\tan x} \sec^2 x\,dx$

22   $\int e^{(\sin x + \cos x)^2} \cos 2x\,dx$

23   $\int e^{\sec x} \sec x \tan x\,dx$

24   $\int e^{\ln \sin x} \cot x\,dx$

25   $\int_0^1 \dfrac{3^{\arcsin x}}{\sqrt{1 - x^2}}\,dx$

26   $\int_0^{\pi/3} 4^{\sec x} \sec x \tan x\,dx$

27   $\int \dfrac{7^{1/x^2}}{x^3}\,dx$

28   $\int_0^{\pi/2} 4^{\sin x} \cos x\,dx$

29   $\int \sec x\,dx$, multiply and divide by $\sec x + \tan x$
30   $\int \sec x\,dx$, multiply and divide by $\sec x - \tan x$
31   $\int \csc x\,dx$, multiply and divide by $\csc x + \cot x$
32   $\int \csc x\,dx$, multiply and divide by $\csc x - \cot x$
33   $\int \cot x \ln \sin x\,dx$
34   $\int \tan x \ln \cos x\,dx$

35   $\int \dfrac{dx}{x \ln x}$

36   $\int \dfrac{\ln x\,dx}{x}$

17.3   Trigonometric forms

If we make use of the definition of the indefinite integral along with the derivative of each of the trigonometric functions, and the suggestions in Problems 11, 12, 29, and 31 of Exercise 17.1, we can obtain more integration formulas. They are:

$$\int \cos u \, du = \sin u + c \tag{17.3}$$

$$\int \sin u \, du = -\cos u + c \tag{17.4}$$

$$\int \sec^2 u \, du = \tan u + c \tag{17.5}$$

integrals of the
derivatives of
the trigonometric
functions and
of trigonometric
functions

$$\int \csc^2 u \, du = -\cot u + c \tag{17.6}$$

$$\int \sec u \tan u \, du = \sec u + c \tag{17.7}$$

$$\int \csc u \cot u \, du = -\csc u + c \tag{17.8}$$

$$\int \cot x \, dx = \ln|\sin x| + c \tag{17.9}$$

$$\int \tan x \, dx = -\ln|\cos x| + c \tag{17.10}$$

$$\int \sec x \, dx = \ln|\sec x + \tan x| + c \tag{17.11}$$

$$\int \csc x \, dx = -\ln|\csc x + \cot x| + c \tag{17.12}$$

Consequently, we should now be able to integrate the usual six trigonometric functions and their derivatives.

Example 1
$$\int \cos 2x \, dx = \tfrac{1}{2} \int (\cos 2x) 2 \, dx \qquad d\, 2x = 2\, dx$$
$$= \tfrac{1}{2} \sin 2x + c \qquad by\ (17.3)$$

Example 2 $\int (\sec x - \tan x)^2 \, dx = \int (\sec^2 x - 2 \sec x \tan x + \tan^2 x) \, dx$
$$\qquad\qquad\qquad\qquad\qquad\qquad\qquad expanding$$
$$= \int (\sec^2 x - 2 \sec x \tan x + \sec^2 x - 1) \, dx$$
$$\qquad\qquad\qquad\qquad\qquad\qquad \tan^2 x = \sec^2 x - 1$$
$$= \int (2 \sec^2 x - 2 \sec x \tan x - 1) \, dx$$
$$\qquad\qquad\qquad\qquad\qquad\qquad collecting$$
$$= 2\int \sec^2 x \, dx - 2\int \sec x \tan x \, dx - \int dx$$
$$= 2 \tan x - 2 \sec x - x + c \qquad by\ (17.5)\ and\ (17.7)$$

Example 3 $\displaystyle \int \frac{\cos x \, dx}{\sec x + \tan x} = \int \frac{\cos x}{\sec x + \tan x} \frac{\sec x - \tan x}{\sec x - \tan x} \, dx$ *multiplying by 1*
$$= \int \frac{\cos x \sec x - \cos x \tan x}{\sec^2 x - \tan^2 x} \qquad expanding$$
$$= \int (1 - \sin x) \, dx \qquad simplifying$$
$$= x + \cos x + c \qquad by\ (17.4)$$

Example 4
$$\int_0^{\pi/4} \tan^2 x \sec^2 x \, dx = \frac{\tan^3 x}{3} \Big|_0^{\pi/4} = \frac{1}{3}$$

EXERCISE 17.2

Perform the indicated integrations.

1 $\displaystyle \int \cos 4x \, dx$

2 $\displaystyle \int \frac{dx}{\sec \tfrac{1}{2} x}$

3    $\int \cot 3x \sin 3x \, dx$

4    $\int_0^{0.5} \pi \sin \pi x \cos \pi x \, dx$

5    $\int 3 \sin \pi x \, dx$

6    $\int \sin 2x \, dx$

7    $\int_{\pi/2}^{\pi} (\sin x + \cos x)^2 \, dx$

8    $\int \sin x \cos 2x \, dx$

9    $\int \sec^2 (3x + 1) \, dx$

10    $\int_0^{\pi/8} \dfrac{2 \, dx}{\cos^2 2x}$

11    $\int_0^{\pi/4} \tan x \csc 2x \, dx$

12    $\int (\sin^2 3x + \cos^2 3x + \tan^2 2x) \, dx$

13    $\int \csc^2 (2x - 1) \, dx$

14    $\int \dfrac{dx}{4 \sin^2 x \cos^2 x}$

15    $\int \dfrac{\csc 2x \cot 2x \, dx}{\cos^2 x - \sin^2 x}$

16    $\int x \csc^2 (x^2 + 1) \, dx$

17    $\int_{\pi/6}^{\pi/3} \sec 4x \tan 4x \, dx$

18    $\int \dfrac{\sec 3x}{\cot 3x} \, dx$

19    $\int \dfrac{dx}{\sin 2x \tan 2x}$

20    $\int_0^{\pi/6} \dfrac{2 \cos x}{\sin^2 x} \, dx$

21    $\int \dfrac{dx}{\cos \pi x}$

22    $\int \sec (3x + 2) \, dx$

23    $\int \dfrac{\sin 2x \, dx}{\cos^2 x \sin x}$

24    $\int \dfrac{\cos 2x}{\cos x} \, dx$

25    $\int \csc (2x - 1) \, dx$

26    $\int_{\pi/4}^{\pi/2} \dfrac{2 \cos x \, dx}{\sin 2x}$

27    $\int x \csc x^2 \, dx$

28    $\int \dfrac{\cos 2x}{\sin x} \, dx$

29    $\int \dfrac{\sec x}{\csc x} \, dx$

30    $\int \sec^2 x \tan 2x \, dx$

31    $\int_{\pi/12}^{\pi/4} \dfrac{\csc 2x}{\sec 2x} \, dx$

32    $\int \left(\csc x \sin 2x + \dfrac{1}{\sin x \sec x} \right) dx$

33    $\int_0^{\pi/4} \cos 2x \, dx$

34    $\int_0^{\pi/6} \sin 3x \, dx$

35    $\int_{\pi/6}^{\pi/3} (\sec x + \tan x)^2 \, dx$

36    $\int_{\pi/6}^{\pi/4} \dfrac{1 - \cos x}{\sin^2 x} \, dx$

37    $\int_0^{\pi/3} \dfrac{\sin 2x}{2 \cos^2 x} \, dx$

38    $\int_{\pi/4}^{\pi/2} \dfrac{\cot x}{\sin x} \, dx$

39    $\int_{\pi/6}^{\pi/2} \dfrac{\sin x + \cos x}{\sin x} \, dx$

40    $\int_{\pi/6}^{\pi/4} \dfrac{2(1 - \cos^2 x)}{\sin 2x} \, dx$

17.4   Products of sines and cosines

An integral of the type $\int \sin 5x \cos 3x$ cannot be evaluated by use of any of the formulas previously discussed. We can, however, show that $\sin 5x \cos 3x = \frac{1}{2}(\sin 8x + \sin 2x)$, by employing one of the identities that we shall next develop. Thus we have $\int \sin 5x \cos 3x \, dx = \frac{1}{2}\int \sin 8x \, dx + \frac{1}{2}\int \sin 2x \, dx$. Then we can complete the integration by use of formula (17.4).

We shall also develop identities that enable us to deal with integrands of the type $\cos ax \cos bx$ and $\sin ax \sin bx$. For this purpose, we shall employ the identities for the sine and the cosine of the sum of two angles and proceed as below.

$$\sin x \cos y + \cos x \sin y = \sin (x + y)$$
$$\sin x \cos y - \cos x \sin y = \sin (x - y)$$

Now, if we equate the sums of the corresponding members of these equations, we get

sin x cos y
$$2 \sin x \cos y = \sin (x + y) + \sin (x - y) \tag{1}$$

Similarly, from the identities

$$\cos x \cos y - \sin x \sin y = \cos (x + y)$$
$$\cos x \cos y + \sin x \sin y = \cos (x - y)$$

we obtain

cos x cos y
$$2 \cos x \cos y = \cos (x + y) + \cos (x - y) \tag{2}$$

sin x sin y
$$2 \sin x \sin y = \cos (x - y) - \cos (x + y) \tag{3}$$

Now, by use of (1), (2), or (3), we can express integrands of the type $\sin ax \cos bx$, $\cos ax \cos bx$, and $\sin ax \sin bx$ as the sum of two sines, the sum of two cosines, and the difference of two cosines. Then formula (17.3) or (17.4) or both can be applied.

Example 1
$$\int 2 \sin 3x \cos x \, dx = \int [\sin (3x + x) + \sin (3x - x)] \, dx \qquad \textit{by (1)}$$
$$= \int (\sin 4x + \sin 2x) \, dx$$
$$= -\frac{1}{4} \cos 4x - \frac{1}{2} \cos 2x + C$$

Example 2
$$\int 2 \cos x \cos 2x \, dx = \int [\cos (x + 2x) + \cos (x - 2x)] \, dx \qquad \textit{by (2)}$$
$$= \int (\cos 3x + \cos x) \, dx \qquad \qquad \textit{since } \cos (-x)$$
$$= \cos x$$
$$= \frac{1}{3} \sin 3x + \sin x + C$$

Example 3 $\qquad \int 2 \sin 4x \sin 3x\, dx = \int (-\cos 7x + \cos x)\, dx \qquad$ *by* (3)

$$= -\frac{1}{7} \sin 7x + \sin x + C$$

17.5 Hyperbolic functions

If we employ the definition of the indefinite integral along with each of the Eqs. (13.6) to (13.11), we can obtain six more integration formulas. Thus,

integrals of the derivatives of the hyperbolic functions

$$\int \cosh u\, du = \sinh u + c \tag{17.13}$$

$$\int \sinh u\, du = \cosh u + c \tag{17.14}$$

$$\int \operatorname{sech}^2 u\, du = \tanh u + c \tag{17.15}$$

$$\int \operatorname{csch}^2 u\, du = -\coth u + c \tag{17.16}$$

$$\int \operatorname{sech} u \tanh u\, du = -\operatorname{sech} u + c \tag{17.17}$$

$$\int \operatorname{csch} u \coth u\, du = -\operatorname{csch} u + c \tag{17.18}$$

We shall now derive formulas for the integrals of $\tanh u$ and $\coth u$. Thus,

$$\int \tanh u = \int \frac{\sinh u}{\cosh u}\, du = \int \frac{d \cosh u}{\cosh u} = \ln |\cosh u| + c$$

The $\coth u$ can be treated similarly, and we then have

$\int \tanh u\, du$ $\qquad\qquad\qquad$ $\int \tanh u\, du = \ln |\cosh u| + c \qquad (17.19)$

$\int \coth u\, du$ $\qquad\qquad\qquad$ $\int \coth u\, du = \ln |\sinh u| + c \qquad (17.20)$

EXERCISE 17.3

Perform the indicated integrations.

1. $\int 2 \sin 5x \cos x\, dx$ $\qquad$ 2. $\int 6 \sin x \cos 3x\, dx$
3. $\int 2 \sin (3x - 1) \cos (3x + 1)\, dx$
4. $\int 2 \sin (1 - 3x) \cos (x - 1)\, dx$
5. $\int 2 \cos 2x \cos x\, dx$
6. $\int 2 \cos 6x \cos (-4x)\, dx$
7. $\int 4 \cos (3x - \pi) \cos (\pi - 3x)\, dx$
8. $\int 8 \cos (x - \pi/3) \cos (x + \pi/6)\, dx$
9. $\int 6 \sin 5x \sin x\, dx$
10. $\int 4 \sin (2x + 3) \sin (1 - 2x)\, dx$
11. $\int 2 \sin (2x - \pi) \sin (3\pi - 2x)\, dx$
12. $\int 2 \sin (x + \pi/3) \sin (x - \pi/6)\, dx$

13   $\int \cosh 2x \, dx$

14   $\int \cosh (3 - x) \, dx$

15   $\int \sinh (1 - 3x) \, dx$

16   $\int x \sinh x^2 \, dx$

17   $\int \tanh 2x \, dx$

18   $\int \dfrac{\tanh x^{1/2}}{x^{1/2}} \, dx$

19   $\int \dfrac{x \coth \sqrt{x^2 + 1}}{\sqrt{x^2 + 1}} \, dx$

20   $\int (3x^2 + 2) \coth (x^3 + 2x) \, dx$

21   $\int \operatorname{sech}^2 (2x + 1) \, dx$

22   $\int \operatorname{sech}^2 mx \, dx$

23   $\int \operatorname{csch}^2 nx \, dx$

24   $\int x \operatorname{csch}^2 x^2 \, dx$

25   $\int x \operatorname{sech} x^2 \tanh x^2 \, dx$

26   $\int 9 \operatorname{sech} (3x + 2) \tanh (3x + 2) \, dx$

27   $\int 4 \operatorname{csch} 2x \coth 2x \, dx$

28   $\int \dfrac{6x^2}{\sqrt{x^3 + 1}} \operatorname{csch} \sqrt{x^3 + 1} \coth \sqrt{x^3 + 1} \, dx$

29   $\int \cosh^3 3x \sinh 3x \, dx$

30   $\int \sqrt{\sinh x} \cosh x \, dx$

31   $\int \tanh^3 2x \operatorname{sech}^2 2x \, dx$

32   $\int \coth^{n-1} x \operatorname{csch}^2 x \, dx$

33   $\int \operatorname{sech}^5 x \tanh x \, dx$

34   $\int \operatorname{csch}^4 x \coth x \, dx$

35   $\int \cosh 2x \sinh 2x \, dx$

36   $\int \tanh^5 x \operatorname{sech}^2 x \, dx$

17.6   Powers of sines and cosines

If we are to integrate a function of x, we must be able to express the function in one of the forms that we have learned to integrate, or we must develop a new integration formula. We can do the former if we are given the product of an integral power of sin x and an integral power of cos x. These powers may be equal or unequal, both even, both odd, or one even and the other odd. In each situation, we use *integrating powers of trigonometric functions* the trigonometric identities and *express the given integrand as a power of a trigonometric function times the derivative of that function or as the sum of powers of a function times the derivative of the function.*

We shall now see how to perform the details under specified conditions, but we must not lose sight of the general principle that we are following.

$\int \sin^m x \cos^n x \, dx$ *with m or n or both odd integers.* If n is an odd integer, we save cos x as the derivative of sin x and then express $\sin^m x \cos^{n-1} x$ as the sum of powers of sin x. Since n is odd, it follows that $n - 1$ is even; hence $(n - 1)/2$ is an integer. Therefore,

$$\cos^{n-1} x = (\cos^2 x)^{(n-1)/2} \qquad n - 1 = 2(n - 1)/2$$

$$= (1 - \sin^2 x)^{(n-1)/2} \qquad \cos^2 x = 1 - \sin^2 x$$

can be expanded by the binomial theorem into a sum of powers of sin x. Consequently, $\sin^m x \cos^{n-1} x$ can also be expressed as a sum of powers of sin x. If m is an odd integer, a procedure similar to the one just followed enables us to keep sin x as a constant times $D_x \cos x$ and to express $\sin^{m-1} x \cos^n x$ as a sum of powers of cos x.

Example 1   Evaluate $\int \sin^3 x \cos^5 x \, dx$.

Solution: Since both exponents are odd, we can keep either sin x or cos x and express the remaining factor of the integrand as a sum of powers of the other. If we save cos x, we must express $\sin^3 x \cos^4 x$ as the sum of powers of sin x. Thus,

$$\cos^4 x = (\cos^2 x)^2 = (1 - \sin^2 x)^2 = 1 - 2 \sin^2 x + \sin^4 x$$

Therefore,

$$\int \sin^3 x \cos^5 x \, dx = \int \sin^3 x (1 - 2 \sin^2 x + \sin^4 x) \cos x \, dx$$
$$= \int (\sin^3 x - 2 \sin^5 x + \sin^7 x) \, d \sin x$$
$$= \frac{\sin^4 x}{4} - \frac{2 \sin^6 x}{6} + \frac{\sin^8 x}{8} + c$$

If we save sin x, we then have

$$\int \sin^3 x \cos^5 x \, dx = \int \sin^2 x (\cos^5 x) \sin x \, dx$$
$$= \int (1 - \cos^2 x) \cos^5 x \sin x \, dx$$
$$= \int (\cos^5 x - \cos^7 x) \sin x \, dx$$
$$= -\frac{\cos^6 x}{6} + \frac{\cos^8 x}{8} + k$$

This is certainly not the same form as the first form we obtained for the integral. The two answers can differ by a constant and the reader might find it to be an interesting task to show that they do.

$\int \sin^m x \cos^n x \, dx$ *with both m and n even integers.* The general principle to be followed is, as before, that of expressing the integrand as the sum of powers of a trigonometric function times the derivative of that function. For this purpose, we make use of the half-angle identities

$$\sin^2 \frac{x}{2} = \frac{1}{2} (1 - \cos x) \qquad \text{and} \qquad \cos^2 \frac{x}{2} = \frac{1}{2} (1 + \cos x)$$

if $m \neq n$. These same identities can be used if $m = n$, but at times the identity

$$\sin x \cos x = \frac{1}{2} \sin 2x$$

for the sine of twice an angle can be used to greater advantage.

Example 2 Evaluate $\int \sin^4 x \cos^2 x \, dx$.

Solution: If we use the half-angle identities, we have

$\int \sin^4 x \cos^2 x \, dx$

$$= \int \left[\frac{(1 - \cos 2x)}{2} \right]^2 \frac{(1 + \cos 2x) \, dx}{2}$$

$$= \frac{1}{8} \int [1 - \cos 2x - \cos^2 2x + \cos^3 2x] \, dx \qquad \textit{expanding}$$

$$= \frac{1}{8} \int [1 - \cos 2x - \frac{(1 + \cos 4x)}{2} + (1 - \sin^2 2x) \cos 2x] \, dx$$

where we replace $\cos^2 2x$ by means of a half-angle identity and apply the usual procedure to the odd power $\cos^3 2x$.

Now integrating, we see that

$\int \sin^4 x \cos^2 x \, dx$

$$= \frac{1}{8} \left[x - \frac{\sin 2x}{2} - \frac{x}{2} - \frac{\sin 4x}{8} + \frac{\sin 2x}{2} - \frac{\sin^3 2x}{6} \right] + c$$

$$= \frac{1}{8} \left(\frac{x}{2} - \frac{\sin 4x}{8} - \frac{\sin^3 2x}{6} \right) + c \qquad \textit{collecting}$$

$$= \frac{1}{192} (12x - 3 \sin 4x - 4 \sin^3 2x) + c \qquad \begin{array}{l} \textit{getting a common} \\ \textit{denominator} \end{array}$$

Example 3 Evaluate $\int \sin^2 x \cos^2 x \, dx$.

Solution: If we apply the identity for the sine of twice an angle, we get

$$\int \sin^2 x \cos^2 x \, dx = \frac{1}{4} \int \sin^2 2x \, dx = \frac{1}{4} \int \frac{(1 - \cos 4x)}{2} \, dx$$

$$= \frac{1}{8} \left(x - \frac{\sin 4x}{4} \right) + c = \frac{1}{32} (4x - \sin 4x) + c$$

EXERCISE 17.4

Perform the indicated integration in each of Problems 1 to 36.

1	$\int \sin^3 x \, dx$	2	$\int \sin^7 x \, dx$
3	$\int \sin^5 2x \, dx$	4	$\int \sin^3 (x/2) \, dx$
5	$\int \cos^7 x \, dx$	6	$\int \cos^3 3x \, dx$
7	$\int \cos^5 2x \, dx$	8	$\int \cos^5 (x/2) \, dx$
9	$\int \cos^2 x \, dx$	10	$\int \sin^4 x \, dx$
11	$\int \sin^6 x \, dx$	12	$\int \cos^4 2x \, dx$
13	$\int \sin^3 x \cos^5 x \, dx$	14	$\int \cos^3 x \sin^7 x \, dx$
15	$\int \sin^5 2x \cos^5 2x \, dx$	16	$\int \cos^5 x \sin^7 x \, dx$

17 $\int \sin^2 x \cos^3 x \, dx$	18 $\int \sin^4 x \cos^3 x \, dx$
19 $\int \cos^4 x \sin^3 x \, dx$	20 $\int \cos^6 x \sin^5 x \, dx$
21 $\int \sin^2 x \cos^4 x \, dx$	22 $\int \sin^6 x \cos^2 x \, dx$
23 $\int \sin^4 x \cos^2 x \, dx$	24 $\int \sin^6 x \cos^4 x \, dx$
25 $\int \sin^2 x \cos^2 x \, dx$	26 $\int \sin^4 x \cos^4 x \, dx$
27 $\int \tan x \sec^2 x \, dx$	28 $\int \tan^2 x \cos x \, dx$
29 $\int \tan^3 x \, dx$	30 $\int \cot^5 x \, dx$
31 $\int \cos x \csc^2 x \, dx$	32 $\int \tan x \sec^3 x \, dx$
33 $\int (\sin x - \cos x)^2 \, dx$	34 $\int (1 + \cos 2x)^2 \, dx$
35 $\int (\sin x + \cos 2x)^2 \, dx$	36 $\int (\cos x + 2\sqrt{\sin x})^2 \, dx$

17.7   A power of $\tan x$ times an even power of $\sec x$

We shall first consider $\int \tan^p x \, dx$ for p an integer greater than 1 by factoring $\tan^p x$ into $\tan^{p-2} x \tan^2 x$ and replacing $\tan^2 x$ by $\sec^2 x - 1$. Thus,

a reduction formula

$$\int \tan^p x \, dx = \int \tan^{p-2} x (\sec^2 x - 1) \, dx$$

$$= \int \tan^{p-2} x \sec^2 x \, dx - \int \tan^{p-2} x \, dx$$

$$= \frac{\tan^{p-1} x}{p - 1} - \int \tan^{p-2} x \, dx \qquad p > 1 \qquad (1)$$

This replaces the problem of integrating one power of $\tan x$ by that of integrating another power that is two smaller than the original one. Consequently, if p is an integer, we can reduce the problem to that of integrating $\tan x$ for p odd and integrating $(\tan x)^0 = 1$ for p even, by repeated use of (1). Since (1) enables us to reduce the exponent of the power of $\tan x$ that is to be integrated, it is called a *reduction formula*.

Example 1   Evaluate $\int \tan^5 x \, dx$.

Solution:

$$\int \tan^5 x \, dx = \int \tan^3 x \tan^2 x \, dx \qquad \text{\textit{factoring with} } \tan^2 x \text{ \textit{as a factor}}$$

$$= \int \tan^3 x (\sec^2 x - 1) \, dx \qquad \tan^2 x = \sec^2 x - 1$$

$$= \int \tan^3 x \sec^2 x \, dx - \int \tan^3 x \, dx$$

$$= \frac{\tan^4 x}{4} - \int \tan x (\sec^2 x - 1) \, dx$$

$$= \frac{\tan^4 x}{4} - \frac{\tan^2 x}{2} + \int \tan x \, dx$$

$$= \frac{\tan^4 x}{4} - \frac{\tan^2 x}{2} - \ln |\cos x| + c$$

We can express an even power of sec x as a sum of powers of tan x by use of $\sec^2 x = 1 + \tan^2 x$. Therefore, we can change $\sec^e x \tan^p x$, e an even integer, to a sum of powers of tan x and then use (1). There is, however, another procedure that is often preferable.

$\int \sec^e x \tan^p x \, dx$
for e even

If we have $\sec^e x \tan^p x$ as an integrand, we can reduce it to the sum of powers of tan x times $\sec^2 x = D_x \tan x$ by use of the identity $\sec^2 x = 1 + \tan^2 x$. Thus,

$$\sec^e x \tan^p x = (\sec^2 x)\sec^{e-2} x \tan^p x \qquad \textit{factoring out } \sec^2 x$$
$$= (\sec^2 x)(\sec^2 x)^{(e-2)/2} \tan^p x$$
$$= \sec^2 x(1 + \tan^2 x)^{(e-2)/2} \tan^p x$$
$$= (\text{sum of powers of tan } x)\sec^2 x$$

Consequently, we are able to integrate $\sec^e x \tan^p x$ by use of $\int u^n \, du$.

Example 2   Evaluate $\int \sec^6 x \tan^3 x \, dx$.

Solution: We shall express the integrand as $\sec^2 x$ times a sum of powers of tan x by use of the binomial theorem and the identity $\sec^2 x = 1 + \tan^2 x$. For this purpose, we write

$$\int \sec^6 x \tan^3 x \, dx = \int \sec^2 x (\sec^4 x)\tan^3 x \, dx \qquad \textit{factoring out } \sec^2 x$$
$$= \int \sec^2 x (1 + \tan^2 x)^2 \tan^3 x \, dx \qquad \sec^4 x = (1 + \tan^2 x)^2$$
$$= \int \sec^2 x (1 + 2\tan^2 x + \tan^4 x)\tan^3 x \, dx \qquad \textit{expanding}$$
$$= \int (\tan^3 x + 2\tan^5 x + \tan^7 x)\sec^2 x \, dx$$
$$= \frac{\tan^4 x}{4} + \frac{\tan^6 x}{3} + \frac{\tan^8 x}{8} + c$$

A power of cot x times an even power of csc x can be integrated in a manner similar to that used for a power of tan x times an even power of sec x.

17.8   A power of sec x times a positive odd integral power of tan x

We shall express the situation described in the heading of this section in symbols as $\sec^p x \tan^q x$ with p an integer and q a positive odd integer. We shall now express this as the sum of powers of sec x times $\sec x \tan x = D_x \sec x$. This can be done by writing

$$\int \sec^p x \tan^q x \, dx = \int \sec^{p-1} x \tan^{q-1} x \sec x \tan x \, dx$$
$$\textit{factoring out } \sec x \tan x$$
$$= \int \sec^{p-1} x(\tan^2 x)^{(q-1)/2} \sec x \tan x \, dx$$
$$q - 1 \textit{ is even}$$
$$= \int \sec^{p-1} x(\sec^2 x - 1)^{(q-1)/2} \, d \sec x$$
$$\textit{since } d \sec x = \sec x \tan x \, dx$$
$$= \int (\text{sum of powers of sec } x) \, d \sec x$$

$\int \sec^p x \tan^q x \, dx$
for q odd

by use of the binomial theorem. Therefore, we can integrate the product of a power of sec x and an odd integral power of tan x by use of $\int u^n \, du$. A product of a power of csc x and an odd power of cot x can be integrated in a similar manner.

Example

Evaluate $\int \csc^4 x \cot^3 x \, dx$.

Solution: We shall save csc x cot $x = -D_x$ csc x and express the other factor of the integrand as the sum of powers of csc x. Thus,

$$\int \csc^4 x \cot^3 x \, dx = \int \csc^3 x \cot^2 x \csc x \cot x \, dx$$

$$= \int \csc^3 x (\csc^2 x - 1) \csc x \cot x \, dx$$

$$= \int (\csc^5 x - \csc^3 x) \csc x \cot \, dx$$

$$= -\frac{\csc^6 x}{6} + \frac{\csc^4 x}{4} + c$$

This problem can also be solved by the method of Sec. 17.7 since it is a power of cot x times a positive even integral power of csc x. If this is done, we obtain the answer in terms of powers of cot x. Thus,

$$\int \cot^3 x \csc^4 x \, dx = \int \cot^3 x \csc^2 x \csc^2 x \, dx$$

$$= \int \cot^3 x (1 + \cot^2 x) \csc^2 x \, dx$$

$$= -\frac{\cot^4 x}{4} - \frac{\cot^6 x}{6} + K$$

The reader might find it an interesting exercise to show that these two answers differ by a constant.

EXERCISE 17.5

Evaluate each of the integrals in Problems 1 to 32.

1   $\int \tan^2 x \, dx$
2   $\int \tan^4 x \, dx$
3   $\int \cot^4 2x \, dx$
4   $\int \cot^2 (x/2) \, dx$
5   $\int \cot^3 x \, dx$
6   $\int \cot^5 2x \, dx$
7   $\int \tan^5 x \, dx$
8   $\int \tan^3 (x/4) \, dx$

9   $\int_0^{\pi/4} \sec^2 x \, dx$
10   $\int_{\pi/8}^{\pi/6} \sec^2 2x \, dx$

11   $\int \csc^6 2x \, dx$
12   $\int_{\pi/3}^{\pi/2} \csc^4 (x/2) \, dx$

13   $\int_0^{\pi/4} \tan^2 x \sec^2 x \, dx$
14   $\int \tan^2 x \sec^4 x \, dx$

15   $\int \cot^6 x \csc^4 x \, dx$
16   $\int \cot^4 x \csc^6 x \, dx$

Use the method of Sec. 17.7 in Problems 17 to 20.

17   $\int \tan^3 x \sec^4 x \, dx$ 18 $\int \tan^5 x \sec^6 x \, dx$

19   $\int \cot^7 2x \csc^4 2x \, dx$ 20 $\int \cot^3 (x/2) \csc^6 (x/2) \, dx$

Use the method of Sec. 17.8 in Problems 21 to 24.

21   $\int \cot^3 x \csc^4 x \, dx$ 22 $\int \cot^5 x \csc^6 x \, dx$

23   $\int \tan^7 2x \sec^4 2x \, dx$ 24 $\int \tan^3 (x/2) \sec^6 (x/2) \, dx$

25   $\int_0^{\pi/4} \sec^3 x \tan x \, dx$ 26 $\int_0^{\pi/3} \sec x \tan^3 x \, dx$

27   $\int \csc^3 x \cot^3 x \, dx$ 28 $\int \csc^5 x \cot^5 x \, dx$

29   $\int \sin 2x \sec^4 x \, dx$ 30 $\int \cos^2 x \sin^4 x \, dx$

31   $\int \sqrt{\tan x} \sec^6 x \, dx$ 32 $\int \sqrt{\sec x} \sec^2 x \tan x \, dx$

17.9   Compound interest law or law of growth

If y is a differentiable function of x and if the rate of change of y with respect to x is proportional to y with $k \neq 0$ the constant of pro-

exponential growth and decay

portionality or variation, we say that y follows the law of *exponential growth* if $k > 0$ and the law of *exponential decay* if $k < 0$. Lord Kelvin called this law the *compound-interest law* since it describes

compound-interest law

the way in which a sum of money increases if placed at interest which is compounded continuously.

A symbolic statement of the law is given by

$$\frac{dy}{dx} = ky$$

If we multiply through by dx and divide by y, this becomes

$$\frac{dy}{y} = k \, dx$$

Hence,

$$\int \frac{dy}{y} = \int k \, dx$$

and we have

$$\ln y = kx + c \qquad \text{or} \qquad y = e^{kx+c} = e^{kx}e^c = Ce^{kx}$$

Consequently,

$$\text{If} \quad \frac{dy}{dx} = ky \quad \text{then} \quad y = Ce^{kx}$$

Example 1   A substance is in the process of dissolving, and the rate of decomposition at any time is known to be proportional to the amount of the substance present at that instant. If 27 g is present initially, that is for $t = 0$, and if 8 g is present after 3 hr, how much will be left after another hour?

Solution: If Q is the amount present at any time t, then the statement of the problem can be put in the form

$$\frac{dQ}{dt} = kQ \qquad \text{where } Q = 27 \text{ for } t = 0, Q = 8 \text{ for } t = 3$$

If we state the law symbolically, we have

$$\frac{dQ}{dt} = kQ$$

$$\frac{dQ}{Q} = k\,dt \qquad\qquad \textit{multiplying each member by } dt/Q$$

$$\ln Q = kt + c \qquad\qquad \textit{integrating each member}$$

$$Q = e^{kt+c} = e^c e^{kt} = Ce^{kt} \tag{1}$$

If we now make use of the given fact that $Q = 27$ for $t = 0$, (1) becomes $27 = Ce^0$; hence, $27 = C$ and

$$Q = 27e^{kt} \tag{2}$$

We can now find the value of k by making use of the given fact that $Q = 8$ for $t = 3$. Thus, (2) becomes

$$8 = 27e^{3k}$$

$$e^{3k} = \tfrac{8}{27}$$

$$e^k = \tfrac{2}{3}$$

Therefore, (2) becomes

$$Q = 27(\tfrac{2}{3})^t \qquad \textit{since } e^{kt} = (e^k)^t = (\tfrac{2}{3})^t$$

Consequently, the number of grams of material present at the end of 4 hr is

$$Q = 27(\tfrac{2}{3})^4 = 5\tfrac{1}{3}$$

Example 2   A sum of \$100 is invested at a rate of 6 percent per year compounded continuously. What is the accumulated value at the end of 10 years. Compare this value with the accumulated value at the end of 10 years at a rate of 6 percent compounded annually and compounded semi-annually.

Solution: If S is the accumulated value at any time t, then $S = 100$ for $t = 0$, and we are to find S for $t = 10$. Symbolically, we are given that

$$\frac{dS}{dt} = 0.06S$$

Therefore,

$$\frac{dS}{S} = 0.06 \, dt \qquad \textit{multiplying each member by dt/S}$$

$$\ln S = 0.06t + c \qquad \textit{integrating}$$

Now, using $S = 100$ for $t = 0$ as given, we get

$$\ln 100 = (0.06)0 + c$$

$$c = \ln 100$$

Hence,

$$\ln S = 0.06t + \ln 100$$

$$S = e^{0.06t + \ln 100}$$

$$= 100e^{0.06t}$$

is the relation between S and t for any t. The value of S for $t = 10$ is

$$S = 100e^{0.06(10)} = 100e^{0.6} \qquad \textit{by use of Table V}$$

$$= \$182.20$$

In order to make the comparisons called for in the problem, we need

$$S = (\$100)(1.06^{10}) = \$179.08 \quad \text{and} \quad S = (\$100)(1.03^{20}) = \$180.61$$

Thus, we see that the accumulated value if compounded semiannually is between the value if compounded annually and the value if compounded continuously.

EXERCISE 17.6

1   If one chemical substance S is being changed into another A at a rate that is proportional to the amount of S present at any time, find the amount of S that remains after 4 min if 36 lb is reduced to 25 lb in 2 min.

2   Assume that the rabbit population of Central Australia for one long period of years doubled every 3 years. If the rabbit population at a certain time is R, after how many years will it be $6R$?

3   How many years are required for $500 to increase to $1,000 if invested at 3 percent per year compounded continuously?

4   To what sum will $2,000 accumulate in 8 years if invested at 4 percent per year compounded continuously? Compare this with the accumulated value if compounded semiannually.

5   Show that the rate on U.S. Government Series E bonds was just under 2.9 percent per year compounded continuously during the period in which a $75 bond accumulated to $100 in 10 years.

6   Assume that the half-life of a substance is 1,200 years, and find the percentage that remains after 240 years. The time required for half
half-life   of a substance to decompose is called the *half-life*.

7   An upright tank is in the shape of a right circular cylinder and contains 100 gal of a liquid. Through a hole in the bottom, the liquid escapes at a rate that is proportional to the square root of the number of gallons present. How much will escape during the second day and during the fifth day if 10 gal leaks out the first day?

8   There are 800 lb of salt in 1,000 gal of saline solution. If the mixture is kept uniform by an agitator, how many pounds of salt remains after 25 hr if 16 gal of solution is withdrawn and 16 gal of water added uniformly each hour?

9   How many hours are required to reduce the salt concentration in Problem 8 to ¼ lb per gal?

10   Assume that the rate at which a body cools in air is proportional to the difference between the temperatures of it and the air. If a body at 80° cools down to 60° in 20 min, what will be its temperature in another 40 min if the air temperature is 40°?

11   The temperature of a body drops from 100 to 80° in 3 hr in an air temperature of 20°. Assume the law of cooling stated in Problem 10, and find how long is required for the temperature to drop from 100 to 30°.

12   The population of the United States was 150 million in 1950 and 180 million in 1960. If the population growth follows the exponential law, find the expected population in 1970. If you work this problem after the 1970 census, compare your result with the census figure.

13   If sugar in solution decomposes into other substances at a rate proportional to the amount present and if 10 lb reduces to 3 lb in 4 hr, when was half of it decomposed?

14   Find i in terms of t if a current i diminishes according to the exponential decay law $di/dt = -50i$ and $i = 20$ for $t = 0$. Evaluate the expression for $t = 2$.

In Problems 15 and 16, assume that the compound-interest law holds; thus, assume that the rate of change of value with respect to time or rate of depreciation is proportional to the value at the time under consideration.

15   If a machine costs $10,000 and is worth only the scrap value of $100 at the end of 10 years, find the value at the end of 5 years. Use the statement preceding Problem 15.

16   If a car costs $3,000 and is worth $600 after 5 years, find its value after 2 years. Use the statement preceding Problem 15.

18

Methods of integration

18.1   Introduction

If the integrand is not in a form that appears in some one of the standard integration formulas, it frequently can be converted to a form that is integrable by use of one of the standard formulas. Sometimes this conversion can be accomplished by replacing the variable in the integrand by some function of another variable. If the integrand involves trigonometric functions, it might be converted to an integrable form by use of the trigonometric identities. Some integrands can be expressed in a different algebraic form to which one or more of the standard formulas can be applied. If, in the process of integration, a new variable is introduced, the final integral should be expressed in terms of the original variable. An example of this procedure follows.

Example   Evaluate $\int \dfrac{dx}{\sqrt{x^2 - 9}}$ by replacing x by 3 sec θ.

Solution: If we let $x = 3 \sec \theta$, then $dx = 3 \sec \theta \tan \theta \, d\theta$ and $x^2 - 9 = 9 \sec^2 \theta - 9 = 9 \tan^2 \theta$. See Fig. 18.1. Consequently, the given integral becomes

$$\int \frac{3 \sec \theta \tan \theta \, d\theta}{3 \tan \theta} = \int \sec \theta \, d\theta = \ln |\sec \theta + \tan \theta| + C$$

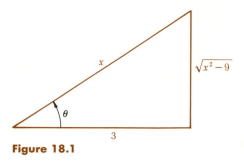

Figure 18.1

Now, since $\sec \theta = x/3$,

$$\tan \theta = \sqrt{\left(\frac{x}{3}\right)^2 - 1} = \frac{\sqrt{x^2 - 9}}{3}$$

and we have

$$\int \frac{dx}{\sqrt{x^2 - 9}} = \ln \left| \frac{x + \sqrt{x^2 - 9}}{3} \right| + C$$

$$= \ln |x + \sqrt{x^2 - 9}| - \ln 3 + C$$

$$= \ln |x + \sqrt{x^2 - 9}| + K$$

In the subsequent sections of this chapter we shall discuss other methods for transforming an integrand by a change of variable and shall explain several other devices that are useful for evaluating many types of integrals.

18.2   Trigonometric substitutions

Integrands that contain $a^2 - x^2$, $a^2 + x^2$, or $x^2 - a^2$ are often encountered. Each of these expressions can be replaced by a single term by means of a properly chosen trigonometric substitution, and it may happen that the integral is thereby changed to a form that enables us to use one of the standard integration formulas.

If $x = a \sin \theta$ then $a^2 - x^2 = a^2 - a^2 \sin^2 \theta = a^2 \cos^2 \theta$

If $x = a \tan \theta$ then $a^2 + x^2 = a^2 + a^2 \tan^2 \theta = a^2 \sec^2 \theta$

If $x = a \sec \theta$ then $x^2 - a^2 = a^2 \sec^2 \theta - a^2 = a^2 \tan^2 \theta$

We must be certain to express dx in terms of $d\theta$ and a trigonometric function of θ rather than erroneously by using $d\theta$ alone.

The example of Sec. 18.1 illustrates the procedure to be followed, but we shall give another example.

Example

Evaluate $\int x^3 \sqrt{4 + x^2}\, dx$ by use of a trigonometric substitution.

Solution: Since $4 + x^2 = 2^2 + x^2$ is the sum of two squares, we shall use the tangent substitution to reduce it to a single term. See Fig. 18.2. If

$$x = 2 \tan \theta$$

then

$$dx = 2 \sec^2 \theta\, d\theta$$

$$4 + x^2 = 4 + 4 \tan^2 \theta = 4 \sec^2 \theta$$

and

$$\frac{\sqrt{4 + x^2}}{2} = \sec \theta$$

Therefore,

$$
\begin{aligned}
\int x^3 \sqrt{4 + x^2}\, dx &= \int (8 \tan^3 \theta)(2 \sec \theta)(2 \sec^2 \theta)\, d\theta \\
&= 32 \int \tan^3 \theta \sec^3 \theta\, d\theta \\
&= 32 \int (\sec^2 \theta - 1) \tan \theta \sec^2 \theta \sec \theta\, d\theta \\
&= 32 \int (\sec^4 \theta - \sec^2 \theta) \sec \theta \tan \theta\, d\theta \\
&= 32 \left(\frac{\sec^5 \theta}{5} - \frac{\sec^3 \theta}{3} \right) + c \\
&= 32 \left[\frac{(4 + x^2)^{5/2}}{160} - \frac{(4 + x^2)^{3/2}}{24} \right] + c \qquad \text{\textit{replacing} sec } \theta \\
& \qquad\qquad\qquad\qquad\qquad\qquad\qquad\qquad \text{\textit{by} } \sqrt{4 + x^2}/2 \\
&= 32(4 + x^2)^{3/2} \left(\frac{4 + x^2}{160} - \frac{1}{24} \right) + c \\
&= \frac{32}{480}(4 + x^2)^{3/2}(12 + 3x^2 - 20) + c \\
&= \frac{1}{15}(4 + x^2)^{3/2}(3x^2 - 8) + c
\end{aligned}
$$

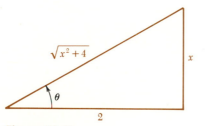

Figure 18.2

EXERCISE 18.1

Evaluate each of the following integrals by use of a trigonometric substitution.

1 $\int x\sqrt{x^2 + 9}\, dx$

2 $\int x\sqrt{x^2 + 4}\, dx$

3 $\int \dfrac{x\, dx}{\sqrt{x^2 + 16}}$

4 $\int \dfrac{x^3\, dx}{\sqrt{x^2 + 1}}$

5 $\int \dfrac{25\, dx}{x^2\sqrt{x^2 + 25}}$

6 $\int \dfrac{81\, dx}{x^4\sqrt{x^2 + 9}}$

7 $\int \dfrac{\sqrt{x^2 + 16}}{x^2}\, dx$

8 $\int \dfrac{4\, dx}{x^2\sqrt{x^2 + 4}}$

9 $\int x^2\sqrt{16 - x^2}\, dx$

10 $\int x^3\sqrt{9 - x^2}\, dx$

11 $\int 64x^2\sqrt{9 - 4x^2}\, dx$

12 $\int \dfrac{\sqrt{9 - 4x^2}}{x^2}\, dx$

13 $\int \dfrac{dx}{x^2\sqrt{16 - x^2}}$

14 $\int \dfrac{x^3\, dx}{\sqrt{25 - 4x^2}}$

15 $\int \dfrac{dx}{x\sqrt{4 - x^2}}$

16 $\int \dfrac{54\, dx}{x^3\sqrt{9 - x^2}}$

17 $\int \dfrac{\sqrt{x^2 - 1}}{x}\, dx$

18 $\int \dfrac{\sqrt{x^2 - 9}}{x^2}\, dx$

19 $\int \dfrac{\sqrt{4x^2 - 25}}{8x^3}\, dx$

20 $\int \dfrac{\sqrt{9x^2 - 1}}{27x^4}\, dx$

21 $\int x^3\sqrt{x^2 - 1}\, dx$

22 $\int \dfrac{\sqrt{x^2 - 4}}{x^3}\, dx$

23 $\int x^3\sqrt{4x^2 - 9}\, dx$

24 $\int \dfrac{x^3\, dx}{\sqrt{16x^2 - 25}}$

25 $\int \dfrac{dx}{x\sqrt{x^2 - 4}}$

26 $\int \dfrac{3\, dx}{x\sqrt{x^2 - 9}}$

27 $\int \dfrac{dx}{x^2\sqrt{4x^2 - 9}}$

28 $\int \dfrac{dx}{x^2\sqrt{x^2 - 6}}$

29 $\int \dfrac{x^2\, dx}{(x^2 + a^2)^{3/2}}$

30 $\int \dfrac{dx}{\sqrt{x^2 + 4}}$

31 $\int \dfrac{x\, dx}{\sqrt{x^2 + 1}}$

32 $\int \dfrac{dx}{x\sqrt{x^2 + 9}}$

33 $\int x(9 - x^2)^{3/2}\, dx$

34 $\int x^3(9 - 4x^2)^{3/2}\, dx$

35 $\int \dfrac{x^2\, dx}{(a^2 - x^2)^{3/2}}$

36 $\int \dfrac{dx}{x\sqrt{a^2 - x^2}}$

18.3   Five standard formulas

There are five integration formulas that are needed sufficiently often that it may be worthwhile to learn them, even though integrals of the types involved can be evaluated by use of the trigonometric substitution discussed in Sec. 18.2. The integrals referred to are

$$\int \frac{du}{\sqrt{a^2 - u^2}} = \text{Arcsin} \frac{u}{a} + C \tag{18.1}$$

$$\int \frac{du}{a^2 + u^2} = \frac{1}{a} \text{Arctan} \frac{u}{a} + C \tag{18.2}$$

integrals that involve the sum and difference of two squares

$$\int \frac{du}{u^2 - a^2} = \frac{1}{2a} \ln \left| \frac{u - a}{u + a} \right| + C \tag{18.3}$$

$$\int \frac{du}{\sqrt{u^2 - a^2}} = \ln |(u + \sqrt{u^2 - a^2})| + C \tag{18.4}$$

$$\int \frac{du}{\sqrt{u^2 + a^2}} = \ln |(u + \sqrt{u^2 + a^2})| + C \tag{18.5}$$

Each of these formulas can be obtained readily by use of the appropriate trigonometric substitution. Furthermore, the first two come immediately from the definition of an indefinite integral and the derivatives of Arcsin (u/a) and Arctan (u/a).

18.4   Quadratic integrands

quadratic integrands   In order to evaluate an integral if the integrand contains a quadratic trinomial, we make use of the fact that any quadratic trinomial can be expressed as the sum or difference of two squares. After so expressing the quadratic, we can then use the trigonometric substitution method of Sec. 18.2.

Example   Evaluate $$\int \frac{dx}{\sqrt{x^2 - 2x - 8}}$$

Solution: Our first task is to complete the square of the quadratic in x, since we can then express the integrand as the difference of two squares. Thus, we write

$$x^2 - 2x - 8 = x^2 - 2x + 1 - 1 - 8 = (x - 1)^2 - 3^2$$

Hence, $$\int \frac{dx}{\sqrt{x^2 - 2x - 8}} = \int \frac{dx}{\sqrt{(x - 1)^2 - 3^2}}$$

Consequently, we make the substitution

$$x - 1 = 3 \sec \theta \tag{1}$$

Hence, $dx = 3 \sec \theta \tan \theta \, d\theta$

and $(x - 1)^2 - 3^2 = 9(\sec^2\theta - 1) = 9 \tan^2 \theta$

Therefore, the given integral becomes

$$\int \frac{3 \sec \theta \tan \theta \, d\theta}{3 \tan \theta} = \int \sec \theta \, d\theta$$

$$= \ln |\sec \theta + \tan \theta| + C$$

$$= \ln \left| \frac{x - 1}{3} + \frac{\sqrt{(x - 1)^2 - 3^2}}{3} \right| + C \qquad \begin{array}{l} \textit{replacing } \theta \textit{ in} \\ \textit{terms of } x \\ \textit{by use of (1)} \end{array}$$

$$= \ln |x - 1 + \sqrt{x^2 - 2x - 8}| + C_1$$

since $(x - 1)^2 - 3^2 = x^2 - 2x - 8$, the logarithm of a quotient is the difference of the logarithms, and $C - \ln 3 = C_1$.

EXERCISE 18.2

Derive the integration formulas in Problems 1 to 4 by use of trigonometric substitutions. They are (18.1) to (18.5).

1 $\int \dfrac{du}{\sqrt{a^2 - u^2}} = \text{Arcsin} \dfrac{u}{a} + C$

2 $\int \dfrac{du}{a^2 + u^2} = \dfrac{1}{a} \text{Arctan} \dfrac{u}{a} + C$

3 $\int \dfrac{du}{u^2 - a^2} = \dfrac{1}{2a} \ln \left| \dfrac{u - a}{u + a} \right| + C$

4 $\int \dfrac{du}{\sqrt{u^a \pm a^2}} = \ln |(u + \sqrt{u^2 \pm a^2})| + C$

Perform the following integrations by use of trigonometric substitutions and by use of the formulas given in Problems 1 to 4.

5 $\int \dfrac{dx}{\sqrt{9 - x^2}}$ 6 $\int \dfrac{dx}{9x^2 + 4}$

7 $\int \dfrac{x \, dx}{x^4 - 16}$ 8 $\int \dfrac{\sin x \, dx}{\sqrt{\cos^2 x + 16}}$

9 $\int \dfrac{e^x \, dx}{e^{2x} + 4}$ 10 $\int \dfrac{x \, dx}{(x^2 + 4)^2 - 9}$

11 $\int \dfrac{dx}{\sqrt{x^2 + 2x + 7}}$ 12 $\int \dfrac{dx}{\sqrt{10 - 4x^2}}$

13 $\int \dfrac{dx}{4x^2 - 4x - 3}$ 14 $\int \dfrac{dx}{\sqrt{4x^2 - 12x + 13}}$

15 $\displaystyle\int\frac{dx}{\sqrt{a^2 - b^2 x^2}}$ 16 $\displaystyle\int\frac{dx}{x(\ln^2 x + 9)}$

17 $\displaystyle\int\frac{dx}{\sqrt{16x^2 - 24x - 7}}$ 18 $\displaystyle\int\frac{dx}{\sqrt{8 - 4x - 4x^2}}$

19 $\displaystyle\int\frac{dx}{x^2 + 2x + 5}$ 20 $\displaystyle\int\frac{dx}{4x^2 - 12x + 5}$

21 $\displaystyle\int\frac{(18x - 3)\,dx}{\sqrt{9x^2 - 12x}}$ 22 $\displaystyle\int\frac{dx}{\sqrt{2x - x^2}}$

23 $\displaystyle\int\frac{dx}{4x^2 - 4x + 5}$ 24 $\displaystyle\int\frac{e^x\,dx}{e^{2x} - 1}$

25 $\displaystyle\int\frac{x\,dx}{x^4 - 4x^2 + 13}$ 26 $\displaystyle\int\frac{(8x - 1)\,dx}{4x^2 - 4x - 3}$

27 $\displaystyle\int\frac{(4x + 1)\,dx}{\sqrt{4x^2 - 12x + 5}}$ 28 $\displaystyle\int\frac{dx}{\sqrt{6x - x^2 - 5}}$

29 $\displaystyle\int\frac{(2x + 6)\,dx}{x^2 + 4x + 8}$ 30 $\displaystyle\int\frac{(x + 3)\,dx}{x^2 + 4x - 5}$

31 $\displaystyle\int\frac{(4x + 7)\,dx}{\sqrt{x^2 - 3x + 2}}$ 32 $\displaystyle\int\frac{x\,dx}{\sqrt{23 + 12x^2 - 4x^4}}$

33 $\displaystyle\int\frac{dx}{9x^2 - 6x - 8}$ 34 $\displaystyle\int\frac{dx}{\sqrt{9x^2 + 6x + 10}}$

35 $\displaystyle\int\frac{2^x\,dx}{\sqrt{9 - 4^x}}$ 36 $\displaystyle\int\frac{(2x + 1)\,dx}{4x^2 + 12x + 13}$

37 $\displaystyle\int\frac{(x - 6)\,dx}{\sqrt{x^2 - 5x + 4}}$ 38 $\displaystyle\int\frac{(2x - 7)\,dx}{\sqrt{3 - 6x - 9x^2}}$

39 $\displaystyle\int\frac{(3x + 5)\,dx}{x^2 + x + 1}$ 40 $\displaystyle\int\frac{(7x + 2)\,dx}{15 + 6x - 9x^2}$

18.5   Algebraic substitutions

The transformations discussed in Secs. 18.2 to 18.4 are those most often used, but others are needed from time to time. In this section we shall discuss three algebraic substitutions, two of which enable us to transform some integrands that involve irrationalities into rational integral functions, and the third will transform some irrational integrands into others that can be integrated by one of the standard formulas.

An integrand that contains two or more radicals of different orders but with the same radicand can frequently be transformed to a rational integral function of another variable by the method illustrated below.

If in

$$\frac{\sqrt[r]{ax + b} + \sqrt[s]{ax + b}}{\sqrt[t]{ax + b}} dx \tag{1}$$

we let
$$y = (ax + b)^{1/rst}$$

then
$$ax + b = y^{rst}$$

and
$$dx = \frac{rst}{a}(y^{rst-1}) dy$$

If we substitute these values in (1), we get

$$\frac{rst(y^{st} + y^{rt})y^{rst-1}}{ay^{rs}} dy$$

Since r, s, and t are integers, this function contains no irrationalities. We shall illustrate the procedure with two examples.

Example 1   Evaluate $\int x\sqrt{2x - 1}\, dx$.

Solution: If we let $\sqrt{2x - 1} = z$

then
$$2x - 1 = z^2$$

$$x = \frac{z^2 + 1}{2}$$

and
$$dx = z\, dz$$

Consequently,

$$\int x\sqrt{2x - 1} = \int \frac{(z^2 + 1)z^2}{2}\, dz = \frac{1}{2}\int (z^4 + z^2)\, dz$$

$$= \frac{1}{2}\left(\frac{z^5}{5} + \frac{z^3}{3}\right) + c = \frac{z^3(3z^2 + 5)}{30} + c$$

Now if we replace z by $\sqrt{2x - 1}$, we have

$$\int x\sqrt{2x - 1} = \frac{(2x - 1)^{3/2}[3(2x - 1) + 5]}{30} + c$$

$$= \frac{(2x - 1)^{3/2}(3x + 1)}{15} + c$$

Example 2   Evaluate
$$\int \frac{dx}{\sqrt[3]{x} + \sqrt[4]{4}}$$

Solution: We shall let $y = \sqrt[12]{x}$. Then we have

$$x = y^{12} \qquad dx = 12y^{11}\, dy \qquad \sqrt[3]{x} = y^4 \qquad \sqrt[4]{x} = y^3$$

If we substitute these values in the given integral, we have

$$\int \frac{dx}{\sqrt[3]{x} + \sqrt[4]{x}} = \int \frac{12y^{11}\,dy}{y^4 + y^3}$$

$$= 12\int \frac{y^8\,dy}{y + 1}$$

$$= 12\int \left(y^7 - y^6 + y^5 - y^4 + y^3 - y^2 + y - 1 \right.$$
$$\left. + \frac{1}{y + 1}\right) dy$$

$$= 12\left(\frac{y^8}{8} - \frac{y^7}{7} + \frac{y^6}{6} - \frac{y^5}{5} + \frac{y^4}{4} - \frac{y^3}{3} + \frac{y^2}{2} \right.$$
$$\left. - y + \ln|y + 1|\right) + c$$

$$= 12\left(\frac{x^{2/3}}{8} - \frac{x^{7/12}}{7} + \frac{x^{1/2}}{6} - \frac{x^{5/12}}{5} + \frac{x^{1/3}}{4} - \frac{x^{1/4}}{3} \right.$$
$$\left. + \frac{x^{1/6}}{2} - x^{1/12} + \ln|x^{1/12} + 1|\right) + c \qquad \textit{replacing y by } x^{1/12}$$

Sometimes an integrand that is not in an integrable form can be converted to one that is integrable by replacing the variable by the reciprocal of another. That is, if the variable in the first integrand is x we replace x by $1/y$.

Example 3   Evaluate

$$\int \frac{dx}{x^3\sqrt{a^3x^4 - x}}$$

Solution: We shall let $x = 1/y$, then $dx = -dy/y^2$. If we substitute these values in the given integrand, we get

$$\int \frac{dx}{x^2\sqrt{a^3x^4 - x}} = -\int \frac{dy/y^2}{(1/y^2)\sqrt{a^3/y^4 - 1/y}}$$

$$= -\int \frac{y^2dy}{\sqrt{a^3 - y^3}}$$

$$= \frac{1}{3}\int \frac{-3y^2dy}{\sqrt{a^3 - y^3}}$$

$$= \frac{2}{3}\sqrt{a^3 - y^3} + c$$

$$= \frac{2}{3}\sqrt{a^3 - \frac{1}{x^3}} + c \qquad \textit{replacing y by } 1/x$$

$$= \frac{2\sqrt{a^3x^4 - x}}{3x^2} + c$$

Considerable thought and ingenuity are necessary for a judicious choice of a substitution that will transform an integrand into one of the standard forms. If no one of the above substitutions accomplishes this purpose, we try replacing a troublesome term in the integrand by a single variable, or by some expression that involves this variable. Several attempts may be made before a suitable substitution is found, and, in some cases, no substitution will accomplish the desired objective.

18.6   Another trigonometric substitution

An integrand of the type $f(\sin x)\, dx$, $g(\cos x)\, dx$, or $h(\sin x, \cos x)\, dx$ frequently can be transformed to an integrand of the type $F(z)\, dz$ where $F(z)$ is a rational integral function. The transformation employed for this purpose is $x = 2\,\text{Arctan}\, z$ or $z = \tan(x/2)$. We shall presently illustrate the method with an example, but we shall first derive some pertinent relations between x and z. Since $x = 2\,\text{Arctan}\, z$, it follows that

$$dx = \frac{2\, dz}{1 + z^2} \tag{1}$$

Furthermore, since

$$z = \tan \frac{x}{2} = \pm \sqrt{\frac{1 - \cos z}{1 + \cos z}}$$

it follows that

$$z^2 = \frac{1 - \cos x}{1 + \cos x}$$

and if we solve the latter equation for $\cos x$, we get

$$\cos x = \frac{1 - z^2}{1 + z^2} \tag{2}$$

Now we use the identity $\sin x = \sqrt{1 - \cos^2 x}$ and get

$$\sin x = \frac{2z}{1 + z^2} \tag{3}$$

Example   Evaluate $$\int \sqrt{\frac{1 + \sin x}{1 + \cos x}}\, dx$$

Solution: We shall replace $\sin x$, $\cos x$, and dx by their values in (3), (2), and (1), respectively, and then complete the evaluation as indicated below.

$$\int\left[\sqrt{\frac{1 + \sin x}{1 + \cos x}}\right] dx = \int\left[\sqrt{\frac{1 + 2z/(1 + z^2)}{1 + (1 - z^2)/(1 + z^2)}}\ \frac{2\ dz}{1 + z^2}\right]$$

$$= \int\left[\sqrt{\frac{(1 + z)^2/(1 + z^2)}{2/(1 + z^2)}}\ \frac{2\ dz}{1 + z^2}\right]$$

$$= \sqrt{2}\int\frac{1 + z}{1 + z^2}\ dz$$

$$= \sqrt{2}\int\frac{dz}{1 + z^2} + \sqrt{2}\int\frac{z\ dz}{1 + z^2}$$

$$= \sqrt{2}\ \text{Arctan}\ z + \frac{\sqrt{2}}{2}\ \ln|1 + z^2| + c$$

$$= \frac{\sqrt{2}}{2}\ x + \frac{\sqrt{2}}{2}\ \ln\left|1 + \tan^2\frac{x}{2}\right| + c \qquad \begin{array}{l}\text{since } x = 2 \\ \text{Arctan } x\end{array}$$

EXERCISE 18.3

Evaluate each of the following integrals by use of a substitution.

1   $\int 4x\sqrt{1 + 2x}\ dx$ 2 $\int 8x\sqrt{1 - 2x}\ dx$

3   $\int(x + 2)\sqrt{2 - x}\ dx$ 4 $\int(2x - 3)\sqrt{x + 3}\ dx$

5   $\int\dfrac{(2x + 1)\ dx}{\sqrt{3x - 2}}$ 6 $\int\dfrac{(3x - 2)\ dx}{\sqrt{2x + 1}}$

7   $\int\dfrac{(3x + 1)\ dx}{\sqrt{4 - x}}$ 8 $\int\dfrac{(x - 1)\ dx}{\sqrt{x + 1}}$

9   $\int x\sqrt[3]{x + 2}\ dx$ 10 $\int(3x - 5)\sqrt[3]{3x + 2}\ dx$

11   $\int\dfrac{x\ dx}{\sqrt[3]{x - 4}}$ 12 $\int\dfrac{(1 - x)\ dx}{\sqrt[3]{4x + 3}}$

Use $x = 1/z$ in Problems 13 to 16.

13   $\int\dfrac{dx}{x^2\sqrt{x^2 + 4}}$ 14 $\int\dfrac{dx}{x^2\sqrt{1 - x^2}}$

15   $\int\dfrac{dx}{x^3\sqrt{x^4 + 1}}$ 16 $\int\dfrac{dx}{x^4\sqrt[3]{1 - x^6}}$

17   $\int\dfrac{dx}{\sqrt{x} - \sqrt[3]{x}}$, use $\sqrt[6]{x} = y$.

18   $\int\dfrac{dx}{x\sqrt{x^2 - 2x}}$, use $x = 2\ \sec^2\theta$.

19   $\int\dfrac{dx}{x\sqrt{x}\ \sqrt{4 - x}}$, use $x = 4\ \sin^2\theta$.

20 $\displaystyle\int \frac{\sqrt{x+4}\,dx}{x\sqrt{x}}$, use $x = 4\tan^2\theta$.

21 $\int \sqrt{1 - \sqrt{x}}\,dx$ 22 $\int \sqrt{4 + \sqrt{x}}\,dx$

23 $\displaystyle\int \frac{e^x(e^x - 2)\,dx}{e^x + 1}$ 24 $\displaystyle\int \frac{e^{2x}\,dx}{e^x - 1}$

Use the substitution $x = 2\,\mathrm{Arctan}\,z$ *in Problems 25 to 32.*

25 $\displaystyle\int \frac{dx}{1 + \sin x}$ 26 $\displaystyle\int \frac{dx}{2 + \cos x}$

27 $\displaystyle\int \frac{dx}{1 + \cos x - \sin x}$ 28 $\displaystyle\int \frac{\cos dx}{1 + \cos x}$

29 $\displaystyle\int \frac{2\cos x\,dx}{\cos x + \sin x}$ 30 $\displaystyle\int \frac{\sin x\,dx}{1 + \cos x}$

31 $\displaystyle\int \frac{dx}{\csc x - \cot x}$ 32 $\displaystyle\int \frac{dx}{1 - \sin x}$

18.7   Integration by parts

If the integrand is not in the form of any of the standard forms and if none of the usual substitutions make it so, we can often integrate by resorting to a procedure that is known as *integration by parts*. We shall now derive a formula for use in this method of integration.
    If u and v are differentiable functions, then

$$d(uv) = u\,dv + v\,du$$

and $$uv = \int u\,dv + \int v\,du \qquad \text{\textit{integrating}}$$

If we solve this equation for $\int u\,dv$, we have

formula for integration by parts

$$\int u\,dv = uv - \int v\,du \qquad\qquad (18.6)$$

as the formula for integrating by parts.
    In order to use this formula, we must regard the integrand as the product of a function u and the differential dv of another function v.

some general principles No general rule for separating the integrand into two factors can be given. We usually take as much as we can integrate for dv and call the other factor u. If the integrand has an integral power of x as a factor, we usually take that as u provided that we can integrate the other factor.

Example 1 Perform the integration indicated by $\int xe^x\,dx$.

Solution: We shall resort to integration by parts since none of the methods studied earlier enable us to perform the indicated integration.

Since we can integrate $e^x \, dx$, we shall let $dv = e^x \, dx$. Consequently, we must use x as u so that $xe^x \, dx$ is $u \, dv$. Now we obtain $v = \int e^x dx = e^x$ and $du = dx$. Therefore,

$$\int xe^x \, dx = xe^x - \int e^x \, dx = xe^x - e^x + c$$

An unfortunate choice for dv can lead to an integral that is more complicated than the given one. Thus, if we let $dv = x \, dx$, then we would have $u = e^x$; hence, $v = x^2/2$ and $du = e^x \, dx$. Consequently,

$$\int xe^x \, dx = \frac{x^2 e^x}{2} - \frac{1}{2} \int x^2 e^x \, dx$$

and this involves a more complicated integrand than that of the given integral.

Frequently it is necessary to apply the method of integration by parts more than once in the evaluation of a given integral. For example

$$\int x^2 e^x \, dx = x^2 e^x - 2 \int xe^x \, dx \qquad \textit{obtained by letting } u = x^2 \textit{ and } dv = e^x \, dx$$

$$= x^2 e^x - 2xe^x + 2e^x + c$$

obtained by letting $u = x$ and $dv = e^x \, dx$ in the integral on the right. Occasionally a repeated application of the method of integration by parts will yield an integral that differs from the original integral by only a constant factor. In such cases we apply the procedure illustrated in Example 2.

Example 2   Evaluate $\int e^{-x} \cos x \, dx$.

Solution: If we let $dv = \cos x \, dx$ with $u = e^{-x}$, then

$$v = \sin x \qquad \text{and} \qquad du = -e^{-x} \, dx$$

Therefore,

$$\int e^{-x} \cos x \, dx = e^{-x} \sin x + \int e^{-x} \sin x \, dx \qquad (1)$$

The form of the integral on the right suggests the use of the method of integration by parts again. Accordingly, in this integral we shall let $dv = \sin x \, dx$ with $u = e^{-x}$, then

$$v = -\cos x \qquad \text{and} \qquad du = -e^{-x} \, dx$$

Consequently, (1) becomes

$$\int e^{-x} \cos x \, dx = e^{-x} \sin x - e^{-x} \cos x - \int e^{-x} \cos x \, dx \qquad (2)$$

Now we notice that the integrals on the left and right in (2) have the same form but differ in sign. Consequently, if we add $\int e^{-x} \cos x \, dx$

to each member of (2), we get

$$2\int e^{-x} \cos x \, dx = e^{-x} \sin x - e^{-x} \cos x + c$$

Hence

$$\int e^{-x} \cos x \, dx = \frac{e^{-x}(\sin x - \cos x)}{2} + K$$

where $K = c/2$.

EXERCISE 18.4

Perform the indicated integration.

1	$\int xe^{-x} \, dx$	2	$\int x^2 e^x \, dx$
3	$\int xa^x \, dx$	4	$\int xa^{2x} \, dx$
5	$\int e^x \cos x \, dx$	6	$\int e^{-x} \sin x \, dx$
7	$\int e^{-2x} \sin 3x \, dx$	8	$\int e^{3x} \cos 2x \, dx$
9	$\int \ln x \, dx$	10	$\int x \ln x \, dx$
11	$\int x^2 \ln x \, dx$	12	$\int \ln^2 x \, dx$
13	$\int x \sin x \, dx$	14	$\int x \cos x \, dx$
15	$\int x^2 \cos 2x \, dx$	16	$\int x \sin x \cos x \, dx$
17	$\int \text{Arccos } x \, dx$	18	$\int \text{Arcsin } 2x \, dx$
19	$\int x \, \text{Arcsin } x \, dx$	20	$\int x \, \text{Arccos } x \, dx$
21	$\int \text{Arctan } x \, dx$	22	$\int x \, \text{Arctan } x \, dx$
23	$\int \sin (\ln x) \, dx$	24	$\int x \sec^2 x/2 \, dx$
25	$\int \sec^3 x \, dx$	26	$\int \csc^3 x \, dx$
27	$\int x \sinh x \, dx$	28	$\int \sin 2x \cos x \, dx$

Replace x by y^2 in Problems 29 to 32.

29	$\int \sin \sqrt{x} \, dx$	30	$\int \cos \sqrt{x} \, dx$
31	$\int \text{Arcsin } \sqrt{x} \, dx$	32	$\int \text{Arctan } \sqrt{x} \, dx$

33 $\int \cos x \sqrt{1 + \sin^2 x} \, dx$; let $\sin x = \tan y$.
34 $\int \sqrt{\sec^2 x + 1} \, \sec x \tan x \, dx$; let $\sec x = \tan y$.

Expand before integrating in Problems 35 and 36.

35 $\int (x + \sin x)^2 \, dx$ 36 $\int (x - e^{-x})^2 \, dx$

37 Show that $\int x^n \ln x \, dx = x^{n+1} [(n + 1) \ln x - 1]/(n + 1)^2$.
38 Show that $\int e^{ax} \sin bx \, dx = e^{ax}(a \sin bx - b \cos bx)/(a^2 + b^2)$.
39 Show that $\int e^{ax} \cos bx \, dx = e^{ax}(a \cos bx + b \sin bx)/(a^2 + b^2)$.
40 Show that

$$\int e^{ax} P(x) \, dx = e^{ax} \left[\frac{P(x)}{a} - \frac{P'(x)}{a^2} + \frac{P''(x)}{a^3} - \cdots + (-1)^n \frac{P^{(n)}(x)}{a^{n+1}} \right]$$

for $P(x)$ a polynomial of degree n.

18.8   Integration of rational fractions

rational fraction A *rational fraction* is a fraction in which the numerator is a monomial or a polynomial and the denominator is a polynomial. If an integrand is a rational fraction of such form that no one of the standard formulas of integration can be applied, we can express the integrand as the sum of two or more fractions, each of which can be integrated by one of the methods previously discussed. We shall employ the following theorem from college algebra:

A polynomial can be expressed as the product of distinct or repeated linear and irreducible quadratic factors with real coefficients.

The factors of the polynomial will be in one of the classifications described below.

1. Each of the factors is linear and none is repeated as in
$$x^3 - 2x^2 - 5x + 6 = (x - 1)(x + 2)(x - 3)$$

2. The factors are all linear but one or more is repeated as in
$$x^3 + 3x^2 - 4 = (x - 1)(x + 2)^2$$

3. One or more of the factors is an irreducible quadratic trinomial as in
$$x^3 - x^2 - x - 2 = (x - 2)(x^2 + x + 1)$$

4. At least one of the factors is a repeated irreducible quadratic trinomial as in
$$x^5 - x^4 + 2x^3 - 2x^2 + x - 1 = (x - 1)(x^2 + 1)^2$$

It is proved in college algebra that a rational fraction with the numerator of lower degree than the denominator[°] can be expressed as the sum of two or more fractions. Each of the fractions in the sum has a power of each of the factors of the given denominator *form for* as a denominator, and each of these fractions is called a *partial* *partial* fraction. Furthermore, these partial fractions must satisfy the follow- *fractions* ing requirements:

I.   To every factor $ax + b$ of the denominator that appears without repetition, there corresponds a partial fraction $A/(ax + b)$ where A is a constant.

II. To every factor $(ax + b)^k$ of the denominator, there correspond the fractions

$$\frac{A_1}{ax + b} + \frac{A_2}{(ax + b)^2} + \cdots + \frac{A_k}{(ax + b)^k}$$

where $A_1, A_2, \ldots, A_k$ are constants.

[°] A fraction of this type is called a *proper rational* fraction.

III.   To every irreducible quadratic factor $ax^2 + bx + c$ of the denominator that appears without repetition, there corresponds a partial fraction $(Ax + B)/(ax^2 + bx + c)$ where A and B are constants.

IV.   If $ax^2 + bx + c$ is irreducible, then to every factor $(ax^2 + bx + c)^k$ of the denominator, there correspond the partial fractions

$$\frac{A_1x + B_1}{ax^2 + bx + c} + \frac{A_2x + B_2}{(ax^2 + bx + c)^2} + \cdots + \frac{A_kx + B_k}{(ax^2 + bx + c)^k}$$

where $A_1, A_2, \ldots, A_k$ and $B_1, B_2, \ldots, B_k$ are constants.

We shall now illustrate the way in which the above information can be employed to express a proper rational fraction as the sum of partial fractions.

Case I.   *All factors of the denominator are linear and none are repeated.*

Example 1   Express
$$\frac{2x^2 - 5x - 9}{(x - 1)(x + 1)(x + 2)}$$

as the sum of partial fractions.

Solution: According to statement I above, each linear factor of the given denominator must appear as the denominator of one of the partial fractions in the sum, and the numerator of each partial fraction is a constant. Hence we have

$$\frac{2x^2 - 5x - 9}{(x - 1)(x + 1)(x + 2)} = \frac{A}{x - 1} + \frac{B}{x + 1} + \frac{C}{x + 2} \qquad (1)$$

We shall next determine A, B, and C so that (1) is an identity. If we combine the fractions on the right in (1) into a single fraction, we have

$$\frac{2x^2 - 5x - 9}{(x - 1)(x + 1)(x + 2)}$$
$$= \frac{A(x + 1)(x + 2) + B(x - 1)(x + 2) + C(x - 1)(x + 1)}{(x - 1)(x + 1)(x + 2)} \qquad (2)$$

Since the two denominators in (2) are identical, (2) is an identity if the equation

$$2x^2 - 5x - 9$$
$$= A(x + 1)(x + 2) + B(x - 1)(x + 2) + C(x - 1)(x + 1)$$
$$= (A + B + C)x^2 + (3A + B)x + 2A - 2B - C \qquad (3)$$

is an identity. Furthermore, (3) is an identity if the coefficients of equal powers of x on the left and right are equal. Consequently, we have the following system of equations for evaluating A, B, and C.

$$A + B + C = 2 \qquad \textit{equating the coefficients of } x^2$$
$$3A + B = -5 \qquad \textit{equating the coefficients of } x$$
$$2A - 2B - C = -9 \qquad \textit{equating the constant terms}$$

If we solve this system, we find that $A = -2$, $B = 1$, and $C = 3$. Consequently,

$$\frac{2x^2 - 5x - 9}{(x - 1)(x + 1)(x + 2)} = \frac{-2}{x - 1} + \frac{1}{x + 1} + \frac{3}{x + 2} \tag{4}$$

There is another method that can be employed for evaluating A, B, and C in (1). It is relatively simple to use and is based on the fact that if two polynomials of degree n are equal for more than n replacements for the variable, they are equal for all values of the variable. We now consider Eq. (3) and shall determine A, B, and C so that (3) is an identity. If we replace x by -2 in (3), we get $9 = 3C$, and it follows that $C = 3$. Similarly, if we replace x by -1, we have $-2 = -2B$, so that $B = 1$; and if we replace x by 1, we get $-12 = 6A$ and have $A = -2$. Now if we replace A, B, and C in (3) by -2, 1, and 3, respectively, we get

$$2x^2 - 5x - 9$$
$$= -2(x + 1)(x + 2) + (x - 1)(x + 2) + 3(x - 1)(x + 2) \tag{5}$$

The expression on the right is a polynomial of degree 2, and so is the polynomial on the left. Furthermore, it is readily verified that these two polynomials are equal for the *three* values -1, 1, and 2 of x. Consequently, (5) is an identity. Furthermore, if x is not equal to -1, 1, or 2, we can divide each member of (5) by $(x - 1)(x + 1)(x + 2)$ and get Eq. (4). Consequently, (4) is an identity since the two members are equal for all replacements for x for which each member is a real number.

Now since the integral of the fraction on the left in (4) is equal to the sum of the integrals of the fractions on the right, we have

$$\int \frac{(2x^2 - 5x - 9)\, dx}{(x - 1)(x + 1)(x + 2)}$$
$$= \int \frac{-2\, dx}{x - 1} + \int \frac{dx}{x + 1} + \int \frac{3\, dx}{x + 2}$$
$$= -2 \ln |x - 1| + \ln |x + 1| + 3 \ln |x + 2| + c$$
$$= \ln \left| \frac{(x + 1)(x + 2)^3}{(x - 1)^2} \right| + c$$

Case II.   Factors of the denominator all linear but some repeated. If a linear factor $px + q$ occurs to the power r in the denominator of the given fraction, we write r fractions corresponding to it. These fractions have $px + q$, $(px + q)^2$, $(px + q)^3$, ..., $(px + q)^r$ as denominators. For example, if $(x - 4)^2$ is a factor of the denominator of the given fraction, we use $x - 4$ and $(x - 4)^2$ as denominators in expressing the given fraction as the sum of partial fractions and use

$$\frac{A}{x - 4} \quad \text{and} \quad \frac{B}{(x - 4)^2}$$

as the corresponding partial fractions, where A and B are to be determined.

Example 2   Express

$$\frac{3x^2 + 18x + 15}{(x - 1)(x + 2)^2}$$

as the sum of partial fractions and then integrate.

Solution: Since the denominator has the linear factor $x - 1$ that is not repeated and the linear factor $x + 2$ that enters to the power 2, we write

$$\frac{3x^2 + 18x + 15}{(x - 1)(x + 2)^2} = \frac{A}{x - 1} + \frac{B}{x + 2} + \frac{C}{(x + 2)^2}$$

$$= \frac{A(x + 2)^2 + B(x - 1)(x + 2) + C(x - 1)}{(x - 1)(x + 2)^2}$$

Therefore, we must determine A, B, and C such that

$$3x^2 + 18x + 15$$

$$\equiv A(x + 2)^2 + B(x - 1)(x + 2) + C(x - 1) \tag{1}$$

$$\equiv (A + B)x^2 + (4A + B + C)x + 4A - 2B - C \tag{2}$$

We shall use a combination of the methods considered in Case I in evaluating A, B, and C. Thus, from (1),

For $x = -2$ we get $-9 = -3C$

For $x = 1$ we get $36 = 9A$

Therefore, $A = 4$ and $C = 3$. We shall now determine B by equating coefficients of x^2 and then solving the resulting equation. Thus, we get

$$A + B = 3$$

$$B = 3 - A$$

$$= 3 - 4 \qquad \textit{since } A = 4$$

$$= -1$$

Now, using the values found for the undetermined coefficients, we see that

$$\int\frac{3x^2 + 18x + 15}{(x - 1)(x + 2)^2} dx \equiv \int\frac{4\,dx}{x - 1} - \int\frac{dx}{x + 2} + \int\frac{3\,dx}{(x + 2)^2}$$

$$\equiv 4\ln|x - 1| - \ln|x + 2| + 3\int(x + 2)^{-2}\,dx + C$$

$$\equiv \ln\frac{(x - 1)^4}{|x + 2|} + \frac{3(x + 2)^{-1}}{-1} + C$$

$$\equiv \ln\frac{(x - 1)^4}{|x + 2|} - \frac{3}{x + 2} + C$$

Cases III *and* IV.   *Denominator contains one or more irreducible quadratic factors.* If an irreducible quadratic factor $ax^2 + bx + c$ occurs in the denominator, we must have a fraction of the form $(Ax + B)/(ax^2 + bx + c)$ as one of the partial fractions. If the quadratic factor occurs to the power r in the denominator, we must have r partial fractions corresponding to it. The denominators of these partial fractions are $ax^2 + bx + c, (ax^2 + bx + c)^2, \ldots, (ax^2 + bx + c)^r$; a linear function with undetermined coefficients is the numerator of each.

Example 3   Evaluate $$\int\frac{x^4 + 9x^3 + 31x^2 + 43x + 16}{(x - 1)(x^2 + 4x + 5)^2} dx$$

Solution: In keeping with the procedure mentioned above, we write

$$\frac{x^4 + 9x^3 + 31x^2 + 43x + 16}{(x - 1)(x^2 + 4x + 5)^2}$$

$$= \frac{A}{x - 1} + \frac{Bx + C}{x^2 + 4x + 5} + \frac{Ex + F}{(x^2 + 4x + 5)^2} \qquad (6)$$

If we multiply each member of (6) by $(x - 1)(x^2 + 4x + 5)^2$ and equate the products, we get

$$x^4 + 9x^3 + 31x^2 + 43x + 16$$

$$= A(x^2 + 4x + 5)^2 + (Bx + C)(x - 1)(x^2 + 4x + 5)$$

$$+ (Ex + F)(x - 1)$$

$$= (A + B)x^4 + (8A + 3B + C)x^3 + (26A + B + 3C + E)x^2$$

$$+ (40A - 5B + C - E + F)x + 25A - 5C - F$$

We now equate the coefficients of the equal powers of x and get the following system of equations in A, B, C, E, and F:

$$A + B \qquad\qquad\quad = 1 \tag{7}$$

$$8A + 3B + C \qquad\qquad = 9 \tag{8}$$

$$26A + B + 3C + E \qquad = 31 \tag{9}$$

$$40A - 5B + C - E + F = 43 \tag{10}$$

$$25A \qquad\; - 5C \qquad - F = 16 \tag{11}$$

We solve this system of equations by the following procedure:

First, we replace B by $A - 1$, obtained from (7), in Eqs. (8) to (11) and get

$$11A + C \qquad\qquad = 12 \tag{12}$$

$$27A + 3C + E \qquad = 32 \tag{13}$$

$$35A + C - E + F = 38 \tag{14}$$

$$25A - 5C \qquad - F = 16 \tag{15}$$

Next, we equate the sums of the members of (14) and (15) and obtain

$$60A - 4C - E = 54 \tag{16}$$

Similarly, we equate the sums of the members of (13) and (16) and have

$$87A - C = 86 \tag{17}$$

Finally, we solve (12) and (17) for A and C and get $A = 1$ and $C = 1$. Now by substitution, we obtain $B = 0$ from (7), $E = 2$ from (13), and $F = 4$ from (15).

We now replace A, B, C, E, and F by the above values in (6), multiply by dx, and get

$$\int \frac{x^4 + 9x^3 + 31x^2 + 43x + 16}{(x-1)(x^2 + 4x + 5)^2}\, dx$$

$$= \int \frac{dx}{x-1} + \int \frac{dx}{x^2 + 4x + 5} + \int \frac{(2x+4)\, dx}{(x^2 + 4x + 5)^2}$$

$$= \int \frac{dx}{x-1} + \int \frac{dx}{(x+2)^2 + 1} + \int \frac{(2x+4)\, dx}{(x^2 + 4x + 5)^2}$$

$$= \ln |x - 1| + \operatorname{Arctan} (x + 2) - \frac{1}{x^2 + 4x + 5} + c$$

EXERCISE 18.5

Perform the integrations indicated in Problems 1 to 40.

$$1 \quad \int \frac{x}{x-3}\, dx \qquad\qquad 2 \quad \int \frac{2x+5}{x-1}\, dx \qquad\qquad 3 \quad \int \frac{7-2x}{3x-5}\, dx$$

4 $\displaystyle\int \frac{x^3 + 1}{x - 1}\, dx$ **5** $\displaystyle\int \frac{x + 7}{x^2 + 2x - 8}\, dx$ **6** $\displaystyle\int \frac{3x + 4}{x^2 + 5x + 6}\, dx$

7 $\displaystyle\int \frac{x^2 + x + 1}{x^2 - 7x + 10}\, dx$ **8** $\displaystyle\int \frac{x - 6}{x^2 - x}\, dx$

9 $\displaystyle\int \frac{6x^2 - 23x + 9}{x^3 - 4x^2 + 3x}\, dx$ **10** $\displaystyle\int \frac{dx}{x^3 - 3x^2 + 2x}$

11 $\displaystyle\int \frac{x^2 - 17x + 22}{(x - 1)(x - 3)(x + 2)}\, dx$ **12** $\displaystyle\int \frac{3x + 15}{(x - 4)(2x + 1)(x - 1)}\, dx$

13 $\displaystyle\int \frac{x^3\, dx}{(x + 1)(x^2 - 4)}$ **14** $\displaystyle\int \frac{2x^4 - x^3 - 2x^2 - 4x + 3}{x(x - 1)(2x - 3)}\, dx$

15 $\displaystyle\int \frac{4x^4 - 8x^3 + 3x^2 - 11x + 1}{(x - 2)(2x - 1)(x + 1)}\, dx$

16 $\displaystyle\int \frac{6x^4 - 3x^3 - 4x^2 - 11x - 6}{(x + 1)(x - 1)(2x + 1)}\, dx$

17 $\displaystyle\int \frac{-2x^2 + x - 1}{(x - 1)^2(x - 3)}\, dx$ **18** $\displaystyle\int \frac{3x + 4}{(x + 2)^2(x - 6)}\, dx$

19 $\displaystyle\int \frac{dx}{x^2(x^2 - 4)}$ **20** $\displaystyle\int \frac{x^5 - 2}{x^4 - 2x^3}\, dx$

21 $\displaystyle\int \sec\theta\, d\theta = \int \frac{\cos\theta\, d\theta}{1 - \sin^2\theta}$. Let $\sin\theta = x$.

22 $\displaystyle\int \csc\theta\, d\theta$

23 $\displaystyle\int \sec^3\theta\, d\theta = \int \frac{\cos\theta\, d\theta}{(1 - \sin^2\theta)^2}$. Let $\sin\theta = x$.

24 $\displaystyle\int \csc^3\theta\, d\theta$ **25** $\displaystyle\int \frac{dx}{x^3 - 10x^2 + 33x - 36}$

26 $\displaystyle\int \frac{-3x^2 + 7x - 16}{x^3 - 5x^2 + 7x - 3}\, dx$ **27** $\displaystyle\int \frac{11x^2 - 7x}{2x^3 - x^2 - 2x + 1}\, dx$

28 $\displaystyle\int \frac{11x^2 - 12x - 5}{2x^3 - x^2 - 7x + 6}\, dx$ **29** $\displaystyle\int \frac{2x^2 + 6x - 1}{x^3 + x^2 + x}\, dx$

30 $\displaystyle\int \frac{x^2 + 9x + 29}{(x - 4)(x^2 + 2x + 3)}\, dx$ **31** $\displaystyle\int \frac{dx}{x^3 - 8}$

32 $\displaystyle\int \frac{6x^2 + 33x + 62}{3x^3 + 16x^2 + 18x - 20}\, dx$

33 $\displaystyle\int \frac{6x^3 - 19x^2 + 23x - 28}{(x - 1)(x - 4)(x^2 + x + 4)}\, dx$

34 $\displaystyle\int \frac{-x^3 + 10x^2 - 5x + 15}{x^2(x^2 + 5)}\, dx$

35 $\int \dfrac{dx}{x^4 - 16}$

36 $\int \dfrac{4x^3 + 23x^2 - 14x + 52}{x^4 + 6x^3 + 14x^2 + 36x + 48} \, dx$

37 $\int \dfrac{2x^3 + 5x^2 + 11x + 13}{(x^2 + 4)(x^2 + 2x + 5)} \, dx$

38 $\int \dfrac{2x^3 + 7x^2 + 14x + 11}{(x^2 + 2x + 5)(x^2 + x + 3)} \, dx$

39 $\int \dfrac{3x^4 - 5x^3 + 4x^2 - 2x + 1}{x(x^2 - x + 1)^2} \, dx$

40 $\int \dfrac{x^4 - x^3 + 5x^2 - 3x + 2}{(x - 1)(x^2 + 1)^2} \, dx$

18.9   More improper integrals

We studied improper integrals in Sec. 8.7 but have learned to integrate a variety of forms since then. The reader might do well to reread Sec. 8.7 before proceeding further since the methods and types of situations covered here are the same as there. Only the forms to be integrated will differ in this section.

Example   Evaluate $\int_2^3 \dfrac{dx}{x - 2}$ or show that it diverges.

Solution: Since the integrand is infinite for the lower limit 2, we write

$$\int_2^3 \frac{dx}{x - 2} = \lim_{\varepsilon \to 0} \int_{2+\varepsilon}^3 \frac{dx}{x - 2} = \lim_{\varepsilon \to 0} \ln |x - 2| \Big|_{2+\varepsilon}^3$$

$$= \lim_{\varepsilon \to 0} [\ln (3 - 2) - \ln (2 + \varepsilon - 2)]$$

$$= \ln 1 - \lim_{\varepsilon \to 0} \ln \varepsilon = -\infty$$

Hence the integral diverges.

EXERCISE 18.6

Evaluate each integral in Problems 1 to 20 or show that it diverges.

1 $\int_4^\infty \dfrac{dx}{3x - 2}$

2 $\int_0^\infty e^{-x/2} \, dx$

3 $\int_0^\infty x e^{-x} \, dx$

4 $\int_3^\infty \dfrac{4x \, dx}{x^2 - 3}$

5   $\displaystyle\int_{-\infty}^{0} e^x \, dx$

6   $\displaystyle\int_{-\infty}^{0} \frac{4x}{x^2 + 3} \, dx$

7   $\displaystyle\int_{-\infty}^{-1} \frac{dx}{x(x - 1)}$

8   $\displaystyle\int_{-\infty}^{0} \frac{dx}{x^2 + 1}$

9   $\displaystyle\int_{-\infty}^{\infty} 4xe^{-x^2} \, dx$

10   $\displaystyle\int_{-\infty}^{\infty} \frac{dx}{x^2 + 4}$

11   $\displaystyle\int_{-\infty}^{\infty} \frac{dx}{x^2 - 4}$

12   $\displaystyle\int_{-\infty}^{\infty} 3^x \, dx$

13   $\displaystyle\int_{1}^{5} \frac{dx}{x - 1}$

14   $\displaystyle\int_{-2}^{2} \frac{dx}{\sqrt[3]{2x + 4}}$

15   $\displaystyle\int_{0}^{\pi/2} \frac{\cot x \, dx}{\ln \sin x}$

16   $\displaystyle\int_{\pi/2}^{0} \tan x \, dx$

17   $\displaystyle\int_{0}^{4} \frac{x \, dx}{\sqrt{4 - x}}$

18   $\displaystyle\int_{0}^{a} \frac{dx}{\sqrt{a^2 - x^2}}$

19   $\displaystyle\int_{0}^{a} \frac{x^2 \, dx}{\sqrt{a^2 - x^2}}$

20   $\displaystyle\int_{3}^{2} \frac{dx}{x - 2}$

21   Compute the area "bounded" by $y = 12/(x^2 + 6)$ and the X axis.

22   Sketch the curve $y^2(4 - x) = x^3$, and find the area "bounded" by it and $x = 4$.

23   Find the area of a circle by using the parametric equations $x = a \cos \theta$ and $y = a \sin \theta$.

24   Find the area enclosed by the four-cusped hypocycloid whose parametric equations are $x = a \cos^3 \theta$ and $y = a \sin^3 \theta$.

19

Applications of the definite integral

19.1   More areas

In Sec. 8.5 and 8.6 we explained the method for obtaining certain areas by use of the definite integral, but at that time we employed integrals that could be evaluated only by the power formula. We are now in position to evaluate a greater variety of integrals, and in this chapter we shall discuss several additional applications of the definite integral.

Example 1   Find the area of the region bounded by the curve $y = 12x/(x^2 + 4)$, the X axis, $x = 1$, and $x = 4$.

Solution: The desired region is shown in Fig. 19.1 and is given by

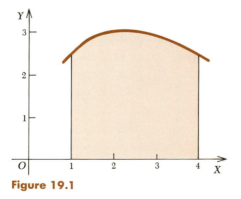

Figure 19.1

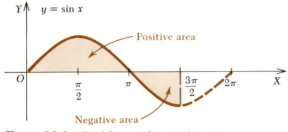

Figure 19.2 **Positive and negative area**

$$A = \int_1^4 \frac{12x}{x^2 + 4}\, dx = 6 \ln (x^2 + 4)\Big|_1^4$$

$$= 6 \ln 20 - 6 \ln 5$$

$$= 6 \ln \text{²⁰⁄₅} = 6 \ln 4$$

Example 2 Find the area of the region bounded by $y = \sin x$, the X axis, $x = 0$, and $x = 3\pi/2$.

Solution: The desired area is the shaded portion in Fig. 19.2. Since a part of it is above the X axis and the remainder below it and since $y = \sin x$ crosses the X axis at $x = \pi$, we find the area by evaluating

$$A = \int_0^\pi \sin x\, dx - \int_\pi^{3\pi/2} \sin x\, dx = -\cos x\Big|_0^\pi + \cos x\Big|_\pi^{3\pi/2}$$

$$= -\cos \pi + \cos 0 + \cos \frac{3\pi}{2} - \cos \pi$$

$$= 1 + 1 + 0 + 1 = 3$$

Example 3 Find the area enclosed by $y = 8/(x^2 + 4)$ and $y = 1$.

Solution: The two curves are shown in Fig. 19.3. Their intersections were found by solving the pair of equations simultaneously and are $(-2, 1)$ and $(2, 1)$ as indicated in the figure.

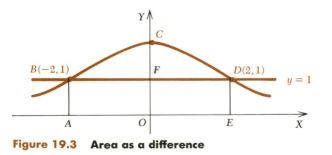

Figure 19.3 **Area as a difference**

The desired area is the difference between the area of $ABCDE$ and that of $ABFDE$; hence

$$\text{Area} = \int_{-2}^{2} \frac{8\,dx}{x^2 + 4} - \int_{-2}^{2} 1\,dx$$

$$= 4\,\text{Arctan}\,\frac{x}{2}\,\Big|_{-2}^{2} \;-\; x\,\Big|_{-2}^{2}$$

$$= 4\left[\frac{\pi}{4} - \left(-\frac{\pi}{4}\right)\right] - [2 - (-2)] \qquad \text{since } -\pi/2 < \text{Arctan}\,N < \pi/2$$

$$= 2\pi - 4$$

EXERCISE 19.1

Sketch the curves in Problems 1 to 28, and find the area of the region bounded by the given curve, the specified ordinates, and the X axis.

1 $y = e^x$, $x = 1$, $x = 2$
2 $y = xe^{x^2}$, $x = 0$, $x = 1$
3 $y = xe^x$, $x = 1$, $x = 3$
4 $y = 4xe^{-x}$, $x = 0$, $x = 2$
5 $y = \ln x$, $x = 1$, $x = e$
6 $y = \ln(2x - 1)$, $x = 1$, $x = (e + 1)/2$
7 $y = x \ln x$, $x = e$, $x = e^2$
8 $y = \ln^2 x$, $x = 1$, $x = e^2$
9 $y = 2 \sin 3x$, $x = 0$, $x = \pi/3$
10 $y = \sin^2 x$, $x = 0$, $x = \pi$
11 $y = 4 \cos^2(x/2)$, $x = 0$, $x = \pi$
12 $y = 2 \sin^3(x/2)$, $x = 0$, $x = \pi$
13 $y = \sin x + \cos x$, $x = 0$, $x = \pi/2$
14 $y = \sin x + \tan x$, $x = 0$, $x = \pi/4$
15 $y = \sec x - \tan x$, $x = \pi/6$, $x = \pi/2$
16 $y = \tan x + \cot x$, $x = \pi/4$, $x = \pi/3$
17 $xy = 4$, $x = 1$, $x = 3$
18 $x^2 y = 4$, $x = 1$, $x = 3$
19 $xy^2 = 4$, $x = 1$, $x = 4$
20 $xy = -1$, $x = 1$, $x = 3$
21 $y = x\sqrt{16 - x^2}$, $x = 0$, $x = 4$
22 $y = \sqrt{25 - x^2}$, $x = 0$, $x = 5$
23 $y = \sqrt{x^2 - 25}$, $x = 5$, $x = 5\sqrt{2}$
24 $y = \sqrt{x^2 + 36}$, $x = 0$, $x = 6$
25 $x^2 y + y - 5 = 0$, $x = 0$, $x = \sqrt{3}$

26 $x^2y + 4y = 8x, x = 0, x = 2$
27 $x^2y - 4y = 15, x = 3, x = 5$
28 $y = 6/\sqrt{4 - x^2}, x = 0, x = 1$

In each of Problems 29 to 36, find the area of the entire region bounded by the two given curves after sketching the curves and shading the desired area.

29 $2y = x^2, y^2 = 16x$ **30** $x^2 = 6y, x^2 = 12y - 9$
31 $y = 8x - x^2, y = 2x$ **32** $y = 8x - x^2, 3y = x^2$
33 $x^2y + 2y = 3x, 4y = x^2$
34 $x^2y + 2y = 6x, 3y = x$
35 $x^2y + 3y + 6x = 0, 2y = 5x - 2x^2$
36 $4y = x^2(5 - x), y = (x - 2)^2$

37   Find the area of the region bounded by $y = e^x$, $y = e^{-x}$, and $x = 1$.
38   Find the area of the region bounded by $y = \ln x$, $y = 1$, and $x = e^2$.
39   Find the area of the region bounded by $y = \sin x$ and $y = \cos x$ between two consecutive points of intersection.
40   Find the area of the first-quadrant region bounded by $y = \sin 2x$ and $y = \tan x$.

19.2   Area in polar coordinates

We considered the problem in rectangular coordinates of finding the area bounded by the curve $y = f(x)$, $x = a$, $x = b$, and the X axis. We shall now consider the corresponding problem in polar coordinates. Thus, we shall determine the area of the region bounded by $r = f(\theta)$ and the two lines $\theta = \theta_1$ and $\theta = \theta_2$. This area is shown in Fig. 19.4 as the area $OABO$.

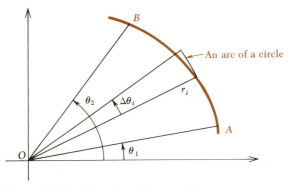

Figure 19.4   Area in polar coordinates

The procedure for finding the area of this region is very similar to that used in dealing with rectangular coordinates. We begin by dividing the angle from θ_1 to θ_2 into n subintervals $\Delta\theta_1, \Delta\theta_2, \ldots, \Delta\theta_n$. We then draw the corresponding radius vectors, denote their lengths by $r_1, r_2, \ldots, r_n$, and draw the circular arcs as shown.

We shall make use of the fact that the area of a sector of a circle of radius a and central angle α is $\frac{1}{2}a^2\alpha$ for α in radians; hence, the sum of the areas of the circular sectors is

$$\frac{r_1^2}{2}\Delta\theta_1 + \frac{r_2^2}{2}\Delta\theta_2 + \cdots + \frac{r_n^2}{2}\Delta\theta_n$$

We now define the area of the region bounded by $\theta = \theta_1$, $r = f(\theta)$, and $\theta = \theta_2$ to be the limit of this sum as the largest of the subintervals $\Delta\theta_1, \Delta\theta_2, \ldots, \Delta\theta_n$ approaches zero. If we use a regular partition, all the $\Delta\theta_i$ are of the same magnitude, and we can write the definition of area as

$$A = \lim_{n \to \infty} \sum_{i=1}^{n} \frac{r_i^2}{2}\Delta\theta$$

If we make use of the fundamental theorem, we see that *the area bounded by $\theta = \theta_1$, $r = f(\theta)$, and $\theta = \theta_2$ is*

area in polar coordinates

$$A = \frac{1}{2}\int_{\theta_1}^{\theta_2} r^2 \, d\theta \tag{19.1}$$

where $r = f(\theta)$ and $\theta_1 < \theta_2$. If $\theta_1 > \theta_2$, then (19.1) gives the negative of the area.

If we do not use a regular partition, we must then require that the largest $\Delta\theta_i$ approach zero; hence, all subintervals approach zero; n becomes infinite, and we again have (19.1).

Example 1   Find the area enclosed by $r = 4\cos\theta$.

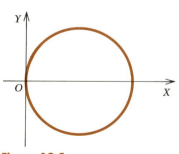

Figure 19.5

Solution: The graph of the given equation is shown in Fig. 19.5. It was drawn by assigning values to θ from 0 to $\pi/2$, drawing the corresponding curve, and then making use of the symmetry with respect to the polar axis. Because of the symmetry, we can integrate from $\theta = 0$ to $\theta = \pi/2$ and multiply by 2 to get the desired area. Thus,

$$A = 2\left(\frac{1}{2}\right)\int_0^{\pi/2} r^2\ d\theta$$

$$= \int_0^{\pi/2} 16\ \cos^2\theta\ d\theta \qquad\qquad since\ r = 4\cos\theta$$

$$= \int_0^{\pi/2} 16\left(\frac{1 + \cos 2\theta}{2}\right) d\theta$$

$$= 8\theta + 4\sin 2\theta\ \Big|_0^{\pi/2}$$

$$= 4\pi$$

Example 2   Find the area of the region that is inside the circle $r = 3\sin\theta$ and also inside the cardioid $r = 1 + \sin\theta$.

Solution: Both curves are symmetric with respect to the normal axis; they are shown in Fig. 19.6. If we solve the equations simultaneously, we find that the curves intersect at $A(\frac{3}{2}, \pi/6)$ and $D(\frac{3}{2}, 5\pi/6)$. Half of the required area is that bounded by the circular arc OA and the line $\theta = \pi/6$ plus that bounded by the line $\theta = \pi/6$, the arc AC of the cardioid, and $\theta = \pi/2$ since both curves are symmetric with respect to the normal axis. Consequently, the desired area is

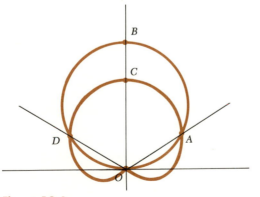

Figure 19.6

$$A = 2\left(\frac{1}{2}\right)\left[\int_0^{\pi/6}(3\sin\theta)^2\,d\theta + \int_{\pi/6}^{\pi/2}(1+\sin\theta)^2\,d\theta\right]$$

$$= \int_0^{\pi/6}9\sin^2\theta\,d\theta + \int_{\pi/6}^{\pi/2}\left(1+2\sin\theta + \frac{1-\cos 2\theta}{2}\right)d\theta$$

$$= 9\int_0^{\pi/6}\frac{1-\cos 2\theta}{2}\,d\theta + \int_{\pi/6}^{\pi/2}\left(1+2\sin\theta + \frac{1-\cos 2\theta}{2}\right)d\theta$$

$$= \frac{9}{2}\left(\theta - \frac{1}{2}\sin 2\theta\right)\Big|_0^{\pi/6} + \left(\theta - 2\cos\theta + \frac{\theta}{2} - \frac{1}{4}\sin 2\theta\right)\Big|_{\pi/6}^{\pi/2}$$

$$= \frac{9}{2}\left(\frac{\pi}{6} - \frac{1}{2}\sin\frac{\pi}{3}\right) + \left(\frac{3}{2}\frac{\pi}{2} - 2\cos\frac{\pi}{2} - \frac{1}{4}\sin\pi\right)$$

$$\quad - \left(\frac{3}{2}\frac{\pi}{6} - 2\cos\frac{\pi}{6} - \frac{1}{4}\sin\frac{\pi}{3}\right)$$

$$= \frac{9}{2}\left(\frac{\pi}{6} - \frac{\sqrt{3}}{4}\right) + \left(\frac{3\pi}{4}\right) - \left(\frac{\pi}{4} - \sqrt{3} - \frac{\sqrt{3}}{8}\right) = \frac{5\pi}{4}$$

EXERCISE 19.2

1 Sketch the curve whose equation is $r = 4\sin 3\theta$ and indicate the area obtained by evaluating $\frac{1}{2}\int_0^{\pi/2}r^2\,d\theta$.

2 Sketch the curve whose equation is $r = a\cos 3\theta$ and indicate the area obtained by evaluating $\int_0^{\pi/2}r^2\,d\theta$.

3 Which of the following integrals gives the entire area enclosed by the graph of $r = 2\cos 3\theta$?

$$2\left(\frac{1}{2}\right)\int_0^{\pi/2}r^2\,d\theta \qquad 2\left(\frac{1}{2}\right)\int_0^{\pi}r^2\,d\theta \qquad 6\left(\frac{1}{2}\right)\int_0^{\pi/3}r^2\,d\theta$$

4 Sketch the cardioid whose equation is $r = a(1+\cos\theta)$. Which of the following integrals give its entire area?

$$4\left(\frac{1}{2}\right)\int_0^{\pi/2}r^2\,d\theta \qquad 2\left(\frac{1}{2}\right)\int_0^{\pi}r^2\,d\theta \qquad \left(\frac{1}{2}\right)\int_0^{2\pi}r^2\,d\theta$$

Find the area of the region enclosed by the graph of the equation in each of Problems 5 to 16.

5 $r = 4\sin\theta$ 6 $r = 2\cos 3\theta$

7 $r = a\sin 2\theta$ 8 $r = 3\sin 3\theta$

9 $r = 2\sqrt{\cos\theta}$ 10 $r = 4\sqrt{\sin\theta}$

11 $r^2 = 4\cos 2\theta$ 12 $r^2 = a^2\sin 2\theta$

13 $r = 3 - \cos \theta$ **14** $r = 2 + \sin \theta$
15 $r = a(1 + \sin \theta)$ **16** $r = 1 + \cos \theta$

17   Compute the area of the region that is inside the small loop of the limaçon $r = 1 - 2 \cos \theta$.
18   Sketch the lemniscate $r^2 = a^2 \cos 2\theta$, and find the area of the region enclosed by it.
19   Find the area of the region that is common to the circles $r = a$ and $r = 2a \sin \theta$.
20   Find the area of the region that is common to the circle $r = 3 \sin \theta$ and the cardioid $r = 1 + \sin \theta$.
21   Find the area of the region that is inside the lemniscate $r^2 = 8 \cos 2\theta$ and outside the circle $r = 2$.
22   Find the area of the region bounded by $r \sin^2 \theta = \cos \theta$ and the line $r = \sec \theta$.
23   Find the area of the region that is inside the cardioid $r = 4(1 + \cos \theta)$ and outside the circle $r = 8 \cos \theta$.
24   Find the area of the region that is inside the circle $r = a \sin \theta$ and outside the cardioid $r = a(1 - \sin \theta)$.

19.3   Volumes of solids of revolution by the disk method

If the area bounded by the curve $y = f(x)$, the X axis, the line $x = a$, and the line $x = b$ is revolved about the X axis, a solid is generated. A quarter of one such solid is shown in Fig. 19.7 and the magnitude of the entire volume of the solid may be formulated as follows:

1.   We partition the interval $[a, b]$ into n subintervals and denote the points of division successively by $x_0 = a, x_1, x_2, x_3, \ldots,$ $x_n = b$. Then at each x_i, $i = 1, \ldots, n$, we construct a line segment perpendicular to the X axis that terminates on the graph of $y = f(x)$. Finally, at the upper terminus of the ith perpendicular, $i = 1, \ldots, n$, we construct a line segment parallel to the X axis that terminates on the $(i + 1)$st perpendicular. Thus, we obtain a set of rectangles, and the ith is shown in Fig. 19.7.
2.   We rotate each of these rectangles with base Δx_i and height y_i about the X axis and thereby generate cylindrical disks of radius y_i and thickness Δx_i. The volume of the ith disk is $\pi y_i^2 \, \Delta x_i$.
3.   We form the sum of these volumes and thus get

$$\sum_{i=1}^{n} \pi y_i^2 \, \Delta x_i = \pi y_1 \, \Delta x_1^2 + \pi y_2^2 \, \Delta x_2 + \cdots + \pi y_n^2 \, \Delta x_n$$

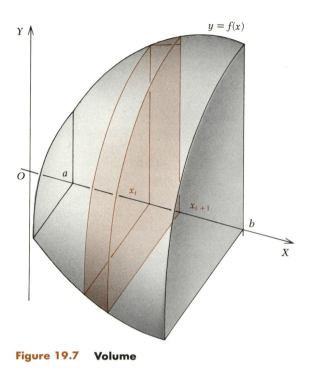

Figure 19.7   Volume

4.   Then we define the volume V of the solid of revolution to be

$$V = \lim_{\sigma \to 0} \pi \sum_{i=1}^{n} y_i^2 \, \Delta x_i$$

where $\sigma \geq \Delta x_i$, $i = 1, \ldots, n$, that is, where σ is the largest Δx_i.

By use of the fundamental theorem, this limit of a sum may be replaced by $\int_{x=a}^{x=b} \pi y^2 \, dx$. Consequently, we know that *the volume of the solid generated by revolving about the X axis the region bounded by $y = f(x)$, the X axis, $x = a$, and $x = b$ is given by*

volume of a solid formed by rotation of an area about the X axis

$$V = \pi \int_{a}^{b} y^2 \, dx \qquad (19.2)$$

If y is real throughout the interval $[a, b]$, then the integrand y^2 is positive and V is positive or negative according as each Δx_i in the sum used to obtain (19.2) is positive or negative.

We can show in a similar manner that if the region bounded by $x = g(y)$, the Y axis, $y = c$, and $y = d$ is revolved about the Y axis, the volume of the solid generated is given by

$$V = \pi \int_c^d x^2 \, dy \qquad (19.3)$$

Example 1   Find the volume of the solid generated by revolving the area bounded by the X axis, the parabola $x^2 - 4x + 4y - 12 = 0$, and the lines $x = -1$ and $x = 3$ about the X axis.

Solution: The region to be revolved is the region $ABCD$ in Fig. 19.8. We first partition the interval $[-1, 3]$ into n subintervals of length $\Delta x_i, i = 1, \ldots, n$, where $\Delta x_i \le \sigma$, and then at each of the subdivision points we erect a perpendicular to the X axis that terminates on the curve. Two such perpendiculars are EG and FH in the figure. If the region $ABCD$ is revolved about the X axis, then $EFGH$ generates a volume that is approximately equal to $\pi y_i^2 \Delta x_i$. Consequently, the volume V generated by revolving $ABCD$ about the X axis is approximately equal to

$$\pi y_1^2 \, \Delta x_1 + \pi y_2^2 \, \Delta x_2 + \cdots + \pi y_n^2 \, \Delta x_n = \pi \sum_{i=1}^{n} y_i^2 \, \Delta x_i$$

and the exact volume V is equal to $\lim_{\sigma \to 0} \pi \sum_{i=1}^{n} y_i^2 \, \Delta x_i$. Therefore, by the fundamental theorem of calculus we have

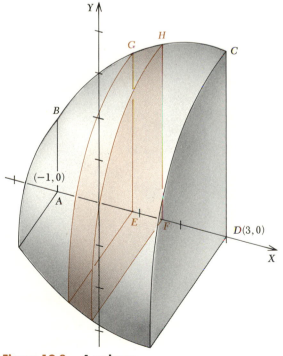

Figure 19.8   A volume

$$V = \pi \int_{-1}^{3} y^2 \, dx = \frac{\pi}{16} \int_{-1}^{3} (12 + 4x - x^2)^2 \, dx$$

$$= \frac{\pi}{16} \int_{-1}^{3} (144 + 96x - 8x^2 - 8x^3 + x^4) \, dx$$

$$= \frac{\pi}{16} \left(144x + 48x^2 - \frac{8x^3}{3} - 2x^4 + \frac{x^5}{5} \right)\Big|_{-1}^{3} = \frac{2903\pi}{60}$$

If a region is bounded in part by a straight line that is parallel to one of the coordinate axes and if the region is revolved about that *rotation about* line, then the volume cannot be obtained by use of (19.2) or (19.3). *a line parallel* It can, however, be obtained by making use of the same principles *to an axis* and concepts used in deriving those formulas.

Example 2   Find the volume of the solid generated by revolving about $y = 2$ the region bounded by $y^2 = 2x$, $x = 8$, and $y = 2$.

Solution: A sketch of the region that is to be revolved along with a typical element of it is shown in Fig. 19.9. A typical strip of the area that is to be rotated is of length $y_i - 2$. Since the interval from $x = 2$ to $x = 8$ is to be covered, the desired volume is given by

$$V = \lim_{n \to \infty} \sum_{i=1}^{n} \pi (y_i - 2)^2 \, \Delta x_i$$

$$= \int_{2}^{8} \pi (y - 2)^2 \, dx$$

$$= \pi \int_{2}^{8} (y^2 - 4y + 4) \, dx$$

$$= \pi \int_{2}^{8} (2x - 4\sqrt{2} \sqrt{x} + 4) \, dx$$

$$= \pi [x^2 - 4\sqrt{2} \, x^{3/2} \, (\tfrac{2}{3}) + 4x]\Big|_{2}^{8}$$

$$= \pi \left[\left(64 - \frac{8\sqrt{2}(8)^{3/2}}{3} + 32 \right) - \left(4 - \frac{8\sqrt{2}(2)^{3/2}}{3} + 8 \right) \right]$$

since $y^2 = 2x$

$$= \frac{28\pi}{3}$$

19.4   Volumes of solids of revolution by use of "washers"

Heretofore, we have considered volumes of solids of revolution if and only if the axis of rotation is one of the boundaries of the generating area. We shall now consider situations in which the rotation is about

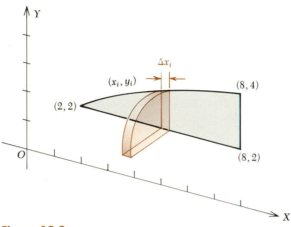

Figure 19.9

a line that is not a boundary of the area that is being rotated. Such

volume as the situations are handled by expressing the desired volume as the differ-
difference of ence of two volumes each of which can be obtained by rotating
two volumes a region about a line that forms one of its boundaries. For example,
in the situation that is pictured in Fig. 19.10, we want to find the
volume of the solid generated by rotating the area *CDEF* about the
X axis. We do this by subtracting the volume of the solid generated
by rotating the area *ACFB* about the *X* axis from that obtained by
rotating the area *ADEB* about the *X* axis. We thus find that *the*

volume as the *volume obtained by rotating about the X axis the area bounded by*
difference of $y_1 = f(x)$, $y_2 = g(x)$, $x = a$ *and* $x = b$ *is given by*
two volumes

$$V = \pi \int_a^b (y_1{}^2 - y_2{}^2) \, dx \tag{19.4}$$

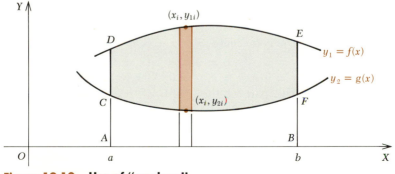

Figure 19.10   Use of "washers"

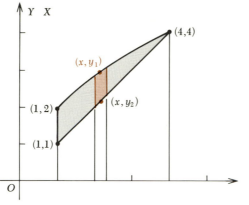

Figure 19.11

Example

Find the volume generated by rotating the region bounded $y_2 = x$, $x = 1$ and $y_1{}^2 = 4x$ about the X axis.

Solution: The specified area and a typical strip of each area that is to be rotated is shown in Fig. 19.11. We shall subtract the volume of the solid generated by rotating about the X axis the region bounded by $x = 1$, $y_2 = x$, $x = 4$ and the X axis from that generated by rotating about the X axis the region bounded by $x = 1$, $y_1{}^2 = 4x$, $x = 4$ and the X axis and thus find that the desired volume is

$$V = \pi \int_1^4 (y_1{}^2 - y_2{}^2)\, dx = \pi \int_1^4 (4x - x^2)\, dx = \pi \left(2x^2 - \frac{x^3}{3} \right)\Big|_1^4 = 9\pi$$

EXERCISE 19.3

In Problems 1 to 4 use the disk method to find the volume of the solid that is generated by revolving the region bounded by the given curves about the line whose equation is given after the semicolon.

   1   The coordinate axes, $x = 3$, $y = x + 2$; the X axis
   2   $y = 2\sqrt{3}x$, $y = 0$, $x = 3$; $x = 3$
   3   $y = \ln x$, $x = e$, $y = 0$; X axis
   4   The part of the graph of $x = a \sec t$, $y = a \cos t$ between $x = a$ and $x = 2a$; X axis

Use the washer method to find the volume of the solid generated by revolving the region bounded by the curves about the line given after the semicolon in each of Problems 5 to 8.

5   $y = x^2$, $y = 0$, $x = 2$; Y axis

6   $y = \sqrt{t + 2}$, $x = 0.5(t + 2)$, $x = 1$; Y axis

7   $y = 3\sqrt{x}$, X axis, $x = 4$; $y = -1$

8   $y = x + 1$, $x = 2$, $y = 3$; X axis

9   Find the volume of the sphere generated by revolving about the X axis the upper half of the circle whose equation is $x^2 + y^2 = 25$.

10   Find the volume of the solid generated by revolving about the X axis the first-quadrant region bounded by the X axis, $y = x$, and $x^2 + y^2 = 16$.

11   Find the volume of the solid generated by revolving about the Y axis the region bounded by the Y axis, the line $y = 9$, and the curve $x = 2\sqrt{y}$.

12   Find the volume of the solid generated by revolving about the X axis the region bounded by $2y^2 = 9x$ and $x = 6$.

13   Find the volume of the solid generated by revolving about $x = 4$ the region bounded by $y^2 = 9x$ and $x = 4$.

14   Find the volume of the solid generated by revolving the right-hand half of the region enclosed by $b^2x^2 + a^2y^2 = a^2b^2$ about the Y axis.

15   Find the volume of the solid generated by revolving the upper half of the region enclosed by $b^2x^2 + a^2y^2 = a^2b^2$ about the X axis.

16   Find the volume of the cone generated by revolving about the Y axis the region bounded by $y = hx/r$, $y = h$, and the Y axis.

17   The region bounded by one arch of $y = \cos x$ and the X axis is rotated about the X axis. Find the volume of the solid that is generated.

18   The region under $y = \sin x$ from $x = \pi/2$ to $x = \pi$ is rotated about $x = \pi/2$. Find the volume of the solid that is generated.

19   The region under $y = \cot x$ in the interval from $\pi/4$ to $\pi/2$ is rotated about the X axis. Find the volume of the solid that is generated.

20   The region under $y = \cos x/2$ from $x = 0$ to $x = \pi$ is rotated about the X axis. Find the volume of the solid generated.

21   The region under $y = e^{-x}$ in $[0, \infty)$ is rotated about the X axis. Find the volume of the solid that is generated.

22   The region bounded by the X axis and the parabola $y = 4x - x^2$ is rotated about the Y axis through an arc of $2\pi/3$. Find the volume of the solid that is generated.

23   The region enclosed by the hypocyloid whose equations are $x = a \cos^3 \theta$, $y = a \sin^3 \theta$ is rotated about the Y axis. Find the volume of the solid that is generated.

24   The region under $y = 8/(x^2 + 4)$ and above the X axis in $[0, \infty)$ is rotated about the X axis. Find the volume of the solid that is generated.

19.5   Volumes of solids of revolution by use of cylindrical shells

In this section we shall discuss a second method for obtaining the volume of a solid of revolution; we shall consider the solid generated by revolving the region $ABCD$ in Fig. 19.12 about the Y axis. We partition the interval from $A(a, 0)$ to $B(b, 0)$ into n subintervals of length Δx and construct rectangles such as $EFGH$ over each subinterval. As the region $ABCD$ is revolved about the Y axis these rectangles generate a set of concentric cylindrical shells. The sum of the volumes of these shells is approximately equal to the volume V of the solid, and the limit of this sum as n approaches infinity is equal to V.

We shall now consider the volume ΔV_i of the shell generated by the rectangle $EFGH$. This volume is equal to the difference between the volumes generated by the rectangles $OEHP$ and $OFGP$ as each is revolved about the Y axis. If the coordinates of E and H are $(x_i, 0)$ and (x_i, y_i) respectively, then the coordinates of F and G are $(x_i + \Delta x, 0)$ and $(x_i + \Delta x, y_i)$. The radius of the first cylinder is x_i and the radius of the second is $x_i + \Delta x$. Consequently, since the altitude of each cylinder is y_i, we have

$$\Delta V_i = \pi y_i[(x_i + \Delta x)^2 - x_i^2] = \pi y_i(x_i + \Delta x + x_i)(x_i + \Delta x - x_i)$$
$$= \pi y_i(2x_i + \Delta x)\,\Delta x = 2\pi(x_i y_i)\,\Delta x + \pi(y_i\,\Delta x)\,\Delta x$$

Therefore, since

$$V = \lim_{n\to\infty}\sum_{i=1}^{n}\Delta V_i$$

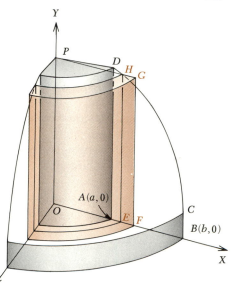

Figure 19.12   Case of cylindrical shells

we have

$$V = \lim_{n \to \infty} 2\pi \sum_{i=1}^{n} y_i x_i \, \Delta x + \left(\lim_{n \to \infty} \pi \sum_{i=1}^{n} y_i \, \Delta x \right) \Delta x$$

$$= 2\pi \int_a^b xy \, dx + \left(\pi \int_a^b y \, dx \right) \left(\lim_{n \to \infty} \Delta x \right)$$

$$= 2\pi \int_a^b xy \, dx + \pi(\text{area of } ABCD)(0)$$

since $\int_a^b y \, dx = $ area of $ABCD$ and $\lim_{n \to \infty} \Delta x = 0$. Consequently, we have

$$V = 2\pi \int_a^b xy \, dx$$

Since x represents the radius of a typical cylindrical shell and y its height, we might do well to write the formula for volume as

volume by use of cylindrical shells

$$V = 2\pi \int_a^b rh \, dr \qquad (19.5)$$

Example 1   Find the volume of the solid generated by rotating about the Y axis the area bounded by $y = \sin x$ and the X axis from $x = 0$ to $x = \pi$.

Solution: The area under consideration and a typical strip of it are shown in Fig. 19.13. A typical element of the volume is generated by revolving this strip about the Y axis. The shell thus formed has a radius x_i and a height y_i and a thickness Δx_i. Hence, as seen from (19.5), the volume is

$$V = 2\pi \int_0^\pi xy \, dx$$

$$= 2\pi \int_0^\pi x \sin x \, dx \qquad\qquad y = \sin x$$

$$= 2\pi(-x \cos x + \sin x) \Big|_0^\pi \qquad \textit{integrating by parts}$$

$$= 2\pi^2$$

since $\cos \pi = -1$ and $\sin \pi = \sin 0 = 0$.

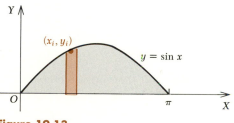

Figure 19.13

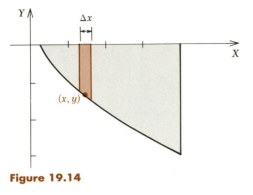

Figure 19.14

Example 2   Find the volume of the solid generated by revolving the smaller area bounded by the parabola $y = t + 1$, $x = t^2/4$, the X axis, and the line $x = 4$ about $x = 4$.

Solution: A sketch of the area that is being rotated and a typical strip of it are shown in Fig. 19.14. A typical element of the volume is obtained by rotating the strip about the line $x = 4$. Therefore, the radius of rotation is $4 - x$, the height of the cylindrical shell is $-y$ and its thickness is Δx. Furthermore, x varies from ¼ to 4 for the area under consideration. Consequently, the desired volume is

$$V = -\int_{1/4}^{4} 2\pi(4 - x)y \, dx$$

Since $y = t + 1 < 0$, $x = t^2/4$, $dx = t \, dt/2$, $t = -1$ for $x = $ ¼, and $t = -4$ for $x = 4$, it follows that

$$V = -\int_{-1}^{-4} 2\pi \left(4 - \frac{t^2}{4}\right)(t + 1)\left(\frac{t}{2}\right) dt$$

$$= -\frac{\pi}{4}\int_{-1}^{-4}(-t^4 - t^3 + 16t^2 + 16t) \, dt \quad \textit{expanding}$$

$$= -\frac{\pi}{4}\left(-\frac{t^5}{5} - \frac{t^4}{4} + \frac{16t^3}{3} + 8t^2\right)\Big|_{-1}^{-4}$$

$$= \frac{1,503\pi}{80}$$

EXERCISE 19.4

By use of the cylindrical-shell method, in Problems 1 to 8 find the volume of the solid that is generated by revolving the region bounded by the given curves about the line indicated after the semicolon.

1   $x = 1 - \cos 2\alpha, y = 2 \sin \alpha, x = 1;$ Y axis
2   First-quadrant part of $y = x, y = x^3;$ X axis
3   $y = e^{-2x}, y = 0, x = 0, x = 2;$ X axis
4   $(x - 1)^2 = -y + 4, y = 3; y = 3$
5   $x^{2/3} + y^{2/3} = a^{2/3};$ either coordinate axis
6   $b^2x^2 + a^2y^2 = a^2b^2;$ X axis
7   $x^2 + y^2 = a^2; x = b > a$
8   $x = \sin \alpha, y = 1 - \cos 2\alpha, x = 1, y = 0;$ Y axis

Solve the following problems by any applicable method.

9   Find the volume of the sphere that is generated by revolving the lower half of the region enclosed by $x^2 + y^2 = a^2$ about the X axis.

10   The region under $y = \cot x$ in $[\pi/4, \pi/2]$ and above the X axis is rotated about the X axis. Find the volume of the solid that is generated.

11   The region under the graph of $y = 2 \sin (x/2)$ in $[0, 2\pi]$ and above the X axis is rotated about the Y axis. Find the volume that is generated.

12   The region under $y = e^{-x}$ in $[0, \infty)$ and above the X axis is revolved about the Y axis. Find the volume generated.

13   The region under $y = \ln x$ in $[1, e]$ and above the X axis is rotated about the Y axis. Find the volume generated.

14   The region under $y = e^x$ in $[-1, 1]$ and above the X axis is rotated about $x = -1$. Find the volume generated.

15   Find the volume generated by revolving the circle $x^2 + y^2 = 2rx$ about the Y axis.

16   The region in the first quadrant bounded by $x^2 + y^2 = r^2$, the Y axis, and $y = x$ is revolved about the Y axis. Find the volume generated.

17   Find the volume of the solid that is generated by rotating the region bounded by the X axis and the upper half of $25x^2 + 9y^2 = 225$ about the X axis.

18   The region under the curve $xy = 1$ from $x = 0$ to $x = 2$ and above the X axis is rotated about the Y axis. Find the volume generated.

19   Find the volume of the solid generated by revolving the region bounded by the Y axis and the parabola $y^2 = 2x + 4$ about the Y axis.

20   The region inside both $x^2 + y^2 = 8x$ and $x^2 + y^2 = 16$ is revolved about their common chord. Find the volume generated.

21   Find the volume of the solid generated by revolving the region bounded by $xy = 4$ and $x + y = 5$ about the Y axis.

22   Find the volume of the solid generated if the area bounded by one arch of $x = a(\theta - \sin \theta), y = a(1 - \cos \theta)$ is rotated about the X axis.

23   The arch of $x = a(\theta - \sin \theta)$, $y = a(1 - \cos \theta)$ in $[0, 2\pi]$ is rotated about the Y axis. Find the volume generated.

24   Find the volume generated by revolving about the Y axis the region bounded by $x^2 + y^2 = 6$ and $y = x^2$.

25   The region bounded by $y = -1$ and $y = 2 \sin x$ in $[7\pi/6, 11\pi/6]$ is rotated about $y = -1$. Find the volume generated.

26   The region of Problem 25 is rotated about the X axis. Find the volume generated.

27   Find the volume of the solid generated by revolving the region under $x^2y + 4y = 8$ in $[0, 2]$ and above the X axis about the Y axis.

28   Find the volume of the solid generated by revolving the region under $6y = 9 + x^2y$ in $[0, 2]$ and above the X axis about the Y axis.

19.6   Volumes of solids with sections of known area

The disk method of Sec. 19.3 amounts to regarding the solid of revolution as made up of slices into which it can be cut by means of a set of parallel planes. These slices are circular disks. The method *volume with* can be extended so as to include all solids that can be cut into parallel *known cross-* slices of known cross-sectional area. If the sections of a solid parallel *sectional area* to a given plane are of area A where A is a continuous function of the distance h from the given plane, we find the volume of the solid as the limit of the sum of the volumes of the slices and have

$$V = \int_{h_1}^{h_2} A(h)\, dh$$

This integral is the same as that used for the disk method except the circular cross-sectional area is replaced by the more general expression $A(x)$.

Example   A wooden wedge is made from a rectangular block that is 5 by 8 by 9 in. by cutting from the lower front edge to the upper back one as indicated in Fig. 19.15. Find the volume by use of integration.

Solution: We can think of the solid as consisting of slices that are parallel to the base. Therefore, we must find the volume of a typical slice. It is 8 in. long, w in. wide, and a distance h in. above the base. Now from similar triangles, we have

$$\frac{w}{9 - h} = \frac{5}{9}$$

hence, $w = 5(9 - h)/9$. Consequently, the cross-sectional area is

$$A(h) = 8w = \frac{40(9 - h)}{9}$$

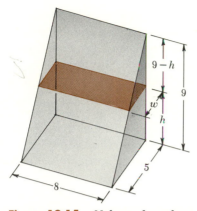

Figure 19.15 Volume from known cross section

Therefore, the volume is

$$V = \int_0^9 A(h)\,dh = \frac{40}{9} \int_0^9 (9 - h)\,dh = \frac{40}{9}\left(9h - \frac{h^2}{2}\right)\Big|_0^9 = 180 \text{ cu in.}$$

This value can be checked since the wedge is half of a rectangular parallelepiped that is 5 by 8 by 9 in., hence has a volume of 360 cu in.

EXERCISE 19.5

1 The base of a right pyramid is a rectangle with sides 3 and 6 in., and its altitude is 9 in. Find the volume after expressing the area of a section a distance h above the base in terms of h.

2 The base of a triangular pyramid is a right triangle with legs $AC = 6$ and $BC = 8$. Its vertex is directly above C and the altitude is 16 in. Express the area of the section at a distance z above the base in terms of z. Find the volume.

3 Every horizontal section of a tower h ft from the top is a square with side $0.3h$. If the steeple is 9 ft high, find its volume.

4 Find the volume of a right pyramid of altitude h and with a square base of side s.

5 A solid has the circle $x^2 + y^2 = 25$ as a base. The sections perpendicular to a specified diameter are squares. Sketch the solid and find its volume.

6 The base of a solid is the circle $x^2 + y^2 = 36$. Find the volume if each cross section perpendicular to a specified diameter is an equilateral triangle.

7 The base of a solid is the circle $x^2 + y^2 = 36$, and each section perpendicular to a fixed diameter is an isosceles triangle of altitude 9. Find the volume.

8   The base of a solid is the circle $x^2 + y^2 = 16$, and each section perpendicular to a fixed diameter is a semicircle with ends of the diameter on the circumference of the base. Find the volume.

9   The base of a solid is bounded by the parabola $y^2 = 2px$ and $x = 2p$. Find the volume if each section perpendicular to the X axis is a semicircle.

10   Use the base of Problem 9 and find the volume if each section perpendicular to the X axis is an equilateral triangle.

11   The base of a solid is the ellipse $b^2x^2 + a^2y^2 = a^2b^2$. Each cross section perpendicular to the X axis is a square with ends of a side on the ellipse. What is the volume?

12   The ellipse $b^2x^2 + a^2y^2 = a^2b^2$ is the base of a solid. The cross sections perpendicular to the X axis are squares with opposite ends of a diagonal on the ellipse. Find the volume of the solid.

13   The area bounded by $y^2 = x$ and $x = 2y$ is the base of a solid. Find the volume if each section perpendicular to the X axis is a square with an end of a side on each bounding curve.

14   The loop of $y^2 = 5x^2 - x^3$ is the base of a solid. Find the volume if each section perpendicular to the X axis is a circle with ends of the diameter on the loop.

15   The region between $y = \sin x$, $x = \pi/2$, and the X axis is the base of a solid. Find its volume if each section perpendicular to the X axis is an equilateral triangle with the ends of one side in the base of the solid.

16   Use the same base as in Problem 15, but let a square with ends of a side in the base of the solid be the cross section perpendicular to the X axis. Find the volume.

17   The axes of two circular cylinders of radius a intersect at right angles. Find the common volume after showing that the section h units above the XY plane is a square with side $2\sqrt{a^2 - h^2}$.

18   A wedge is cut from a tree 6 ft in diameter by making a horizontal cut half through the tree and then making another cut inclined at $45°$ to the horizontal until it meets the first cut along a diameter. Find the volume of the wedge.

19   The radius of a hemispherical vat is 5 ft, and it contains a liquid to a depth of 4 ft. Find the volume of the liquid.

20   Find the volume of the segment cut from a sphere of radius 6 in. by two parallel planes whose distances from the center are 2 and 4 in.

19.7   The length of a curve

In this section we shall discuss a method for obtaining the length of an arc of a curve defined by the equation $y = f(x)$. We shall assume that $f(x)$ is differentiable in the interval $[a, b]$ and shall partition $[a, b]$

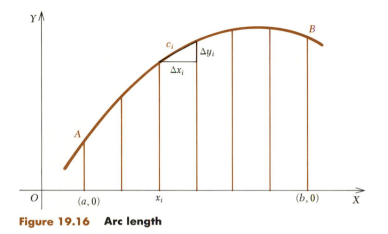

Figure 19.16   Arc length

into n subintervals Δx_i, $i = 1, 2, \ldots, n$, such that $\Delta x_i \leq \sigma$. We shall also connect the points where the ordinates through the points of subdivision intersect the curve. We thus obtain a chord of the curve above each Δx_i as in Fig. 19.16. If we let c_i stand for the length of the chord over Δx_i, $i = 1, \ldots, n$, then $c_i = \sqrt{(\Delta x_i)^2 + (\Delta y_i)^2}$. We now define the length s of the arc of the curve between $A[a, f(a)]$ and $B[b, f(b)]$ as follows:

$$s = \lim_{\sigma \to 0} \sum_{i=1}^{n} \sqrt{(\Delta x_i)^2 + (\Delta y_i)^2}$$

We shall now convert the above limit to a form to which the fundamental theorem of calculus can be applied. By the mean value theorem,

$$\Delta y_i = f(x_i + \Delta x_i) - f(x_i) = f'(x) \, \Delta x_i$$

where $x_i + \Delta x_i \leq x \leq x_i$. Consequently,

$$s = \lim_{\sigma \to 0} \sum_{i=1}^{n} \sqrt{(\Delta x_i)^2 + [f'(x)]^2 (\Delta x_i)^2}$$

$$= \lim_{\sigma \to 0} \sum_{i=1}^{n} \sqrt{1 + [f'(x)]^2} \, \Delta x_i$$

$$= \int_a^b \sqrt{1 + [f'(x)]^2} \, dx$$

arc length   Then, since $f'(x) = dy/dx$, we have

$$s = \int_a^b \sqrt{1 + \left(\frac{dy}{dx}\right)^2} \, dx \qquad (19.6)$$

An alternate formula for obtaining the length of an arc when the equation of the curve is $x = F(y)$ can be derived in a similar manner.

It is

$$s = \int_c^d \sqrt{1 + \left(\frac{dx}{dy}\right)^2}\, dy \qquad (19.6')$$

more arc-length formulas

If $x = g(t)$ and $y = h(t)$, we can express the formula (19.6) in terms of the parameter t by the following procedure.

$$\frac{dy}{dx} = \frac{dy}{dt}\frac{dt}{dx} \qquad \text{and} \qquad dx = \frac{dx}{dt}\, dt$$

Then, if we substitute these values in (19.6), we get

$$s = \int_{t_1}^{t_2} \sqrt{1 + \left(\frac{dy}{dt}\right)^2 \left(\frac{dt}{dx}\right)^2}\, \frac{dx}{dt}\, dt$$

and if we insert dx/dt into the radicand, we have

$$s = \int_{t_1}^{t_2} \sqrt{\left(\frac{dx}{dt}\right)^2 + \left(\frac{dy}{dt}\right)^2}\, dt \qquad (19.6'')$$

where $t_1 = g(a)$ and $t_2 = g(b)$.

Finally, the arc-length formula for use when the equation of the curve is given in polar form can be obtained by use of the usual relations between rectangular and polar coordinates. If

$$x = r \cos \theta \qquad \text{and} \qquad y = r \sin \theta$$

then

$$dx = -r \sin \theta\, d\theta + \cos \theta\, dr$$

$$dy = r \cos \theta\, d\theta + \sin \theta\, dr$$

where r is a differentiable function of θ.

If we put these expressions for dx and dy in (19.6) and collect coefficients of $(dr)^2$ and $(d\theta)^2$, we have

$$s = \int \sqrt{r^2 (d\theta)^2 + (dr)^2}$$

Consequently, dividing the radicand by $(d\theta)^2$ and multiplying the integrand by $d\theta$, we see that

$$s = \int_{\theta_1}^{\theta_2} \sqrt{r^2 + \left(\frac{dr}{d\theta}\right)^2}\, d\theta \qquad (19.6''')$$

is a formula for arc length if the equation is in polar form.

Example 1 Find the length of the arc of the curve defined by $y = (x + 1)^{3/2}$ between the points $(-1, 0)$ and $(4, 5^{3/2})$.

Solution: If

$$y = (x + 1)^{3/2}$$

then

$$\left(\frac{dy}{dx}\right)^2 = \frac{9(x + 1)}{4}$$

Hence by (19.6), the desired length is

$$s = \int_{-1}^{4} \sqrt{1 + \frac{9(x+1)}{4}} \, dx = \frac{1}{2} \int_{-1}^{4} \sqrt{13 + 9x} \, dx$$

$$= \frac{1}{18} \int_{-1}^{4} \sqrt{13 + 9x} \, (9) \, dx = \frac{1}{27} (13 + 9x)^{3/2} \Big|_{-1}^{4}$$

$$= \frac{1}{27} (49^{3/2} - 4^{3/2}) = \frac{1}{27} (343 - 8) = \frac{335}{27} \text{ square units}$$

Example 2 Find the length of the curve defined by $x = 2t^3$, $y = 3t^2$ from $t = 0$ to $t = 4$.

Solution: From the given expressions for x and y, we have $dx = 6t^2 \, dt$ and $dy = 6t \, dt$. Consequently,

$$s = \int_{t_1}^{t_2} \sqrt{\left(\frac{dx}{dt}\right)^2 + \left(\frac{dy}{dt}\right)^2} \, dt$$

becomes $s = \int_{0}^{4} 6t \sqrt{t^2 + 1} \, dt = 3(t^2 + 1)^{3/2} \cdot \frac{2}{3} \Big|_{0}^{4} = 2(17\sqrt{17} - 1)$

Example 3 Find the entire length of the cardioid $r = a(1 - \cos \theta)$.

Solution: A sketch of the upper half of the curve is shown in Fig. 19.17. Since the curve is symmetric with respect to the polar axis and since the upper half is traced out as θ varies from 0 to π, we shall obtain the entire length by taking twice the length from $x = 0$ to $x = \pi$.

From the given equation, we find that

$$\frac{dr}{d\theta} = a \sin \theta$$

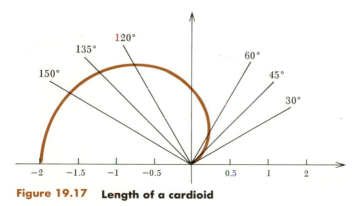

Figure 19.17   Length of a cardioid

Consequently,

$$s = 2 \int_0^{\pi} a \sqrt{(1 - \cos \theta)^2 + \sin^2 \theta} \, d\theta \qquad \text{by (19.6''')}$$

$$= 2a \int_0^{\pi} \sqrt{2} \, \sqrt{1 - \cos \theta} \, d\theta$$

$$= 4a \int_0^{\pi} \sin \frac{\theta}{2} \, d\theta \qquad \text{since } \sqrt{1 - \cos \theta} = \sqrt{2} \sin (\theta/2)$$

$$= 8a \left(-\cos \frac{\theta}{2} \right) \Big|_0^{\pi}$$

$$= 8a(0 + 1)$$

$$= 8a$$

19.8   Area of a surface of revolution

If a curve $y = f(x)$ along with n chords as shown in Fig. 19.18 is revolved about the X axis, the curve generates a surface of revolution, and each chord generates the lateral surface of a frustum of a cone. Since the lateral area of a frustum of a cone is the average circumference times the slant height, we see that if the rotation is about the X axis, the area of the ith frustum is

$$2\pi(y_{i-1} + \tfrac{1}{2}\Delta y_{i-1}) \sqrt{(\Delta x_{i-1})^2 + (\Delta y_{i-1})^2}$$

Consequently, the sum of the areas of the frustums as the norm σ of the subintervals into which $[a, b]$ is divided approaches zero is

$$S = \lim_{\sigma \to 0} \sum_{i=1}^{n} 2\pi \left(y_{i-1} + \frac{\Delta y_{i-1}}{2} \right) \sqrt{(\Delta x_{i-1})^2 + (\Delta y_{j-1})^2}$$

$$= \lim_{\sigma \to 0} \sum_{i=1}^{n} 2\pi \left(y_{i-1} + \frac{\Delta y_{i-1}}{2} \right) \sqrt{1 + \left(\frac{\Delta y_{i-1}}{\Delta x_{i-1}} \right)^2} \, \Delta x_{i-1}$$

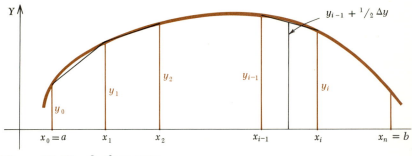

Figure 19.18   Surface area

Therefore, since by the mean value theorem we have

$$\frac{\Delta y_{i-1}}{\Delta x_{i-1}} = \frac{dy}{dx}\bigg|_x$$

where $x_{i-1} \leq x \leq x_i$, it follows from the fundamental theorem of calculus that the limit of the sum of the lateral areas of the frustum as σ approaches zero is

$$S = 2\pi \int_a^b y \sqrt{1 + \left(\frac{dy}{dx}\right)^2} \, dx \tag{19.7}$$

surface area This, by definition, is the area of the surface of revolution obtained by rotating $y = f(x)$ from $x = a$ to $x = b$ about the X axis.

We can find similarly that if the curve $x = g(y)$ is rotated about the Y axis, then

$$S = 2\pi \int_a^b x \sqrt{1 + \left(\frac{dy}{dx}\right)^2} \, dx \tag{19.7'}$$

The expression $\sqrt{1 + (dy/dx)^2}\, dx$ in Eqs. (19.7) and (19.7′) is ds and can be replaced by $\sqrt{1 + (dx/dy)^2}\, dy$, $\sqrt{(dx/dt)^2 + (dy/dt)^2}\, dt$, or $\sqrt{r^2 + (dr/d\theta)^2}\, d\theta$, as indicated by the situation, provided that appropriate changes are made in the limits of integration.

Example Find the curved surface of the solid generated by revolving the part of $y = x^2$ from $(0, 0)$ to $(\sqrt{6}, 6)$ about the Y axis.

Solution: If we differentiate, we find that $dy/dx = 2x$; hence,

$$S = 2\pi \int_a^b x \sqrt{1 + \left(\frac{dy}{dx}\right)^2} \, dx$$

becomes

$$S = 2\pi \int_0^{\sqrt{6}} x\sqrt{1 + 4x^2} \, dx = \frac{2\pi}{8}(1 + 4x^2)^{3/2}\left(\frac{2}{3}\right)\bigg|_0^{\sqrt{6}}$$

$$= \frac{2\pi}{12}[25^{3/2} - 1^{3/2}] = \frac{62\pi}{3}$$

EXERCISE 19.6

In each of Problems 1 to 10, find the length of arc of the curve in the specified interval.

1 $y = x$ in $[1, 4]$ 　　　　　2 $y = 2x^{3/2}/3$ in $[0, 3]$
3 $y^3 = x^2$ in $[0, 4]$ 　　　　4 $y = x^3/3 + 1/4x$ in $[2, 4]$
5 $y = \ln \cos x$ in $[0, \pi/3]$ 　　6 $y = \ln(x^2 - 1)$ in $[2, 5]$

7 $y = \ln \dfrac{e^x + 1}{e^x - 1}$ in $[a, b]$ **8** $y = a(e^{x/a} + e^{-x/a})/2$ in $[0, a]$

9 $y^2 = x^3$; $x = 0$ to $x = 4$

10 $y = \ln x$; $x = 1$ to $x = \sqrt{3}$

11 Find the perimeter of the four-cusped hypocycloid $x^{2/3} + y^{2/3} = a^{2/3}$.

12 Set up an integral whose value is equal to the length of one arc of the curve $y = \sin x$.

13 Find the perimeter of the cardioid $r = a(1 + \cos \theta)$.

14 Sketch the graph of $r = 4 \sin^3(\theta/3)$ for $\theta = 0$ to $3\pi/2$, and compute the length of arc.

15 Set up an integral whose value is equal to the perimeter of the ellipse $b^2x^2 + a^2y^2 = a^2b^2$.

16 Find the perimeter of $r = 4 \cos \theta$ by integration.

17 Solve Problem 11 by using $x = a \cos^3 \theta$, $y = a \sin^3 \theta$ as parametric equations of the curve.

18 Find the length of one arc of $x = a(\theta - \sin \theta)$, $y = a(1 - \cos \theta)$.

19 Find the length of $r = 2/(\cos \theta + \sin \theta)$ from $\theta = 0$ to $\theta = \pi/2$.

20 Find the circumference of the circle $x = 2 \sin t$, $y = 2 \cos t$ by integration.

21 Compute the area of the surface generated by revolving the arc of $y = x^2$ from $x = 0$ to $x = \sqrt{2}$ about the Y axis.

22 If a parabolic reflector of an automobile head light is 12 in. in diameter and 4 in. deep, what is its area?

23 Compute the area of the surface generated by revolving the four-cusped hypocycloid $x^{2/3} + y^{2/3} = a^{2/3}$ about the X axis.

24 Compute the area of the surface generated by revolving one arc of $y = \sin x$ about the X axis.

25 Find the area of the surface generated by revolving $y = \ln x$ from $x = 0$ to $x = 1$ about the Y axis.

26 The part of $y = e^x$ for which $x < 0$ is revolved about the X axis. Find the area of the surface.

27 The part of $y = \sqrt{9 - x^2}$ from $x = 0$ to $x = 3$ is revolved about the X axis. Find the area of the surface.

28 Find the area of the surface generated by revolving the part of $y = x + 2$ from $x = 1$ to $x = \sqrt{3}$ about the line $y = 1$.

29 Solve Problem 23 by using $x = a \cos^3 \theta$, $y = a \sin^3 \theta$ as parametric equations of the curve.

30 Find the area of the surface generated by revolving one arc of $x = a(\theta - \sin \theta)$, $y = a(1 - \cos \theta)$ about the X axis.

31 Find the area of the surface generated by revolving $x = a \cos \theta$, $y = a \sin \theta$ from $\theta = 0$ to $\theta = \pi/2$ about the Y axis.

32 Find the area of the surface generated by revolving $x = t^3$, $y = t$ from $t = 0$ to $t = 1$ about the Y axis.

33   Find the area of the surface generated by revolving the upper half of $r = 2a \cos \theta$ about the polar axis.

34   What is the area of the surface generated by revolving the upper half of $r = a(1 + \cos \theta)$ about the polar axis?

35   The circle $r = 2a \cos \theta$ is revolved about the normal axis. Find the area of the surface generated.

36   Find the surface area of the torus generated by revolving the circle $r = 2a \cos \theta$ about the line $r \cos \theta = b, b > 2a$.

19.9   Liquid pressure

It is shown in physics that if a liquid is at rest the pressure at any point within the liquid is the same in all directions. Furthermore, it is shown that the pressure is proportional to the depth below the surface and that at a depth h ft below the free surface of a liquid, the pressure (that is, force per unit area) is

$$p = \delta h \text{ lb per sq ft}$$

where δ is the weight in pounds of a cubic foot of the liquid.

We shall now consider the problem of computing the force exerted on one side of a flat plate, as in Fig. 19.19, by a liquid in which the plate is submerged vertically. The procedure is essentially that of setting up a definite integral as the limit of a sum. Consequently, we divide the submerged area into n strips parallel to the liquid surface and denote their areas by $\Delta A_1, \Delta A_2, \ldots, \Delta A_n$. We multiply the area of each strip by the pressure at a point in it, add the quantities so obtained as indicated by

$$\sum_{i=1}^{n} \delta h_i \, \Delta A_i$$

and find the limit of this sum as the width of each strip approaches zero and n approaches infinity. This limit, by definition, is the force

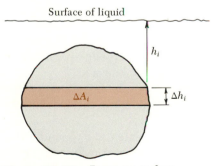

Surface of liquid

h_i

ΔA_i

Δh_i

Figure 19.19   Force on a dam

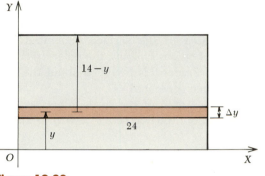

Figure 19.20

force of a   against the area. Therefore,
liquid

$$F = \lim_{\Delta h_i \to 0} \sum_{i=1}^{n} \delta h_i\, \Delta A_i = \delta \int h\, dA \tag{19.8}$$

where δ is the density of the liquid and appropriate limits are used.

Example   Find the force against the face of a rectangular dam that is 24 ft long if the water behind the dam is 14 ft deep and the weight of water is 62.5 lb per cu ft.

Solution: A sketch of the situation is shown in Fig. 19.20. The X axis is taken at the bottom of the dam and the Y axis along one end. The typical strip is of length 24, thickness Δy, and of depth $14 - y$. Therefore,

$$F = \delta \int_0^{14}(14 - y)24\, dy$$

$$= \delta(336y - 12y^2)\Big|_0^{14} = 2352\,\delta$$

$$= 147{,}000 \text{ lb} \qquad\qquad \delta = 62.5 \text{ lb per cu ft}$$

EXERCISE 19.7

1   A rectangular plate is 4 ft long and 2 ft wide. It is submerged vertically in water with the upper 4 ft edge parallel to, and 2 ft below, the surface. Find the force against one side.

2   If the plate of Problem 1 is submerged vertically with the upper 2-ft edge parallel to and 3 ft below the surface, find the force on one side.

3   The edges of a triangular plate are 5, 5, and 6 ft. Find the force against one side if it is submerged vertically with the 6-ft edge in the surface of the water.

4 The plate of Problem 3 is submerged vertically with the altitude to the longest side horizontal and 5 ft below the surface. Find the force on one side.

5 The edges of a cubical box are each 5 ft, and the box contains liquid concrete that weighs 250 lb per cu ft. Find the force on one end if the concrete is 4 ft deep.

6 A cylindrical tank 4 ft in diameter and 6 ft long has its axis horizontal and is half full of oil that weighs 50 lb per cu ft. Find the force exerted on one end.

7 Find the force on one end of the tank of Problem 6 if it is full of oil.

8 A circular gate in the vertical face of a dam is 4 ft in diameter. Find the force on it if the water level is 3 ft above the top of the gate.

9 The cross section of an oil tank is an ellipse with semiaxes 6 and 4 ft. The tank has its axis and the major axis of the ends horizontal. Find the force on the upper half and on the lower half of one end if the tank is full of oil that weighs 50 lb per cu ft.

10 An elliptic gate with semiaxes 3 and 4 ft is submerged vertically in water with the minor axis horizontal and 5 ft below the surface. Find the force on one side of the gate.

11 A gate is in the form of the parabola $x^2 = 8y$. Find the force on one side if it is submerged vertically with the focal chord along the surface.

12 Find the force on one side of a gate bounded by $x^2 = -8y$ and its focal chord if the gate is submerged vertically with the vertex in the surface.

13 A trapezoidal gate is of altitude 4 ft and the parallel sides are 3 and 5 ft. Find the force on one side if the gate is submerged vertically and if the longest side is in the surface.

14 Find the force on one side of the gate in Problem 13 if the shorter of the parallel sides is in the surface.

15 A gate is in the form of the letter T with the vertical 2 ft wide and 4 ft high. The horizontal cross piece is 6 ft long and 1 ft high. Find the force on one side if the gate is submerged vertically with the top of the cross piece in the surface.

16 A gate has the shape of one arch of the cycloid $y = a(1 - \cos \theta)$, $x = a(\theta - \sin \theta)$ and is inverted and submerged vertically with the straight side in the surface. Find the force on one side.

17 A trough with a trapezoidal cross section is 6 ft long; furthermore, it is 5 ft wide at the top and 3 ft wide at the bottom. Find the force on one end if the trough is full of water and if the altitude is 3 ft.

18 The sloping sides of a triangular trough that is 12 ft long are inclined at 45° to the horizontal. Find the force on each end if the water is 2 ft deep.

19   A dam is 10 ft wide at the bottom, 4 ft wide at the top, and 8 ft high. Find the force on the vertical face if water is flowing over the top and the dam is 20 ft long.

20   Find the force on the dam in Problem 19 if the water is 6 ft deep.

19.10   Work

If an object is moved over a distance by the continuous application of a force, we say that work is done. More precisely, if a force F acts on a body and if the point of application moves d ft in the direction of the force, then the product of F and d is the measure of the work done by the force. If the measure of the work is denoted by W, then

$$W = Fd \text{ ft-lb}$$

Thus, if one lifts a 30-lb weight upward a distance of 2 ft, the work done is 60 ft-lb.

In this section we shall consider the work done by a variable force as it moves a particle over a given distance. We shall prove that if a force moves a particle from $(a, 0)$ to $(b, 0)$ and the magnitude of the force at the point $(x, 0)$ is $F(x)$, then the work done is equal to $\int_a^b F(x)\, dx$. In order to prove this statement, we shall first assume that $F(x)$ is a continuous, increasing function of x over the interval $[a, b]$. We next partition the interval $[a, b]$ into n subintervals where the norm of the partition is σ, and we shall consider the subinterval $[x_i, x_i + \Delta x_i]$, as in Fig. 19.21. If we let W_i stand for the work done by the force in moving the particle over this interval, then, since the force increases continuously over the interval, we have $F(x_i)\, \Delta x_i \leq$

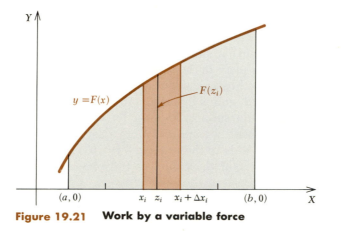

Figure 19.21   **Work by a variable force**

$W_i \le F(x_i + \Delta x_i)\, \Delta x_i$. Therefore, since $F(x)$ is continuous, there is a number z_i between x_i and $x_i + \Delta x_i$ such that $W_i = F(z_i)\, \Delta x_i$.°* Furthermore, the work W done in moving the particle over $[a, b]$ is equal to

$$\lim_{\sigma \to 0} \sum_{i=1}^{n} W_i = \lim_{\sigma \to 0} \sum_{i=1}^{n} F(z_i)\, \Delta x_i = \int_a^b F(x)\, dx$$

by the fundamental theorem of calculus. Consequently, if x is an element of the interval $[a, b]$ and the force $F(x)$ moves a particle over this interval, then the work done is given by the formula

formula for work

$$W = \int_a^b F(x)\, dx \qquad (19.9)$$

It can be proved by a similar argument that if $F(x)$ is a decreasing function over $[a, b]$, formula (19.9) holds. If $F(x)$ is an increasing function over certain subintervals of $[a, b]$ and a decreasing function over others, we can prove that (19.9) also holds. For example, if $F(x)$ increases over $[a, c]$, decreases over $[c, d]$, and increases over $[d, b]$, then

$$W = \int_a^c F(x)\, dx + \int_c^d F(x)\, dx + \int_d^b F(x)\, dx = \int_a^b F(x)\, dx$$

Example 1   Find the work done in stretching a spring of natural length 8 in. from 10 to 13 in. Assume a force of 6 lb is needed to hold it at a length of 11 in.

Solution: We shall make use of the fact that the force required to stretch a spring beyond its natural length and within its elastic limit is proportional to the amount it is stretched. Consequently,

$$f(x) = kx$$

where x is the increase in length. Hence,

$$f(3) = k \cdot 3 = 6$$

since a force of 6 lb is required to hold the 8-in. spring stretched $11 - 8 = 3$ in. Therefore, $k = 2$ and $f(x) = 2x$. Furthermore, the limits are $10 - 8 = 2$ and $13 - 8 = 5$. Consequently,

$$W = \int_2^5 2x\, dx = x^2 \Big|_2^5 = 21$$

° This statement seems obvious, but it is not self-evident. Rigorous proofs of the statement can be found in most texts dealing with the theory of functions of a real variable. For example, see Graves, "Theory of Functions of a Real Variable," p. 64, McGraw-Hill Book Company, New York, 1946.

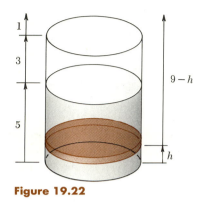

Figure 19.22

Example 2   A tank is a right circular cylinder 3 ft in radius and 8 ft tall. It is standing upright on one end and has 5 ft of water in it. How much work is required to lift the water to a point that is 1 ft above the tank?

Solution: A sketch of the tank is shown in Fig. 19.22. The radius of the tank is 3 ft; hence, the cross-sectional area is $\pi(3)^2 = 9\pi$ and the volume of a typical film is $9\pi\,\Delta h$. The weight is 62.5 lb per cu ft, the film must be lifted a distance of $9 - h$ ft, and there is water from $h = 0$ to $h = 5$. Hence the amount of work required is

$$W = 62.5 \int_0^5 9\pi(9 - h)\,dh = 562.5\pi\left(9h - \frac{h^2}{2}\right)\Big|_0^5 = 18{,}281.25\pi$$

Hence, $18{,}281.25\pi$ ft-lb of work is needed.

EXERCISE 19.8

1   The natural length of a spring is 8 in. and a force of 10 lb will hold it at 12 in. How much work is done in stretching it from 9 to 13 in.?

2   How much work is done in stretching a spring from 6 to 8 ft if its natural length is 6 ft and a force of 40 lb will hold it at 7 ft?

3   A force of 18 lb will hold a spring of natural length 9 ft at 10 ft. How much work is done in stretching it from 9.5 to 11 ft?

4   Prove that the same amount of work is not required to stretch a spring of natural length 11 in. from 12 to 14 in. as from 11 to 13 in.

5   Find the work done in overcoming a force of $f(x) = 1/(1 + x^2)$ from $x = 0$ to $x = \sqrt{3}$.

6   A force of $f(x) = 1 + 2x$ is required to move a load along a line. How much work is done in moving the load from $x = 1$ to $x = 5$?

7   A force of $f(x) = 6 \cos (\pi x/2)$ just keeps a load from moving

along the direction in which the force is applied. How much work is done if x varies from 3 to 5?

8   A force of $f(x) = 2x/\sqrt{x + 2}$ is required to move a pay load from $x = 2$ to $x = 7$. How much work is done?

9   How much work is done in winding a 32-ft uniform cable on a windlass if the cable weighs 3 lb per ft?

10   A 60-ft cable that weighs 4 lb per ft has a 500-lb weight attached to the end. How much work is done in winding up the last 20 ft of the cable?

11   How much work is done in winding up the middle 20 ft of the cable in Problem 10?

12   A uniform chain that weighs 4 oz per ft has a leaky 4-gal bucket attached to it. If the bucket is full of a liquid when 30 ft of chain is out and half full when no chain is out, how much work was done in winding the chain on a windlass. Assume that the liquid leaks out at a uniform rate and weighs 7.5 lb per gal.

13   How much work is required to pump all of the water from a right circular cylindrical tank that is 4 ft in radius and 9 ft tall if it is emptied at a point 1 ft above the tank? Assume that the tank is on one end and upright and that it is initially full of water.

14   Find the work done in pumping the water from the tank in Problem 13 to a height 1 ft above the tank if the tank is on its side and horizontal.

15   The cross section of a cylindrical tank 10 ft long is an ellipse with axes 8 and 6 ft. The tank has its axis and the major axis of the end horizontal. It is full of oil that weighs 48 lb per cu ft. How much work is done in emptying the tank if the oil is pumped to a point 8 ft above the bottom of the tank?

16   If the tank in Problem 15 is only half full of oil, how much work is done in pumping it to a height of 8 ft above the bottom of the tank?

17   A hemispherical tank that is 12 ft in diameter is filled with a liquid that weighs 100 lb per cu ft. Find the work done in lowering the liquid 4 ft if it is expelled at a point 3 ft above the top of the tank.

18   A conical tank that is 5 ft high has a radius of 2 ft and is filled with a liquid that weighs 50 lb per cu ft. How much work is done in discharging the liquid at a point 3 ft above the top of the tank?

19   If there is only 2 ft of liquid in the tank of Problem 18, how much work is done in discharging it at a point 3 ft above the top of the tank?

20   A unit positive charge repels a positive charge of e with a force of e/r^2 if the two are r units apart. Find the work done when e moves along a line that connects the unit charge and e, from a distance of r_1 from the unit charge to a distance of r_2 from that charge.

20 Approximate integration

20.1 Introduction

The function $f(x)$ may be of such a nature that $\int_a^b f(x)\, dx$ cannot be evaluated by the methods discussed in this book. In such cases, we must resort to methods that enable us to obtain an approximation to the value. A geometric interpretation of $\int_a^b f(x)\, dx$ is the area of the region bounded by the graph of $y = f(x)$, $x = a$, $x = b$, and $y = 0$. We can obtain an estimate of the value of the integral by sketching the boundaries of the region and estimating the area of the enclosed region. Methods exist, however, that enable us to obtain an approximation to the value of a definite integral that will be correct to a reasonable degree of accuracy. We shall discuss three of these methods in this chapter.

20.2 The trapezoidal rule

We shall obtain an approximation to $\int_a^b f(x)\, dx$ by finding the sum of the areas of a set of trapezoids. Thus, we begin as indicated in Fig. 20.1 by dividing $[a, b]$ into n *equal* subintervals and constructing

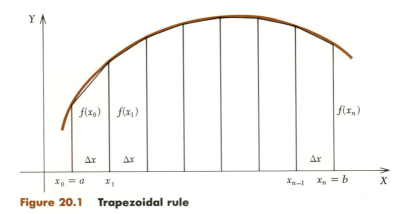

Figure 20.1 Trapezoidal rule

a trapezoid on each base. If we designate the lengths of the ordinates drawn at the points of subdivision by $f(x_0)$, $f(x_1)$, . . . , $f(x_{n-1})$, and $f(x_n)$ and the width of each trapezoid by $\Delta x = (b - a)/n$, we find that the sum of the area of the trapezoids is

$$A = \tfrac{1}{2}\,[\,f(x_0) + f(x_1)]\,\Delta x + \tfrac{1}{2}\,[\,f(x_1) + f(x_2)]\,\Delta x + \cdots$$
$$+ \tfrac{1}{2}\,[\,f(x_{n-1}) + f(x_n)]\,\Delta x$$

Consequently, we have

trapezoidal rule $\displaystyle\int_a^b f(x)\,dx \doteq \frac{\Delta x}{2}\,[\,f(x_0) + 2f(x_1) + 2f(x_2) + \cdots + 2f(x_{n-1}) + f(x_n)]$

$$(20.1)$$

where $\doteq$ is read "is approximately equal to." This approximation is known as the *trapezoidal rule*. The accuracy of the approximation depends on the function $f(x)$ and on the number of subintervals used.

Example Use the trapezoidal rule with $n = 4$ to find an approximation to

$$\int_2^6 \sqrt{x^2 - 2}\; dx$$

Solution: If we use $n = 4$, then $\Delta x = (6 - 2)/4 = 1$. Hence, $x_0 = 2$, $x_1 = 3$, $x_2 = 4$, $x_3 = 5$, and $x_4 = 6$. Consequently,

$$\int_2^6 \sqrt{x^2 - 2}\; dx \doteq \tfrac{1}{2}\,[\,\sqrt{2^2 - 2} + 2\sqrt{3^2 - 2} + 2\sqrt{4^2 - 2}$$
$$+ 2\sqrt{5^2 - 2} + \sqrt{6^2 - 2}\,]$$
$$= \tfrac{1}{2}\,[\,\sqrt{2} + 2\sqrt{7} + 2\sqrt{14} + 2\sqrt{23} + \sqrt{34}\,]$$
$$= \tfrac{1}{2}\,[1.41 + 2(2.65) + 2(3.74) + 2(4.80) + 5.83]$$
$$= 14.81$$

EXERCISE 20.1

In each of Problems 1 to 16, find an approximation to the value of the integral by use of the trapezoidal rule with the specified value of n. In each of Problems 1 to 8, obtain the exact value of the integral.

1 $\int_2^7 x^2 \, dx, \ n = 5$

2 $\int_0^4 x^3 \, dx, \ n = 4$

3 $\int_0^3 x\sqrt{16 + x^2} \, dx, \ n = 3$

4 $\int_1^4 \frac{dx}{x}, \ n = 3$

5 $\int_0^{\pi/2} \cos x \, dx, \ n = 4$

6 $\int_0^{\pi/3} \tan x \, dx, \ n = 4$

7 $\int_0^1 \frac{dx}{1 + x^2}, \ n = 5$

8 $\int_1^3 e^x \, dx, \ n = 2$

9 $\int_1^5 \sqrt{0.5x^3 + 4} \, dx, \ n = 4$

10 $\int_1^6 \sqrt{x^2 + 3x} \, dx, \ n = 5$

11 $\int_4^7 \frac{\sqrt{1 + x^2}}{x} \, dx, \ n = 3$

12 $\int_0^3 \sqrt{x^3 + 6} \, dx, \ n = 3$

13 $\int_0^{\pi/2} \sqrt{\cos x} \, dx, \ n = 9$

14 $\int_0^{\pi/4} (2 + \sin x) \, dx, \ n = 5$

15 $\int_0^2 e^{-x^2} \, dx, \ n = 10$

16 $\int_0^{\pi/2} (2 - \cos x) \, dx, \ n = 6$

17 Find an approximation to the length of $y = \sin x$ from $x = 0$ to $x = \pi/2$ with $n = 9$ by use of the trapezoidal rule.

18 Find an approximation to the length of $y = e^x$ from $x = 0$ to $x = 4$ with $n = 4$ by use of the trapezoidal rule.

19 Find an approximation to the length of $y = x^3/6$ from $x = -4$ to $x = -1$ with $n = 6$ by use of the trapezoidal rule.

20 Find an approximation to the length of $y = \cos x$ from $x = 0$ to $x = 2\pi$. Use $n = 9$ in obtaining the arc length from $x = 0$ to $x = \pi/2$ by use of the trapezoidal rule.

20.3 The prismoidal formula

We shall now develop another formula for use in finding an approximation to the value of a definite integral. This formula is called the *prismoidal formula*. It is obtained by approximating the graph of the function $y = f(x)$ to be integrated by an arc of a parabola. The parabola is made to pass through the points on $y = f(x)$ corresponding to $x = a$, $a + h$, and $a + 2h$.

We shall now show that

$$\int_a^{a+2h} P(x)\, dx = \frac{h}{3}(y_0 + 4y_1 + y_2)$$

where $P(x) = C_0 x^2 + C_1 x + C_2$, $y_0 = P(a)$, $y_1 = P(a + h)$, and $y_2 = P(a + 2h)$.

Thus, $y_0 = P(a) = C_0 a^2 + C_1 a + C_2$

$$y_1 = P(a + h) = C_0(a + h)^2 + C_1(a + h) + C_2$$

$$y_2 = P(a + 2h) = C_0(a + 2h)^2 + C_1(a + 2h) + C_2$$

$$y_0 + 4y_1 + y_2 = C_0[a^2 + 4(a + h)^2 + (a + 2h)^2]$$
$$+ C_1[a + 4(a + h) + a + 2h] + 6C_2$$
$$= C_0(6a^2 + 12ah + 8h^2) + C_1(6a + 6h) + 6C_2$$

Furthermore,

$$\int_a^{a+2h} P(x)\, dx = \frac{C_0 x^3}{3} + \frac{C_1 x^2}{2} + C_2 x \Big|_a^{a+2h}$$

$$= \frac{C_0}{3}[(a + 2h)^3 - a^3] + \frac{C_1}{2}[(a + 2h)^2 - a^2]$$
$$+ C_2(a + 2h - a)$$

$$= \frac{C_0}{3}(6a^2 h + 12ah^2 + 8h^3) + C_1(2ah + 2h^2) + 2C_2 h$$

$$= \frac{h}{3}[C_0(6a^2 + 12ah + 8h^2) + C_1(6a + 6h) + 6C_2]$$

$$= \frac{h}{3}(y_0 + 4y_1 + y_2)$$

Consequently,

*prismoidal
formula*
$$\int_a^{a+2h} f(x)\, dx \doteq \frac{h}{3}(y_0 + 4y_1 + y_2) \tag{20.2}$$

where $y_0 = f(a) = P(a)$, $y_1 = f(a + h) = P(a + h)$, and $y_2 = f(a + 2h) = P(a + 2h)$. Equation (20.2) is known as the *prismoidal formula*. It clearly gives the exact value for a polynomial of degree 1 or 2; furthermore, it is also exact for a cubic. Since any cubic can be obtained by adding $K(x - a)^3$ to a polynomial of lower degree, we need only show that (20.2) is exact for $(x - a)^3$ to know that it is exact for all cubics. We have

$$\int_a^{a+2h}(x - a)^3\, dx = \frac{(x - a)^4}{4}\Big|_a^{a+2h} = 4h^4$$

and the right-hand member of (20.2) is

$$\frac{h}{3}[(a-a)^3 + 4(a+h-a)^3 + (a+2h-a)^3]$$

$$= \frac{h}{3}[0 + 4h^3 + 8h^3] = 4h^4$$

Hence, the prismoidal formula gives the exact value for $f(x)$ a cubic.

Example 1   The cross-sectional area of a solid is a square in a plane perpendicular to the Y axis with one side perpendicular to the Y axis and extending from the Y axis to the portion of the graph of $x^2 = y^3 - 5y^2 - 16y + 80$ that is in the first quadrant. Find the volume of the solid by use of (20.2).

Solution: A sketch of the solid is shown in Fig. 20.2. By Sec. 19.6, we have

$$V = \int_0^4 x^2 \, dy = \int_0^4 (y^3 - 5y^2 - 16y + 80) \, dy$$

We shall apply (20.2) with $f(y) = y^3 - 5y^2 - 16y + 80$, $h = 2$, $a = 0$, $a + h = 2$, and $a + 2h = 4$. Hence,

$$V = \tfrac{2}{3}[f(0) + 4f(2) + f(4)] = \tfrac{2}{3}[80 + 4(36) + 0]$$
$$= \tfrac{2}{3}(224) = 149\tfrac{1}{3} \text{ cubic units}$$

Since $f(y)$ is a polynomial of degree 3, this is the exact value of the volume.

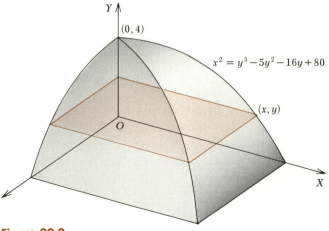

Figure 20.2

Example 2   The region bounded by $x = 0$, $y = 0$, $y = 2$, and the graph of $y = [(25 - x^4)/16]^2$ is revolved about the Y axis. (*a*) Use formula (20.2) to obtain an approximation to the volume generated. (*b*) Obtain another approximation to this volume by applying (20.2) separately to solids that are generated by revolving the portion of the above region that is below $y = 1$ and the portion that is above $y = 1$ about the Y axis.

Solution: (*a*) The definite integral that yields the required volume is $\int_0^2 \pi x^2 \, dy$. If we solve the given equation for x^2, we get $x^2 = \sqrt{25 - 16y^{1/2}}$. Consequently, we have

$$V = \int_0^2 \pi \sqrt{25 - 16y^{1/2}} \, dy$$

We now apply (20.2) with $a = 0$, $h = 1$, $a + h = 1$, $a + 2h = 2$, and $f(y) = \pi\sqrt{25 - 16y^{1/2}}$. Hence, we get

$$V = \frac{\pi}{3}[f(0) + 4f(1) + f(2)]$$

$$= \frac{\pi}{3}\left[\sqrt{25} + 4\sqrt{25 - 16} + \sqrt{25 - 16\sqrt{2}}\right]$$

$$= \frac{\pi}{3}(5 + 12 + 1.54) = 6.18\pi$$

(*b*) In this case we have $V = V_1 + V_2$ where

$$V_1 = \int_0^1 \pi\sqrt{25 - 16y^{1/2}} \, dy \qquad \text{and} \qquad V_2 = \int_1^2 \pi\sqrt{25 - 16y^{1/2}} \, dy$$

Comparing these integrals with (20.2), we have for the first, $a = 0$, $h = \frac{1}{2}$, $a + h = \frac{1}{2}$, and $a + 2h = 1$; and for the second, $a = 1$, $h = \frac{1}{2}$, $a + h = \frac{3}{2}$, and $a + 2h = 2$. In each case,

$$f(y) = \pi\sqrt{25 - 16y^{1/2}}$$

Therefore,

$$V = \frac{\pi}{6}\left[f(0) + 4f\left(\frac{1}{2}\right) + f(1)\right] + \frac{\pi}{6}\left[f(1) + 4f\left(\frac{3}{2}\right) + f(2)\right]$$

$$= \frac{\pi}{6}[(5 + 14.80 + 3) + (3 + 9.32 + 1.54)] = \frac{\pi}{6}(36.66) = 6.11\pi$$

The integral $\int_0^2 \pi\sqrt{25 - 16y^{1/2}} \, dy$ can be evaluated by replacing the radical by z^2, and the evaluation yields 6.06π for the volume accurate to two decimal places. Consequently, the error in the approximations in (*a*) and (*b*) is 0.12π and 0.05π, respectively.

Examples 1 and 2 illustrate the fact that the volume of a solid with the height H and cross-sectional area $A(z)$ is given by the formula

$$V = \frac{H}{6}(A_b + 4A_m + A_t) \tag{20.3}$$

where A_b, A_m, and A_t are the cross-sectional areas at the base, the middle, and the top, respectively, provided that $A(z)$ is a polynomial of degree 3 or less. If $A(z)$ is not such a polynomial, the volume yielded by (20.3) is an approximation to the exact volume.

20.4   Simpson's rule

We shall make use of the prismoidal formula in obtaining another approximation to $\int_a^b f(x)\,dx$. We proceed by dividing the interval from $x = a$ to $x = b$ into an *even* number of equal subintervals, erecting an ordinate at $x = a$, $x = b$, and each point of subdivision, and calling the intersections of these lines and $y = f(x)$ by the names $P_0, P_1, P_2, \ldots, P_n$. Next we pass a parabola through P_0, P_1, and P_2, another through P_2, P_3, and $P_4, \ldots$, and one through P_{n-2}, P_{n-1}, and P_n. We then find the sum of the areas under the parabolas. The situation is shown in Fig. 20.3.

If we apply the prismoidal formula to the area under each parabola, we find that

$$\int_a^b f(x)\,dx \doteq \frac{h}{3}\{[f(x_0) + 4f(x_1) + f(x_2)] + [f(x_2) + 4f(x_3) + f(x_4)]$$

$$+ \cdots + [f(x_{n-2}) + 4f(x_{n-1}) + f(x_n)]\}$$

where $h = (b - a)/n$ is the length of each subinterval. Consequently, we have

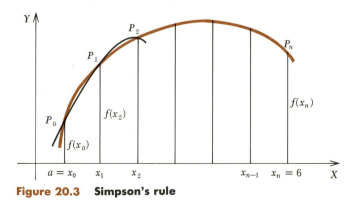

Figure 20.3   Simpson's rule

Simpson's rule $\int_a^b f(x)\,dx \doteq \dfrac{h}{3}[f(x_0) + 4f(x_1) + 2f(x_2) + 4f(x_3) + \cdots + 2f(x_{n-2})$

$$+ 4f(x_{n-1}) + f(x_n)] \quad (20.4)$$

which is known as *Simpson's rule*. The value given by Simpson's rule is exact if $f(x)$ is a polynomial of degree 3 or less since an exact value is given by the prismoidal formula under those conditions.

Example 1 Use Simpson's rule with $n = 4$ to find an approximation to

$$\int_2^6 \sqrt{x^2 - 2}\,dx$$

Solution: If we use $n = 4$, then $h = (6 - 2)/4 = 1$. Hence, $x_0 = 2$, $x_1 = 3$, $x_2 = 4$, $x_3 = 5$, and $x_4 = 6$. Consequently,

$$\int_2^6 \sqrt{x^2 - 2}\,dx$$

$$= \tfrac{1}{3}\left(\sqrt{2^2 - 2} + 4\sqrt{3^2 - 2} + 2\sqrt{4^2 - 2} + 4\sqrt{5^2 - 2} + \sqrt{6^2 - 2}\right)$$

$$= \tfrac{1}{3}\left(\sqrt{2} + 4\sqrt{7} + 2\sqrt{14} + 4\sqrt{23} + \sqrt{34}\right)$$

$$= \tfrac{1}{3}[1.41 + 4(2.65) + 2(3.74) + 4(4.80) + 5.83] = 14.84$$

We shall not give the proof, but it can be proved, that the error E due to the use of Simpson's rule with n subintervals to approximate $\int_a^b f(x)\,dx$ satisfies the inequality

$$E \leq \frac{(b - a)^5 K}{180n^4}$$

provided that $K \geq f^{(4)}(x)$ for every x in the closed interval from a to b.

Example 2 Find an approximation to $\int_1^5 \dfrac{dx}{1 + x}$ by use of Simpson's rule with $n = 4$, and find a value of E.

Solution: By use of Simpson's rule with $n = 4$, we have

$$\int_1^5 \frac{dx}{1 + x} = \frac{1}{3}\left(\frac{1}{1 + 1} + \frac{4}{1 + 2} + \frac{2}{1 + 3} + \frac{4}{1 + 4} + \frac{1}{1 + 5}\right)$$

$$= \frac{1}{3}\left(\frac{1}{2} + \frac{4}{3} + \frac{1}{2} + \frac{4}{5} + \frac{1}{6}\right) = 1.1$$

In order to be able to find a value of E, we must get $D_x^4 1/(1 + x)$. It is $24/(1 + x)^5$. Its largest value in $[1, 5]$ occurs for $x = 1$ and is $24/2^5 = 0.75$. Furthermore, $b - a = 5 - 1 = 4$, and n is given as 4. Consequently,

$$E \leq \frac{4^5(0.75)}{180(4^4)} = \frac{1}{60}$$

EXERCISE 20.2

Evaluate the integrals in Problems 1 to 8 by use of Simpson's rule with the specified value of n.

1 $\int_1^5 \sqrt[3]{126 - x^3}\, dx,\ n = 4$ 2 $\int_3^9 \sqrt[3]{3 + x^2}\, dx,\ n = 6$

3 $\int_0^4 \dfrac{\sqrt{1 + x}\, dx}{x + 2},\ n = 4$ 4 $\int_2^8 \dfrac{\sqrt{1 + x^2}\, dx}{x},\ n = 6$

5 $\int_0^{\pi/2} \sqrt{2 + \sin x}\, dx,\ n = 6$ 6 $\int_0^{\pi/2} \sqrt{\cos x}\, dx,\ n = 6$

7 $\int_1^3 e^{-x^2}\, dx,\ n = 4$ 8 $\int_1^3 x e^{-x^2}\, dx,\ n = 4$

9   The points $(1, 1)$, $(3, 3)$, and $(5, 3)$ determine a parabola. Find the area under it from $x = 1$ to $x = 5$ without finding the equation. Check by finding the equation and integrating.

10   Evaluate $\int_1^5 (x^3 - 2x^2 - 3x + 5)\, dx$ by Simpson's rule with $n = 4$ and by integrating.

11   Find the area under $y = x^3 + 2x^2 - 5x + 3$ in $[1, 5]$ by Simpson's rule with $n = 4$ and by integration.

12   Find the area bounded by the X axis and $y = (x + 1)^2(3 - x)$ by Simpson's rule with $n = 2$ and by integration.

13   Show that the prismoidal formula gives the exact value of volume of a right circular cone.

14   Use the prismoidal formula to find the volume common to two right circular cylinders of radius r whose axes intersect at right angles.

15   By use of Simpson's rule with $n = 4$, approximate the volume generated by revolving the area bounded by $x^2 = 4y$ and $y = 4$ about the Y axis.

16   Approximate the volume generated by revolving about the X axis the area under $y = \sin x$ from $x = 0$ to $x = \pi$. Use $n = 4$.

Use Simpson's rule to approximate the area under the curve determined by the data in each of Problems 17 and 18.

17

x	1	3	5	7	9
y	1.7	2.1	2.6	3.0	3.5

18

x	1	2	3	4	5	6	7
y	4.7	5.0	5.4	5.9	6.6	7.4	8.1

The cross-sectional area for several heights is given in each of Problems 19 and 20. Find the volume by use of Simpson's rule.

19

Height	0	2	4	6	8
Area	2.3	3.1	4.0	4.8	4.1

20

Height	0	1	2	3	4	5	6
Area	5.9	5.7	5.3	4.9	4.4	4.8	5.2

Find an approximation to the integral in each of Problems 21 to 24 by use of Simpson's rule with the given value of n and also find E.

21 $\int_2^{10} \dfrac{dx}{x-1}$, $n = 4$

22 $\int_0^4 (x^3 - x)\,dx$, $n = 2$

23 $\int_0^{\pi/3} \sin x\,dx$, $n = 4$

24 $\int_0^6 (x^4 - 2x)\,dx$, $n = 6$

21

Moments, centroids

21.1   Introduction

The product obtained by multiplying the magnitude of a force by the length of the perpendicular drawn from the point O to the line of action of the force is called the *first moment* of the force about O. If there is more than one force as in Fig. 21.1, then the sum of their first moments is called their *resultant moment*. For the forces and distances, except 30 lb, shown in the figure, the resultant moment in foot-pounds is

first moment

resultant moment

$$5(15) + 8(9) + 17(6) = 249$$

If we multiply the sum of the forces, 30 lb, by a distance x ft from O, we find that the first moment is $30x$ ft-lb. Equating this and the resultant moment 249 ft-lb, we find that a single force equal to the sum of the three forces, if acting at a distance 8.3 ft from O, has the same first moment about O as the given system of three forces. This distance is a weighted average of the distances 5, 8, and 17 of the individual forces from O.

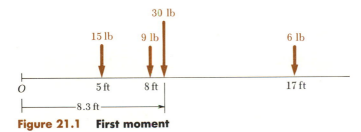

Figure 21.1   First moment

21.2   First moment and centroid of plane areas and arcs

We can obtain the first moment of a plane area with respect to a line λ in its plane by dividing the area into n strips that are parallel to the line, denoting the area of the ith strip by ΔA_i (see Fig. 21.2), multiplying this area by its distance l_i from λ, forming the sum $\sum\limits_{i=1}^{n} l_i \, \Delta A_i$ of these products, and taking the limit of this sum as the

first moment as an integral

maximum ΔA_i approaches zero. Now defining the first moment of the area with respect to λ as this limit, we have

$$M_\lambda = \lim_{\max \Delta A_i \to 0} \sum_{i=1}^{n} l_i \, \Delta A_i = \int l \, dA \qquad (21.1)$$

where limits are used so as to extend the integration over the entire area under consideration. In evaluating M_λ, we replace dA in terms of x or y and take limits on the one of them whose differential appears in dA.

In general, we shall be interested in finding the first moment of an area with respect to a coordinate axis and shall see how to do that by use of an example.

Example 1   Find the first moment with respect to the Y axis of the area bounded by the parabola $y = x^2/3$ and the line $y = x$.

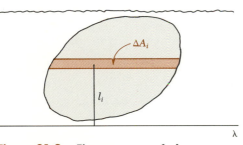

Figure 21.2   First moment of plane area

Solution: The graphs of the given equations and a strip of the area between them and parallel to the Y axis are shown in Fig. 21.3. Since the strip is parallel to the Y axis, its length is $y_l - y_p$, its width is Δx, and its distance from the Y axis is x. Consequently, we need only the limits of integration. Solving the given equations simultaneously, we find that the graphs intersect at $(0, 0)$ and $(3, 3)$. Consequently $dA = (y_l - y_p)\, dx$ and (21.1) becomes

$$M_y = \int_0^3 x(y_l - y_p)\, dx = \int_0^3 x\left(x - \frac{x^2}{3}\right) dx = \frac{x^3}{3} - \frac{x^4}{12}\,\Big|_0^3$$

$$= \frac{27}{3} - \frac{81}{12} = \frac{9}{4} = 2.25$$

centroid   We shall now define the *centroid* of a plane area as that point $(\bar{x}, \bar{y})$ at which the entire area can be imagined as being concentrated without changing the first moment of the area with respect to either coordinate axis. If the entire area could be concentrated at $(\bar{x}, \bar{y})$, then

$$M_y = \bar{x}A = \int x\, dA \tag{1}$$

and $$M_x = \bar{y}A = \int y\, dA \tag{2}$$

where proper limits are used in each integral. If (1) and (2) are solved for $\bar{x}$ and $\bar{y}$, we find that *the coordinates of the centroid of a plane area are*

$$\bar{x} = \frac{\int x\, dA}{A} \qquad \text{and} \qquad \bar{y} = \frac{\int y\, dA}{A} \tag{21.2}$$

where proper limits are used after expressing dA in terms of x or y.

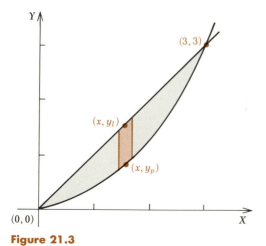

Figure 21.3

Example 2   Find the coordinates of the centroid of the area bounded by $y = x$ and $y = x^2/3$.

Solution: This is the area used in Example 1. A strip with edges parallel to the Y axis is shown in Fig. 21.3. The first moment about the Y axis was found to be 2.25. The area is

$$A = \int_0^3 (y_l - y_p)\, dx = \int_0^3 \left(x - \frac{x^2}{3} \right) dx = \frac{x^2}{2} - \frac{x^3}{9} \Big|_0^3 = 1.5$$

Consequently,
$$\bar{x} = \frac{2.25}{1.5} = 1.5$$

If we take a strip of the area with edges parallel to the X axis, we find that

$$M_x = \int_0^3 y(x_p - x_l)\, dy = \int_0^3 y(\sqrt{3}y^{1/2} - y)\, dy$$

$$= \left[\sqrt{3}y^{5/2}\left(\frac{2}{5}\right) - \frac{y^3}{3} \right] \Big|_0^3 = \frac{54}{5} - 9 = 1.8$$

Therefore,
$$\bar{y} = \frac{1.8}{1.5} = \frac{6}{5}$$

Hence the centroid $(\bar{x}, \bar{y})$ is $(1.5, 1.2)$

moment of an arc   By following a procedure similar to that used in finding a formula for the first moment of a plane area with respect to a line, we can find that the first moment of an arc relative to a line is

$$M_\lambda = \int l\, ds \qquad (21.3)$$

where l is a general expression for the distance from the line to the arc and limits are chosen so as to extend the integration over the entire arc under consideration after expressing ds in terms of x or y and dx or dy.

Example 3   Find the first moment of the arc $y = \sqrt{a^2 - x^2}$, $x > 0$, with respect to the X axis.

Solution: By differentiation, we find that

$$y' = \frac{-x}{\sqrt{a^2 - x^2}}$$

Hence,
$$ds = \sqrt{1 + y'^2}\, dx = \frac{a\, dx}{\sqrt{a^2 - x^2}}$$

and
$$M_x = \int y\, ds = \int_0^a \frac{a\sqrt{a^2 - x^2}\, dx}{\sqrt{a^2 - x^2}} = ax \Big|_0^a = a^2$$

By analogy to the centroid of a plane area, we define the centroid of an arc $(\bar{x}, \bar{y})$ to be determined by the equations

centroid
of an arc

$$M_y = s\bar{x} = \int x \, ds \qquad \text{and} \qquad M_x = s\bar{y} = \int y \, ds \qquad (21.4)$$

where limits are chosen so as to extend the integration over the desired arc after expressing ds in terms of x or y and dx or dy.

Example 4   Find the y coordinate of the centroid of the arc $y = \sqrt{a^2 - x^2}$, $x > 0$.

Solution: In Example 3, we found that $M_x = a^2$. If we use (21.4), we have

$$M_x = \bar{y} \int ds = \bar{y} a \int_0^a \frac{dx}{\sqrt{a^2 - x^2}} = \bar{y} a \sin^{-1} \frac{x}{a} \Big|_0^a = \frac{\bar{y} a \pi}{2}$$

Consequently, equating the two expressions for M_x, we have

$$\frac{\bar{y} a \pi}{2} = a^2$$

Hence $\bar{y} = 2a/\pi$.

21.3   Computation of first moment by use of strips perpendicular to the axis

Heretofore, we have found the first moment by using a strip or element with edges parallel to the axis about which the moment is being found. At times, we are led to a simpler integration by taking a strip with edges perpendicular to the axis. Consequently, we shall now develop a method of procedure and a formula for use in such cases.

We shall consider a rectangle of width w and altitude h that abuts on the X axis as shown in Fig. 21.4 and shall find its first moment about that axis in terms of w and h. To do this, we begin by taking a typical strip of the rectangle with edges parallel to the X axis and finding the first moment as in Sec. 21.2. Thus,

$$M_x = \int_0^h wy \, dy = \frac{wy^2}{2} \Big|_0^h = w \frac{h^2}{2} = A \frac{h}{2}$$

element
perpendicular
to the axis

since $A = wh$. Consequently, we see that *the first moment of a rectangular element that abuts on and is perpendicular to the axis about which the moment is taken is equal to the area of the strip times one-half of its height.*

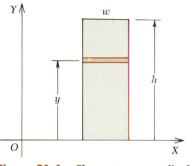

Figure 21.4 **Element perpendicular to axis**

Example

Find M_x for the area under $y = \cos x$ from $x = 0$ to $x = \pi/2$.

Solution: The area under consideration is shown in Fig. 21.5. If we use elements parallel to the X axis, we have

$$M_x = \int_0^1 yx\, dy = \int_0^1 y \operatorname{Arccos} y\, dy$$

and if the elements are taken abutting on and perpendicular to the X axis, we get

$$M_x = \int_0^{\pi/2} \frac{y}{2}\, y\, dx = \frac{1}{2}\int_0^{\pi/2} \cos^2 x\, dx$$

We now can choose which of the two forms of M_x to evaluate. If we choose the second and replace $\cos^2 x$ by $(1 + \cos 2x)/2$, we obtain

$$M_x = \frac{1}{4}\int_0^{\pi/2}(1 + \cos 2x)\, dx = \frac{1}{4}\left(x + \frac{1}{2}\sin 2x\right)\Big|_0^{\pi/2} = \frac{\pi}{8}$$

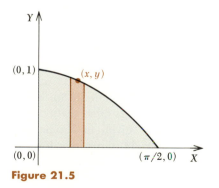

Figure 21.5

EXERCISE 21.1

In each of Problems 1 to 20, find the centroid of the specified area.

1   The area bounded by $y^2 = 8x$ and $x = 4$

2   The first-quadrant area bounded by $y^2 = 4x$, the X axis, and $x = 4$

3   The second-quadrant area enclosed by $x^2 + y^2 = 25$

4   The first-quadrant area bounded by the coordinate axes and $x^2/a^2 + y^2/b^2 = 1$

5   The area enclosed by $y^2 = 4x$ and $x^2 = 4y$

6   The area bounded by $y^2 = 8x$ and $y = 2x$

7   The area bounded by $ax + by = c$ and the coordinate axes

8   The area bounded by $y = x^4$ and $y = x^2$

9   The area bounded by $y = \sqrt{6x}$, $y = 3x - 4$, and $y = 0$

10   The area bounded by the X axis, $x = 1$, and $y = 4x - x^3$

11   The area bounded by $x^{1/2} + y^{1/2} = a^{1/2}$ and the x and y axes

12   The area between the X axis and $4y = 16 - x^4$

13   The area bounded by $y = e^x$, $x = 0$, $x = 1$, and $y = 0$

14   The area under $y = \cos x$ from $x = 0$ to $x = \pi/2$

15   The area under $y = \sin x$ from $x = 0$ to $x = \pi$

16   The area bounded by $y = \cosh x$, $x = -1$, $x = 1$, and $y = 0$

17   The area above $y = 0$ under one arc of the cycloid $x = a(\theta - \sin \theta)$, $y = a(1 - \cos \theta)$

18   The area bounded by the parabola $x = pt^2$, $y = 2pt$ and its latus rectum.

19   The first-quadrant area bounded by $x = a \sin \theta$, $y = a \cos \theta$

20   The upper half of the ellipse $x = a \cos \theta$, $y = b \sin \theta$

In each of Problems 21 to 24 find the centroid of the given arc.

21   The parabola $x = 3t^2$, $y = 6t$ from $t = 0$ to $t = 2$

22   The circle $x = \sin t$, $y = \cos t$ from $t = \pi/2$ to $t = \pi$

23   One arc of the cycloid $x = a(\theta - \sin \theta)$, $y = a(1 - \cos \theta)$

24   The curve $x = e^{-t} \cos t$, $y = e^{-t} \sin t$ from $t = 0$ to $t = 2\pi$.

25   Prove that if an area that lies entirely on one side of a line is revolved about the line, the volume of the solid generated is equal to the product of the area and the distance traveled by its centroid. This *Pappus'* is known as the theorem of Pappus.

theorem   26   Use the theorem of Pappus (Problem 25) to derive a formula for the volume of a torus of radius R if its cross section has radius r.

27   Use the theorem of Pappus given in Problem 25 to derive a formula for the volume of a right circular cone of radius r and altitude h.

28   Make use of the area of a circle, the volume of a sphere, and the theorem of Pappus to locate the centroid of a semicircle.

21.4   Second moment of area and arc

We have studied the first moment of an area with respect to a line
moment of and shall now consider the *second moment*, or *moment of inertia*, of
inertia a plane area with respect to a line. The moment of inertia differs from
the first moment only in that the area of each strip parallel to the
line is multiplied by the second power of the distance of the strip from
the line instead of the first power. If we follow a procedure that is
essentially the same as that in Sec. 21.2, we find that the moment of
inertia of a plane area about the X axis and that about the Y axis are,
respectively,

$$I_x = \int y^2 \, dA \qquad \text{and} \qquad I_y = \int x^2 \, dA \qquad (21.5)$$

where in each case, dA refers to an element whose edges are parallel
to the axis about which the moment of inertia is being taken. In each
case appropriate limits are used after expressing dA in terms of x or y
and dx or dy.

Example 1   Find the moment of inertia of a circle with respect to a diameter.

Solution: We shall choose a circle of radius r and put the axes
through its center as in Fig. 21.6. If the element is parallel to the
Y axis, then

$$dA = 2y \, dx$$

$$x^2 \, dA = x^2 \, 2y \, dx = 2x^2 \sqrt{r^2 - x^2} \, dx$$

and $$I_y = \int x^2 \, dA = \int_{-r}^{r} 2x^2 \sqrt{r^2 - x^2} \, dx = \frac{\pi r^4}{4}$$

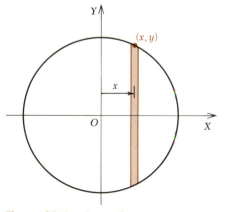

Figure 21.6   Second moment

Example 2

Find the moment of inertia of the area of a rectangle with respect to its base.

Solution: We shall consider a rectangle with base b and altitude h in the position shown in Fig. 21.7 relative to the coordinate axes. Since the base is b, the element of area with sides parallel to the base is

$$dA = b\,dy$$

and
$$I_x = \int y^2\,dA = \int_0^h y^2 b\,dy = \frac{y^3}{3}b\bigg|_0^h$$

$$= \frac{bh^3}{3} = bh\left(\frac{h^2}{3}\right)$$

If we make use of the fact that the area of a parallelogram with base b and altitude h is bh, we see from the value of I_x in Example 2 that *the moment of inertia of a rectangular element that abuts on the axis about which the moment is taken is equal to the area of the element times one-third of the square of its altitude.* This enables us to set up the integral for moment of inertia in a second, and at times more readily integrated, form.

element perpendicular to the axis

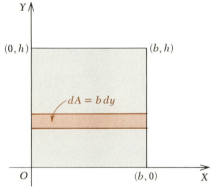

Y

$(0,h)$ (b,h)

$dA = b\,dy$

O $(b,0)$ X

Figure 21.7

Example 3

Find the second moment with respect to the X axis of the area bounded by $y = e^x$, the coordinate axes, and $x = 1$.

Solution: A sketch of the area and an element perpendicular to the X axis are shown in Fig. 21.8. If we now use the italized statement that is given just before this example, we have

$$dA = y\,dx$$

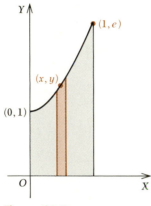

Figure 21.8

and the second moment of the element with respect to the X axis is

$$\tfrac{1}{3}\,y^2\,dA = \tfrac{1}{3}\,y^2 y\,dx$$

$$= \tfrac{1}{3}\,e^{3x}\,dx \qquad y = e^x$$

Therefore, $\qquad I_x = \dfrac{1}{3}\displaystyle\int_0^1 e^{3x}\,dx = \dfrac{1}{9}\,e^{3x}\Big|_0^1 = \dfrac{1}{9}\,(e^3 - 1)$

We define the second moments with respect to the coordinate axes of a plane arc length by the equations

$$I_x = \textstyle\int y^2\,ds \qquad \text{and} \qquad I_y = \textstyle\int x^2\,ds \qquad\qquad (21.6)$$

where the limits of integration are so chosen as to extend the integration over the desired arc after expressing ds in terms of x or y and dx or dy.

Example 4 Find the moment of inertia with respect to each coordinate axis of the part of the arc $y = \sqrt{r^2 - x^2}$ which is in the first quadrant.

Solution: Since $y = f(x) = \sqrt{r^2 - x^2}$, it follows that

$$dy/dx = -x/\sqrt{r^2 - x^2} \qquad \text{and} \qquad \sqrt{1 + (dy/dx)^2} = r/\sqrt{r^2 - x^2}$$

Consequently by (21.6)

$$I_x = \int_0^r r\sqrt{r^2 - x^2}\,dx$$

and

$$I_y = \int_0^r \frac{rx^2\,dx}{\sqrt{r^2 - x^2}}$$

We can evaluate each of these integrals by replacing x by $r \sin \theta$ and get

$$I_x = r^3 \int_0^{\pi/2} \cos^2 \theta \, d\theta = r^3 \int_0^{\pi/2} \left(\frac{1}{2} + \frac{1}{2} \cos 2\theta \right) d\theta$$

$$= \frac{r^3}{2} \left[\theta + \frac{1}{2} \sin 2\theta \right]_0^{\pi/2} = \frac{\pi r^3}{4}$$

$$I_y = r^3 \int_0^{\pi/2} \sin^2 \theta \, d\theta = r^3 \int_0^{\pi/2} \left(\frac{1}{2} - \frac{1}{2} \cos 2\theta \right) d\theta$$

$$= \frac{r^3}{2} \left[\theta - \frac{1}{2} \sin 2\theta \right]_0^{\pi/2} = \frac{\pi r^3}{4}$$

21.5   The parallel-axis theorem

parallel-axis theorem

We shall now give a theorem which enables us to find the moment of inertia of a plane area with respect to any line if the moment of inertia with respect to a parallel line through the centroid is known. The statement of the theorem is: *If I_c is the moment of inertia of a plane area A with respect to a line through its centroid and if I_λ is the second moment with respect to a line λ parallel to and s units from the line through the centroid, then*

$$I_\lambda = I_c + As^2$$

We shall use Fig. 21.9 in proving this theorem.

$$I_\lambda = \int y^2 \, dA = \int (y_1 + s)^2 \, dA = \int (y_1{}^2 + 2y_1 s + s^2) \, dA$$

$$= \int y_1{}^2 \, dA + 2s \int y_1 \, dA + s^2 \int dA = \int y_1{}^2 \, dA + 0 + s^2 A$$

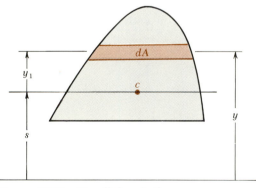

Figure 21.9   Parallel-axis theorem

where $\int y_1 \, dA = 0$ because it is the first moment of the area about an axis through the centroid. Hence

$$I_\lambda = I_c + s^2 A$$

Example   Find the second moment with respect to a tangent of the area of a circle of radius r.

Solution: A sketch of the situation is shown in Fig. 21.10. In Example 1 of Sec. 21.4, we found the moment of inertia of a circle of radius r about a diameter to be $\pi r^4/4$. Since a diameter is a line through the centroid and since a tangent parallel to a diameter is a distance r from the diameter, it follows by use of the parallel-axis theorem that the second moment about the tangent is

$$I_\lambda = I_c + A r^2$$

$$= \frac{\pi r^4}{4} + \pi r^2 r^2 \qquad \textit{since } A = \pi r^2$$

$$= \frac{5 \pi r^4}{4}$$

21.6   Radius of gyration

If I is the moment of inertia of a mass M with respect to a line λ, then the equation

$$I = MR^2$$

defines a number $\qquad R = \sqrt{\dfrac{I}{M}}$

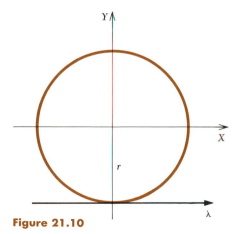

Figure 21.10

radius of   that is called the *radius of gyration*. Consequently, a particle of
gyration   mass M located R units from the line λ has the same moment of
inertia with respect to λ as the given mass has. As a special case, if
M is area, we find that *the radius of gyration R of an area A with
respect to a line L is*

$$R = \sqrt{\frac{I_L}{A}}$$

Example   Find the radius of gyration R of the area of a circle with respect to a
tangent.

Solution: We found in the example of Sec. 21.5 by use of the
parallel-axis theorem that the moment of inertia of a circle about a
tangent is $5\pi r^4/4$, and we know that the area of a circle is πr^2.
Consequently, the radius of gyration is

$$R = \sqrt{\frac{I}{A}} = \sqrt{\frac{5\pi r^4}{4} \div \pi r^2} = \sqrt{\frac{5r^2}{4}} = \frac{r}{2}\sqrt{5}$$

EXERCISE 21.2

*Find the moment of inertia and radius of gyration of the area in each
of Problems 1 to 8 with respect to the X and Y axes.*

1   The area bounded by $y = 2x$, $x = 2$, and $y = 0$
2   The area bounded by $y = 2x$, $y = 4$, and $x = 0$
3   The area bounded by $y^2 = 5x$ and $x = 5$
4   The area bounded by $x^2/a^2 + y^2/b^2 = 1$
5   The area bounded by $y = 4 - x^2$ and the X axis
6   The area bounded by $y = x^3$, $x = 2$, and the X axis
7   The area bounded by $y = e^x$, $x = -1$, and the coordinate axes
8   The area under $y = \cos x$ from $x = 0$ to $x = \pi/2$

9   Find I_y for the area under $y = 8/(x^2 + 4)$ from $x = 0$ to $x = 2$.
10 Compute I_y for the area bounded by $y = 4x - x^2$ and the
X axis.
11 Find I_x for the triangle with vertices at $(0, 0)$, $(b, 0)$, and $(0, h)$.
12 Find I_x for the area bounded by $x^2 = 2py$, $y = 0$, and $x = 2p$.
13 Find I_x for the area under one arc of $y = \sin x$.
14 Find I_x for the area under $y = \tan x$ from $x = 0$ to $x = \pi/3$.
15 Find I_x for the area under $y = \sec x$ from $x = 0$ to $x = \pi/4$.
16 Find I_x for the area under $y = \cosh x$ for $x = -1$ to $x = 1$.
17 Find I_x for the first-quadrant area bounded by $y = x$ and $y = x^3$.
18 Find I_x for the area bounded by $y = 2x^2$ and $y = 2x$.
19 Find I_y for the area bounded by $x = e^y$ and $x = 1 + (e - 1)y$.

20 Find I_y for the area bounded by $x = \cos y$ and $x = 1 - 2y/\pi$.

21 Find the second moment about the Y axis of the arc of one branch of $y^2 = 4x^3/9a$ from $x = 0$ to $x = 3a$.

22 Find the moment of inertia about the X axis of the arc of $y = e^x$ from $x = 0$ to $x = 1$.

23 Find the moment of inertia about the Y axis of the hypocycloid $x = a \sin^3 \theta$, $y = a \cos^3 \theta$.

24 Find the second moment with respect to the X axis of one arc of $x = a(\theta - \sin \theta)$, $y = a(1 - \cos \theta)$.

Use the parallel-axis theorem in finding the values called for in Problems 25 to 28.

25 The moment of inertia of the area bounded by $x^2 + y^2 = a^2$ with respect to $x = 2a$

26 The second moment of the area bounded by $b^2x^2 + a^2y^2 = a^2b^2$ with respect to $y = 3b$

27 The second moment of the area bounded by $y = \cosh x$, $x = -1$, $x = 1$, and the X axis with respect to $x = 2$

28 The moment of inertia of the area bounded by $y^2 = 5x$ and $x = 5$ axis with respect to $y = 6$

21.7 First moment and centroid of a solid of revolution

We shall consider the solid generated by revolving about the X axis the area bounded by $y = f(x)$, $x = a$, $x = b$, and the X axis and shall find the first moment and centroid of the solid. In order to do this, we partition the interval $[a, b]$ into n subintervals of length $\Delta x_i \leq \delta$, $i = 1, 2, \ldots, n$. We then divide the solid into slices by planes that are perpendicular to the X axis at the points of subdivision and construct the approximating cylindrical disks. The ith one of these disks is shown in Fig. 21.11. We denote its volume by $\Delta V_i = \pi[f(z_i)]^2 \Delta x_i$ where z_i is a value of x in the closed interval $[x_{i-1}, x_i]$; we form the product $z_i \Delta V_i$ and take the sum $\sum_{i=1}^{n} z_i \Delta V_i = \sum_{i=1}^{n} z_i \pi f^2(z_i) \Delta x_i$ of all such products over the solid. If the limit of this sum exists as δ approaches zero, it is $\int_a^b x\pi y^2 \, dx$. We now give a definition by saying that

$$M = \lim_{\delta \to 0} \sum_{i=1}^{n} z_i \Delta V_i = \pi \int_a^b xy^2 \, dx$$

first moment of a solid of revolution *is called the first moment of volume of the solid of revolution with respect to a plane through the origin and perpendicular to the X axis.*

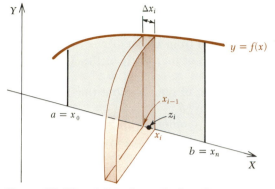

Figure 21.11   Solid of revolution, first moment

The volume of the solid was obtained in Sec. 19.3 and is

$$V = \pi \int_a^b y^2 \, dx$$

The centroid of the solid of revolution is the point $(\bar{x}, 0)$ where

*centroid of a
solid of
revolution*

$$\bar{x} = \frac{M}{V} = \frac{\pi \int_a^b xy^2 \, dx}{\pi \int_a^b y^2 \, dx}$$

The centroid for the solid under consideration is on the X axis, but, in general, such a centroid is on the axis of revolution.

Example

A solid is generated by revolving the area bounded by $xy = 1$, $y = 0$, $x = 1$, and $x = 3$ about the X axis. Find the first moment of the volume of the solid about the plane through the origin and perpendicular to the X axis, and find the x coordinate of the centroid.

Solution: A sketch of the area that is being revolved and a typical element of the volume is shown in Fig. 21.12. By use of the definition of M and for $xy = 1$, we have

$$M = \pi \int_1^3 x \left(\frac{1}{x}\right)^2 dx \qquad \textit{since } y = 1/x$$

$$= \pi \int_1^3 \frac{dx}{x} = \pi \ln 3$$

Furthermore, $V = \pi \int_1^3 \left(\frac{1}{x}\right)^2 dx = \pi \left(-\frac{1}{x}\right)\Big|_1^3 = \frac{2\pi}{3}$

Consequently, $\bar{x} = \dfrac{M}{V} = \pi \ln 3 \div \dfrac{2\pi}{3} = \dfrac{3 \ln 3}{2} = \ln 3\sqrt{3}$

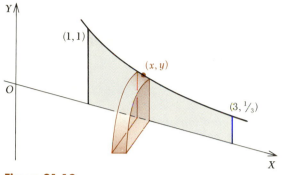

Figure 21.12

21.8   Second moment of a solid of revolution with respect to the axis of revolution

For a solid of revolution, the value of the moment of inertia with respect to the axis of revolution can be obtained as follows. We shall let the solid V be generated by revolving the area bounded by $x = a$, $y = f(x)$, $x = b$, and the X axis about the Y axis as indicated in Fig. 21.13. As usual, we divide $[a, b]$ into subintervals of width $\Delta x_i < \delta$, let z_i be a value of x in the ith subinterval, and let ΔA_i designate an element of the area with Δx_i as a base. The volume of the cylindrical shell generated by revolving the element of area about the Y axis is $\Delta V_i = 2\pi z_i f(z_i)\, \Delta x_i$. We now define the moment of inertia of V with respect to the Y axis as

$$\sum_{i=1}^{n} z_i^2\, \Delta V_i = \sum_{i=1}^{n} z_i^2 (2\pi) z_i f(z_i)\, \Delta x_i$$

If the limit of this sum exists it is $\int_a^b x^2(2\pi)xy\, dx$.

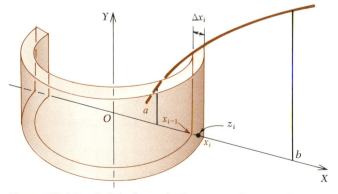

Figure 21.13   Solid of revolution, second moment

second
moment   Consequently, the *second moment*, or *moment of inertia*, with respect to its axis of the volume of the solid of revolution obtained by revolving about the Y axis the area bounded by $y = f(x)$, $x = a$, $x = b$, and the X axis is

$$I = \lim_{\delta \to 0} \sum_{i=1}^{n} z_i{}^2 \, \Delta V_i = 2\pi \int_a^b x^3 y \, dx$$

radius of
gyration   The *radius of gyration* of the solid of revolution with respect to its axis of revolution is, by definition, the positive number R such that $R^2 V = I$ where I and V are the moment of inertia and volume of the solid.

Example   Find the moment of inertia about the Y axis of the solid generated by revolving the first-quadrant area bounded by $y = \sqrt{a^2 - x^2}$ about the Y axis. Find the radius of gyration.

Solution: The area that is to be revolved and a typical strip of it are shown in Fig. 21.14. If we use the formula developed in this section, we get

$$I = 2\pi \int_0^a x^3 y \, dx = 2\pi \int_0^a x^3 \sqrt{a^2 - x^2} \, dx$$

In order to evaluate this integral, we put $x = a \sin \theta$ and find that

$$I = 2\pi a^5 \int_0^{\pi/2} \sin^3 \theta \cos^2 \theta \, d\theta = 2\pi a^5 \int_0^{\pi/2} (1 - \cos^2 \theta) \cos^2 \theta \sin \theta \, d\theta$$

$$= 2\pi a^5 \left(-\frac{\cos^3 \theta}{3} + \frac{\cos^5 \theta}{5} \right) \Big|_0^{\pi/2} = \frac{4\pi a^5}{15}$$

Consequently, the radius of gyration is

$$R = \sqrt{\frac{I}{V}} = \sqrt{\frac{4\pi a^5}{15} \frac{3}{2\pi a^3}} = \frac{a}{5} \sqrt{10}$$

since the volume of a hemisphere is $2\pi a^3/3$.

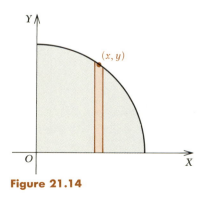

Y

(x, y)

O

X

Figure 21.14

EXERCISE 21.3

In each of Problems 1 to 16 a solid is generated by revolving about the X axis the area determined, or bounded, by the X axis and the given curves. In each problem find the first moment about a plane through the origin and perpendicular to the X axis and also the x-coordinate of the centroid.

1   $by = ax, x = a$ 2 $y = x^2, x = 2$
3   $x = y^3, x = 8$ 4 $y = 2x, x = 1, x = 4$
5   First-quadrant part of $b^2x^2 + a^2y^2 = a^2b^2$
6   First-quadrant part of $y^2 = 4ax, x = 2a$
7   First-quadrant part of $b^2x^2 - a^2y^2 = a^2b^2, x = 2a$
8   First-quadrant part of $x^{2/3} + y^{2/3} = a^{2/3}$
9   $y = e^x, x = 0, x = -1$
10   $y = \sin x, x = \pi/2$
11   $y = \ln x, x = 3$ 12 $y = \tan x, x = \pi/4$
13   First-quadrant part of $x = a \sin^3 \theta, y = a \cos^3 \theta$
14   The part of $x = a \sin \theta, y = a \cos \theta$ for $\theta = 0$ to $\theta = \pi/2$
15   The part of $x = a \sec t, y = a \tan t$ for $t = 0$ to $t = \pi/3$
16   The part of $x = \sin^2 t, y = \cos^2 t$ for $t = 0$ to $t = \pi/2$

A solid is generated in each of Problems 17 to 24 by rotating the area bounded by the given curves about the Y axis. Find the moment of inertia and radius of gyration with respect to the Y axis for each solid.

17   $xy = 1, x = 1, x = 3, y = 0$
18   $y = x^2, y = 0, x = 2$
19   $y = x^3, x = 2, y = 0$
20   $y = 2\sqrt{x}, y = 0, x = 4$
21   $y = e^x, x = 0, y = 0$
22   $y = \cos x, x = 0, y = 0$
23   $y = \ln x, x = e, y = 0$
24   $y = 2 \cosh x, x = -1, x = 1, y = 0$

Vectors, planes, lines

Coordinate systems are needed in three-dimensional space just as they are in two-dimensional space. We shall introduce the rectangular-coordinate system, express it in terms of vectors, and then use vectors in investigating some three-dimensional configurations.

22.1 Coordinates

There are several types of coordinate systems that may be used in three-dimensional space. Others will be given later, but we shall begin by studying the system of rectangular coordinates. In this system a point is located by giving its distance and direction from each of three mutually perpendicular planes that divide space into eight *octants* parts called *octants*. The planes are shown in Fig. 22.1 as XOY, XOZ, and YOZ and are referred to as the XY, XZ, and YZ planes, respectively. The line $X'X$ is called the X axis, YY' is the Y axis, and $Z'Z$ is the Z axis. The coordinates of a point are written as (x, y, z) where x, y, and z are the directed distances from the YZ, XZ, and XY planes, respectively. Thus the coordinates of P in Fig. 22.1 are $x = OM$, $y = MN$, and $z = NP$. The *positive directions* are indicated by the arrows; the direction opposite each positive direction is *negative*. The point $(-1, 3, 2)$ is located in Fig. 22.1.

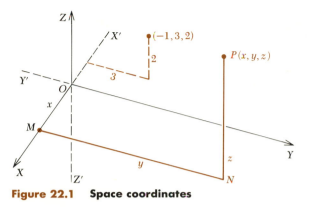

Figure 22.1   Space coordinates

22.2   Length and direction of the radius vector

radius vector   The directed segment OP from the origin to any point $P(x, y, z)$ is called the *radius vector* of P. In order to find the length of the radius vector, we make use of the fact, as seen from Fig. 22.2, that OP is the diagonal of a rectangular parallelepiped whose edges are $OA = x$, $OB = y$, and $OC = z$. Consequently, we have

$$\overline{OP}^2 = \overline{OD}^2 + \overline{DP}^2 = \overline{OA}^2 + \overline{AD}^2 + \overline{DP}^2$$
$$= \overline{OA}^2 + \overline{OB}^2 + \overline{OC}^2 = x^2 + y^2 + z^2$$

Hence we know that the length ρ of the radius vector is

$$\rho = \sqrt{x^2 + y^2 + z^2} \tag{22.1}$$

The angles $AOP = \alpha$, $BOP = \beta$, and $COP = \gamma$ are called the direction angles of OP. Each of them is a nonnegative angle less

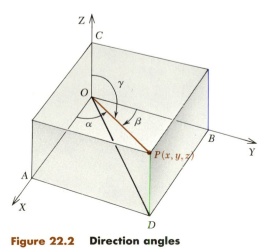

Figure 22.2   Direction angles

than or equal to $180°$ and their cosines are called the *direction cosines* of *OP*. Thus,

direction cosines

$$\cos \alpha = \frac{OA}{OP} = \frac{x}{\rho}$$

$$\cos \beta = \frac{OB}{OP} = \frac{y}{\rho}$$

$$\cos \gamma = \frac{OC}{OP} = \frac{z}{\rho}$$

If we square each member of each of these equations and add corresponding members of the new equations, we see that

$$\cos^2 \alpha + \cos^2 \beta + \cos^2 \gamma = \frac{x^2 + y^2 + z^2}{\rho^2} = 1$$

sum of the squares of the direction cosines

Consequently, we know that *the sum of the squares of the direction cosines of OP is one.*

If *P* is in the first octant, its coordinates are all positive; hence its direction cosines are all positive and its direction angles are all acute. If *P* has one or more negative coordinates, then at least one of its direction cosines is negative and at least one of its direction angles is in the interval $(90°, 180°)$.

22.3   The distance formula

We shall now find the distance between any two points in three-dimensional space. Two such points are shown as $P_1(x_1, y_1, z_1)$ and $P_2(x_2, y_2, z_2)$ in Fig. 22.3. Since P_1P_2 is the hypotenuse of a right triangle with P_1Q and QP_2 as sides, we have

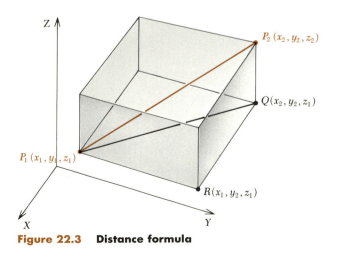

Figure 22.3   Distance formula

$$\overline{P_1P_2}^2 = \overline{P_1Q}^2 + \overline{QP_2}^2 = \overline{P_1R}^2 + \overline{QR}^2 + \overline{QP_2}^2 = \overline{QR}^2 + \overline{P_1R}^2 + \overline{QP_2}^2$$

The coordinates of R and Q are as shown in Fig. 22.3 since P_1R, QR, and QP_2 are parallel to the Y, X, and Z axes, respectively. Therefore, putting in the lengths of the segments in terms of the coordinates of their end points, we see that

$$P_1P_2 = \sqrt{(x_1 - x_2)^2 + (y_1 - y_2)^2 + (z_1 - z_2)^2}$$

is the distance between $P_1(x_1, y_1, z_1)$ *and* $P_2(x_2, y_2, z_2)$.

Example   Find the distance between $P_1(3, -1, 2)$ and $P_2(5, -4, 8)$.

Solution: If we substitute the coordinates of P_1 and P_2 in the distance formula, we find that

$$P_1P_2 = \sqrt{(3 - 5)^2 + [-1 - (-4)]^2 + (2 - 8)^2}$$
$$= \sqrt{(-2)^2 + (+3)^2 + (-6)^2} = 7$$

22.4   Direction cosines and direction numbers of a directed line

In Sec. 22.2 we discussed the direction of the radius vector. Here we shall begin our discussion of the direction cosines of the directed line from P_1 to P_2 by stating that the direction angles and direction cosines of a directed line that does not pass through the origin are the same as those of the line through the origin parallel to, and directed as, the given line. Therefore, if we use d as the length of P_1P_2, we have

$$\cos \alpha = \frac{x_2 - x_1}{d}$$

formulas for the direction cosines

$$\cos \beta = \frac{y_2 - y_1}{d} \qquad (22.2)$$

$$\cos \gamma = \frac{z_2 - z_1}{d}$$

as the direction cosines of the line through P_1 and P_2. Each of the direction cosines has the same sign as its numerator since d is positive. The effect of changing the positive direction on the line is to change the signs of all of the direction cosines, hence, to replace the direction angles by their supplements.

Any three numbers that are proportional to the direction cosines of a *direction* line are called *direction numbers* of the line. Hence, if we multiply the *numbers* right members of Eq. (22.2) by d, we see that

$$x_2 - x_1 \qquad y_2 - y_1 \qquad z_2 - z_1$$

are direction numbers of the line through $P_1(x_1, y_1, z_1)$ *and* $P_2(x_2, y_2, z_2)$.

Any three numbers such as a, b, and c that are not all zero can be used as direction numbers of a line. The corresponding values of the direction cosines are obtained by dividing each number by $\pm \sqrt{a^2 + b^2 + c^2}$. Consequently, we have

$$\cos \alpha = \frac{\pm a}{\sqrt{a^2 + b^2 + c^2}}$$

$$\cos \beta = \frac{\pm b}{\sqrt{a^2 + b^2 + c^2}}$$

$$\cos \gamma = \frac{\pm c}{\sqrt{a^2 + b^2 + c^2}}$$

where the signs with the numerators depend on which direction on the line is chosen as positive.

Example   If the positive direction on the line through $P(6, 2, 1)$ and $Q(3, 8, 3)$ is from P to Q, find the direction cosines.

Solution: Since the positive direction is from P to Q, we must use a coordinate of Q minus the corresponding one of P as the numerator of each direction cosine. We must use

$$PQ = \sqrt{(3 - 6)^2 + (8 - 2)^2 + (3 - 1)^2} = 7$$

for each denominator. Thus, we get

$$\cos \alpha = \frac{3 - 6}{7} = -\frac{3}{7}$$

$$\cos \beta = \frac{8 - 2}{7} = \frac{6}{7}$$

$$\cos \gamma = \frac{3 - 1}{7} = \frac{2}{7}$$

These values can be checked by seeing if the sum of their squares is one.

EXERCISE 22.1

In each of Problems 1 to 4 draw the radius vector OP, and find its length and direction cosines.

  1 $P(6, 3, 2)$ **2** $P(-4, 0, 3)$
  3 $P(6, 4, -3)$ **4** $P(2, -1, 5)$

Locate each pair of points in Problems 5 to 8; find the distance between each pair and the direction cosines of the line from the first point to the second in each case.

5   $(2, 7, 8)$, $(1, 3, 6)$ 6 $(5, 9, 11)$, $(2, 3, 8)$
7   $(-3, 2, -6)$, $(0, 4, -8)$ 8 $(7, 5, -3)$, $(1, 4, 2)$

9   Show that $(4, 6, 12)$, $(2, 7, 6)$, and $(-2, 5, 7)$ can be used as the vertices of a right triangle.
10   Show that $(2, -3, 5)$, $(4, 5, -1)$, and $(-6, 3, 3)$ can be used as the vertices of an isosceles triangle.
11   Show that the segment between $(2, -1, 4)$ and $(5, 2, 1)$ is parallel to that between $(-3, -7, 6)$ and $(0, -4, 3)$.
12   Show in two ways that $(2, -5, 0)$, $(5, -1, -2)$, and $(-4, -13, 4)$ lie on a line.

In each of Problems 13 to 16 the values of two direction cosines of a line are given; find all values of the other one.
13   $\cos \alpha = \frac{2}{3}$, $\cos \beta = \frac{1}{3}$
14   $\cos \beta = 1/\sqrt{2}$, $\cos \gamma = \frac{1}{2}$
15   $\cos \alpha = -1/\sqrt{2}$, $\cos \gamma = 1/\sqrt{2}$
16   $\cos \beta = -\frac{1}{3}$, $\cos \gamma = -\frac{2}{3}$

17   Find the point on the Z axis that is equidistant from $(-2, 6, 1)$ and $(5, 3, 4)$.
18   Find the Z coordinates of the points on the Z axis that are five units from $(3, 1, 4)$.
19   Find the X coordinates of the points on the X axis that are seven units from $(-2, 4, 5)$.
20   Find the equation of all points in the XY plane that are six units from $(2, -4, 3)$.
21   If $\alpha = \pi/4$ and $\beta = \pi/3$, find an acute angle γ.
22   If $\beta = \pi/4$ and $\gamma = 3\pi/4$, find α.
23   If $\alpha = \pi/3$ and $\beta = 2\pi/3$, find an obtuse angle γ.
24   If $\alpha = \pi/2$ and $\gamma = \pi/3$, find β.

In Problems 25 to 28 direction numbers of a line are given; find the specified direction cosine.
25   $a = 2$, $b = -6$, $c = 3$, find $\cos \alpha$ for α acute.
26   $a = -3$, $b = 2$, $c = 6$, find $\cos \beta$ for β obtuse.
27   $a = 2$, $b = 2$, $c = -1$, find $\cos \gamma$ for γ obtuse.
28   $a = 1$, $b = -2$, $c = 2$, find $\cos \beta$ for β acute.

22.5   Vectors

Much of the material in this section is a review of the material in Secs. 16.7 and 16.8. If a quantity is determined by a magnitude *vector* and a direction, it is called a *vector quantity*. Such a quantity can be *quantity* represented by a directed line segment called a *vector*. Thus AB

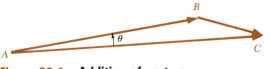

Figure 22.4   Addition of vectors

represents the vector from A to B; whereas BA represents a vector from B to A. The two vectors are equal in magnitude but oppositely directed; hence we write $AB = -BA$. We indicate geometrically that the vector BC is added to the vector AB by drawing BC from the end of AB as in Fig. 22.4. We then say that AC is the sum of AB and BC and write

$$AB + BC = AC$$

If the vector AC can be given the same direction as AB by rotating it through an angle θ, we say that the angle from AC to AB is θ. We add three vectors by adding one vector to the sum of the other two; the order of adding is immaterial since the associative law holds for vectors. We subtract a vector by adding its negative.

equal vectors   We say that two vectors are *equal* if they have the same magnitude and direction.

A vector is often represented by a single letter written in boldface as **A** or with an arrow over it. We shall use **A**. A number that *scalar* specifies magnitude only is called a *scalar* and is represented by a letter in italics as m. The product of the scalar m and the vector **A** is indicated by $m\mathbf{A}$ and is a vector with the same direction as **A** and a magnitude m times that of **A**.

unit vector   If a vector is one unit long, it is called a *unit vector*.

If a vector **V** is obtained by adding two or more other vectors, we *components* say that they are *components* of **V**. We shall be primarily interested in finding components that are parallel to the coordinate axes. For this purpose we lay off a unit vector along each coordinate axis as shown in Fig. 22.5 and then express the components in terms of these unit

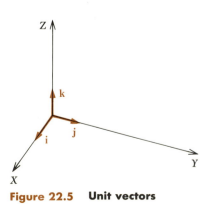

Figure 22.5   Unit vectors

vectors. The unit vectors along the X, Y, and Z axes are represented by $\mathbf{i}$, $\mathbf{j}$, and $\mathbf{k}$, respectively. Consequently, if $AB = \mathbf{V} = a\mathbf{i} + b\mathbf{j} + c\mathbf{k}$, it is a vector originating at A and with components a units in length parallel to the X axis, b units in length parallel to the Y axis, and c units in length parallel to the Z axis.

absolute value   The length of a vector $\mathbf{V} = a\mathbf{i} + b\mathbf{j} + c\mathbf{k}$ is called its *absolute value* and is represented by $|\mathbf{V}|$. Consequently, by use of the distance formula, we have

$$|\mathbf{V}| = \sqrt{a^2 + b^2 + c^2}$$

If we represent the vector from the origin to $P(x, y, z)$ by

$$\mathbf{R} = \mathbf{i}x + \mathbf{j}y + \mathbf{k}z$$

we say that $\mathbf{R}$ is a position vector. If $P(x, y, z)$ traces out or moves along a curve in three-dimensional space and if its coordinates are functions of t that have first and second derivatives with respect to t, then

velocity
$$\mathbf{v} = \frac{d\mathbf{R}}{dt} = \mathbf{i}\frac{dx}{dt} + \mathbf{j}\frac{dy}{dt} + \mathbf{k}\frac{dz}{dt}$$

acceleration   and
$$\mathbf{a} = \frac{d\mathbf{v}}{dt} = \frac{d^2\mathbf{R}}{dt^2} = \mathbf{i}\frac{d^2x}{dt^2} + \mathbf{j}\frac{d^2y}{dt^2} + \mathbf{k}\frac{d^2z}{dt^2}$$

represent, respectively, the velocity and acceleration of $\mathbf{R}$.

We should notice that the components of $\mathbf{R}$ are the coordinates of P and that the components of $\mathbf{v}$ and $\mathbf{a}$ are the first and second derivative, respectively, of the coordinates of P.

22.6   The dot product of two vectors

If $\mathbf{A}$ and $\mathbf{B}$ are any two vectors with common initial points, then their *scalar product*   dot, or *scalar, product* is written as $\mathbf{A} \cdot \mathbf{B}$ and is defined by

$$\mathbf{A} \cdot \mathbf{B} = |\mathbf{A}||\mathbf{B}| \cos \theta \tag{22.3}$$

where θ is the angle from $\mathbf{A}$ to $\mathbf{B}$ when their initial points coincide as shown in Fig. 22.6. Since the cosine of an angle and of its negative are equal, it follows that

$$\mathbf{B} \cdot \mathbf{A} = \mathbf{A} \cdot \mathbf{B}$$

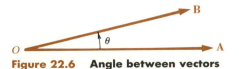

Figure 22.6   Angle between vectors

Consequently, we say that scalar, or dot, multiplication of two vectors is *commutative*.

The scalar product can be evaluated in terms of the components or coefficients of **i**, **j**, and **k** in the two vectors, and we shall now do that. Thus, if

$$\mathbf{A} = a_1\mathbf{i} + b_1\mathbf{j} + c_1\mathbf{k} \qquad \text{and} \qquad \mathbf{B} = a_2\mathbf{i} + b_2\mathbf{j} + c_2\mathbf{k}$$

then
$$\mathbf{A} + \mathbf{B} = (a_1 + a_2)\mathbf{i} + (b_1 + b_2)\mathbf{j} + (c_1 + c_2)\mathbf{k}$$

and as shown in Fig. 22.7,

$$\mathbf{C} = \mathbf{B} - \mathbf{A} = (a_2 - a_1)\mathbf{i} + (b_2 - b_1)\mathbf{j} + (c_2 - c_1)\mathbf{k}$$

We now apply the law of cosines to the triangle with sides **A**, **B**, and **C** and θ the angle from **A** to **B** and get

$$|\mathbf{C}|^2 = |\mathbf{A}|^2 + |\mathbf{B}|^2 - 2|\mathbf{A}||\mathbf{B}| \cos \theta$$

Therefore, $\qquad |\mathbf{A}||\mathbf{B}| \cos \theta = \frac{1}{2}(|\mathbf{A}|^2 + |\mathbf{B}|^2 - |\mathbf{C}|^2)$

The left member of this equation is $\mathbf{A} \cdot \mathbf{B}$ and the right member can be evaluated by use of the absolute value of a vector and the expressions for **A**, **B**, and **C** in terms of their components. If this is done, we find that

$$\mathbf{A} \cdot \mathbf{B} = a_1a_2 + b_1b_2 + c_1c_2 \qquad\qquad (22.4)$$

Example

If vectors are drawn from the origin to the position vectors $\mathbf{A}(-2, 5, 1)$ and $\mathbf{B}(4, 6, -3)$, find the angle between them.

Solution: If we equate the expressions for $\mathbf{A} \cdot \mathbf{B}$ as given by (22.3) and (22.4), we have

$$|\mathbf{A}||\mathbf{B}| \cos \theta = a_1a_2 + b_1b_2 + c_1c_2$$

Now, since the position vectors can be written as

$$\mathbf{A} = -2\mathbf{i} + 5\mathbf{j} + 1\mathbf{k} \qquad \text{and} \qquad \mathbf{B} = 4\mathbf{i} + 6\mathbf{j} - 3\mathbf{k}$$

we have

$$|\mathbf{A}||\mathbf{B}| \cos \theta = |-2\mathbf{i} + 5\mathbf{j} + 1\mathbf{k}||4\mathbf{i} + 6\mathbf{j} - 3\mathbf{k}| \cos \theta$$
$$= (-2)(4) + (5)(6) + (1)(-3) = 19$$

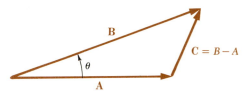

Figure 22.7   Scalar product

Hence, since $|\mathbf{A}| = \sqrt{4 + 25 + 1} = \sqrt{30}$ and $|\mathbf{B}| = \sqrt{16 + 36 + 9} = \sqrt{61}$ it follows that

$$\sqrt{30} \sqrt{61} \cos \theta = 19$$

$$\theta = \arccos \frac{19}{\sqrt{1830}} = \arccos 0.4442 = 63°40'$$

22.7   Perpendicular and parallel vectors

perpendicular vectors

If two vectors $\mathbf{A}$ and $\mathbf{B}$ are perpendicular, the angle between them is $90°$; hence $\cos \theta = \cos 90° = 0$ and $\mathbf{A} \cdot \mathbf{B} = 0$. Conversely, if the dot product of two vectors is zero, then either they are perpendicular or one of them is zero. If we adopt the convention that a zero vector is perpendicular to all vectors, we can say that *two vectors are perpendicular if and only if their dot product is zero.*

Example 1   Show that $\mathbf{A} = 2\mathbf{i} + 3\mathbf{j} + 5\mathbf{k}$ and $\mathbf{B} = 9\mathbf{i} - 1\mathbf{j} - 3\mathbf{k}$ are perpendicular.

Solution: By use of (22.4)

$$\mathbf{A} \cdot \mathbf{B} = (2)(9) + (3)(-1) + (5)(-3) = 0$$

Therefore the vectors are perpendicular.

parallel vectors

If two vectors $\mathbf{A}$ and $\mathbf{B}$ are parallel, the angle between them is zero or $180°$; hence $\cos \theta = \pm 1$ and $\mathbf{A} \cdot \mathbf{B} = \pm |\mathbf{A}| |\mathbf{B}|$. Conversely, if the dot product of two vectors is $\pm |\mathbf{A}| |\mathbf{B}|$, then they are parallel. We can now say that *two vectors $\mathbf{A}$ and $\mathbf{B}$ are parallel if and only if their dot product is $\pm |\mathbf{A}| |\mathbf{B}|$.*

Example 2   Show that $\mathbf{A} = 2\mathbf{i} - 2\mathbf{j} - 3\mathbf{k}$ and $\mathbf{B} = -4\mathbf{i} + 4\mathbf{j} + 6\mathbf{k}$ are parallel.

Solution: If we use the values of the coefficients of $\mathbf{A}$ and $\mathbf{B}$ in (22.4), we find that

$$\mathbf{A} \cdot \mathbf{B} = (2)(-4) + (-2)(4) + (-3)(6) = -34$$

and $|\mathbf{A}| |\mathbf{B}| = \sqrt{2^2 + (-2)^2 + (-3)^2} \sqrt{(-4)^2 + 4^2 + 6^2} = 34$

Consequently, the vectors are parallel since $\mathbf{A} \cdot \mathbf{B} = -|\mathbf{A}| |\mathbf{B}|$.

22.8   The cross product of two vectors

If $\mathbf{A}$ and $\mathbf{B}$ are any two vectors with common initial point and if $\mathbf{n}$ is a unit vector perpendicular to the plane determined by them and pointing in the direction that a right-hand screw would advance if its

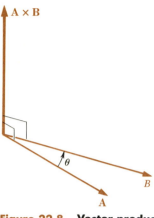

Figure 22.8   Vector product

head were turned from **A** to **B**, then we write

$$\mathbf{A} \times \mathbf{B} = \mathbf{n}|\mathbf{A}| \, |\mathbf{B}| \sin \theta \qquad (22.5)$$

*vector
product*   where θ is the angle from **A** to **B** and say that the *cross, or vector, product* of **A** and **B** is equal to $\mathbf{n}|\mathbf{A}| \, |\mathbf{B}| \sin \theta$ (see Fig. 22.8).

Since the sine of an angle and the sine of its negative are of equal magnitudes but have opposite signs, it follows that

$$\mathbf{B} \times \mathbf{A} = -\mathbf{A} \times \mathbf{B} \qquad (1)$$

Consequently, we see that cross, or vector, multiplication of two vectors is not commutative. Hence we must pay attention to the order of factors in multiplication of this type.

If we make use of Fig. 22.9, we see that $|\mathbf{B}| \sin \theta = h$ is the altitude from the terminal end **B** of the parallelogram that has **A** and **B** as a pair of consecutive sides. Hence, *the absolute value of*

*geometric
interpretation
of* **A** × **B**
$$\mathbf{A} \times \mathbf{B} = \mathbf{n}|\mathbf{A}| \, |\mathbf{B}| \sin \theta$$

is equal to the area of the parallelogram with **A** *and* **B** *a pair of consecutive sides.*

If we apply the definition of vector multiplication to **i** × **j**, we find that **i** × **j** is a unit vector perpendicular to both **i** and **j** times

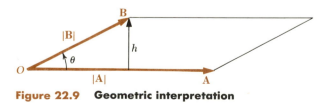

Figure 22.9   Geometric interpretation

$$|\mathbf{i}|\,|\mathbf{j}|\,\sin 90° = (1)(1)(1) = 1$$

Consequently, $\qquad\qquad\qquad \mathbf{i} \times \mathbf{j} = \mathbf{k}$ $\qquad\qquad\qquad$ (2a)

since $\mathbf{k}$ is a unit vector perpendicular to both $\mathbf{i}$ and $\mathbf{j}$. Hence, by (1) we have

$$\mathbf{j} \times \mathbf{i} = -\mathbf{k} \qquad\qquad (2b)$$

It can be shown similarly that

products of unit vectors
$$\mathbf{j} \times \mathbf{k} = \mathbf{i} \qquad \mathbf{k} \times \mathbf{j} = -\mathbf{i} \qquad\qquad (3)$$
$$\mathbf{k} \times \mathbf{i} = \mathbf{j} \qquad \mathbf{i} \times \mathbf{k} = -\mathbf{j} \qquad\qquad (4)$$
$$\mathbf{i} \times \mathbf{i} = \mathbf{j} \times \mathbf{j} = \mathbf{k} \times \mathbf{k} = 0 \qquad\qquad (5)$$

Furthermore, it follows immediately from the definition that

$$(\alpha\mathbf{A}) \times (\beta\mathbf{B}) = \alpha\beta(\mathbf{A} \times \mathbf{B}) \qquad\qquad (6)$$

where α and β are scalars.

It is possible to express the cross product of two vectors in terms of *distributive* their components, but to do so requires the use of the distributive law *law*

$$\mathbf{A} \times (\mathbf{B} + \mathbf{C}) = \mathbf{A} \times \mathbf{B} + \mathbf{A} \times \mathbf{C} \qquad\qquad (7)$$

which we have not developed. We shall leave this development as a problem in the next exercise and point out that

$$(\mathbf{B} + \mathbf{C}) \times \mathbf{A} = \mathbf{B} \times \mathbf{A} + \mathbf{C} \times \mathbf{A} \qquad\qquad (8)$$

is a consequence of (7) and (1) since the order of factors in each member of (8) is opposite to the order in the corresponding member of (7).

We shall now consider the two vectors

$$\mathbf{A} = a_1\mathbf{i} + b_1\mathbf{j} + c_1\mathbf{k}$$

and $\qquad\qquad\qquad \mathbf{B} = a_2\mathbf{i} + b_2\mathbf{j} + c_2\mathbf{k}$

and evaluate $\mathbf{A} \times \mathbf{B}$ in terms of $\mathbf{i}$, $\mathbf{j}$, and $\mathbf{k}$ and the components of $\mathbf{A}$ and $\mathbf{B}$. Thus,

$$\mathbf{A} \times \mathbf{B} = (a_1\mathbf{i} + b_1\mathbf{j} + c_1\mathbf{k}) \times (a_2\mathbf{i} + b_2\mathbf{j} + c_2\mathbf{k})$$

$$= a_1a_2\mathbf{i} \times \mathbf{i} + a_1b_2\mathbf{i} \times \mathbf{j} + a_1c_2\mathbf{i} \times \mathbf{k} + b_1a_2\mathbf{j} \times \mathbf{i} + b_1b_2\mathbf{j} \times \mathbf{j}$$
$$+ b_1c_2\mathbf{j} \times \mathbf{k} + c_1a_2\mathbf{k} \times \mathbf{i} + c_1b_2\mathbf{k} \times \mathbf{j} + c_1c_2\mathbf{k} \times \mathbf{k}$$
$$\textit{by (6), (7), and (8)}$$

A × B in $\qquad = a_1b_2\mathbf{k} - a_1c_2\mathbf{j} - b_1a_2\mathbf{k} + b_1c_2\mathbf{i} + c_1a_2\mathbf{j} - c_1b_2\mathbf{i}$
terms of $\qquad\qquad\qquad\qquad\quad \textit{by (2a), (2b), (3), (4), and (5)}$
components $\quad \mathbf{A} \times \mathbf{B} = (b_1c_2 - b_2c_1)\mathbf{i} - (a_1c_2 - a_2c_1)\mathbf{j} + (a_1b_2 - a_2b_1)\mathbf{k} \qquad (22.5a)$
$$\textit{by the commutative and distributive axioms}$$

Hence, making use of the expansion of a third-order determinant, we see that

$$\mathbf{A} \times \mathbf{B} = \begin{vmatrix} \mathbf{i} & \mathbf{j} & \mathbf{k} \\ a_1 & b_1 & c_1 \\ a_2 & b_2 & c_2 \end{vmatrix} \tag{22.5b}$$

Example

Find a unit vector that is perpendicular to both $\mathbf{A} = 3\mathbf{i} + \mathbf{j} + 2\mathbf{k}$ and $\mathbf{B} = 2\mathbf{i} - 3\mathbf{j} + \mathbf{k}$.

Solution: The cross product of two vectors is a vector perpendicular to both of them. Hence, we shall use (22.5b) with $a_1 = 3$, $b_1 = 1$, $c_1 = 2$ and $a_2 = 2$, $b_2 = -3$, $c_2 = 1$ and get

$$\mathbf{A} \times \mathbf{B} = \begin{vmatrix} \mathbf{i} & \mathbf{j} & \mathbf{k} \\ 3 & 1 & 2 \\ 2 & -3 & 1 \end{vmatrix}$$

$$= 7\mathbf{i} + \mathbf{j} - 11\mathbf{k}$$

Furthermore, $\mathbf{A} \times \mathbf{B}$ is perpendicular to both $\mathbf{A}$ and $\mathbf{B}$, but its absolute value is $\sqrt{7^2 + 1^2 + (-11)^2} = \sqrt{171}$. Therefore, dividing by $\sqrt{171}$, we find that $(7\mathbf{i} + \mathbf{j} - 11\mathbf{k})/\sqrt{171}$ is a unit vector that is perpendicular to both $\mathbf{A}$ and $\mathbf{B}$.

EXERCISE 22.2

Find the dot product of the two vectors given in each of Problems 1 to 4.
1   $\mathbf{A} = 2\mathbf{i} - 3\mathbf{j} + 6\mathbf{k}$, $\mathbf{B} = 3\mathbf{i} + 4\mathbf{j} - \mathbf{k}$
2   $\mathbf{A} = 7\mathbf{i} + 2\mathbf{j} - 3\mathbf{k}$, $\mathbf{B} = \mathbf{i} - 6\mathbf{j} + \mathbf{k}$
3   $\mathbf{A} = -3\mathbf{i} + 6\mathbf{j} + 2\mathbf{k}$, $\mathbf{B} = 4\mathbf{i} + \mathbf{j} + 3\mathbf{k}$
4   $\mathbf{A} = 2\mathbf{i} + 3\mathbf{j}$, $\mathbf{B} = 5\mathbf{i} - 3\mathbf{j} + 8\mathbf{k}$

Find the cross product of the two vectors given in each of Problems 5 to 8.
5   $\mathbf{A} = 3\mathbf{i} + 2\mathbf{j} + 6\mathbf{k}$, $\mathbf{B} = 2\mathbf{i} + 6\mathbf{j} - 3\mathbf{k}$
6   $\mathbf{A} = 2\mathbf{i} - 3\mathbf{j} + 7\mathbf{k}$, $\mathbf{B} = \mathbf{i} + 3\mathbf{j} + 4\mathbf{k}$
7   $\mathbf{A} = 4\mathbf{i} + 5\mathbf{k}$, $\mathbf{B} = 7\mathbf{j} - \mathbf{k}$
8   $\mathbf{A} = 5\mathbf{i} - 4\mathbf{j} - 2\mathbf{k}$, $\mathbf{B} = -4\mathbf{i} + 3\mathbf{j} + 4\mathbf{k}$

In each of Problems 9 to 12 find the area of the triangle with vertices at the given points.
9   $A(2, 1, 3)$, $B(3, 2, 0)$, $C(-1, 4, 6)$
10   $A(5, 0, -2)$, $B(-1, 2, 3)$, $C(3, 4, -5)$
11   $A(0, 0, -7)$, $B(3, -2, 2)$, $C(-4, 0, 5)$

12   $A(6, -1, 3)$, $B(-2, 4, -1)$, the origin
13   Find a unit vector that is perpendicular to both $\mathbf{A} = 3\mathbf{i} - 2\mathbf{j} + \mathbf{k}$ and $\mathbf{B} = 2\mathbf{i} + 6\mathbf{j} + \mathbf{k}$.
14   Find a unit vector that is perpendicular to both $\mathbf{A} = 6\mathbf{i} + 2\mathbf{j} - 3\mathbf{k}$ and $\mathbf{B} = -\mathbf{i} + 3\mathbf{j} + 2\mathbf{k}$.
15   Find a unit vector that is perpendicular to the plane determined by $A(2, 1, -5)$, $B(-3, 2, 4)$, and $C(0, 6, 3)$.
16   Find a unit vector that is perpendicular to the plane determined by $A(1, -2, 6)$, $B(2, 3, 4)$, and $C(-5, 0, 2)$.

Find the angle between the given vectors in each of Problems 17 to 20.
17   $\mathbf{A} = 2\mathbf{i} + 3\mathbf{j} - 6\mathbf{k}$, $\mathbf{B} = 3\mathbf{i} + 2\mathbf{j} + 2\mathbf{k}$
18   $\mathbf{A} = \mathbf{i} - 2\mathbf{j} + 3\mathbf{k}$, $\mathbf{B} = 2\mathbf{i} + 3\mathbf{j} - \mathbf{k}$
19   $\mathbf{A} = -3\mathbf{i} + 5\mathbf{j} - \mathbf{k}$, $\mathbf{B} = -6\mathbf{i} - 10\mathbf{j} + 2\mathbf{k}$
20   $\mathbf{A} = 5\mathbf{i} + 2\mathbf{j} + 3\mathbf{k}$, $\mathbf{B} = \mathbf{i} - 4\mathbf{k}$

Find the angle A of a triangle with the vertices given in each of Problems 21 to 24.
21   $A(2, 1, 5)$, $B(3, 2, 7)$, $C(1, 5, -1)$
22   $A(5, -2, 3)$, $B(6, -2, 4)$, $C(-4, 0, 1)$
23   $A(0, 1, 5)$, $B(-3, 0, 2)$, $C(2, -1, 0)$
24   $A(7, -6, -1)$, $B(5, -8, 1)$, $C(10, -4, 6)$

In each of Problems 25 to 28, $\mathbf{R} = i\mathbf{x} + j\mathbf{y} + k\mathbf{z}$ where x, y, and z are functions of t as given. Find the velocity and acceleration in each case for the specified value of t.
25   $x = t^2$, $y = t + 1$, $z = 3t - 1$, $t = 2$
26   $x = 2t - 1$, $y = \cos t$, $z = e^t$, $t = 0$
27   $x = e^t$, $y = \ln t$, $z = t^2 + t$, $t = 1$
28   $x = t^2 - 1$, $y = \sec t$, $z = \tan t$, $t = 0$
29   Prove that $\mathbf{A} \cdot (\mathbf{B} + \mathbf{C}) = \mathbf{A} \cdot \mathbf{B} + \mathbf{A} \cdot \mathbf{C}$.
30   Prove that $(\mathbf{A} + \mathbf{B}) \cdot (\mathbf{C} + \mathbf{D}) = \mathbf{A} \cdot \mathbf{C} + \mathbf{A} \cdot \mathbf{D} + \mathbf{B} \cdot \mathbf{C} + \mathbf{B} \cdot \mathbf{D}$.
31   Prove that $\mathbf{A} \times (\mathbf{B} + \mathbf{C}) = \mathbf{A} \times \mathbf{B} + \mathbf{A} \times \mathbf{C}$.
32   Prove that $(\mathbf{A} + \mathbf{B}) \cdot \mathbf{C} = \mathbf{A} \cdot \mathbf{C} + \mathbf{B} \cdot \mathbf{C}$.

22.9   Equations of a line

If $P_1(x_1, y_1, z_1)$ is a specified point and if L is a line in three-dimensional space that passes through P_1 and is parallel to the vector

$$\mathbf{v} = A\mathbf{i} + B\mathbf{j} + C\mathbf{k}$$

then L is the set of points $P(x, y, z)$ such that the vector P_1P is parallel to the vector $\mathbf{v}$ as in Fig. 22.10. Consequently, P is on L if and only if there is a scalar t such that

$$P_1P = t\mathbf{v}$$

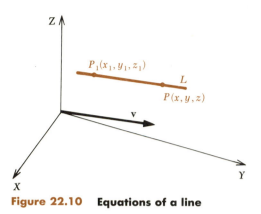

Figure 22.10   Equations of a line

If we equate coefficients of **i**, of **j**, and of **k** from the two members of this equation, we see that

$$x - x_1 = tA \qquad y - y_1 = tB \qquad z - z_1 = tC \qquad (22.6)$$

parametric equations   are *parametric equations* for the line through P_1 parallel to $A\mathbf{i} + B\mathbf{j} + C\mathbf{k}$. The point $P(x, y, z)$ traces out the line if we let t vary over the interval $(-\infty, \infty)$.

If we divide each equation in (22.6) by the coefficient of t in it, we obtain three expressions for t. If we equate these expressions, we find that

$$\frac{x - x_1}{A} = \frac{y - y_1}{B} = \frac{z - z_1}{C} \qquad (22.7)$$

symmetrical equations   are *the cartesian equations of the line.* They are called *symmetrical equations* of the line. If a denominator in (22.7) is zero, then it is clear from (22.6) that the corresponding numerator must also be zero. If it is A that is zero, then $x - x_1$ must be zero, and we use

$$x - x_1 = 0 \qquad \text{along with} \qquad \frac{y - y_1}{B} = \frac{z - z_1}{C}$$

as the cartesian equations of the line. If two denominators are zero, say A and B, then

$$x - x_1 = 0 \qquad \text{and} \qquad y - y_1 = 0 \qquad (22.8)$$

and these are the equations of the line.

22.10   Forms of the equation of a plane

We shall now find the equation of a plane by considering a vector

$$\mathbf{N} = A\mathbf{i} + B\mathbf{j} + C\mathbf{k}$$

that is perpendicular to the plane and a specified point $P_1(x_1, y_1, z_1)$

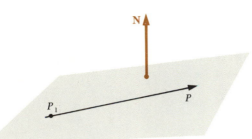

Figure 22.11   Equation of a plane

that is on the plane, as shown in Fig. 22.11. Furthermore, we shall represent any point on the plane by $P(x, y, z)$. Now, since N is perpendicular to the plane and P_1P is in the plane, we have

$$N \cdot P_1P = 0$$

Consequently, by use of (22.4) and Sec. 22.7 we see that

equation of a plane

$$A(x - x_1) + B(y - y_1) + C(z - z_1) = 0 \qquad (22.9)$$

is an *equation of the plane* through $P_1(x_1, y_1, z_1)$ and perpendicular to the vector $N = A\mathbf{i} + B\mathbf{j} + C\mathbf{k}$.

If we remove parentheses and replace $Ax_1 + By_1 + Cz_1$ by D, this equation becomes

$$Ax + By + Cz = D \qquad (22.9a)$$

We now know that the coefficients of $\mathbf{i}$, $\mathbf{j}$, and $\mathbf{k}$ in a vector perpendicular to a plane are the coefficients in the equation of a plane.

Example 1   Find the equation of the plane through $P_1(2, 1, -5)$ if $N = 3\mathbf{i} - 4\mathbf{j} + 7\mathbf{k}$ is perpendicular to it.

Solution: If we substitute the given numbers in (22.9), we find that

$$3(x - 2) - 4(y - 1) + 7(z + 5) = 0$$

is one form for the equation. Another form can be obtained by removing parentheses and collecting like terms, that is,

$$3x - 4y + 7z + 33 = 0$$

or
$$3x - 4y + 7z = -33$$

Example 2   Find a vector that is parallel to the line of intersection of the planes $2x + 6y - 3z = 14$ and $x - 3y + 5z = 9$.

Solution: The vectors $N_1 = 2\mathbf{i} + 6\mathbf{j} - 3\mathbf{k}$ and $N_2 = 1\mathbf{i} - 3\mathbf{j} + 5\mathbf{k}$ are perpendicular to the first and the second of the given planes, respectively, as seen by the statement immediately preceding Example 1. Hence they are perpendicular to the line of intersection of the two planes. Consequently, any perpendicular to the two vectors is parallel to the line of intersection of the planes. Therefore, the cross product of $\mathbf{N_1}$ and $\mathbf{N_2}$ is the desired vector. Thus, by (22.5a)

$$\mathbf{V} = \mathbf{N_1} \times \mathbf{N_2} = \begin{vmatrix} \mathbf{i} & \mathbf{j} & \mathbf{k} \\ 2 & 6 & -3 \\ 1 & -3 & 5 \end{vmatrix} = 21\mathbf{i} - 13\mathbf{j} - 12\mathbf{k}$$

is a vector that is parallel to the line of intersection of the given planes.

EXERCISE 22.3

In each of Problems 1 to 4 find the angle between the given pair of planes.

1 $2x - 3y + 6z = 21,\ 3x + 6y + 2z = 35$
2 $2x + y - 2z = 9,\ x - 2y + 2z = 27$
3 $x + 2y + z = 3,\ 4x + 3y - z = 7$
4 $x - 2y - 2z = -9,\ 6x - 3y - 2z = -6$

In each of Problems 5 to 8 find the coordinates of the intersection of the line and the plane.

5 $\dfrac{x-1}{2} = \dfrac{y-2}{2} = \dfrac{z-3}{1},\ 3x - y + 2z = 7$

6 $\dfrac{x-2}{3} = \dfrac{y-3}{2} = \dfrac{z+1}{2},\ 2x - 3y - 3z = 4$

7 $\dfrac{x+2}{4} = \dfrac{y-1}{1.5} = \dfrac{z+3}{2},\ x - 2y + 4z = 2$

8 $\dfrac{x+5}{-2} = \dfrac{y+3}{3} = \dfrac{z-1}{2},\ 4x + 2y - z = -15$

Find the equation of the plane through the points given in each of Problems 9 to 12.

9 $(1, 5, 2),\ (2, -1, 3),\ (4, 0, 6)$
10 $(7, 6, 2),\ (3, 2, -2),\ (-1, 1, 3)$
11 $(-4, 1, 3),\ (6, -2, 5),\ (3, -1, 5)$
12 $(7, -4, -2),\ (-2, 3, -1),\ (1, 0, 0)$

Find the equations of the line that connects the two points in each of Problems 13 to 16.

13 $(2, 4, 5),\ (3, -1, 0)$ 14 $(7, 6, -2),\ (5, 3, 1)$
15 $(3, 0, -5),\ (-1, 2, 3)$ 16 $(0, -3, 4),\ (2, 1, 8)$

Find the equation of each of the following planes.

17 Through $(2, 3, 5)$, parallel to $6x + 2y - 3z - 5 = 0$

18   Through $(4, 0, -1)$, parallel to $2x - 5y + \sqrt{7}z + 5 = 0$
19   Through $(2, 1, -1)$ and $(3, 0, 1)$, perpendicular to $3x - 4y + z = 5$
20   Through $(4, 2, 0)$ and $(1, 2, 3)$, perpendicular to $x + y - z = 3$
21   Through $P_1(4, 5, -3)$ and perpendicular to the vector OP_1
22   Through $P_1(3, -6, 2)$ and perpendicular to the vector OP_1
23   Through $(4, 2, -3)$ and perpendicular to the vector through $(4, 1, 6)$ and $(2, 3, 5)$
24   Through $(-1, 2, -3)$ and perpendicular to the vector through $(0, 6, 1)$ and $(3, 0, 2)$
25   Show that the equation $ax + by + cz + d = 0$ can be put in the form $x \cos \alpha + y \cos \beta + z \cos \gamma - p = 0$ where α, β, and γ are direction angles of the normal to the plane and p is the distance from the origin to the plane. This is called the *normal form* of the equation of the plane.

Put the following equations in normal form by use of Problem 25.
26   $3x - 2y + 6z - 21 = 0$
27   $2x + 6y + 3z - 35 = 0$
28   $x + 2y - 2z - 18 = 0$

29   Show that the distance between $P_1(x_1, y_1, z_1)$ and the plane $ax + by + cz + d = 0$ is

$$\frac{|ax_1 + by_1 + cz_1 + d|}{\sqrt{a^2 + b^2 + c^2}}$$

Use the formula given in Problem 29 to find the distance between the point and the plane in each of Problems 30 to 32.
30   $3x - 2y + 6z + 13 = 0$, $P_1(2, -3, 4)$
31   $2x + y - 2z - 10 = 0$, $P_1(4, -5, 7)$
32   $5x + 4y + 2\sqrt{2}z - 35 = 0$, $P_1(-3, 1, \sqrt{2})$

23

Surfaces and curves

In the first two sections of this chapter we shall discuss some of the concepts that assist in visualizing and sketching surfaces and curves in three-dimensional space. We shall next consider the equations of some of the more usual surfaces and then see how to eliminate the product terms. Finally, we shall discuss curves in space and the use of parametric equations.

23.1   Sections, traces, intercepts, and symmetry

If we have an equation in x, y, and z, we may assign real values to any two of them and thereby determine at least one value for the third. Hence, at least one ordered triple of real numbers which satisfies the given equation is determined provided that there is a real value of the third variable corresponding to the assigned values of the others. This ordered triple of real numbers determines a point, and the set of all such ordered triples which satisfy the given equation is called *surface*   a *surface*.

The remainder of this section will be devoted to concepts that simplify sketching surfaces and curves. The curve that is determined *section*   by the intersection of a surface and a plane is called a *section* of the surface and is represented by the simultaneous use of the equation of the surface and the equation of the plane. The section made by a coordinate plane is called the *trace* on that plane, and its equation can be obtained by setting one variable in the equation of the surface equal to zero. Thus, the trace on the YZ plane is obtained by replacing x by zero. The nonzero coordinate of a point of intersection of a surface and a coordinate axis is an *intercept;* furthermore, zero is an intercept if $(0, 0, 0)$ is on the surface.

trace

intercept

Example 1   Consider the surface $F(x, y, z) = x^2 + 2y^2 - 4y + 3z - 7 = 0$. Find the section made by $x = 2$, the trace on the XY plane, and the z intercept.

Solution: The section made by $x = 2$ is the parabola

$$F(2, y, z) = 2y^2 - 4y + 3z - 3 = 0$$

in the plane $x = 2$ and is often represented by simultaneous use of the plane and the surface. The trace in the XZ plane is obtained by putting $y = 0$ in the equation of the surface, that is,

$$F(x, 0, z) = x^2 + 3z - 7 = 0$$

hence, it is a parabola in the XZ plane. The z intercept is obtained by solving $F(0, 0, z) = 3z - 7 = 0$ for z; hence it is ⅔.

We can obtain additional worthwhile information about a surface by examining it for symmetry. We studied symmetry in a plane relative to a line and relative to a point in Sec. 2.15. The definitions of symmetry with respect to a point and relative to a line as given there apply here. We extend the concept of symmetry by saying that two points are *symmetric* with respect to a plane if the line segment joining them is bisected by the plane and is perpendicular to the plane. The three tests that are most widely used for symmetry in three-dimensional space are given below but will not be proved. The proofs are similar to those in Sec. 2.15. The statements used in testing are:

symmetry

1. A surface $s(x, y, z) = 0$ is symmetric with respect to the origin if and only if $s(-x, -y, -z) = \pm s(x, y, z)$.
2. A surface $s(x, y, z) = 0$ is symmetric with respect to the X axis if and only if $s(x, -y, -z) = \pm s(x, y, z)$, to the Y axis if and only if $s(-x, y, -z) = \pm s(x, y, z)$, and to the Z axis if and only if $s(-x, -y, z) = \pm s(x, y, z)$.
3. A surface $s(x, y, z) = 0$ is symmetric with respect to the XY plane if and only if $s(x, y, -z) = \pm s(x, y, z)$, to the XZ plane if and only if $s(x, -y, z) = \pm s(x, y, z)$, and to the YZ plane if and only if $s(-x, y, z) = \pm s(x, y, z)$.

Example 2   Test the surface $s(x, y, z) = x^2 - y^2 + z^2 + 2x - 5 = 0$ for symmetry with respect the origin, the X axis, and XZ plane.

Solution: If we use the tests given above, we find that

$$s(-x, -y, -z) = x^2 - y^2 + z^2 - 2x - 5 \neq \pm s(x, y, z)$$
$$s(x, -y, -z) = x^2 - y^2 + z^2 + 2x - 5 = s(x, y, z)$$
$$s(x, -y, z) = x^2 - y^2 + z^2 + 2x - 5 = s(x, y, z)$$

Consequently, the surface is not symmetric with respect to the origin but is symmetric with respect to the X axis and the XZ plane.

23.2   Parallel sections in sketching

We used plane sections in Example 1 of Sec. 23.1 and shall now consider them further. If a sequence of sections by planes parallel to one of the coordinate planes does not give a clear enough picture, then we should use sections parallel to one or more other coordinate planes in order to sketch the surface. Before making the sketch of a surface, we should:

1.   Determine the type of section parallel to each coordinate plane.
2.   Determine the traces on the coordinate planes.
3.   Find the intercepts on the coordinate axes.
4.   Test for symmetry.

Example   Discuss and sketch

$$x^2 + 2yz - 6 = 0$$

Solution: The sections parallel to the XY plane are made by the plane $z = k$ and are the parabolas $x^2 + 2ky - 6 = 0$. Sections made by $y = k$ are the parabolas $x^2 + 2kz - 6 = 0$ and those made by $x = k$ are the hyperbolas $k^2 + 2yz - 6 = 0$. The traces in the XY and XZ planes are the lines $x = \pm \sqrt{6}$ and that in the YZ plane is the hyperbola $yz = 3$. The x intercepts are $x = \pm \sqrt{6}$ and there are no y or z intercepts.

The sketch in Fig. 23.1 makes use of the sections cut by the plane $y = k$. The first step in drawing it is to put in the traces. We lay off a distance OA on the Y axis because that axis is perpendicular to sections we are going to use, and we divide the interval OA into subintervals. At each point of subdivision, Q for example, we draw the traces QP and QR of the plane $y = k$; they intersect the traces of the surface at P and R. Then, the parabolic arc PR is the desired section.

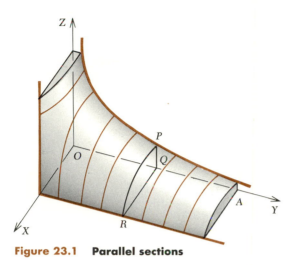

Figure 23.1 Parallel sections

EXERCISE 23.1

Find the sections parallel to the coordinate planes of the surfaces in Problems 1 to 20. Also find the traces and intercepts. Finally test for symmetry and sketch the surface by use of parallel sections.

1	$x^2 + y^2 + z^2 = 16$	2	$x^2 + y^2 + z^2 = 9$
3	$x^2 + y^2 + 2z^2 = 8$	4	$x^2 + 2y^2 + 4z^2 = 8$
5	$x^2 - y^2 + z^2 = 9$	6	$x^2 + y^2 - z^2 = 4$
7	$x^2 - y^2 - 2z^2 = 2$	8	$x^2 - 3y^2 - z^2 = 9$
9	$x^2 + xy = z$	10	$y^2 - yz = x$
11	$z = xy$	12	$z = xy + y$
13	$x^2 + z^2 = 4y$	14	$x^2 - z^2 = y$
15	$z^2 = xy$	16	$z^2 = xy - y$
17	$x^2 + 4y^2 = 4$	18	$x^2 - 2y = 1$
19	$y = 2xz - z$	20	$x = 2yz + z$

23.3 Cylindrical surfaces

We shall begin our discussion of a rather common class of surfaces by saying that *the surface generated by a moving line that remains parallel to its original position and intersects a given plane curve at all times is called a cylindrical surface.*

cylindrical surface

generator, generating curve

The line, or any position of it, is called a *generator* of the surface and the given curve is known as the *generating curve*. Any curve obtained by cutting the surface with a plane perpendicular to a generator is called a *right section*. The line through the center, if any, of a right section and parallel to the generators is called the *axis* of the surface.

right section

axis

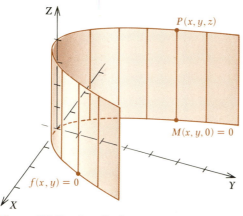

Figure 23.2   A cylinder

We shall restrict our study of cylindrical surfaces to those whose generators are parallel to one of the coordinate axes and whose generating curve is in the plane that is perpendicular to the generators. If the generating curve is in the XY plane, it can be represented by $f(x, y) = 0$ and the generators are perpendicular to that plane.

We shall begin our study of the surface by erecting a generator at any point $M(x, y, 0)$ in Fig. 23.2 on the generating curve and shall then consider any point $P(x, y, z)$ on the generator. The coordinates of M satisfy the equation $f(x, y) = 0$ of the generating curve. Hence, *equation of a* $f(x, y) = 0$ is satisfied by the coordinates of P since z does not enter in *cylinder* the equation and the x and y of P are the same as those of M. Consequently, we know that *any equation in two variables, if considered in three-dimensional space, represents a cylindrical surface that is perpendicular to the plane of the two variables and whose generating curve is the plane curve whose equation is given.* If the generating curve of the cylindrical surface is a circle, a parabola, *types of* an ellipse, or a hyperbola, the surface is called *circular, parabolic, cylinders elliptic,* or *hyperbolic,* respectively.

23.4   The sphere

Any surface whose equation is a quadratic in three variables is called a *quadric* *quadric surface* or *conicoid.* The first name is generally used and was *surface* introduced because the equation is a quadratic. One of the more common quadric surfaces is a sphere; it is characterized by saying that a *sphere* *sphere* consists of the set of those and only those points in three-dimensional space which are at a given distance from a given point. *center* The given point is called the *center* of the sphere and the given *radius* distance is the *radius.*

We shall now derive an equation of the sphere with center at (h, k, l)

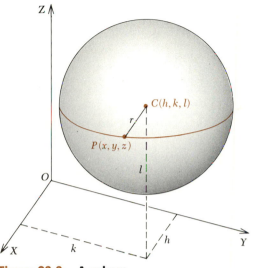

Figure 23.3   A sphere

and radius r (see Fig. 23.3). If $P(x, y, z)$ is any point on the sphere, then $CP = r$ or $CP - r = 0$ from the definition of the sphere; furthermore, $CP + r \neq 0$ since CP and r are both nonnegative. Therefore $(CP - r)(CP + r) = \overline{CP}^2 - r^2 = 0$ or $\overline{CP}^2 = r^2$ is the equation of the required locus. Consequently, by use of the distance formula, we see that

$$(x - h)^2 + (y - k)^2 + (z - l)^2 = r^2 \tag{23.1}$$

is the equation of the sphere with center at (h, k, l) and radius r.

standard form   Equation (23.1) is called the *standard form* of the equation of a sphere. It reduces to

$$x^2 + y^2 + z^2 = r^2 \tag{23.1a}$$

if the center is at the origin. Furthermore, if (23.1) is expanded, it takes the form

$$x^2 + y^2 + z^2 + Hx + Ky + Lz + R = 0 \tag{23.1b}$$

general   This is called the *general equation* of a sphere. Since it contains four
equation   essential constants, a sphere is determined by four consistent and independent conditions.

Example 1   Find the equation of the sphere that has center at $C(2, 1, -3)$ and passes through $P_1(-4, -2, -5)$.

Solution: Since the coordinates of the center are known, we need only find the radius in order to write the equation of the sphere. If we use the distance formula to evaluate CP_1, we find that the radius is

$$r = \sqrt{(-4-2)^2 + (-2-1)^2 + [-5-(-3)]^2}$$
$$= \sqrt{36+9+4} = 7$$

Therefore, by use of (23.1), the equation of the sphere is

$$(x-2)^2 + (y-1)^2 + (z+3)^2 = 7^2$$

The equation of a sphere can be put in standard form by completing the squares of the three quadratic expressions in the equation and collecting constants.

Example 2   Put $x^2 + y^2 + z^2 + 2x - 8y + 6z = 10$ in standard form and find the center and radius.

Solution: We complete the square of $x^2 + 2x$ by adding 1 to the left member of the equation, and we offset this by adding the same amount to the right member. If this is done and the other squares are similarly completed, we have

$$x^2 + 2x + 1 + y^2 - 8y + 16 + z^2 + 6z + 9 = 10 + 1 + 16 + 9$$
$$(x+1)^2 + (y-4)^2 + (z+3)^2 = 6^2$$

Consequently, the center is at $(-1, 4, -3)$ and the radius is 6.

EXERCISE 23.2

Sketch and classify the cylinder in each of Problems 1 to 12.

1	$x^2 + y^2 = 9$	2	$y^2 + z^2 = 4$
3	$x^2 + 2x + z^2 - 4z = 4$	4	$x^2 + 6x + y^2 - 8y = 11$
5	$y^2 = 4x$	6	$z^2 = 8y$
7	$y^2 - 2y + 4z + 9 = 0$	8	$x^2 + 4x = 4y$
9	$9x^2 + 4y^2 = 36$	10	$9y^2 + 16z^2 = 144$
11	$x^2 - 4z^2 = 4$	12	$16y^2 - 25x^2 = 400$

Find the equation of the sphere determined by the data in each of Problems 13 to 20.

13   Center at $(2, 1, 3)$, radius 4
14   Center at $(3, 5, -2)$, radius 7
15   Center at $(-4, 0, 6)$, through $(2, 2, 3)$
16   Center at $(5, 1, -4)$, through $(3, -5, -1)$
17   Through $(0, 6, 0)$, $(0, -6, 4)$, $(-7, -1, 0)$, and $(9, -1, 0)$
18   Through $(4, 0, 7)$, $(0, -3, 2)$, $(-5, 5, -1)$, $(-8, 5, 2)$
19   With $(-4, 5, 1)$ and $(2, 3, -7)$ as ends of a diameter
20   With $(7, 6, -2)$ and $(-1, 2, 8)$ as ends of a diameter

Change the equation in each of Problems 21 to 24 to standard form.
21   $x^2 + y^2 + z^2 - 2x - 4y - 6z = 2$
22   $x^2 + y^2 + z^2 - 4x + 6y + 8z = -4$
23   $x^2 + y^2 + z^2 + 6x - 2y - 8z = 10$
24   $x^2 + y^2 + z^2 - 4x + 12y + 6z = 0$

23.5   The ellipsoid

The surface determined by

$$\frac{x^2}{a^2} + \frac{y^2}{b^2} + \frac{z^2}{c^2} = 1$$

an ellipsoid   is called an *ellipsoid* and is shown in Fig. 23.4. It reduces to an ellipsoid of revolution if two of the denominators are equal and to a sphere if all three are equal. If real, the sections parallel to the coordinate planes are ellipses and the traces are ellipses. The x, y, and z intercepts are $\pm a$, $\pm b$, and $\pm c$, respectively.

23.6   The elliptic hyperboloid of one sheet

The surface represented by

$$\frac{x^2}{a^2} + \frac{y^2}{b^2} - \frac{z^2}{c^2} = 1$$

elliptic hyperboloid of one sheet   is called an *elliptic hyperboloid of one sheet* and is shown in Fig. 23.5. If $b = a$, the surface is a hyperboloid of revolution of one sheet. By putting $z = 0$, we see that the trace in the XY plane is an ellipse. Similarly the traces in the other coordinate planes are hyperbolas.

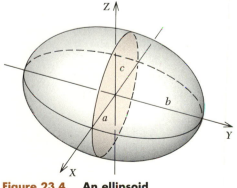

Figure 23.4   An ellipsoid

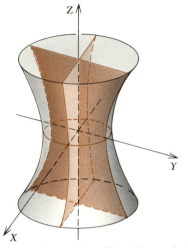

Figure 23.5   An elliptic hyperboloid of one sheet

Furthermore, the sections parallel to the XY plane are ellipses and those parallel to the other coordinate planes are hyperbolas including the degenerate cases for $x = \pm a$ and $y = \pm b$. The x and y intercepts are $\pm a$ and $\pm b$, respectively. There are no z intercepts. The surface is symmetric with respect to each coordinate plane.

23.7   The elliptic hyperboloid of two sheets

The surface represented by

$$\frac{y^2}{a^2} - \frac{x^2}{b^2} - \frac{z^2}{c^2} = 1$$

elliptic hyperboloid of two sheets

is an *elliptic hyperboloid of two sheets* and is shown in Fig. 23.6. If $b = c$, it becomes a hyperboloid of revolution of two sheets. The surface consists of two distinct parts; one of them is in the region $y \geq a$ and the other is in $y \leq -a$. The trace in the XY plane is a hyperbola. Real sections parallel to the XZ plane are ellipses, and sections parallel to the other coordinate planes are hyperbolas. The y intercepts are $\pm a$ and there are no x or z intercepts. The surface is symmetric with respect to each coordinate plane.

23.8   The elliptic cone

The surface represented by

$$\frac{x^2}{a^2} + \frac{y^2}{b^2} - \frac{z^2}{c^2} = 0$$

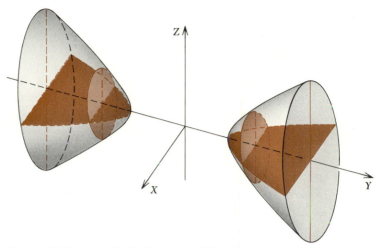

Figure 23.6   An elliptic hyperboloid of two sheets

elliptic cone   is called an *elliptic cone* (see Fig. 23.7). It is a cone of revolution if $a = b$ and can be generated by revolving a line about the Z axis. The traces in the XZ and YZ planes are intersecting straight lines, and the sections parallel to those planes are hyperbolas. The trace in the XY plane is the origin and the sections parallel to that plane are ellipses. All intercepts are zero and the surface is symmetric with respect to each coordinate plane.

If any one of the signs is negative and the other two are positive, the surface is still an elliptic cone but the traces and sections are not as stated above. They are, however, readily determined.

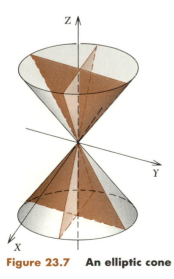

Figure 23.7   An elliptic cone

23.9   Removal of linear terms

If an equation in three variables contains all three variables to the second power and one or more linear terms but no product term, it can be put into a standard form of the equation of an ellipsoid, a hyperboloid, or an elliptic cone by completing the squares of the quadratics in x, y, and z and then replacing the linear terms that are squared by x', y', and z'. This procedure is essentially a translation of the axes. The origin is translated to (h, k, l) by replacing $x - h$, $y - k$, and $z - l$ by x', y', and z', respectively.

Example   Remove the linear terms from

$$x^2 + 9y^2 - 4z^2 + 2x - 36y + 32z = 63$$

by a translation.

Solution: We complete the square of the quadratic in x by adding 1 to each member and the others by adding 36 for y and subtracting 64 for z. Thus, we get

$$x^2 + 2x + 1 + 9(y^2 - 4y + 4) - 4(z^2 - 8z + 16)$$
$$= 63 + 1 + 36 - 64$$
$$(x + 1)^2 + 9(y - 2)^2 - 4(z - 4)^2 = 36$$

Hence, after translating the origin to $(-1, 2, 4)$, the equation becomes

$$x'^2 + 9y'^2 - 4z'^2 = 36$$

Consequently, it represents an elliptic hyperboloid of one sheet whose equation in standard form is

$$\frac{x'^2}{6^2} + \frac{y'^2}{2^2} - \frac{z'^2}{3^2} = 1$$

EXERCISE 23.3

1   Show that $Ax^2 + By^2 + Cz^2 = D$ represents an ellipsoid if no coefficients are negative and $D > 0$.

2   Show that $Ax^2 + By^2 + Cz^2 = D$ represents an elliptic hyperboloid of one sheet if one coefficient is negative and $D > 0$.

3   Show that $Ax^2 + By^2 + Cz^2 = D$ represents an elliptic hyperboloid of two sheets if two coefficients are negative and $D > 0$.

4   Show that $Ax^2 + By^2 + Cz^2 = D$ represents an elliptic cone if $D = 0$ and not all coefficients have the same sign.

By means of the statements in Problems 1 to 4, classify the surfaces in Problems 5 to 20. Sketch each surface after putting the equation in standard form.

5   $16x^2 + 36y^2 + 9z^2 = 144$
6   $9x^2 + 16y^2 + 144z^2 = 144$
7   $36x^2 + 225y^2 + 100z^2 = 900$
8   $x^2 + y^2 + 9z^2 = 9$
9   $4x^2 - 16y^2 + z^2 = 16$
10   $4x^2 + 36y^2 - 9z^2 = 36$
11   $225x^2 + 100y^2 - 36z^2 = 900$
12   $16x^2 - 400y^2 + 25z^2 = 400$
13   $x^2 - 4y^2 - z^2 = 4$
14   $4x^2 - y^2 - 9z^2 = 36$
15   $9x^2 + 144y^2 - 16z^2 = -144$
16   $100x^2 - 225y^2 + 36z^2 = -900$
17   $36x^2 + 9y^2 - 4z^2 = 0$
18   $16x^2 + 100y^2 - 25z^2 = 0$
19   $4x^2 - 36y^2 + 81z^2 = 0$
20   $225x^2 - 400y^2 + 144z^2 = 0$

Remove the linear terms from the equations in Problems 21 to 28, and sketch each surface relative to the new axes.

21   $\dfrac{(x-2)^2}{3^2} + \dfrac{(y+1)^2}{2^2} + \dfrac{(z-3)^2}{5^2} = 1$

22   $\dfrac{(x+3)^2}{2^2} - \dfrac{(y-2)^2}{3^2} + \dfrac{(z-1)^2}{2^2} = 1$

23   $\dfrac{(x-1)^2}{5^2} - \dfrac{(y+1)^2}{2^2} - \dfrac{(z-2)^2}{3^2} = 1$

24   $\dfrac{(x-3)^2}{3^2} + \dfrac{(y-2)^2}{2^2} - \dfrac{(z-4)^2}{4^2} = 0$

25   $9x^2 + 4y^2 - 9z^2 - 18x - 8y - 18z = 32$
26   $4x^2 - 36y^2 - 9z^2 - 16x + 72y - 36z = 92$
27   $4x^2 + 4y^2 - 9z^2 - 8y - 36z = 32$
28   $4x^2 + 25y^2 + 100z^2 - 32x - 50y + 200z = -89$

23.10   The elliptic paraboloid

The surface represented by

$$\frac{x^2}{a^2} + \frac{y^2}{b^2} = 4cz$$

elliptic paraboloid is called an *elliptic paraboloid* and is a paraboloid of revolution

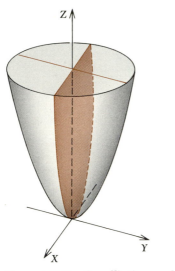

Figure 23.8   An elliptic paraboloid

if $a = b$ (see Fig. 23.8). Sections parallel to and traces in the XZ and YZ planes are parabolas. Those parallel to the XY plane are ellipses and are real for sections above or below the XY plane according as $c > 0$ or $c < 0$. The surface is symmetric with respect to the XZ and YZ planes, and all intercepts are zero.

23.11   The hyperbolic paraboloid

The surface represented by

$$\frac{x^2}{a^2} - \frac{y^2}{b^2} = 4cz$$

hyperbolic is called a *hyperbolic paraboloid* and is shown in Fig. 23.9. The
paraboloid sections parallel to and the traces on the XZ and YZ planes are parabolas. Sections parallel to the XY plane are hyperbolas. The trace on the XY plane is a pair of intersecting lines. The surface is symmetric with respect to the XZ and YZ planes, and all three intercepts are zero.

23.12   Curves

We have used the fact that a surface and a plane intersect in a plane curve in discussing sections and traces. We have used the equation of the surface and the equation of the plane considered simultaneously

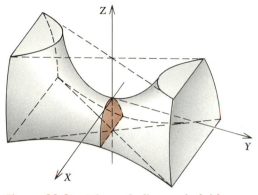

Figure 23.9   A hyperbolic paraboloid

as the equations of the curve. We shall now extend the concept of a curve by considering the equations

$$f_1(x, y, z) = 0 \qquad \text{and} \qquad f_2(x, y, z) = 0 \tag{1}$$

of two surfaces as the simultaneous equations of the curve determined by the intersection of the two surfaces. If the curve lies in a plane, it is called a *plane curve*, and if it does not lie in a plane, it is called *skew curve*   a *skew*, or *twisted, curve.*

The equation obtained by eliminating one of the variables between the two equations of a curve represents a cylinder since any equation in two variables, if considered in three-dimensional space, is a cylinder *projecting*   and is called a *projecting cylinder.* The elements of the projecting *cylinder*   cylinder are perpendicular to the plane of the two variables that are in its equation. It is often desirable, useful, and comparatively easy to represent a curve in space by showing it as the intersection of two of its projecting cylinders.

Example   Find the equations of the projecting cylinders of the curve whose equations are

$$x^2 + 2y^2 + z^2 = 4 \qquad x^2 + y^2 - 2z^2 = -1$$

and sketch the curve.

Solution: If, in turn, we eliminate z, y, and x between the given equations, we see that the projecting cylinders are represented by

$$3x^2 + 5y^2 = 7 \qquad x^2 - 5z^2 = -6 \qquad y^2 + 3z^2 = 5$$

The curve of intersection of the surfaces in Fig. 23.10 is the intersection of the XY and YZ projecting cylinders. It is constructed by drawing rays parallel to the X and Z axes from any point Q on the Y axis and

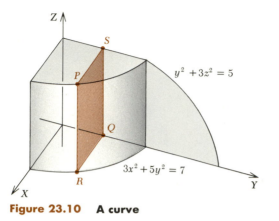

Figure 23.10   **A curve**

extending them until they intersect the directing curves of the projecting cylinders at R and S, then drawing generators of the cylinders perpendicular to the XY and YZ planes until they meet at a point P. This determines a point of the curve of intersection; as many points as desired may be determined in this manner.

23.13   Parametric equations of a space curve

We saw in Sec. 22.9 that the equations of a line through $P_1(x_1, y_1, z_1)$ and with A, B, and C as direction numbers are

$$x = x_1 + At \qquad y = y_1 + Bt \qquad z = z_1 + Ct$$

In general, the three equations

$$x = f_1(t) \qquad y = f_2(t) \qquad z = f_3(t)$$

are parametric equations of a *space* or *skew curve*. If t is eliminated between any two of the equations, we obtain a projecting cylinder of the curve.

Example   Sketch the graph of

$$x = 3 \sin t \qquad y = 3 \cos t \qquad z = \frac{2t}{3}$$

and find the equation of the *xy* projecting cylinder.

Solution: If $t = 0$, we find that $x = 0$, $y = 3$, $z = 0$. Hence, $(0, 3, 0)$ is a point on the curve and it is indicated in Fig. 23.11 by A. If $t = \pi/2$, we have $x = 3$, $y = 0$, $z = \pi/3$. The corresponding point on the curve is $B(3, 0, \pi/3)$.

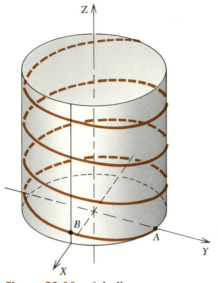

Figure 23.11   **A helix**

If we eliminate t between the first two equations as can be done by squaring each member of each equation and then adding corresponding members of the new equations, we see that

$$x^2 + y^2 = 9$$

is the equation of the xy projecting cylinder. The curve lies on this cylinder and is called a *helix*.

EXERCISE 23.4

Sketch the surface whose equation is given in each of Problems 1 to 12. If desirable, translate before sketching.

1   $x^2/2^2 + y^2/3^2 = 4z$ 2 $x^2/3^2 + y^2/4^2 = 8z$
3   $x^2/3^2 + y^2/2^2 = -8z$ 4 $x^2/1^2 + y^2/2^2 = -4z$
5   $x^2/3^2 - y^2/4^2 = 16z$ 6 $x^2/5^2 - y^2/2^2 = 8z$
7   $y^2/5^2 - x^2/4^2 = 4z$ 8 $y^2/2^2 - x^2/4^2 = 12z$
9   $9x^2 + 4y^2 - 18x + 16y - 144z + 25 = 0$
10   $x^2 + 4y^2 + 6x + 8y - 16z + 13 = 0$
11   $4x^2 - 9y^2 + 8x + 36y - 288z - 32 = 0$
12   $-x^2 + 4y^2 + 6x - 16y - 16z + 7 = 0$

Sketch the curves whose equations are given in Problems 13 to 24.

13   $x^2 + y^2 + z^2 = 6,\ x^2 - y^2 + 2z^2 = 4$

14   $x^2 + y^2 + 3z^2 = 9,\ x^2 - y^2 + z^2 = 6$

15   $x^2 + y^2 - z^2 = 0,\ 2x^2 - y^2 - z^2 = 2$

16   $x^2 + 2y^2 - 3z^2 = 6,\ 2x^2 + y^2 - z^2 = -2$

17   $x^2 + y^2 + z^2 = 25,\ x^2 + y^2 = 9$

18   $x^2 + y^2 + z^2 = 32,\ x^2 + y^2 = 4z$

19   $xy = z,\ x^2 + y^2 + z^2 = 9$

20   $xz = y,\ yz = x$

21   $x = 5 \sin t,\ y = 4 \cos t,\ z = t/2$

22   $x = 4 \cos^2 t,\ y = 4 \sin^2 t,\ z = t^2$

23   $x = \cos t,\ y = \sec t,\ z = t$

24   $x = \sec t,\ y = \tan t,\ z = 3t$

24 Partial derivatives

24.1   Functions of two or more variables

Heretofore we have been concerned with functions of a single variable such as $f(x)$ where the domain was a set of admissible values of x and the range was the set of corresponding function values. In this chapter we shall deal with functions of two or more variables and shall first define a function of two variables.

If, for each element in a set of ordered pairs $\{(x, y)\}$ that belong to a region R, there exists a rule that associates one and only one number z with each ordered pair (x, y) in R, then z is a function of (x, y). The region R is the *domain* of definition of the function and the *range* is the set $\{z\}$.

function of two variables

domain

range

According to this definition, if the rule that determines the correspondence is $z = f(x, y)$, then a function of two variables is the set of ordered triples

$$\{(x, y, z) \mid z = f(x, y) \text{ and } (x, y) \text{ belongs to}$$

$$\text{the set } R \text{ of ordered pairs } (x, y)\}$$

In three-dimensional space, a function of two variables determines a set of points $\{(x, y, z)\}$ where (x, y) belongs to a region of the XY plane. For example, $\sqrt{r^2 - x^2 - y^2}$ determines the set of points (x, y, z)

on the hemisphere above the XY plane with the center at the origin, and the domain is the circle $x^2 + y^2 = r^2$ in the XY plane. This situation is illustrated in Fig. 24.1.

Example

If the domain of a function f is the set of all ordered pairs of real numbers (x, y) and the rule that specifies exactly one number z for each pair (x, y) is $z = x^2 - y$, find the function values $f(3, 1)$, $f(-2, 5)$ and $f(4, -3)$.

Solution: If we substitute the specified values of x and y in $z = f(x, y) = x^2 - y$, we have

$$f(3, 1) = 3^2 - 1 = 8$$
$$f(-2, 5) = (-2)^2 - 5 = -1$$
$$f(4, -3) = 4^2 - (-3) = 19$$

We say that the limit of $f(x, y)$ as (x, y) approaches (x_1, y_1) is L if $|f(x, y) - L|$ can be made less than any preassigned positive number by taking (x, y) sufficiently near to (x_1, y_1). Furthermore, we say that $z = f(x, y)$ is continuous at (x_1, y_1) if the limit of $f(x, y)$ is $f(x_1, y_1)$ as (x, y) approaches (x_1, y_1). This can be stated more precisely by saying that

If for each $\varepsilon > 0$ there exists a $\delta > 0$ such that

continuous function

$$|f(x, y) - f(x_1, y_1)| < \varepsilon \text{ for } |x - x_1| < \delta \text{ and } |y - y_1| < \delta$$

then $f(x, y)$ is continuous at (x_1, y_1).

continuity in a region

A function is *continuous in a region* if it is continuous at every point (x, y) in the region. When we say that a function is continuous, we mean that it is continuous in the region involved in the discussion.

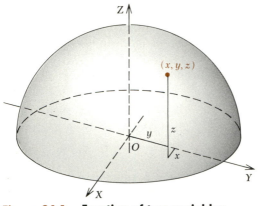

Figure 24.1   Function of two variables

We shall assume that all functions and their derivatives used in this chapter are continuous unless there is a statement to the contrary.

It can be proved that if two functions are continuous at a point (x_1, y_1) and are defined in a region about the point, then their sum, difference, and product are continuous at the point. Furthermore, their quotient is also continuous at the point if the divisor is not zero there.

24.2 Partial derivatives

We shall consider the function $z = f(x, y)$, let x change by an increment Δx, and keep y fixed. If the corresponding change in z is denoted by Δz, then we have

$$z + \Delta z = f(x + \Delta x, y)$$
$$\Delta z = f(x + \Delta x, y) - f(x, y)$$
$$\frac{\Delta z}{\Delta x} = \frac{f(x + \Delta x, y) - f(x, y)}{\Delta x}$$

If we let $\Delta x \to 0$ and denote $\lim\limits_{\Delta x \to 0} (\Delta z / \Delta x)$ by $\partial z / \partial x$, we have

$$\frac{\partial z}{\partial x} = \lim_{\Delta x \to 0} \frac{f(x + \Delta x, y) - f(x, y)}{\Delta x}$$

partial
derivative

If this limit exists, it is called the *partial derivative of z with respect to x*. It is the instantaneous rate of change of z relative to x with y held constant. Therefore, in obtaining the partial derivative of a function of (x, y), we consider y as a constant. Similarly, if x is held constant and if

$$\frac{\partial z}{\partial y} = \lim_{\Delta y \to 0} \frac{f(x, y + \Delta y) - f(x, y)}{\Delta y}$$

exists, it is called the *partial derivative of z with respect to y*. At times we use z_x, f_x, or $\partial f / \partial x$ instead of $\partial z / \partial x$, and z_y, f_y, or $\partial f / \partial y$ for $\partial z / \partial y$.

Example 1   Find the partial derivative of z with respect to each variable if $z = xy^2 + x^3$.

Solution: We first treat y as a constant, differentiate with respect to x, and get

$$\frac{\partial z}{\partial x} = \frac{\partial}{\partial x} xy^2 + \frac{d}{dx} x^3 = y^2 + 3x^2$$

Similarly, treating x as a constant and differentiating with respect to y, we have

$$\frac{\partial z}{\partial y} = \frac{\partial}{\partial y} xy^2 + \frac{\partial}{\partial y} x^3 = 2xy$$

Example 2 If $P(1, 8)$ is in the domain of $z = \sqrt{x^2 + 3y}$, find z_x and z_y at $(1, 8)$.

Solution: By use of the appropriate differentiation formulas, we have

$$z_x = \frac{2x}{2\sqrt{x^2 + 3y}}$$

$$= \frac{2}{2\sqrt{1 + 24}} \qquad at\ P(1, 8)$$

$$= \frac{1}{5}$$

and

$$z_y = \frac{3}{2\sqrt{x^2 + 3y}}$$

$$= \frac{3}{10} \qquad at\ P(1, 8)$$

24.3   Geometric interpretation

geometric interpretation

The points whose coordinates in space satisfy the equation $z = f(x, y)$ lie on a surface S, and we say that the equation represents the surface. In order to obtain a geometric interpretation for z_x, we choose a point $P_1(h, k, l)$ on S and pass a plane $y = k$ through P_1 parallel to the xz plane. The intersection of S and the plane $y = k$ is the curve CP_1D in Fig. 24.2. If we choose a second point $P_2(h + \Delta x, k, l + \Delta z)$ on the intersection of S and $y = k$, then P_2 is on the curve CP_1D, and $\Delta z/\Delta x$ is the slope of the secant P_1P_2. As Δx approaches zero, the secant P_1P_2 approaches the tangent T. Consequently, $\lim\limits_{\Delta x \to 0} (\Delta z/\Delta x) = z_x$ for $x = h$ and $y = k$ is the slope of the tangent to CP_1D in the plane $y = k$. Similarly, z_y for $x = h$, $y = k$ is the slope of the tangent in the plane $x = h$ to the curve of intersection $AP_1 B$ of $x = h$ and the surface $z = f(x, y)$.

Example Find the slope of the tangent to the curve of intersection of the plane $y = 3$ and the surface $z = xy + 3x^2$ at $(2, 3, 18)$.

Solution: Since $\partial z/\partial x$ is the slope of the tangent to the curve of intersection of $y = k$ and $z = f(x, y)$, we must find $\partial z/\partial x$ and evaluate it for $x = 2$ and $y = 3$. Thus

$$\frac{\partial z}{\partial x} = y + 6x = 3 + 12 = 15$$

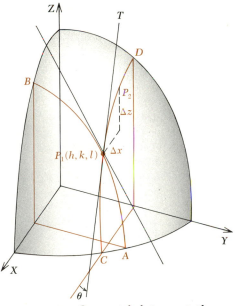

Figure 24.2   Geometric interpretation

EXERCISE 24.1

Find $\partial z/\partial x$ and $\partial z/\partial y$ in each of Problems 1 to 12.

1   $z = x^3 - 4x^2y^2 + 8y^2$ 2 $z = x^3 + 5x^2y + 6$

3   $z = \dfrac{y^2}{y - 4x}$ 4 $z = \dfrac{x - y}{x^2 + y^2}$

5   $z = \ln(x^2 + y^2)$ 6 $z = \text{Arctan}\,(y/x)$

7   $z = x \cos xy$ 8 $z = xe^{y/x}$

9   $z = xe^y + ye^x$ 10 $z = e^{x-y} - x^2y$

11   $z = x \sin y + y \cos x$ 12 $z = y \,\text{Arcsin}\, x + x\sqrt{1 - y^2}$

13   The volume of a surface is $v = x^2y/3$. Find the value of v_x and of v_y for $x = 3$ and $y = 6$.

14   The area of a triangle is $A = 0.5xy \sin \theta$ where θ is the angle between the sides x and y. Find A_x and A_θ for $x = 3$, $y = 4$, and $\theta = \pi/4$.

15   If the temperature at any point (x, y, z) in a body is determined by

$$T = x^2 + xy + yz$$

find $\partial T/\partial x$, $\partial T/\partial y$, and $\partial T/\partial z$ at $(6, -1, 2)$. Here T is a function of three variables and we hold the other two constant in finding the partial derivative with respect to either.

16 The pressure, volume, and absolute temperature of a gas are related by the equation $p = ct/v$ where c is a constant. Find the rate of change of pressure with respect to temperature and with respect to volume if the other variable is held constant, and evaluate these derivatives for $c = 2$, $t = 300°$, and $v = 15$.

Find z_x and z_y in each of Problems 17 to 24 by use of implicit differentiation.

17 $x^2 + y^2 + z^2 = 16$ **18** $xy + z^2 = 6$

19 $z^2 = x^2/4 + y^2/9$ **20** $x^2/a^2 + y^2/b^2 + z^2/c^2 = 1$

21 $x^2z + 4z^2 = 8y$ **22** $x^2y - 2z^2 = 4x$

23 $x^2z + yz^2 = xy$ **24** $xz^2 - y^2z = x$

In each of Problems 25 to 28, find the slope of the curve of intersection of the surface and each given plane at the point of intersection of the surface and the planes.

25 $x^2/9 + y^2/16 + z = 3$, $x = 3$, $y = 4$

26 $x^2 + y^2 = z$, $x = 5$, $y = 6$

27 $x^2/2 + y^2/6 + z^2/3 = 1$, $x = 1$, $y = 1$

28 $2x^2 + 3y^2 = z$, $x = 1$, $y = 2$

29 Show that $xz_x + yz_y = 0$ if $z = (x^2 - y^2)/xy$.

30 Show that $xz_x + yz_y = z$ if $z = (x^3 - y^3)/xy$.

31 Show that $xu_x + yu_y + zu_z = 0$ if $u = (xy + z^2)/yz$.

32 Show that $xz_y - yz_x = 0$ if $z = k\sqrt{x^2 + y^2}$.

24.4   The tangent plane and normal line

The equation

$$z - z_1 = A(x - x_1) + B(y - y_1) \tag{1}$$

represents a plane since it is a first-degree equation in three variables, and the plane passes through the point $P_1(x_1, y_1, z_1)$ since those values of x, y, and z satisfy the equation. We now propose to determine A and B so that (1) represents the plane that is tangent to the surface $z = f(x, y)$ at P_1. We shall use Fig. 24.3 in doing this. The intersection of the surface and the plane $y = y_1$ is the curve CP_1D, and z_x is the slope of this curve at P_1. Furthermore, the plane $y = y_1$ intersects the plane represented by (1) in the line LP_1Q. The equations of this line are $y = y_1$ and $z - z_1 = A(x - x_1)$; hence, its slope is A. In order for the plane (1) to be tangent to the surface, the line represented by $y = y_1$ and $z - z_1 = A(x - x_1)$ must be tangent to the curve CP_1D at P_1. Consequently, A must be equal to the value of z_x at P_1.

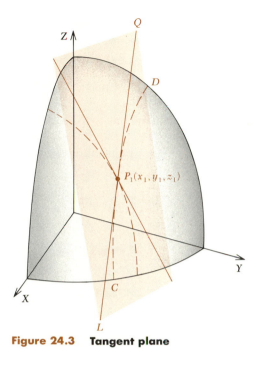

Figure 24.3   Tangent plane

If we take a plane section made by $x = x_1$, we can show in a similar manner that B must be equal to the value of z_y at P_1. *tangent* Therefore, *the equation of the plane that is tangent to the surface at* *plane* P_1 *is*

$$z - z_1 = \left.\frac{\partial z}{\partial x}\right|_{P_1}(x - x_1) + \left.\frac{\partial z}{\partial y}\right|_{P_1}(y - y_1) \qquad (24.1a)$$

We shall now find the equations of the line that is perpendicular to $(24.1a)$ at P_1. It is called the *normal line*. We know from Sec. 22.10, that the coefficients in the equation $Ax + By + Cz + D = 0$ of a plane are components of a vector perpendicular to the plane. Consequently,

$$\left.\frac{\partial z}{\partial x}\right|_{P_1}, \quad \left.\frac{\partial z}{\partial y}\right|_{P_1}, \quad \text{and} \quad -1$$

are direction numbers of any perpendicular to the plane represented by $(24.1a)$. Therefore, the equations of the normal line to the surface at P_1 are

normal
line
$$\frac{x - x_1}{\left.\dfrac{\partial z}{\partial x}\right|_{P_1}} = \frac{y - y_1}{\left.\dfrac{\partial z}{\partial y}\right|_{P_1}} = \frac{z - z_1}{-1} \qquad (24.1b)$$

Since the components of a vector, or directed line segment, are proportional to its direction cosines, they constitute a set of direction numbers of the line. Consequently, it is often said that

$$\left.\frac{\partial z}{\partial x}\right|_{P_1}, \quad \left.\frac{\partial z}{\partial y}\right|_{P_1}, \quad \text{and} \quad -1$$

are direction numbers of the normal line to the surface $z = f(x, y)$ at P_1.

Example   Find the equation of the tangent plane and equations of the normal line to $z = 3x^2 + 2y^2 - 27$ at $P_1(1, 2, 4)$.

Solution: We need and shall begin by finding the values of the partial derivatives of z at P_1. We obtain

$$z_x = 6x \qquad \text{and} \qquad z_y = 4y$$

Hence,
$$z_x = 6x|_{P_1} = 6$$

and
$$z_y = 4y|_{P_1} = 8$$

If we now put these values and the coordinates of P_1 in (24.1a) and in (24.1b), we find that

$$z - 4 = 6(x - 1) + 8(y - 2)$$

is the equation of the tangent plane and that

$$\frac{x - 1}{6} = \frac{y - 2}{8} = \frac{z - 4}{-1}$$

are the equations of the normal line to the given surface at $P_1(1, 2, 4)$.

24.5   A fundamental increment formula

We shall now develop a formula that is of fundamental importance in our work on partial derivatives. It is a formula for the total change of a function of several variables when all of the variables on which it depends change. We shall consider a continuous function $u = f(x, y)$ of two variables with initial values x_1 and y_1 and shall let them assume increments Δx and Δy. Therefore, the corresponding increment of the function is

$$\Delta u = f(x_1 + \Delta x, y_1 + \Delta y) - f(x_1, y_1)$$
$$= [f(x_1 + \Delta x, y_1 + \Delta y) - f(x_1, y_1 + \Delta y)]$$
$$+ [f(x_1, y_1 + \Delta y) - f(x_1, y_1)] \quad (1)$$

by subtracting and adding $f(x_1, y_1 + \Delta y)$. In the first bracket, the value of x changes and that of y remains constant; whereas, in the second bracket, the value of x remains fixed while the value of y

changes. We shall now assume that the first partial derivatives of u are continuous and represent them by $f_x(x, y)$ and $f_y(x, y)$. If we now apply the law of the mean as given in Sec. 14.2 to the expression in each bracket, we have

$$f(x_1 + \Delta x, y_1 + \Delta y) - f(x_1, y_1 + \Delta y) = f_x(\xi, y_1 + \Delta y) \, \Delta x$$

and
$$f(x_1, y_1 + \Delta y) - f(x_1, y_1) = f_y(x_1, \eta) \, \Delta y \tag{2}$$

where ξ is a properly chosen number in $(x_1, x_1 + \Delta x)$ and η is an appropriately chosen number in $(y_1, y_1 + \Delta y)$. If we now substitute from (2) in (1), we get

$$\Delta u = f_x(\xi, y_1 + \Delta y) \, \Delta x + f_y(x_1, \eta) \, \Delta y \tag{3}$$

Now we make use of the assumption that $f_x(x, y)$ and $f_y(x, y)$ are continuous functions of x and y; hence, $f_x(\xi, y_1 + \Delta y)$ approaches $f_x(x_1, y_1)$ and $f_y(x_1, \eta)$ approaches $f_y(x_1, y_1)$ as Δy and Δx approach zero. Consequently,

$$f_x(\xi, y_1 + \Delta y) = f_x(x_1, y_1) + \varepsilon_1$$
$$f_y(x_1, \eta) = f_y(x_1, y_1) + \varepsilon_2 \tag{4}$$

where ε_1 and ε_2 approach zero with Δx and Δy. Equations (3) and (4) now enable us to say that *if $u = f(x, y)$ and its partial derivatives $\partial u / \partial x$ and $\partial u / \partial y$ are continuous and if x and y start at $x = x_1$* *and $y = y_1$ and are given increments Δx and Δy, then the resulting increment in the function is*

fundamental increment formula

$$\Delta u = \frac{\partial u}{\partial x} \Delta x + \frac{\partial u}{\partial y} \Delta y + \varepsilon_1 \, \Delta x + \varepsilon_2 \, \Delta y \tag{24.2a}$$

where $\partial u / \partial x$ and $\partial u / \partial y$ are evaluated at (x_1, y_1) and ε_1 and ε_2 approach zero as Δx and Δy do.

Example 1   If $u = x^2 - xy + y^2$, express Δu as in (24.2a).

Solution: If we replace x by $x + \Delta x$ and y by $y + \Delta y$, we see that

$$u + \Delta u = [(x + \Delta x)^2 - (x + \Delta x)(y + \Delta y) + (y + \Delta y)^2]$$

Hence, expanding and subtracting u from each member, we have

$$\Delta u = 2x \, \Delta x + \overline{\Delta x}^2 - x \, \Delta y - y \, \Delta x - \Delta x \, \Delta y + 2y \, \Delta y + \overline{\Delta y}^2$$
$$= (2x - y) \, \Delta x + (-x + 2y) \, \Delta y + \Delta x \, \Delta x + (-\Delta x + \Delta y) \, \Delta y$$
$$= u_x \, \Delta x + u_y \, \Delta y + \varepsilon_1 \, \Delta x + \varepsilon_2 \, \Delta y$$

since $u_x = 2x - y$ and $u_y = -x + 2y$.

A formula for Δu where $u = F(x, y, z)$ can be derived in a similar manner to that used in obtaining (24.2a). The formula is

another
increment
formula

$$\Delta u = \frac{\partial u}{\partial x}\,\Delta x + \frac{\partial u}{\partial y}\,\Delta y + \frac{\partial u}{\partial z}\,\Delta z + \varepsilon_1\,\Delta x + \varepsilon_2\,\Delta y + \varepsilon_3\,\Delta z \quad (24.3a)$$

If we let Δx, Δy, and Δz approach zero, (24.2a) and (24.3a) become

formulas
for du

$$du = u_x\,dx + u_y\,dy \qquad\qquad\qquad (24.2b)$$

and

$$du = u_x\,dx + u_y\,dy + u_z\,dz \qquad\qquad (24.3b)$$

where du is called the total differential of u.

If the values of x and y are changed by small amounts in $u = f(x, y)$, then du is the approximate change in u.

Example 2   The metal used in making a closed right circular cylindrical can is ¹⁄₆₄ in. thick. Find, approximately, the volume of metal used if the inside dimensions of the can are $r = 4$ in. and $h = 8$ in.

Solution: The volume of metal is equal to the increase in volume if the radius changes from 4 to 4¹⁄₆₄ in. and the height from 8 to 8¹⁄₃₂ in. The volume of a cylinder is

$$V = \pi r^2 h$$

Hence, $$dV = V_r\,dr + V_h\,dh = 2\pi rh\,dr + \pi r^2\,dh$$

is the approximate change in V. Consequently, for $r = 4$, $h = 8$, $dr = ¹⁄₆₄$, and $dh = ¹⁄₃₂$, we have

$$dV = 64\pi(¹⁄₆₄) + 16\pi(¹⁄₃₂) = 1.5\pi$$

Therefore, approximately 1.5π cu in. of metal is used.

If we are working with approximations of measurements, then du is the approximate change (error) in u due to changes (errors) in the variables on which u depends. The fraction du/u is called the *relative error* in u.

relative
error

Example 3   Find the relative error in the area of an ellipse if there is a possible error of 1 percent in measuring each semiaxis.

Solution: If the semiaxes of an ellipse are a and b, its area is $A = \pi ab$. Consequently,

$$dA = \pi(a\,db + b\,da) = \pi[a(0.01b) + b(0.01a)]$$
$$= \pi ab(0.02)$$

Hence, $$\frac{dA}{A} = \frac{\pi ab(0.02)}{\pi ab} = 0.02$$

and the relative error is 2 percent.

EXERCISE 24.2

Find the equation of the tangent plane and the equations of the normal line to the surface at the specified point in each of Problems 1 to 8.

1   $z = x^2 + y^2 - 9$, $(1, 3, 1)$
2   $z = 2x^2 + y^2 - 5$, $(2, 1, 4)$
3   $z = x^2 - 2y^2 + 1$, $(3, 2, 2)$
4   $z = x^2 - xy + 3$, $(-1, 3, 7)$
5   $x^2 + y^2 + z^2 = 54$, $(6, 3, -3)$
6   $xyz = 24$, $(2, 4, 3)$
7   $x^2y - z = 5$, $(2, 2, 3)$
8   $xy - z = 2$, $(3, 2, 4)$

Find the total differential of the function in each of Problems 9 to 16.

9    $z = x^2 - 2xy + 3y^2$ 10 $z = \sqrt{x^2 - y}$
11   $z = \ln xy$ 12 $z = xy/\sqrt{x^2 + y^2}$
13   $u = xyz - xz$ 14 $u = e^x - \ln yz$
15   $w = \ln x + \cos y - e^z$ 16 $w = \ln(x^2 + y^2) + \tan z$

17   Show that if $z = xy$, then $dz = x\,dy + y\,dx$; make use of this differential to find an approximation to $(1.98)(2.01)$.

18   Use the differential of xy^2 to find an approximation to $(4.02)(3.97)^2$.

19   Approximately how much does the volume of a rectangular parallelepiped that is 14 by 17 by 22 cm change if each dimension is increased by 0.1 cm?

20   Find an approximation to the amount of material needed to put a coat 0.03 cm thick on a right circular cylinder that is 10 cm in radius and 8 cm high.

21   Find approximately the volume of wood in a rectangular box whose inside dimensions are 4 by 6 by 5 in. if the material is ⅛ in. thick.

22   A certain type of brick is 2 by 4 by 8 in. A rectangular pile of these bricks is measured to be 6 by 8 by 5 ft, and the number of bricks estimated by dividing the volume of the pile by the volume of a brick. What is the greatest error in the estimated number of bricks if the error in each dimension of the pile can be off as much as 0.1 ft?

23   What is the maximum error in the area of a triangle if two sides and the included angle are measured to be 15 ft, 20 ft, and $\pi/3$ but may be off as much as 0.1 ft and 0.01 radian.

24   If $\cos A = b/c$ and if b and c are measured to be 27 and 35 cm. with a possible error in each measurement of 0.1 cm, find $\cos A$ from the measurements, and find the possible error in it.

25   Show that the relative error in the calculated volume of a rectangular box is equal to the sum of the relative errors in the measurements of the edges. If the dimensions are w, b, and h, the problem is to show that

$$\frac{dV}{V} = \frac{dw}{w} + \frac{db}{b} + \frac{dh}{h}$$

26   Show that the relative error in the volume of a cube is three times the relative error in an edge.

27   Show that the relative error in the volume of a right circular cylinder is three times the common value of the relative error in the height and radius.

28   If the value of $z = kx^m y^n$ is determined from approximations to x and y, show that the relative error in z is m times the relative error in x plus n times the relative error in y.

24.6   Chain rules

We have studied derivatives in connection with the equation $u = f(x, y)$ where x and y are independent variables and shall now consider the situation in which $u = f(x, y)$ and y is a function of x, say $y = g(x)$. Under these circumstances u is a function of the single independent variable x, and we shall develop a formula for du/dx.

If x is given an increment Δx, then y assumes an increment Δy since y is a function of x and the corresponding change in u, as in Sec. 24.5, is

$$\Delta u = \frac{\partial u}{\partial x} \Delta x + \frac{\partial u}{\partial y} \Delta y + \varepsilon_1 \Delta x + \varepsilon_2 \Delta y \tag{1}$$

Consequently, dividing by Δx, taking the limit as Δx approaches zero, and noting that Δy, ε_1, and ε_2 approach zero with Δx, we find that

total derivative
$$\frac{du}{dx} = \frac{\partial u}{\partial x} + \frac{\partial u}{\partial y} \frac{dy}{dx} \tag{24.3}$$

This is called the *total derivative* of u with respect to x.

Example 1   Find du/dx if

$$u = f(x, y) = x^2 y^3 \qquad \text{and} \qquad y = g(x) = 4x - 1$$

Solution: Since u is a function of two variables and one of them is a function of the other, we use (24.3) and have

$$\frac{du}{dx} = \frac{\partial x^2 y^3}{\partial x} + \frac{\partial x^2 y^3}{\partial y} \frac{d(4x - 1)}{dx}$$

$$= 2xy^3 + 3x^2 y^2 \cdot 4 = 2xy^3 + 12x^2 y^2$$

This result agrees with du/dx as found from $u = x^2(4x - 1)^3$.

We shall now consider the situation in which u is a function of two variables and each of them a function of a third variable. We shall represent the situation symbolically by

$$u = f(x, y) \qquad x = g(t) \qquad y = h(t)$$

and develop a formula for du/dt. In order to do this, we begin by dividing each member of (1) by Δt and thus obtain

$$\frac{\Delta u}{\Delta t} = \frac{\partial u}{\partial x} \frac{\Delta x}{\Delta t} + \frac{\partial u}{\partial y} \frac{\Delta y}{\Delta t} + \varepsilon_1 \frac{\Delta x}{\Delta t} + \varepsilon_2 \frac{\Delta y}{\Delta t}$$

If we let Δt approach zero, we get

total derivative for $u = f(x, y)$, x and y functions of t

$$\frac{du}{dt} = u_x \frac{dx}{dt} + u_y \frac{dy}{dt} \tag{24.4}$$

We shall now consider the situation in which

$$u = f(x, y) \qquad x = g(s, t) \qquad y = h(s, t)$$

and s and t are independent variables. Since x, y, and u are functions of the two independent variables s and t, the derivatives with respect to t in (24.4) become partial derivatives, and we see immediately that

$$\frac{\partial u}{\partial t} = \frac{\partial u}{\partial x} \frac{\partial x}{\partial t} + \frac{\partial u}{\partial y} \frac{\partial y}{\partial t} \tag{24.5a}$$

u_t and u_s for $u = f(x, y)$ and x and y functions of s and t

Similarly,

$$\frac{\partial u}{\partial s} = \frac{\partial u}{\partial x} \frac{\partial x}{\partial s} + \frac{\partial u}{\partial y} \frac{\partial y}{\partial s} \tag{24.5b}$$

The extensions to cases in which $u = f(x, y, z)$ with x, y, and z functions of t or functions of the independent variables s and t are immediate. In the first situation, we have

$$\frac{du}{dt} = \frac{\partial u}{\partial x} \frac{dx}{dt} + \frac{\partial u}{\partial y} \frac{dy}{dt} + \frac{\partial u}{\partial z} \frac{dz}{dt} \tag{24.6}$$

more on total and partial derivatives

If $u = f(x, y, z)$ where x, y and z are functions of s and t, then

$$\frac{\partial u}{\partial t} = \frac{\partial u}{\partial x} \frac{\partial x}{\partial t} + \frac{\partial u}{\partial y} \frac{\partial y}{\partial t} + \frac{\partial u}{\partial z} \frac{\partial z}{\partial t} \tag{24.7a}$$

$$\frac{\partial u}{\partial s} = \frac{\partial u}{\partial x} \frac{\partial x}{\partial s} + \frac{\partial u}{\partial y} \frac{\partial y}{\partial s} + \frac{\partial u}{\partial z} \frac{\partial z}{\partial s} \tag{24.7b}$$

Example 2   If $u = x^2 - xy + 2yz$, $x = 2t - s$, $y = st$, and $z = t + s^2$, find u_s and evaluate it for $s = 2$ and $t = 1$.

Solution: Since u is a function of x, y, and z and each of them is in turn a function of s and t, we must use $(24.7b)$. If this is done, we have

$$u_s = (2x - y)(-1) + (-x + 2z)t + 2y(2s)$$

In order to be able to evaluate this for $s = 2$ and $t = 1$, we must find the corresponding values of x, y, and z or express u_s in terms of s and t above. We shall do the former. Correspondingly,

$$x = 2(1) - 2 = 0 \qquad y = 2(1) = 2 \qquad z = 1 + 2^2 = 5$$

Therefore,

$$u_s = (2 \cdot 0 - 2)(-1) + (-0 + 2 \cdot 5)1 + 2 \cdot 2(2 \cdot 2)$$
$$= (-2)(-1) + 10(1) + 4(4) = 28$$

24.7   Directional derivative and gradient

If $u = f(x, y)$ for some region R of the plane as indicated in Fig. 24.4, then u_x and u_y represent the rate of change of u in the direction of the X axis and of the Y axis, respectively. In some applied problems, it is desirable to be able to find the rate of change of u in an arbitrarily chosen direction. In order to derive a formula for this purpose, we shall let $P_1(x_1, y_1)$ be any point in R and through P_1 draw a line that makes an angle α with the horizontal. In addition we shall let $Q(x_1 + \Delta x, y_1 + \Delta y)$ be another point of R that is on the line, and, finally, we shall represent the distance P_1Q by Δs and the change $f(x + \Delta x, y + \Delta y) - f(x, y)$ in u from P_1 to Q by Δu. Then, *directional derivative defined* we say that *the directional derivative of u at P_1 in the α direction is*

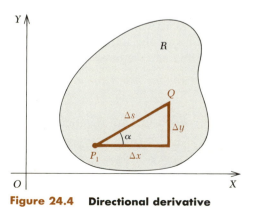

Figure 24.4   Directional derivative

the limit of the ratio $\Delta u / \Delta s$ as Δs approaches P_1 along the line in the α direction and we shall represent it by $(du/ds)\,(\alpha)$.

Now we shall use the fundamental increment formula (24.2a). It is

$$\Delta u = u_x\,\Delta x + u_y\,\Delta y + \varepsilon_1\,\Delta x + \varepsilon_2\,\Delta y$$

Hence,
$$\frac{\Delta u}{\Delta s} = u_x\,\frac{\Delta x}{\Delta s} + u_y\,\frac{\Delta y}{\Delta s} + \varepsilon_1\,\frac{\Delta x}{\Delta s} + \varepsilon_2\,\frac{\Delta y}{\Delta s}$$

$$= u_x \cos \alpha + u_y \sin \alpha + \varepsilon_1 \cos \alpha + \varepsilon_2 \sin \alpha$$

Now, since ε_1 and ε_2 approach zero with Δx and Δy, we can say that *if $u = f(x, y)$, then the directional derivative of u at P_1 in the α direction is*

formula for directional derivative

$$\frac{du}{ds}\,(\alpha) = u_x \cos \alpha + u_y \sin \alpha \qquad (24.8)$$

provided that the derivatives are evaluated at P_1.

If $\alpha = 0$, we thus have $(du/ds)(0) = u_x$, and if $\alpha = \pi/2$, we have

$$\frac{du}{ds}\left(\frac{\pi}{2}\right) = u_y$$

Example 1   If $u = 3xy + y^2$, find $\dfrac{du}{ds}\left(\dfrac{\pi}{6}\right)$ at $P_1(1, 3)$.

Solution: If we take the partial derivatives, we find that $u_x = 3y$ and $u_y = 3x + 2y$. Hence,

$$\frac{du}{ds}\,(\alpha) = 3y \cos \alpha + (3x + 2y) \sin \alpha$$

and for $x = 1$ and $y = 3$, this becomes

$$\frac{du}{ds}\,(\alpha) = 9 \cos \alpha + 9 \sin \alpha$$

Therefore,
$$\frac{du}{ds}\left(\frac{\pi}{6}\right) = \frac{9\sqrt{3}}{2} + \frac{9}{2} = 4.5(\sqrt{3} + 1)$$

If there is a maximum value of the directional derivative at P_1 for some value of α, it is called the *gradient* or *normal derivative* of the function at P_1.

gradient

Example 2   Find the gradient of $u = 3xy + y^2$ at $P_2(1, 3)$.

Solution: The directional derivative at P_1 in the α direction was found in Example 1 and is

$$\frac{du}{ds}\,(\alpha) = 9 \cos \alpha + 9 \sin \alpha$$

As a first step in finding the maximum value, if any, we differentiate $(du/ds)(\alpha)$ with respect to α. Thus,

$$\frac{d}{d\alpha}\left[\frac{du}{ds}(\alpha)\right] = -9\sin\alpha + 9\cos\alpha$$

$$= 0 \qquad\qquad \textit{for } \alpha = \pi/4,\ 5\pi/4$$

Hence if there is a maximum value of the directional derivative at P_1, it occurs for $\alpha = \pi/4$ or $5\pi/4$. To continue the investigation, we find that

$$\frac{d^2}{d^2\alpha}\left[\frac{du}{ds}(\alpha)\right] = -9\cos\alpha - 9\sin\alpha$$

This evaluated for $\alpha = \pi/4$ is negative and for $\alpha = 5\pi/4$ is positive. Consequently, the maximum value of the directional derivative at $P_1(1, 3)$ occurs for $\alpha = \pi/4$. Hence, the value of the gradient is

$$\frac{du}{ds}\left(\frac{\pi}{4}\right) = 9\cos\frac{\pi}{4} + 9\sin\frac{\pi}{4} = 9\sqrt{2}$$

EXERCISE 24.3

Find du/dx in each of Problems 1 to 4.
1    $u = x^2 - xy + y^2,\ y = 3x + 2$
2    $u = x^2 - y^3,\ y = \ln x$
3    $u = \sin x + \cos y,\ y = e^x$
4    $u = \tan 2x + \ln y,\ y = \sin x$

Find du/dt in each of Problems 5 to 8.
5    $u = x^2 + xy - y^2,\ x = 2t - 1,\ y = t - 2$
6    $u = \sin(x + y) + \cos x,\ x = \pi + 2t,\ y = \pi - t$
7    $u = 2\ln x + \ln y^2,\ x = e^{-t},\ y = e^t$
8    $u = \text{Arctan } x + \sqrt{1 - y^2},\ x = t^2,\ y = t - 1$

Find u_r and u_s in each of Problems 9 to 11.
9    $u = x^2 + xy + y^2,\ x = r + s,\ y = r - s$
10    $u = \text{Arctan }(y/x),\ x = r + s,\ y = rs$
11    $u = \cos xy,\ x = r^2s,\ y = e^{rs}$

12   Show that

$$(z_x)^2 + (z_y)^2 = (z_r)^2 + \frac{(z_\theta)^2}{r^2}$$

if $z = f(x, y)$, $x = r\cos\theta$, and $y = r\sin\theta$.

In Problems 13 to 20, find du/dt or u_r and u_s.

13 $u = xy + yz - zx$, $x = \sin t$, $y = \cos t$, $z = \sin 2t$
14 $u = x^2 - y^2 + z^2$, $x = \cos t$, $y = \sin t$, $z = \sin 2t$
15 $u = x/y + y - z$, $x = t^2 - 1$, $y = t - 1$, $z = t + 1$
16 $u = e^x + y^2 - yz$, $x = \ln t$, $y = \sqrt{t^2 + 1}$, $z = t/\sqrt{t^2 + 1}$
17 $u = xy - yz + zx$, $x = rs$, $y = r - s$, $z = r + s$
18 $u = xyz - y + z$, $x = 1/(r^2 - s^2)$, $y = r + s$, $z = r - s$
19 $u = x^2 + y^2 + z^2$, $x = r \cos s$, $y = s \sin r$, $z = rs$
20 $u = \ln \sqrt{x^2 + y^2 + z^2}$, $x = r \cos s$, $y = r \sin s$, $z = r \tan s$

21 If u is a function of x and y in which x and y occur only in the combination y/x, show that $xu_x + yu_y = 0$.
22 Show that if x and y occur only in the combination $x^2 + y^2$ in $u = f(x, y)$, then $yu_x = xy_y$.
23 If u is a function of x and y in which x and y enter only in the combinations $x - y$ and $y - x$, show that $u_x + u_y = 0$.
24 If u is a function of x, y, and z in which these variables occur only in the combination $x - y$, $y - z$, and $z - x$, show that $u_x + u_y + u_z = 0$.

In Problems 25 to 32 find the directional derivative of the given function at the specified point for the direction shown.
25 $u = x^2 + y^2 - 4$, $(2, -1)$, $\pi/4$
26 $u = x^2 - xy + y^2$, $(3, 1)$, $\pi/3$
27 $u = \sqrt{17 - x^2 - y^2}$, $(2, 3)$, $3\pi/4$
28 $u = x^2 - y^2 - 2x$, $(2, -2)$, $\pi/6$
29 $u = e^{x+y} - e^{-x-y}$, $(\ln 3, \ln 2)$, $\pi/4$
30 $u = e^y \cos x$, $(\pi/2, 0)$, $-\pi/6$
31 $u = \tan x + \sec y$, $(\pi/4, \pi/3)$, $\pi/2$
32 $u = x^3 + 3xy - y^3$, $(1, -2)$, π

33 Find the maximum value of du/ds at $(2, 2)$ and the value of α for which it occurs if $u = x^2 + y^2 - 3xy$.
34 Find the gradient of $u = x^2 - 2xy$ at $(1, 0)$.
35 Find the gradient of $u = e^x \sin y$ at $(0, 3\pi/4)$.
36 Find the normal derivative of $u = e^{x-y}$ at $(1, 1)$.
37 The temperature at every point of a heated circular plate is

$$T = \frac{75}{x^2 + y^2 + 11}$$

where T is in degrees and x and y in inches. Find the rate of change of T at $(3, 4)$ for $\alpha = \pi/4$.
38 If $v = \ln \sqrt{x^2 + y^2}$, show that v is constant along any circle with center at the origin. Show also that at any point $P(x, y)$, the direction of the gradient of v is that of the radial line drawn through P and the origin.

39 If $u = f(x, y)$, show that at any point $P(x, y)$, the direction in which $du/ds = 0$ is given by $\tan \alpha = -u_x/u_y$, hence, that the direction in which du/ds is largest is perpendicular to that in which it is zero.

40 Prove that the maximum value of the directional derivative of $u = f(x, y)$ at any point $P_1(x_1, y_1)$ for all directions is in the direction $\alpha = \text{Arctan}\,(u_y/u_x)$ or $\text{Arctan}\,(u_y/u_x) + \pi$. Furthermore, show that its value is $\sqrt{u_x{}^2 + u_y{}^2}$.

24.8 Differentiation of implicit functions

In Chap. 5, we saw that an equation of the form

$$\phi(x, y) = 0 \tag{1}$$

may define y as a differentiable function of x, and we then found dy/dx by use of implicit differentiation. We shall now find a formula for dy/dx in terms of partial derivatives. As a first step, we let

$$u = \phi(x, y)$$

then

$$du = \frac{\partial \phi}{\partial x}\,dx + \frac{\partial \phi}{\partial y}\,dy \tag{2}$$

regardless of whether x and y are independent variables or not. If we restrict x and y to values that satisfy (1), then $u \equiv 0$ and $du \equiv 0$. Consequently, replacing du in (2) by zero and solving for dy/dx, we find that

derivative of an implicit function

$$\frac{dy}{dx} = -\frac{\partial \phi/\partial x}{\partial \phi/\partial y} \qquad \text{if} \qquad \frac{\partial \phi}{\partial y} \neq 0 \tag{24.9}$$

Example 1 Find dy/dx if $\phi(x, y) = x^2 - y^3 + xy - 7 = 0$.

Solution: In applying (24.9), we need ϕ_x and ϕ_y and shall now find them.

$$\phi_x = 2x + y \qquad \text{and} \qquad \phi_y = -3y^2 + x$$

Consequently,

$$\frac{dy}{dx} = \frac{2x + y}{3y^2 - x} \qquad x \neq 3y^2$$

We shall now consider

$$\phi(x, y, z) = 0 \tag{3}$$

If (3) defines z as function of x and y, we can find $\partial z/\partial x$ as follows: Let

$$u = \phi(x, y, z)$$

then
$$du = \phi_x \, dx + \phi_y \, dy + \phi_z \, dz \tag{4}$$

Furthermore, if x, y, and z are restricted to values that satisfy (3), then $u \equiv 0$ and $du \equiv 0$. Hence, (4) becomes

$$\phi_x \, dx + \phi_y \, dy + \phi_z \, dz = 0 \tag{5}$$

and
$$\phi_x \frac{dx}{dx} + \phi_y \frac{dy}{dx} + \phi_z \frac{dz}{dx} = 0 \tag{6}$$

If now x and y are independent variables, then $dy/dx = 0$ and dz/dx becomes $\partial z/\partial x$. Consequently, since $dx/dx = 1$, (6) becomes

$$\phi_x + \phi_z \frac{\partial z}{\partial x} = 0$$

and, solving for $\partial z/\partial x$, we find that

$$\frac{\partial z}{\partial x} = -\frac{\phi_x}{\phi_z} \qquad \phi_z \neq 0 \tag{24.10a}$$

We can find similarly that

$$\frac{\partial z}{\partial y} = -\frac{\phi_y}{\phi_z} \qquad \phi_z \neq 0 \tag{24.10b}$$

If an equation $\phi(x_1, x_2, \ldots, x_n) = 0$ defines x_n as a function of $x_1, x_2, \ldots, x_{n-1}$, then $\partial x_n/\partial x_i$ is determined by an equation such as $(24.10a)$ and $(24.10b)$.

Example 2   Find $\partial z/\partial w$ if

$$\phi(w, x, y, z) = w^2 - wx + x + y^2 + yz^2 = 0$$

Solution: In order to find $\partial z/\partial w$, we need ϕ_w and ϕ_z, and they are

$$\phi_w = 2w - x \qquad \text{and} \qquad \phi_z = 2yz$$

Therefore,
$$\frac{\partial z}{\partial w} = -\frac{\phi_w}{\phi_z} = \frac{x - 2w}{2yz}$$

24.9   Partial derivatives of higher order

Since $\partial z/\partial x$ and $\partial z/\partial y$ where z is a function of x and y are also functions of x and y, we may differentiate each of them with respect to x and y. These derivatives are called second derivatives and are indicated by the following notation:

$$\frac{\partial}{\partial x}\left(\frac{\partial z}{\partial x}\right) = \frac{\partial^2 z}{\partial x^2} = z_{xx}$$

$$\frac{\partial}{\partial y}\left(\frac{\partial z}{\partial y}\right) = \frac{\partial^2 z}{\partial y^2} = z_{yy}$$

$$\frac{\partial}{\partial x}\left(\frac{\partial z}{\partial y}\right) = \frac{\partial^2 z}{\partial x \partial y} = z_{yx}$$

$$\frac{\partial}{\partial y}\left(\frac{\partial z}{\partial x}\right) = \frac{\partial^2 z}{\partial y \partial x} = z_{xy}$$

It is shown in most books on advanced calculus° that $z_{yx} = z_{xy}$ if $z = f(x, y)$ *and both second partial derivatives are continuous.*

Derivatives of order higher than the second can be obtained, and a notation for some of them is as follows:

$$\frac{\partial^3 z}{\partial x^3}, \quad \frac{\partial^3 z}{\partial x^2 \partial y}, \quad \frac{\partial^3 z}{\partial x \partial y^2}, \quad \frac{\partial^3 z}{\partial y^3}, \quad \frac{\partial^4 z}{\partial x^3 \partial y}$$

The first symbol indicates the result of differentiating successively three times with respect to x, and the third symbol indicates the result of differentiating twice with respect to y and then once with respect to x.

Example   Find z_{xx}, z_{xy}, z_{yy} if $z = \sin (2x + 3y) + \ln x$.

Solution: In order to find the second partial derivatives, we begin by finding that

$$z_x = 2 \cos (2x + 3y) + \frac{1}{x}$$

and

$$z_y = 3 \cos (2x + 3y)$$

Consequently,

$$z_{xx} = -4 \sin (2x + 3y) - \frac{1}{x^2}$$

$$z_{xy} = -6 \sin (2x + 3y)$$

$$z_{yy} = -9 \sin (2x + 3y)$$

EXERCISE 24.4

In each of Problems 1 to 8, find dy/dx in two ways without solving for y.

1	$xy + y^3 - 5 = 0$	2	$xy - 5y - 7 = 0$
3	$x^2y + x - y - 3 = 0$	4	$xy^2 - x + 2y + 4 = 0$
5	$x^3 - x^2 + y^4 + 7 = 0$	6	$x^4 + 4xy + 2y^2 - 2 = 0$

° Woods, "Advanced Calculus," p. 68, Ginn, Boston, 1934.

7   $x^3 + x^2y - y^2 - 3 = 0$ 8 $x^2 + xy^2 - xy - y = 0$

Find z_x and z_y in each of Problems 9 to 12.

9   $x^2 + y^2 + 9z^2 = 25$ 10 $x^2 + xy + yz - z^2 = 7$

11   $x^3 - 3xy - 9yz + z^3 = 9$ 12 $x^4 + 2xy^2 - xz^3 = 5$

13   Find $\partial u/\partial x$ if $x^3 + yzu - xu^2 = 3$.

14   Find $\partial v/\partial t$ if $x^2v + ty + vt^2 = 4$.

15   Find $\partial s/\partial y$ if $x^3 + xy^2 + zs^2 - sy = 0$.

16   Find $\partial w/\partial z$ if $wz^2 - zx^2 - xy^2 + yw^2 = 41$.

In each of Problems 17 to 24 find z_{xx}, z_{xy}, and z_{yy}.

17   $z = x^2y^2 - y$ 18 $z = xy^2 - x^2 + y^3$

19   $z = (x - y)/(x + y)$ 20 $z = xy/(x - y)$

21   $z = x \cos (x - y)$ 22 $z = y \sin (2x - y)$

23   $z = x \ln y$ 24 $z = ye^{xy}$

Show that $z_{xy} = z_{yx}$ in each of Problems 25 to 28.

25   $z = 2x^3 - 3xy + y^2$ 26 $z = 2xy^2 - x^3 + 3y$

27   $z = \cos x + xe^y$ 28 $z = \ln (x^2 - y) - x^2 + y$

Show that $z_{xyy} = z_{yxy} = z_{yyx}$ in each of Problems 29 to 32.

29   $z = x^2y^3 + 4x^3y^2 - x^4y$ 30 $z = (x + y)/(x - y)$

31   $z = x \tan y$ 32 $z = e^{xy} + y \ln x$

33   Show that $\partial^2z/\partial x^2 + \partial^2z/\partial y^2 = 0$ if $z = \ln (x^2 + y^2) + $ Arctan (y/x).

34   Show that $\partial^2z/\partial x^2 + \partial^2z/\partial y^2 = 0$ if $z = ke^y \sin x$.

35   Show that $\partial^2z/\partial y^2 - a^2 \; \partial^2z/\partial x^2 = 0$ if $z = K \cos kay \sin kx$.

36   Show that

$$\frac{\partial^2z}{\partial x^2} + \frac{\partial^2z}{\partial y^2} = \frac{\partial^2z}{\partial r^2} + \frac{1}{r^2}\frac{\partial^2z}{\partial \theta^2}$$

if $z = f(x, y)$, $x = r \cos \theta$, and $y = r \sin \theta$.

24.10   Maxima and minima

We studied maxima and minima of a function of a single independent variable in Chap. 10 and shall now consider them for a function of two independent variables. For this purpose, we shall consider the surface $z = f(x, y)$ as shown in Fig. 24.5.

We say that $z = f(x, y)$ has a maximum value for $x = a$ and $y = b$ if the value of $c = f(a, b)$ is greater than the value of $z = f(x, y)$ for all points (x, y) in the neighborhood of (a, b). This can be stated more precisely by saying that if $\delta > 0$ exists such that for all h *definition* in $(a - \delta, a + \delta)$ and all k in $(b - \delta, b + \delta)$ we have $f(a + h, b + k)$ *of maxima* *and minima* $-f(a, b) < 0$, then $c = f(a, b)$ is a *maximum*; furthermore, if $\delta > 0$

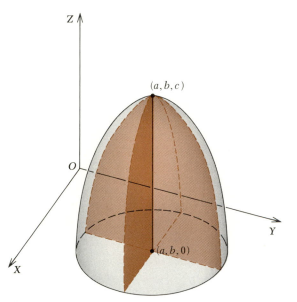

Figure 24.5   Re maxima and minima

exists such that for all h in $(a - \delta, a + \delta)$ and all k in $(b - \delta, b + \delta)$, we have $f(a + h, b + k) - f(a, b) > 0$, then $c = f(a, b)$ is a minimum.

In order for the surface $z = f(x, y)$ to have a maximum or minimum point for $x = a$ and $y = b$, it is necessary for the curve of intersection of the surface and the plane $x = a$ to have a maximum or minimum at that point. Furthermore, the curve of intersection of the surface and $y = b$ must have a maximum or minimum value at the point corresponding to $x = a$, $y = b$. Consequently, *if $z = f(x, y)$ and its first partial derivatives are continuous in a region that includes (a, b),* *a necessary* *then a necessary condition for $c = f(a, b)$ to be a maximum or* *condition* *minimum of the function is that*

$$\frac{\partial f(a, b)}{\partial x} = 0 \qquad \text{and} \qquad \frac{\partial f(a, b)}{\partial y} = 0$$

Example 1   Find a condition that must be satisfied for

$$z = x^2 - y^2 - 2xy - 4x$$

to have a maximum or minimum.

Solution: Since the necessary condition for a maximum or minimum is that both first partial derivatives be zero, we shall find z_x and z_y, set them equal to zero, and solve the resulting pair of equations simultaneously. Thus,

$$z_x = 2x - 2y - 4 = 0 \qquad \text{and} \qquad z_y = -2y - 2x = 0$$

The solution of this pair of equations is $(1, -1)$; hence, if there is a maximum or minimum of the function, it occurs for $x = 1$, $y = -1$.

The condition discussed and illustrated is satisfied if the function has a maximum or a minimum, but there is no assurance that there is a maximum or minimum at a point even though both first partial derivatives are zero there. We shall now state without proof a sufficient condition for a maximum or minimum that is given and proved in most advanced calculus books. The test is:

If, for x = a and y = b,

$$f_x = 0 \qquad and \qquad f_y = 0$$

and if, for these values of x and y,

$$f_{xx}f_{yy} - (f_{xy})^2 > 0$$

then c = f(a, b) is a maximum value if f_{xx} and f_{yy} are negative and is a minimum value if f_{xx} and f_{yy} are positive; furthermore, if

$$f_{xx}f_{yy} - (f_{xy})^2 < 0$$

then c = f(a, b) has a critical value which is neither a maximum nor a minimum; and if

$$f_{xx}f_{yy} - (f_{xy})^2 = 0$$

the test fails to give the desired information.

Example 2   Test $z = x^2 - y^2 - 2xy - 4x$ for maxima and minima.

Solution: This is the function considered in Example 1. We found that $z_x = 2x - 2y - 4$ and $z_y = -2y - 2x$ are simultaneously zero for $x = 1$ and $y = -1$. We now find that $z_{xx} = 2$, $z_{yy} = -2$, and $z_{xy} = -2$. Therefore,

$$z_{xx}z_{yy} - (z_{xy})^2 = (2)(-2) - (-2)^2 = -8 < 0$$

and the function has neither a maximum nor a minimum at $(1, -1)$.

Example 3   Test $z = x^3 - 6x - 6xy + 6y + 3y^2$ for maxima and minima.

Solution: We shall first find the derivatives that are needed. Thus,

$$z_x = 3x^2 - 6 - 6y \qquad z_y = -6x + 6 + 6y$$

$$z_{xx} = 6x \qquad z_{yy} = 6 \qquad z_{xy} = -6$$

Consequently, as seen from solving $z_x = 0$ and $z_y = 0$ simultaneously, the first partial derivatives are both zero at $(0, -1)$ and at $(2, 1)$. Furthermore, $z_{xx}(0, -1) = 0$, $z_{xx}(2, 1) = 12$ and $z_{yy} = 6$ and

$z_{xy} = -6$ at all points. Therefore, at $(0, -1)$,

$$z_{xx}z_{yy} - (z_{xy})^2 = 0 \cdot 6 - (-6)^2 < 0$$

and at $(2, 1)$, $\quad z_{xx}z_{yy} - (z_{xy})^2 = 12 \cdot 6 - (-6)^2 > 0$

We now know that the given function has a minimum for $x = 2$, $y = 1$ since $z_x = z_y = 0$, $z_{xx} > 0$, $z_{yy} > 0$, and $z_{xx}z_{yy} - (z_{xy})^2 > 0$ for $(2, 1)$. Furthermore, the function has a critical value which is neither a maximum nor minimum at $(0, -1)$ since $z_{xx}z_{yy} - (z_{xy})^2 < 0$ there.

EXERCISE 24.5

Test the function in each of Problems 1 to 12 for maxima and minima.

1   $z = xy$ 2 $z = x^2 - y^2$

3   $z = x^2 + y^2 - 2$ 4 $z = (x - h)^2 + (y - k)^2 - a^2$

5   $z = x^3 - 3xy + y^2 + y - 7$

6   $z = x^4 + y^4 - 4xy$ 7 $z = x^3 + y^3 + 3xy$

8   $z = x^3 + y^3 - 3xy$ 9 $z = y \ln x - x$

10   $z = \sin x - 2 \cos y - x - y$

11   $z = y^2 + b^2 + 2by \cos x$

12   $z = e^x + 4 \ln y - x - y$

13   Find the three parts into which a positive number N must be divided so that their product is as large as possible.

14   Find the shortest distance from the origin to $xyz^2 = 1$.

15   Find the shape of the rectangular parallelepiped of greatest volume for a fixed surface area.

16   A rectangular parallelepiped has a given volume. Find its relative dimensions if the surface area is a minimum.

17   Find the shape of the rectangular parallelepiped of given volume if the sum of edges is a minimum.

18   Show that a cube is the largest rectangular parallelepiped that can be inscribed in a sphere.

19   Find the volume of the largest rectangular parallelepiped that can be inscribed in $x^2/a^2 + y^2b^2 + z^2/c^2 = 1$.

20   Prove that if the product of the sines of the angles of a triangle is a maximum, the triangle is equilateral.

Multiple integrals

25.1   Double integrals

In Chap. 24 we defined the second partial derivative $\partial^2 f(x, y)/\partial y \partial x$ as the function obtained by the two steps: first, regarding y as a constant and differentiating with respect to x, obtaining f_x; and second, regarding x as a constant and differentiating f_x with respect to y. In this chapter we shall discuss an analogous procedure in which we perform successive operations of integration on a function of two or more variables, regarding each variable in turn as a constant.

As an illustration we shall consider the pyramid $OABC$ in Fig. 25.1 bounded by the plane

$$z = a - x - y \tag{1}$$

and the coordinate planes. We know from solid geometry that the volume of this pyramid is $a^3/6$, and we shall now obtain this volume by two successive integrations. For this purpose we choose a point $D(0, y, 0)$ on OB and pass a plane through D parallel to the XZ plane and intersecting the pyramid in the right triangle EDH. The point E is on the line AB whose equation is $x + y = a$. Hence the x coordinate of E in terms of y is $a - y$. Furthermore the area of the triangle EDH is given by the definite integral

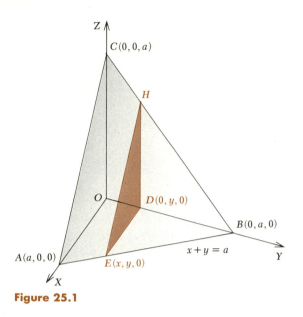

Figure 25.1

$$K = \int_0^{a-y} z \, dx = \int_0^{a-y} (a - x - y) \, dx \tag{2}$$

If for the present we regard y as a constant and evaluate the integral in (2), we have

$$K = \left[ax - \frac{x^2}{2} - yx \right]\Big|_0^{a-y} = \frac{1}{2}(a^2 - 2ay + y^2)$$

Since K is the area of a cross section of the pyramid parallel to the XZ plane, the volume of the pyramid is

$$V = \int_0^a K \, dy \tag{3}$$

$$V = \frac{1}{2} \int_0^a (a^2 - 2ay + y^2) \, dy = \frac{1}{2} \left[a^2 y - ay^2 + \frac{y^3}{3} \right]\Big|_0^a = \frac{a^3}{6}$$

If we replace K in (3) by its value in (2), we have

$$V = \int_0^a \int_0^{a-y} (a - x - y) \, dx \, dy \tag{4}$$

where it is understood that we evaluate the inner definite integral first, regarding y as a constant.

iterated
integral   The integral in (4) is an example of the *iterated* integral

$$\int_a^b \int_{h(y)}^{g(y)} f(x, y) \, dx \, dy$$

double
integral and it may also be expressed as the *double* integral $\int_R \int dA$ where R

is the set of all interior and boundary points of the triangle ABC and $dA = dx\,dy$ is a rectangular element of the area of the triangle.

In this chapter we shall define a multiple integral as the limit of a sum and shall discuss some applications of these integrals.

In order to define the double integral of a function of two variables, we shall begin by considering a finite closed plane region R whose equations are of the form $y = F(x)$ with $F(x)$ and $F'(x)$ continuous. Furthermore, we shall let $f(x, y)$ be a function that is defined throughout R.

We now divide R into n subregions of maximum diameter δ by use of two systems of curves or lines, as in Fig. 25.2, and call the subregions $\Delta A_1, \Delta A_2, \cdots, \Delta A_n$, where the system of subregions is such that δ approaches zero as n approaches infinity. We now take a point (ξ_k, η_k) inside or on the boundary of each subregion ΔA_k and form the sum

$$f(\xi_1, \eta_1)\,\Delta A_1 + f_2(\xi_2, \eta_2)\,\Delta A_2 + \cdots + f_n(\xi_n, \eta_n)\,\Delta A_n$$

$$= \sum_{k=1}^{n} f_k(\xi_k, \eta_k)\,\Delta A_k$$

where ΔA_k denotes the area of the kth subregion.

We now define the double integral of $f(x, y)$ over R by saying that

definition of
$\int_R\!\int f(x, y)\,dA$

if $\displaystyle\sum_{k=1}^{n} f(\xi_k, \eta_n)\,\Delta A_k$ *approaches a limit independent of the mode of subdivision of R and independent of the choice of (ξ_k, η_k) in ΔA_k as δ approaches zero, then this limit is called the double integral of $f(x, y)$ over the region R, is denoted by $\int_R\!\int f(x, y)\,dA$, and we write*

$$\lim_{\delta \to 0} \sum_{k=1}^{n} f(\xi_k, \eta_k)\,\Delta A_k = \int_R\!\int f(x, y)\,dA \qquad (25.1)$$

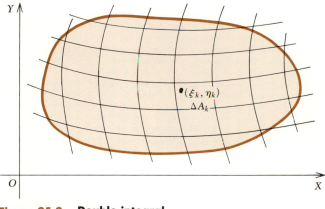

Figure 25.2 **Double integral**

In practice, the double integral is evaluated by evaluating two single integrals successfully. Specifically, the value of the double integral $\int_R \int f(x, y)\, dA$ is calculated by finding the value of either one of the iterated integrals

$$\int_R \int f(x, y)\, dy\, dx \qquad \int_R \int f(x, y)\, dx\, dy \tag{5}$$

The equality of the double integral (25.1) and the iterated integrals (5) is proved in most books on advanced calculus.*

We shall now see how to evaluate the iterated integral

evaluation of a double integral

$$\int_R \int f(x, y)\, dy\, dx$$

We proceed by (a) evaluating $\int f(x, y)\, dy$ with respect to y between $y = f_1(x)$ and $y = f_2(x)$ with x held constant and then (b) integrating the result of (a) with respect to x between $x = a$ and $x = b$. In order to evaluate

$$\int_R \int f(x, y)\, dx\, dy$$

we begin by integrating $\int f(x, y)\, dx$ between $x = g_1(y)$ and $x = g_2(y)$ and then we integrate the result thereof between $y = c$ and $y = d$. In each case, the limits of integration must be chosen so as to extend the integration over the desired region R.

Example 1   Evaluate $\int_0^2 \int_{2x}^{x^2} (x - 2y)\, dy\, dx$.

Solution

$$\int_0^2 \int_{2x}^{x^2} (x - 2y)\, dy\, dx = \int_0^2 (xy - y^2)\Big|_{2x}^{x^2} dx \qquad \text{\textit{since x held constant}}$$

$$= \int_0^2 [x^3 - x^4 - (2x^2 - 4x^2)]\, dx$$

$$= \int_0^2 (x^3 - x^4 + 2x^2)\, dx$$

$$= \frac{x^4}{4} - \frac{x^5}{5} + \frac{2x^3}{3}\Big|_0^2$$

$$= \frac{16}{4} - \frac{32}{5} + \frac{16}{3}$$

$$= \frac{44}{15}$$

* For example, it can be found in Osgood, "Advanced Calculus," p. 262, The Macmillan Company, New York, 1925.

Example 2 Sketch the region covered by the iterated integral in Example 1, and then express and evaluate the integral in the order $dx\, dy$.

Solution: The sketch of the area covered by the integration is shown in Fig. 25.3. It is the area between $y = 2x$ and $y = x^2$ for $x = 0$ to $x = 2$. We want to select the limits of integration so that the same area is covered with the order of integration changed. Thus, x must vary from the x of $y = 2x$ to x of $y = x^2$, hence, from $x = y/2$ to $x = y^{1/2}$. Furthermore, the limits on y must be the values of y that correspond to $x = 0$ and $x = 2$, hence, 0 and 4. Since the order of limits in Example 1 is such that each Δy is negative and each Δx is positive, we choose the order of limits in this example so that each Δy is negative and each Δx is positive. Therefore,

$$\int_0^2 \int_{2x}^{x^2} (x - 2y)\, dy\, dx = \int_4^0 \int_{y/2}^{y^{1/2}} (x - 2y)\, dx\, dy$$

$$= \int_4^0 \left(\frac{x^2}{2} - 2xy\right)\Bigg|_{y/2}^{y^{1/2}} dy \qquad \textit{y held constant}$$

$$= \int_4^0 \left(\frac{y}{2} - 2y^{3/2} - \frac{y^2}{8} + y^2\right) dy$$

$$= \int_4^0 \left(\frac{y}{2} - 2y^{3/2} + \frac{7y^2}{8}\right) dy$$

$$= \frac{y^2}{4} - \frac{2}{1}\frac{2}{5} y^{5/2} + \frac{7}{8}\frac{y^3}{3}\Bigg|_4^0$$

$$= -\left[\frac{16}{4} - \frac{4}{5}(4^{5/2}) + \frac{7}{8}\frac{64}{3}\right]$$

$$= -\left[4 - \frac{4}{5}(32) + \frac{56}{3}\right]$$

$$= \frac{44}{15}$$

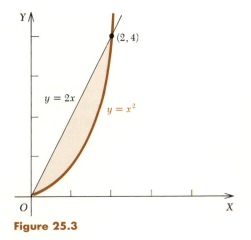

$(2, 4)$

$y = 2x$

$y = x^2$

Figure 25.3

25.2   Triple integrals

Just as the concept of $\int_a^b f(x)\,dx$ was extended to the iterated double integral $\int_a^b \int_{f_1(x)}^{f_2(x)} f(x, y)\,dy\,dx$, the latter can be extended to an iterated triple integral. We use the symbol

$$\int_{z_1}^{z_2} \int_{y_1}^{y_2} \int_{x_1}^{x_2} f(x, y, z)\,dx\,dy\,dz$$

triple integrals

to show that $f(x, y, z)$ is to be integrated with respect to x between the limits x_1 and x_2 with y and z constant; the result of this operation integrated with respect to y between the limits y_1 and y_2 with z constant; and this result integrated with respect to z between the limits z_1 and z_2. All limits of integration are not necessarily constants; in fact, the limits x_1 and x_2 may be functions of y and z and the limits y_1 and y_2 may be functions of z if the order of integration is as above but z_1 and z_2 are constants.

Example 1   Evaluate $\int_3^5 \int_2^4 \int_0^1 (4xy + 3z^2)\,dy\,dx\,dz$.

Solution: Since the first integration is with respect to y, we have

$$\int_3^5 \int_2^4 \int_0^1 (4xy + 3z^2)\,dy\,dx\,dz = \int_3^5 \int_2^4 (2xy^2 + 3z^2 y)\Big|_0^1 dx\,dz$$

$$= \int_3^5 \int_2^4 (2x + 3z^2)\,dx\,dz$$

$$= \int_3^5 (x^2 + 3z^2 x)\Big|_2^4 dz$$

$$= \int_3^5 [(16 + 12z^2) - (4 + 6z^2)]\,dz$$

$$= \int_3^5 (12 + 6z^2)\,dz = (12z + 2z^3)\Big|_3^5$$

$$= (60 + 250) - (36 + 54) = 220$$

We shall now consider an example in which the limits of integration in the first integration are functions of the other two variables and the limits in the second integration are functions of the third variable.

Example 2   Evaluate $\int_0^1 \int_0^{1/y} \int_0^{x+y} (x - y)\,dz\,dx\,dy$.

Solution: Performing the integrations in the order indicated by the positions of the differentials, we have

$$\int_0^1\int_0^{1/y}\int_0^{x+y}(x-y)\,dz\,dx\,dy = \int_0^1\int_0^{1/y}(x-y)z\Big|_0^{x+y}\,dx\,dy$$

$$= \int_0^1\int_0^{1/y}(x^2-y^2)\,dx\,dy$$

$$= \int_0^1\left(\frac{x^3}{3}-xy^2\right)\Big|_0^{1/y}\,dy$$

$$= \int_0^1\left(\frac{1}{3y^3}-y\right)dy = \left(\frac{1}{3}\frac{y^{-2}}{-2}-\frac{y^2}{2}\right)\Big|_0^1$$

$$= -\frac{1}{6}-\frac{1}{2} = -\frac{2}{3}$$

EXERCISE 25.1

Evaluate each of the integrals in Problems 1 to 20.

1 $\displaystyle\int_0^1\int_0^2 y\,dy\,dx$ 2 $\displaystyle\int_0^4\int_0^3 x\,dy\,dx$

3 $\displaystyle\int_1^4\int_1^3(x^2-y)\,dx\,dy$ 4 $\displaystyle\int_1^4\int_{-2}^2 xy\,dx\,dy$

5 $\displaystyle\int_0^1\int_{x^2}^{2x} y\,dy\,dx$ 6 $\displaystyle\int_1^3\int_y^{3y} x\,dx\,dy$

7 $\displaystyle\int_1^3\int_y^{y^2-1}(y-1)\,dx\,dy$ 8 $\displaystyle\int_0^6\int_{y^{1/2}}^{y-2}2x\,dx\,dy$

9 $\displaystyle\int_0^3\int_0^{2t}(x+t)\,dx\,dt$ 10 $\displaystyle\int_0^3\int_0^{\sqrt{9-t^2}}\sqrt{9-t^2}\,dx\,dt$

11 $\displaystyle\int_0^2\int_0^{\sqrt{x+1}}\sqrt{x+1}\,dt\,dx$ 12 $\displaystyle\int_{\sqrt3}^5\int_0^{\sqrt{x^2-1}}\sqrt{x^2-1}\,dt\,dx$

13 $\displaystyle\int_0^{\pi/12}\int_0^{\cos\theta} r\,dr\,d\theta$ 14 $\displaystyle\int_{\pi/2}^{\pi}\int_0^{1-\sin\theta} r\,dr\,d\theta$

15 $\displaystyle\int_0^{\pi/2}\int_{2\sin\theta}^2 r\,dr\,d\theta$ 16 $\displaystyle\int_{\pi/2}^{\pi}\int_0^{4\cos\theta} r\,dr\,d\theta$

17 $\displaystyle\int_0^1\int_0^2\int_{-1}^0 yz\,dx\,dy\,dz$ 18 $\displaystyle\int_{-1}^1\int_0^2\int_{-3}^3(2x+y)\,dx\,dy\,dz$

19 $\displaystyle\int_0^1\int_0^y\int_0^{x+y}6xy\,dz\,dx\,dy$ 20 $\displaystyle\int_0^{\pi/2}\int_{a\sin\theta}^a\int_0^{\sqrt{a^2-r^2}} r\,dz\,dr\,d\theta$

Set up another double iterated integral that is equal to the given integral in each of Problems 21 to 24 but with the order of integration changed. Check your work by evaluating both integrals in each case.

21 $\displaystyle\int_0^1 \int_{x^3}^{x^2} dy \, dx$

22 $\displaystyle\int_0^{\sqrt{2}} \int_{x^2-1}^1 dy \, dx$

23 $\displaystyle\int_0^1 \int_0^{\sqrt{1-y^2}} dx \, dy$

24 $\displaystyle\int_{-2}^2 \int_{y/2}^{-1+\sqrt{y+2}} dx \, dy$

25.3   Area by use of double integration

We may express, as a double iterated integral, the area $PQSR$ bounded by the curves $y = f(x)$ and $y = g(x)$ and the lines $x = a$ and $x = b$. As seen from Fig. 25.4, this area is

$$PQSR = APQB - ARSB = \int_a^b f(x) \, dx - \int_a^b g(x) \, dx$$

$$= \int_a^b [f(x) - g(x)] \, dx$$

$$= \int_a^b \int_{g(x)}^{f(x)} dy \, dx \qquad since \int_{g(x)}^{f(x)} dy = f(x) - g(x)$$

area as a double integral   Hence, $PQSR = \displaystyle\int_a^b \int_{g(x)}^{f(x)} dy \, dx$

We shall now obtain this double interated integral by use of a double summation. To do this, we begin by drawing a set of lines parallel to each coordinate axis as in Fig. 25.5. We thus divide the area into rectangles with dimensions Δx and Δy and parts of such rectangles. One of these rectangles is made to stand out in the figure by shading it in black. If we make use of the fundamental theorem, we can write

$$\lim_{n \to \infty} \sum_{i=1}^n \left(\lim_{m \to \infty} \sum_{j=1}^m \Delta y_j \right) \Delta x_i = \lim_{n \to \infty} \sum_{i=1}^n \left(\int_{g(x)}^{f(x)} dy \right) \Delta x_i = \int_a^b \int_{g(x)}^{f(x)} dy \, dx$$

by a second use of the fundamental theorem. The inner summation indicates the sum of a column of rectangles and leads to a vertical strip

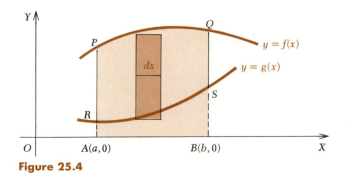

Figure 25.4

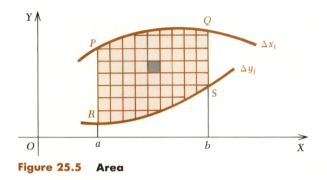

Figure 25.5   Area

as shown in Fig. 25.4. The second or outer summation indicates the sum of a set of vertical rectangular elements as was the case in our first contact with a definite integral in Chap. 7.

The order of integration may have a material effect on the ease of evaluating a double or triple integral; hence, it is desirable to be able to set it up with various orders until one that is relatively easy to evaluate is found.

Example   Find the area bounded by $y^2 = x + 3$ and $y = x - 3$.

Solution: A sketch of the curves and the area bounded by them is shown in Fig. 25.6. We find by solving simultaneously that the curves, intersect at $(1, -2)$ and $(6, 3)$. We shall integrate first with respect to x since integrating with respect to y first would require that we use one set of limits on y to the left of $x = 1$ and another to the right of $x = 1$. We thus have

$$A = \int_{-2}^{3} \int_{x_p}^{x_l} dx \, dy$$

where $x_l = y + 3$ is the x coordinate of a point on the line and

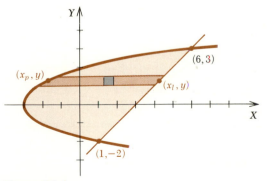

Figure 25.6

$x_p = y^2 - 3$ is the x coordinate of a point on the parabola in terms of y. If we replace the limits on x by these values, we have

$$A = \int_{-2}^{3}\int_{y^2-3}^{y+3} dx\, dy = \int_{-2}^{3} x\Big|_{y^2-3}^{y+3} dy = \int_{-2}^{3}[\, y + 3 - (y^2 - 3)]\, dy$$

$$= \int_{-2}^{3}(-y^2 + y + 6)\, dy = \frac{-y^3}{3} + \frac{y^2}{2} + 6y\Big|_{-2}^{3} = \frac{125}{6}$$

25.4   Area in polar coordinates

The area bounded by the curves $r = f(\theta)$ and $r = g(\theta)$ and the lines $\theta = \theta_1$ and $\theta = \theta_2$ can be expressed as a double iterated integral, and we shall now do so by use of Fig. 25.7. The typical element of area is bounded by two lines and two circular arcs centered at the pole; one such element is shaded in the figure.

The area of the region $ABCD$ is

$$ABCD = OAB - ODC$$

$$= \frac{1}{2}\int_{\theta_1}^{\theta_2} f^2(\theta)\, d\theta - \frac{1}{2}\int_{\theta_1}^{\theta_2} g^2(\theta)\, d\theta \qquad \text{\textit{by Sec. 19.2}}$$

$$= \frac{1}{2}\int_{\theta_1}^{\theta_2}[\, f^2(\theta) - g^2(\theta)]\, d\theta$$

$$= \frac{1}{2}\int_{\theta_1}^{\theta_2}\int_{g(\theta)}^{f(\theta)} 2r\, dr\, d\theta \qquad \int_{g(\theta)}^{f(\theta)} 2r\, dr = f^2(\theta) - g^2(\theta)$$

$$= \int_{\theta_1}^{\theta_2}\int_{g(\theta)}^{f(\theta)} r\, dr\, d\theta$$

Hence, $ABCD = \int_{\theta_1}^{\theta_2}\int_{g(\theta)}^{f(\theta)} r\, dr\, d\theta$

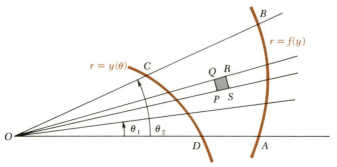

Figure 25.7   Area in polar coordinates

In addition to obtaining the area $ABCD$ as we did, it can be obtained by a double summation process. The typical element is $PQRS$ of Fig. 25.7. If the coordinates of P are $(r_k - \frac{1}{2}\Delta r_k, \theta_k - \frac{1}{2}\Delta\theta_k)$ and those of R are $(r_k + \frac{1}{2}\Delta r_k, \theta_k + \frac{1}{2}\Delta\theta_k)$, then the changes in r and θ of the element are Δr and $\Delta\theta$, and the area of the element is

$$PQRS = OSR - OPQ = \frac{1}{2}\left(r_k + \frac{\Delta r_k}{2}\right)^2 \Delta\theta_k - \frac{1}{2}\left(r_k - \frac{\Delta r_k}{2}\right)^2 \Delta\theta_k$$

$$= r_k\,\Delta r_k\,\Delta\theta_k$$

Therefore, by use of the usual summation process we find that the area

area as a double polar integral

$$ABCD = \int_{\theta_1}^{\theta_2}\int_{g(\theta)}^{f(\theta)} r\,dr\,d\theta$$

25.5   Second and product moments

If we multiply the area of each rectangular element of area by the square of the distance of the element from the X axis, add all of these products, and take the limit of this sum, we have the *second moment*

I_x

$$I_x = \int\int y^2\,dy\,dx$$

where limits of integration are to be so chosen as to cover the area under consideration. In a similar manner, we obtain the

second moments   second moments

$$I_y = \int\int x^2\,dy\,dx$$

$I_y, I_0,$ and

$$I_0 = \int\int (x^2 + y^2)\,dy\,dx$$

I_{xy}

Furthermore, $I_{xy} = \int\int xy\,dy\,dx$

product moment   is called the *product moment* and is used in the theory of bending beams that do not have symmetric cross sections. Each of these four moments can be expressed in terms of polar coordinates by replacing the element $dy\,dx$ of area by the element $r\,dr\,d\theta$ and replacing the coefficient of the element of area in terms of polar coordinates. Thus, we get

$$I_x = \int\int (r\sin\theta)^2 r\,dr\,d\theta = \int\int r^3\sin^2\theta\,dr\,d\theta$$

$$I_y = \int\int r^3\cos^2\theta\,dr\,d\theta$$

$$I_0 = \int\int r^3\,dr\,d\theta$$

$$I_{xy} = \int\int r^3\sin\theta\cos\theta\,dr\,d\theta$$

Example

Find the second moment about the X axis of the first-quadrant region bounded by $r = 5$, $\theta = \pi/2$, and $r = 3 \csc \theta$.

Solution: The region under consideration is shown in Fig. 25.8. If we solve the equation of the line and of the circle simultaneously, we find the value of θ at the first-quadrant intersection to be Arcsin 0.6. If we make use of

$$I_x = \iint r^3 \sin^2 \theta \, dr \, d\theta$$

and put in the limits $3 \csc \theta$ and 5 on r and Arcsin 0.6 and $\pi/2$ on θ, we have

$$I_x = \int_{\text{Arcsin } 0.6}^{\pi/2} \int_{3 \csc \theta}^{5} r^3 \sin^2 \theta \, dr \, d\theta$$

$$= \int_{\text{Arcsin } 0.6}^{\pi/2} \frac{r^4}{4} \Big|_{3 \csc \theta}^{5} \sin^2 \theta \, d\theta$$

$$= \frac{1}{4} \int_{\text{Arcsin } 0.6}^{\pi/2} (625 - 81 \csc^4 \theta) \sin^2 \theta \, d\theta$$

$$= \int_{\text{Arcsin } 0.6}^{\pi/2} (156.25 \sin^2 \theta - \frac{81}{4} \csc^2 \theta) \, d\theta$$

$$= \int_{\text{Arcsin } 0.6}^{\pi/2} (156.25 \frac{1 - \cos 2\theta}{2} - \frac{81}{4} \csc^2 \theta) \, d\theta$$

$$= \frac{156.25}{2} \left(\theta - \frac{1}{2} \sin 2\theta \right) + \frac{81}{4} \cot \theta \Big|_{\text{Arcsin } 0.6}^{\pi/2}$$

$$= \frac{156.25}{2} (\theta - \sin \theta \cos \theta) + \frac{81}{4} \cot \theta \Big|_{\text{Arcsin } 0.6}^{\pi/2}$$

$$= \frac{156.25\pi}{4} - \frac{156.25}{2} \text{Arcsin } 0.6 + \frac{156.25}{2} (0.6)(0.8) - \frac{81}{4} \left(\frac{0.8}{0.6} \right)$$

$$= \frac{156.25}{4} (\pi - 2 \text{ Arcsin } 0.6) + 10.50$$

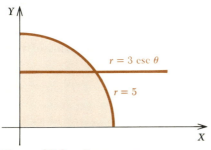

Figure 25.8   A moment

EXERCISE 25.2

Find the area of the region bounded by the curve or curves in each of Problems 1 to 16 by use of a double integral.

1	$y = x^2, y = 2x$	2	$2y = x^2, x + y = 0$
3	$y^2 = 6x, 2x - y = 0$	4	$y^2 = 6 - 5x, y = x$
5	$y = -x^2 + 4x, y = x^2 - x$	6	$y^2 = 6x, x^2 = 6y$
7	$y = 10x/(x^2 + 1), y = 2x$	8	$y = 2x/(x^2 + 1), y = x$
9	$x^2/a^2 + y^2/b^2 = 1$	10	$y = x^2 - 3x^3, y = x - 3x^2$
11	$y = x^2, y = x^3$	12	$x^2/a^2 - y^2/b^2 = 1, x = 2a$
13	$r = 2 \sin 3\theta$	14	$r = a(1 + \cos \theta)$
15	$r = 2 + \sin \theta$	16	$r^2 = a^2 \cos 2\theta$

17 Find the area bounded by $r = \sec \theta \tan \theta$ and $r = \sec \theta (2 - \tan \theta)$.

18 Find the area that is inside the lemniscate $r^2 = 8 \cos 2\theta$ and outside the circle $r = 2$.

19 Find the area that is common to the circle $r = 3 \cos \theta$ and the cardioid $r = 1 + \cos \theta$.

20 Find the area that is inside the circle $r = a \cos \theta$ and outside the cardioid $r = a(1 - \cos \theta)$.

21 Compute I_x for the area of the circle $x^2 + y^2 = a^2$ by use of a double integral.

22 Compute I_x for the area bounded by $y = \sin x, y = 0$, and $x = \pi/2$.

23 Compute I_y for the area in Problem 22.

24 Compute I_y for the area bounded by the X axis, $y = e^x, x = 0$, and $x = 1$.

25 Set up a double integral whose value is equal to I_x for the area inside $r = 3 \cos \theta$ and outside $r = 1 + \cos \theta$.

26 Set up a double integral whose value is equal to I_x for the area common to $r = 3 \cos \theta$ and $r = 1 + \cos \theta$.

27 Set up a double integral whose value is I_y for the first-quadrant area common to $r = \cos \theta$ and $r = 1 - \cos \theta$.

28 Set up a double integral whose value is I_y for the area inside $r^2 = 8 \cos 2\theta$ and outside $r = 2$.

29 Show that for a given area, $I_0 = I_x + I_y$.

30 Find I_0 for the area of the circle $r = 2a \cos \theta$.

31 Find I_0 for the area bounded by $x = \sqrt{y}, x = 0$, and $y = 1$.

32 Calculate I_0 for the first-quadrant area bounded by $x^2 + y^2 = 2$, $y^2 = x$, and $x = 0$.

Find the product moment of the area in each of Problems 33 to 36 with respect to the coordinate axes.

33 One quadrant of the circle $x^2 + y^2 = a^2$.

34   A rectangle of sides 6 and 8 with a vertex at the origin and two sides along the axes.

35   The area bounded by $\theta = 0$, $\theta = \pi/4$, and $r \cos \theta = 1$.

36   The first-quadrant area common to $r = a$ and $r = 2a \cos \theta$.

25.6   Volume by double integration

We shall now consider the volume bounded below by the XY plane, above by the surface $z = f(x, y)$ and laterally by the planes, $y = a$ and $y = b$ and the cylindrical surfaces $x = g(y)$ and $x = h(y)$.

We know from Sec. 19.6 that

$$V = \int A(y)\, dy$$

where $A(y)$ is the area of the section of the solid that is a distance y from the XZ plane. The section $PQRS$ in Fig. 25.9 is such a section. The area of this section, for y constant, is

$$A(y) = \int_{g(y)}^{h(y)} z\, dx = \int_{g(y)}^{h(y)} f(x, y)\, dx$$

volume
as a
Consequently, $$V = \int_a^b \int_{g(y)}^{h(y)} f(x, y)\, dx\, dy$$
double
integral   is the volume of the solid bounded by the XY plane, $z = f(x, y)$, $x = g(y)$, $x = h(x)$, $y = a$, and $y = b$.

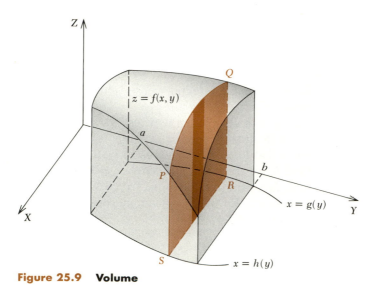

Figure 25.9   **Volume**

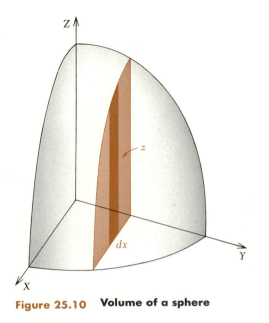

Figure 25.10   Volume of a sphere

A similar expression holds for the volume if $y = G(x)$, $y = H(x)$, $x = A$, and $x = B$ are parts of the boundary of the solid in place of $x = g(y)$, $x = h(y)$, $y = a$, and $y = b$.

This, as other, double integrals can be thought of as a double summation. The element of volume is a vertical rectangular column of base Δx by Δy and height $z = f(x, y)$.

Example   Find the volume of one octant of the sphere whose equation is $x^2 + y^2 + z^2 = a^2$ by use of double integration.

Solution: A sketch of the solid is shown in Fig. 25.10. The area of a cross section that is a distance y from the XZ plane is

$$A(y) = \int_0^{\sqrt{a^2 - y^2}} z\, dx$$

Consequently, the volume is given by

$$V = \int_0^a \int_0^{\sqrt{a^2 - y^2}} \sqrt{a^2 - y^2 - x^2}\, dx\, dy$$

since $z = \sqrt{a^2 - y^2 - x^2}$ is obtained from the equation of the sphere.

The volume of the first octant of the sphere $x^2 + y^2 + z^2 = a^2$ is also shown in Fig. 25.11. If it is used, we obtain the volume as a double summation. Thus

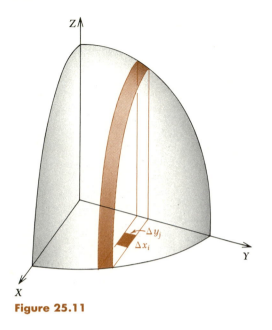

Figure 25.11

$$\Delta V = z\,\Delta x_i\,\Delta y_j$$

volume by double summation

$$V = \lim_{m\to\infty} \sum_{j=1}^{m} \left(\lim_{n\to\infty} \sum_{i=1}^{n} z\,\Delta x_i \right)\Delta y_j$$

$$= \int_0^a \int_0^{\sqrt{a^2-y^2}} z\,dy\,dy \qquad z = \sqrt{a^2 - y^2 - x^2}$$

as found above by considering the solid as one with known cross-sectional area.

The limits of integration are determined by the base of the solid. In the case of the octant of a sphere, they were determined by the quarter circle of intersection of the sphere and the XY plane.

25.7   Volume by triple integration

Just as we used a rectangular base of dimensions Δx and Δy and a height z as the element of volume in finding volume by means of a double integration considered as a double summation, we use a rectangular parallelepiped of dimensions Δx, Δy, and Δz as the element in finding volume by use of a triple integration. Thus,

volume as a triple integral

$$V = \left\{ \lim_{m\to\infty} \sum_{j=1}^{m} \left[\lim_{n\to\infty} \sum_{i=1}^{n} \left(\lim_{p\to\infty} \sum_{k=1}^{p} \Delta z_k \right)\Delta x_i \right]\Delta y_j \right\} = \int_{y_1}^{y_2}\int_{x_1}^{x_2}\int_{z_1}^{z_2} dz\,dx\,dy$$

where the limits are chosen so as to cover the desired volume. If the order of integration is as indicated above, z_1 and z_2 may be functions of x and y, and x_1 and x_2 may be functions of y, but the limits on y must be constant.

If triple integration is used to find the volume of the octant of a sphere of the last section, we have

$$V = \int_0^a \int_0^{\sqrt{a^2-y^2}} \int_0^{\sqrt{a^2-y^2-x^2}} dz\, dx\, dy$$

After the first integration is performed and the limits used, we obtain the expression

$$V = \int_0^a \int_0^{\sqrt{a^2-y^2}} \sqrt{a^2 - y^2 - x^2}\, dx\, dy$$

This is the double iterated integral obtained in the example in Sec. 25.6.

There are six possible orders of performing a triple iterated integral. If one order leads to an unduly complicated integration, another order should be used.

Example

Find the volume bounded by the surface $2z = 4 - x^2 - y^2$ and the XY plane.

Solution: A sketch that shows the first-octant portion of the desired volume is shown in Fig. 25.12. The desired volume is four times that

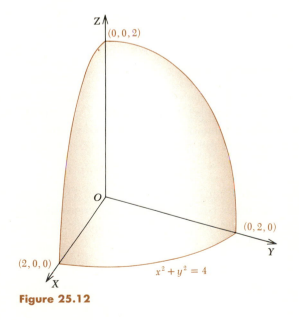

$(0,0,2)$

$(0,2,0)$

$(2,0,0)$

$x^2 + y^2 = 4$

Figure 25.12

shown since the surface is symmetric with respect to the XZ and YZ planes. Thus,

$$\frac{V}{4} = \int_0^2 \int_0^{\sqrt{4-x^2}} \int_0^{(4-x^2-y^2)/2} dz\, dy\, dx = \int_0^2 \int_0^{\sqrt{4-x^2}} z \Big|_0^{(4-x^2-y^2)/2} dy\, dx$$

$$= \frac{1}{2} \int_0^2 \int_0^{\sqrt{4-x^2}} (4-x^2-y^2)\, dy\, dx$$

$$= \frac{1}{2} \int_0^2 \left[(4-x^2)y - \frac{y^3}{3} \right] \Big|_0^{\sqrt{4-x^2}} dx$$

$$= \frac{1}{2} \int_0^2 \left[(4-x^2)\sqrt{4-x^2} - \frac{(\sqrt{4-x^2})^3}{3} \right] dx = \frac{1}{3} \int_0^2 (4-x^2)^{3/2}\, dx$$

This integral can be evaluated by letting $x = 2\sin\alpha$. If this is done, we find that

$$\frac{V}{4} = \frac{16}{3} \int_0^{\pi/2} \cos^4\alpha\, d\alpha$$

$$= \frac{4}{3} \int_0^{\pi/2} (1 + 2\cos 2\alpha + \cos^2 2\alpha)\, d\alpha \qquad \textit{since } \cos^4\alpha = (\cos^2\alpha)^2$$

$$\qquad\qquad\qquad\qquad\qquad\qquad\qquad\qquad = \left(\frac{1 + \cos 2\alpha}{2} \right)^2$$

$$= \frac{4}{3} \int_0^{\pi/2} \left(1 + 2\cos 2\alpha + \frac{1 + \cos 4\alpha}{2} \right) d\alpha$$

$$= \frac{4}{3} \left(\frac{3}{2}\alpha + \sin 2\alpha + \frac{1}{8}\sin 4\alpha \right) \Big|_0^{\pi/2}$$

$$= \pi$$

Consequently, the desired volume is 4π.

25.8   Moments by triple integration

If we multiply each element of volume in Fig. 25.13 by the square, $y^2 + z^2$, of the distance of the element from the X axis, add the resulting products, and take the limit of the sum, we find that the moment of inertia of the volume relative to the X axis is

I_x
$$I_x = \iiint (y^2 + z^2)\, dz\, dy\, dx$$

with appropriate limits on each integral. The moment of inertia relative to the other coordinate axes, the coordinate planes, and the origin can be found similarly. For example, the moment of inertia about the XZ plane is

I_{xz}
$$I_{xz} = \iiint y^2\, dz\, dy\, dx$$

In general, the value of

$$\iiint f(x, y, z)\, dz\, dy\, dx$$

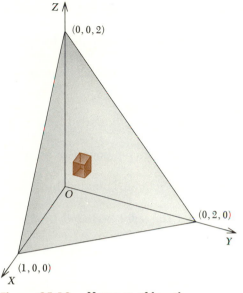

Figure 25.13   Moment of inertia

taken throughout a volume is called the integral of the function $f(x, y, z)$ over the volume. It can have various physical and geometric interpretations depending upon the nature of the function and the physical meanings of the variables.

Example

Find the moment of inertia about the XY plane of the volume bounded by the coordinate planes and $2x + y + z = 2$.

Solution: A sketch of the volume and a typical element thereof is shown in Fig. 25.13. The distance of the element from the XY plane is z. Hence, by using the order $dy \, dz \, dx$, we have

$$I_{xy} = \int_0^1 \int_0^{2-2x} \int_0^{2-2x-z} z^2 \, dy \, dz \, dx = \int_0^1 \int_0^{2-2x} z^2(2 - 2x - z) \, dz \, dx$$

$$= \int_0^1 (2 - 2x) \frac{z^3}{3} - \frac{z^4}{4} \bigg|_0^{2-2x} dx = \int_0^1 \frac{(2 - 2x)^4}{12} \, dx$$

$$= -\frac{2^4}{12} \frac{(1 - x)^5}{5} \bigg|_0^1 = \frac{4}{15}$$

EXERCISE 25.3

Use a double iterated integral in each of Problems 1 to 12.

1   Find the volume bounded by the coordinate planes and $x + y + 2z = 6$.

2   Find the volume under $x^2 + y^2 = 9 - z$, inside $x^2 + y^2 = 1$, and above the XY plane.

3   Find the first-octant volume bounded by $y^2 = x$, $y + z = 1$, $x = 0$, and $z = 0$.

4   Find the first-octant volume bounded by $y^2 + z^2 = 2z$, $2z = 2 - x$, $x = 0$, and $y = 0$.

5   Find the volume of a sphere.

6   Find the first-octant volume bounded by $y = x^2$, $2y + z = 1$, $x = 0$, and $z = 0$.

7   Find the first-octant volume bounded by $x^2 + y^2 = 9$, $y + z = 3$, $z = 0$, $y = 0$, $x = 0$.

8   Find the first-octant volume bounded by $z^2 = e^{-x^2-y^2}$, $x = 0$, $y = 0$, and $z = 0$.

9   Find the volume of the solid bounded by $3x^2 + y^2 = z$, $x^2 + y^2 = 2y$, and $z = 0$.

10   Find the first-octant volume bounded by $x^2 - 2y = 4$, $y + z = 1$, $z = 0$, $y = 0$, and $x = 0$.

11   Find the volume of one of the wedges cut from $x^2 + y^2 = a^2$ by $z = 0$ and $z = mx$.

12   Find the volume that is common to two right circular cylinders of radius a if their axes meet at right angles. This figure is often called a *banister cap*.

Use a triple iterated integral in each of Problems 13 to 16.

13   Find the volume bounded by $z = 4 - y^2$, $z = 0$, $x = 0$, and $y = x$.

14   Find the volume of the solid bounded by $x^2 + y^2 + 4z = 16$ and $x + z = 4$.

15   Find the volume of the solid bounded by $z = ye^x$, $x = y$, $x = 0$, $z = 0$, and $y = 1$.

16   Find the first-octant volume bounded by $z = xe^{-y^3/a^3}$, $y = x$, $x = 0$, and $z = 0$.

In Problems 17 to 20, set up a triple iterated integral whose value is equal to the volume bounded by the given surfaces.

17   The first-octant volume bounded by $x^2 + y^2 = 9 - z$, $x^2 + y^2 = 1$, and the XY plane.

18   The first-octant volume bounded by $x^2 + y^2 = 4az$, $x = 2a$, $y = 0$, $y = x$, and $z = 0$.

19   The first-octant volume bounded by $y = 2x^2$, $2y + z = 1$, $x = 0$, and $z = 0$.

20   The first-octant volume bounded by $z = x^2 + y^2$, $y = x$, and $x = a$.

21   Find I_x for the solid described in Problem 1.

22   Find I_y for the solid described in Problem 3.
23   Find I_x for the solid described in Problem 6.
24   Find I_z for the solid described in Problem 7.
25   Find I_{yz} for the solid described in Problem 1.
26   Find I_{xz} for the solid described in Problem 3.
27   Find I_{xy} for the solid described in Problem 6.
28   Find I_0 for the solid described in Problem 7.

25.9   Cylindrical coordinates

We found in Chap. 15 that the equations of some curves are simpler if we use polar coordinates. Similarly, problems involving some surfaces and skew curves are simplified if we use polar coordinates to locate the point in the XY plane that is directly below or above a desired point in space and use z to indicate the directed distance from the XY plane to the desired point. The coordinates of the point

cylindrical coordinates   are then (r, θ, z) and are called the *cylindrical coordinates* of the point.
The relations between the rectangular and cylindrical coordinates of a point are readily obtained by use of Fig. 25.14. They are

relations between cylindrical and rectangular coordinates

$$x = r \cos \theta \qquad y = r \sin \theta \qquad z = z$$

and $$r = \sqrt{x^2 + y^2} \qquad \theta = \text{Arctan} \frac{y}{x} \qquad z = z$$

The first two equations in each group are the relations between rectangular and polar coordinates as should be expected since r and θ in (r, θ, z) are the polar coordinates of $(r, \theta, 0)$.

Example   Express $x^2 + y^2 = z^2$ in terms of cylindrical coordinates.

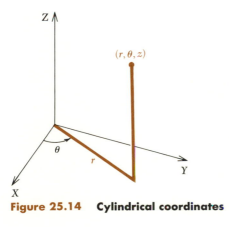

Figure 25.14   Cylindrical coordinates

Solution: If we replace x and y in terms of r and θ, we see that

$$r^2 = z^2$$

is the cylindrical-coordinate form of the equation.

25.10   Spherical coordinates

The spherical-coordinate system employs the spherical coordinates (ρ, θ, ϕ) to determine the position of a point in space. We define these coordinates as follows: ρ is the directed distance from the origin to a point P and is called *the radius vector,* θ is the angle from the positive X axis to the projection of OP on the XY plane, and ϕ is the angle from the positive Z axis to the line OP. This definition is illustrated in Fig. 25.15.

spherical
coordinates

We shall find the relation between the spherical and rectangular coordinates of a point by use of Fig. 25.15. Thus,

$$\sin \theta = \frac{y}{r} \qquad \text{and} \qquad \cos \theta = \frac{OS}{r} = \frac{x}{r} \tag{1}$$

hence, $\qquad\qquad y = r \sin \theta \qquad$ and $\qquad x = r \cos \theta$

We now notice that angle OPQ is equal to ϕ since they are alternate interior angles of parallel lines. Consequently,

$$\sin \phi = \sin OPQ = \frac{r}{\rho}$$

and it follows that $r = \rho \sin \phi$. If we now substitute this expression for r in each of Eqs. (1), we see that

$$x = \rho \sin \phi \cos \theta \qquad \text{and} \qquad y = \rho \sin \phi \sin \theta$$

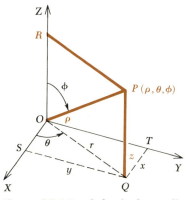

Figure 25.15   Spherical coordinates

Furthermore, $\cos \phi = OR/\rho = z/\rho$ since $OR = QP = z$; hence,

$$z = \rho \cos \phi$$

Therefore,

$$x = \rho \cos \theta \sin \phi \qquad y = \rho \sin \theta \sin \phi \qquad z = \rho \cos \phi$$

relations
between
spherical
and
rectangular
coordinates

are the relations that express the rectangular coordinates of a point in terms of the spherical. If we solve these equations for ρ, θ, and ϕ we find that

$$\rho = \sqrt{x^2 + y^2 + z^2} \qquad \theta = \text{Arctan}\,\frac{y}{x} \qquad \phi = \text{Arccos}\,\frac{z}{\sqrt{x^2 + y^2 + z^2}}$$

express the spherical coordinates of a point in terms of the rectangular.

Example

Express $x^2 + y^2 = z$ in terms of spherical coordinates.

Solution: If we replace x, y, and z in terms of ρ, θ, and ϕ, we get

$$(\rho \cos \theta \sin \phi)^2 + (\rho \sin \theta \sin \phi)^2 = \rho \cos \phi$$

$$\rho^2 \sin^2 \phi(\cos^2 \theta + \sin^2 \theta) = \rho \cos \phi$$

$$\rho^2 \sin^2 \phi = \rho \cos \phi \qquad \textit{since } \cos^2 \theta + \sin^2 \theta = 1$$

Therefore, dividing each member by ρ and multiplying by $\csc^2 \phi$, we find that

$$\rho = \cos \phi \csc^2 \phi$$

is the spherical form of the given equation.

25.11 Volume in cylindrical and spherical coordinates

We shall employ Fig. 25.16 as a basis for the development of a triple integral for the volume of a solid if the bounding surfaces are expressed in terms of cylindrical coordinates. The solid in the figure is bounded by the coordinate planes and the surface $z = f(r, \theta)$. If the coordinates of an interior point $P(r, \theta, z)$ assume the increments Δr, $\Delta \theta$, and Δz, respectively, then an element of the volume ΔV with a vertex at P is equal to $\Delta z\, r\, \Delta r\, \Delta \theta$, since the area of the base is $r\, \Delta r\, \Delta \theta$ by Sec. 25.4 and the altitude is Δz. Then the volume is

$$V = \lim_{\substack{\Delta \theta \to 0 \\ \Delta r \to 0 \\ \Delta z \to 0}} \Sigma\Sigma\Sigma \Delta z\, r\, \Delta r\, \Delta \theta = \iiint_R dz\, r\, dr\, d\theta$$

where R is the region of the XY plane bounded by $f(r, \theta) = 0$, $\theta = 0$, and $\theta = \pi/2$.

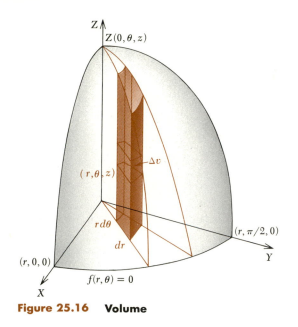

Figure 25.16   Volume

We express the above triple integral as an iterated integral by first observing that the limits for z are from 0 to $f(r, \theta)$. Next, we solve the equation of the trace of the surface in the XY plane, $f(r, \theta) = 0$, for r in terms of θ and get $r = F(\theta)$; thus, the limits for r are from zero to $F(\theta)$. Finally the limits for θ are from zero to $\pi/2$. Thus, we have the iterated integral

$$V = \int_0^{\pi/2} \int_0^{F(\theta)} \int_0^{f(r,\theta)} dz \, r \, dr \, d\theta$$

Example 1   Find the volume cut from the sphere $x^2 + y^2 + z^2 = 16$ by the cylinder $x^2 + y^2 = 4y$ by use of cylindrical coordinates.

Solution: In terms of cylindrical coordinates, the given equations become

$$r^2 + z^2 = 16 \qquad \text{and} \qquad r = 4 \sin \theta$$

Figure 25.17 shows the first-octant part of the sphere and the part of the circle $r = 4 \sin \theta$ in the first quadrant.

We shall find the quarter of the desired volume that is in the first octant. Since the lower boundary of the solid is the XY plane and the upper boundary is the surface $r^2 + z^2 = 16$, the limits for z are 0 and $\sqrt{16 - r^2}$. The coordinate r varies from 0 to a point on the circumference of the circle $r = 4 \sin \theta$ that is in the first quadrant of the XY plane, so the limits for r are 0 and $4 \sin \theta$. Finally, θ varies

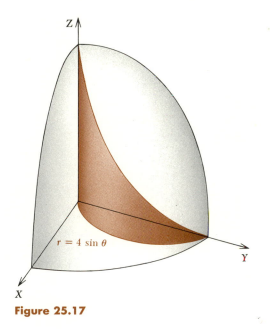

Figure 25.17

from 0 to $\pi/2$. Consequently, the iterated integral for the volume is

$$V = \int_0^{\pi/2} \int_0^{4\sin\theta} \int_0^{\sqrt{16-r^2}} dz\, r\, dr\, d\theta$$

$$= \int_0^{\pi/2} \int_0^{4\sin\theta} z \Big|_0^{\sqrt{16-r^2}} r\, dr\, d\theta$$

$$= \int_0^{\pi/2} \int_0^{4\sin\theta} \sqrt{16-r^2}\, r\, dr\, d\theta$$

$$= \int_0^{\pi/2} -\frac{1}{2}(16-r^2)^{3/2}\left(\frac{2}{3}\right)\Big|_0^{4\sin\theta} d\theta$$

$$= -\frac{1}{3}\int_0^{\pi/2}(64\cos^3\theta - 64)\, d\theta$$

$$= -\frac{64}{3}\left(\sin\theta - \frac{\sin^3\theta}{3} - \theta\right)\Big|_0^{\pi/2} \qquad \textit{since } \cos^3\theta = (1-\sin^2\theta)\cos\theta$$

$$= -\frac{64}{3}\left(1 - \frac{1}{3} - \frac{\pi}{2}\right) = \frac{32(3\pi-4)}{9}$$

Consequently, the required volume is $128(3\pi - 4)/9$.

We shall now consider the solid in Fig. 25.18 bounded by the coordinate planes and the surface whose equation in spherical coordinates is $\rho = f(\theta, \phi)$. If the coordinates $P(\rho, \theta, \phi)$ of an interior point of the solid assume the increments $\Delta\rho$, $\Delta\theta$, and $\Delta\phi$, respectively, we

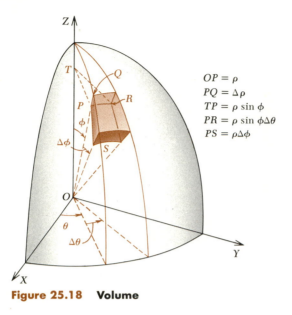

$$OP = \rho$$
$$PQ = \Delta\rho$$
$$TP = \rho \sin\phi$$
$$PR = \rho \sin\phi\Delta\theta$$
$$PS = \rho\Delta\phi$$

Figure 25.18 **Volume**

have an increment ΔV of the volume that is approximately equal to $(PS)(PR)(PQ)$. From the figure we see that $PS = \rho \Delta\phi$, $PR = TP \Delta\theta = \rho \sin\phi \Delta\theta$ and $PQ = \Delta\rho$. Hence

$$V = \rho \Delta\phi \; \rho \sin\phi \; \Delta\theta \; \Delta\rho \qquad \textit{approximately}$$
$$= \rho^2 \sin\phi \; \Delta\rho \; \Delta\phi \; \Delta\theta$$

Then
$$V = \lim_{\substack{\Delta\rho\to 0 \\ \Delta\phi\to 0 \\ \Delta\theta\to 0}} \Sigma\Sigma\Sigma\rho^2 \sin\phi \; \Delta\rho \; \Delta\phi \; \Delta\theta$$

Then by the definition of a triple integral

$$V = \iiint\limits_{R} \rho^2 \sin\phi \; d\rho \; d\phi \; d\theta$$

where R is the region of the XY plane bounded by

$$x = 0 \qquad y = 0 \qquad \text{and} \qquad f(\theta, \phi) = 0$$

Since P is a point on the line OQ, ρ varies from 0 to $f(\theta, \phi)$. Hence, the limits for ρ are 0 and $f(\theta, \phi)$. In this case ϕ varies from 0 to $\pi/2$ and θ varies from 0 to $\pi/2$. Hence, the iterated integral for V is

$$V = \int_0^{\pi/2}\int_0^{\pi/2}\int_0^{f(\theta,\phi)} \rho^2 \sin\phi \; d\rho \; d\phi \; d\theta$$

If the boundaries of a solid involve two or more surfaces that are not either of the coordinate planes, the choice of the limits is somewhat more difficult. For example, the solid in Fig. 25.19 is bounded by the XZ and YZ planes and the surfaces $ATBD$ and $OATB$ whose equations are $\rho = f(\theta, \phi)$ and $\rho = g(\theta, \phi)$, respectively. If $P(\rho, \theta, \phi)$ is a vertex of an element of volume ΔV and the line through O and P intersects the surfaces at S and R, then ρ varies from the ρ coordinate of S to the ρ coordinate of R, so the limits for ρ are $g(\theta, \phi)$ and $f(\theta, \phi)$. The angle ϕ varies from zero to the angle DOT. The latter angle is the ϕ coordinate of the point T on the intersection ATB of the two surfaces and must be expressed in terms of θ by use of the equation $f(\theta, \phi) = g(\theta, \phi)$ obtained by eliminating ρ between the equations of the surfaces. We solve this equation for ϕ and get $\phi = h(\theta)$ and then have the limits 0 and $h(\theta)$ for ϕ. Finally, the limits for θ are 0 and $\pi/2$. Consequently, we have

$$V = \int_0^{\pi/2} \int_0^{h(\theta)} \int_{g(\theta,\phi)}^{f(\theta,\phi)} \rho^2 \sin \phi \, d\rho \, d\phi \, d\theta$$

Example 2   Find the volume of a sphere of radius a.

Solution: The equation of the sphere in cylindrical coordinates is $\rho = a$. We shall find the volume in the first octant and multiply it by 8 in order to find the entire volume. Hence, the limits on ρ are 0 and a, and on both ϕ and θ are 0 and $\pi/2$. Consequently, the volume is

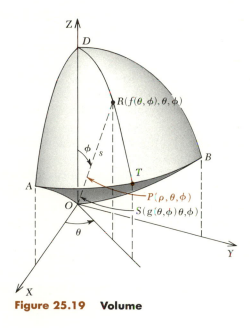

Figure 25.19   Volume

$$V = 8 \int_0^{\pi/2} \int_0^{\pi/2} \int_0^a \rho^2 \sin \phi \, d\rho \, d\theta \, d\phi = 8 \int_0^{\pi/2} \int_0^{\pi/2} \frac{\rho^3}{3} \Big|_0^a \sin \phi \, d\theta \, d\phi$$

$$= \frac{8}{3} \int_0^{\pi/2} \int_0^{\pi/2} a^3 \sin \phi \, d\theta \, d\phi = \frac{8a^3}{3} \int_0^{\pi/2} \theta \Big|_0^{\pi/2} \sin \phi \, d\phi$$

$$= \frac{4\pi a^3}{3} \int_0^{\pi/2} \sin \phi \, d\phi = - \frac{4\pi a^3}{3} \cos \phi \Big|_0^{\pi/2} = \frac{4\pi a^3}{3}$$

EXERCISE 25.4

Express the equation in each of Problems 1 to 12 in terms of two other types of coordinates.

1   $x^2 + y^2 + z^2 = a^2$ 2 $x^2 + y^2 = 2z$
3   $x^2 - y^2 = a^2$ 4 $x^2 + y^2 = 2ax$
5   $r^2 + z^2 = a^2$ 6 $r^2 = 2z$
7   $r^2 \cos 2\theta = a^2$ 8 $r \cos \theta = a$
9   $\rho = a$ 10 $\rho^2 \sin^2 \phi \cos 2\theta = a^2$
11 $\rho \sin^2 \phi \sin 2\theta = \cos \phi$
12 $\rho \sin \phi \cos \theta = a$

Find the volume called for in each of Problems 13 to 28 by use of a triple integral in cylindrical or spherical coordinates.

13   The volume above the XY plane, below the surface $x^2 + y^2 = 3z$, and inside the cylinder $x^2 + y^2 = 9$

14   The volume above the XY plane, below the paraboloid $x^2 + y^2 + z = 12$, and inside the cylinder $x^2 + y^2 = 9$

15   The volume above the XY plane, below the cone $x^2 + y^2 = z^2$, and inside the cylinder $x^2 + y^2 = 9$

16   The volume above the XY plane, below the surface $x^2 + y^2 + 3 = z$, and inside the cylinder $x^2 + y^2 = 9$

17   The volume bounded by the XY plane and the paraboloid $x^2 + y^2 = 4 - z$

18   The volume bounded by the XY plane, the paraboloid $z = x^2 + y^2$, and the cylinder $x^2 + y^2 = 4y$

19   The volume of the smaller of the two pieces into which a sphere of radius 10 in. is cut by a plane that is 8 in. from the center

20   The volume of the ellipsoid $x^2 + y^2 + 9z^2 = 9$

21   The volume bounded by the ellipsoid $x^2 + y^2 + 4z^2 = 4$ and the cylinder $x^2 + y^2 = 2x$

22   The volume inside both the sphere $x^2 + y^2 + z^2 = 4$ and the cone $x^2 + y^2 = z^2$. *Hint:* The limits on r and θ are determined by projecting on the XY plane the intersection of the given surfaces.

23   The volume common to the cone $x^2 + y^2 = z^2$ and the paraboloid $4z = x^2 + y^2 + 4$. *Hint:* The limits on r and θ are determined by the projection on the XY plane of the intersection of the given surfaces.

24   The volume cut from the paraboloid $z = x^2 + y^2$ by the plane $z = x$

25   The volume common to the sphere $r^2 + z^2 = a^2$ and the cylinder $r = a \sin \theta$

26   The volume cut from the sphere $r^2 + z^2 = 16$ by the cylinder $r = 2\sqrt{3}$

27   Find the volume of the smaller piece into which a right circular cylinder of radius 6 in. and height 12 in. is cut by a plane that passes through a diameter of one base and is tangent to the other base.

28   Find the volume that is in the sphere $r^2 + z^2 = 8$ and above $2z = r^2$.

29   Find the second moment of the mass of the homogeneous solid cylinder $r = 2a \cos \theta$ by evaluating $\delta \iiint r^2 (r\,dz\,dr\,d\theta)$ with appropriate limits where δ is mass density.

30   Use cylindrical coordinates to find the second moment of the mass of the homogeneous sphere with respect to a diameter.

31   A torus is generated by rotating a circle of radius a about an axis in its plane and a distance $b > a$ from the center. Find its moment of inertia about the axis of revolution if the mass is m.

32   Find the moment of inertia of a homogeneous cube with respect to an edge if the edge is a and the mass is m.

25.12   Area of a surface

We studied the area of a surface of revolution in Sec. 19.8 and shall now see how to find the area of that part of the surface $z = f(x, y)$ which is bounded by a closed curve as indicated in Fig. 25.20. We shall designate the surface by S and its projection on the XY plane by R. If $\Delta y\, \Delta x$ is a rectangular element of R, then a perpendicular from the XY plane drawn at any point in this rectangular element determines a point P of S. Furthermore, the column erected on the element $\Delta y\, \Delta x$ of the XY plane intersects the tangent plane to S at P in a region whose area we shall call ΔA. Now

$$\Delta A \cos \gamma = \Delta y\, \Delta x \tag{1}$$

where γ is the acute angle between the XY plane and the tangent plane at P. We now make use of the fact that the direction cosines of the normal to the surface $z = f(x, y)$ are proportional to z_x and z_y at

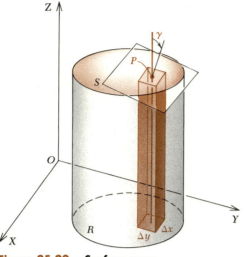

Figure 25.20 **Surface area**

P and -1. Hence, if γ is the angle between the Z axis and the normal to the surface at P, we have

$$\cos \gamma = \frac{1}{\sqrt{1 + (z_x)^2 + (z_y)^2}}$$

We now define the area of S to be the limit of the sum of the areas ΔA as Δx and Δy approach zero, and, from (1), we get

$$S = \lim_{\substack{\Delta x \to 0 \\ \Delta y \to 0}} \Sigma\Sigma\Delta A = \lim_{\substack{\Delta x \to 0 \\ \Delta y \to 0}} \Sigma\Sigma \frac{\Delta y \, \Delta x}{\cos \gamma}$$

$$= \lim_{\substack{\Delta x \to 0 \\ \Delta y \to 0}} \Sigma\Sigma \sqrt{1 + (z_x)^2 + (z_y)^2} \, \Delta y \, \Delta x$$

$$= \int_a^b \int_{y_1(x)}^{y_2(x)} \sqrt{1 + (z_x)^2 + (z_y)^2} \, dy \, dx$$

by two uses of the fundamental theorem.

We can now say that *the area of the part of the surface $z = f(x, y)$ which is bounded by a closed curve whose projection on the XY plane is the region R is*

area of a surface

$$S = \int_R\int \sqrt{1 + (z_x)^2 + (z_y)^2} \, dA = \int_a^b \int_{y_1(x)}^{y_2(x)} \sqrt{1 + (z_x)^2 + (z_y)^2} \, dy \, dx$$

provided that z_x and z_y are continuous in R and the limits are so chosen as to cover R.

Example 1   Find the area of the portion of the surface $2z = x^2$ that lies in the first octant and is intercepted by $x + y = \sqrt{3}$.

Solution: Since we want a part of the surface $2z = x^2$, we have $z_x = x$ and $z_y = 0$. Consequently,

$$S = \int_0^{\sqrt{3}} \int_0^{\sqrt{3}-x} \sqrt{1 + x^2} \, dy \, dx = \int_0^{\sqrt{3}} \sqrt{1 + x^2} \, (\sqrt{3} - x) \, dx$$

$$= \frac{2}{3} + \frac{\sqrt{3}}{2} \ln (2 + \sqrt{3})$$

Quite often it is simpler to find the area of a surface if cylindrical coordinates are used. If we want a portion S of the surface $z = f(r, \theta)$, we must use $r \, \Delta r \, \Delta \theta$ as the element of area in the projection R of S on the XY plane. Furthermore,

$$1 + (z_x)^2 + (z_y)^2 = 1 + (z_r)^2 + \frac{(z_\theta)^2}{r^2}$$

since $x = r \cos \theta$ and $y = r \sin \theta$. Consequently,

surface area in cylindrical coordinates

$$S = \int_R \int \sqrt{1 + (z_r)^2 + \frac{(z_\theta)^2}{r^2}} \, dA$$

$$= \int_{\theta_1}^{\theta_2} \int_{r_1(\theta)}^{r_2(\theta)} \sqrt{1 + (z_r)^2 + \frac{(z_\theta)^2}{r^2}} \, r \, dr \, d\theta$$

provided that the limits on the iterated integral are chosen so as to cover R.

Example 2   Find the area of the portion of the surface of $x^2 + y^2 + z^2 = a^2$ that lies inside the cylinder $x^2 + y^2 = ax$ and above the XY plane.

Solution: We shall use cylindrical coordinates. In terms of them, the equations are $r^2 + z^2 = a^2$ and $r = a \cos \theta$. Since the desired area is a part of the surface whose equation is $r^2 + z^2 = a^2$, we have $z_\theta = 0$ and $2r + 2zz_r = 0$; hence, $z_r = -r/z$. Consequently,

$$1 + (z_r)^2 + \frac{(z_\theta)^2}{r^2} = 1 + \frac{r^2}{z^2} = \frac{z^2 + r^2}{z^2} = \frac{a^2}{z^2} = \frac{a^2}{a^2 - r^2}$$

Therefore,

$$S = 2\int_0^{\pi/2} \int_0^{a \cos \theta} \frac{a}{\sqrt{a^2 - r^2}} r \, dr \, d\theta = -2a \int_0^{\pi/2} (a^2 - r^2)^{1/2} \Big|_0^{a \cos \theta} d\theta$$

$$= -2a^2 \int_0^{\pi/2} (\sin \theta - 1) \, d\theta = -2a^2(-\cos \theta - \theta) \Big|_0^{\pi/2} = a^2(\pi - 2)$$

EXERCISE 25.5

1   Find the area of the first-octant portion of the surface of the cylinder $y^2 + z^2 = a^2$ intercepted by $x = 0$ and $y = x/4$.

2   Find the area of the surface of the plane $3x - 2y + 6z = 12$ intercepted by $2x^2 + y^2 = 2$.

3   Find the area of the portion of the surface $9z^2 = 16x^2 + 16y^2$ that lies in the cylinder $x^2 + y^2 = 2x$.

4   Find the area of the portion of $2x - 3y + 3z = 6$ that is above the part of the XY plane bounded by $x = 0$, $x = 2$, $y = 0$, and $y = 3$.

5   Find the area of the portion of the cylinder $x^2 + z^2 = a^2$ that lies inside the cylinder $x^2 + y^2 = a^2$.

6   Find the area of the first-octant portion of the surface $\sqrt{6}\, z = xy$ intercepted by $x^2 + y^2 = 6$.

7   Find the area of the first-octant portion of the surface $z = x^2 - y^2$ that lies inside the cylinder $x^2 + y^2 = 2$.

8   Find the area of the portion of the surface of the cone $x^2 + y^2 = z^2$ that lies inside the cylinder $x^2 + y^2 = 2x$.

9   Find the area of the first-octant portion of the surface $z = xy$ that lies inside the cylinder $x^2 + y^2 = 3$.

10   Find the curved surface area of a cylinder of radius a and altitude h.

11   Find the area of the portion of the surface of the sphere $x^2 + y^2 + z^2 = 4$ that lies inside the surface $x^2 + y^2 + z = 4$.

12   Find the area of the first-octant portion of the surface $2z = x^2 - y^2$ that lies inside a solid whose projection in the XY plane is $(x^2 + y^2)^2 = x^2 - y^2$.

Infinite series

26.1 Introduction

sequence
series

A set of terms or numbers arranged in an order that is determined according to some law is called a *sequence*. The indicated sum of a sequence is called a *series*. The series is *finite* or *infinite* according as the number of terms is limited or unlimited. In algebra we studied infinite geometric series with ratio less than one in absolute value. We shall devote this chapter to a further study of infinite series. An infinite series is not definitely determined unless a general expression from which any term can be obtained is given. Thus

$$\frac{1}{2} + \frac{1}{4} + \frac{1}{8} + \cdots + \frac{1}{2^n} + \cdots \tag{1}$$

does define or determine an infinite series, but, without the term $\frac{1}{2}^n$, it would not since the terms after $\frac{1}{8}$ could be any numbers. The series with $1/(n^2 - n + 2)$ as general term has the same first three terms as (1) but terms thereafter are different.

At times, we shall indicate a series by use of the sigma notation. If this is used, we would indicate the series (1) by

$$\sum_{n=1}^{\infty} \frac{1}{2^n} \quad \text{or} \quad \sum_{n=0}^{\infty} \frac{1}{2^{n+1}}$$

If we wish to indicate that the terms alternate in sign we can do so by introducing $(-1)^n$ or some other power of -1 as part of the indicated sum. Thus

$$\sum_{n=1}^{\infty} (-1)^{n-1} \frac{1}{n!} = \frac{1}{1!} - \frac{1}{2!} + \frac{1}{3!} - \frac{1}{4!} + \cdots + (-1)^{n-1} \frac{1}{n!} + \cdots$$

$$= 1 - \frac{1}{2} + \frac{1}{6} - \frac{1}{24} + \cdots + (-1)^{n-1} \frac{1}{n!} + \cdots$$

26.2   Convergence and divergence

We cannot find the sum of an infinite series in the sense of adding all of the terms since there is no end to the number of terms. We can, however, find the sum of any number of terms. If the series is

$$a_1 + a_2 + a_3 + \cdots + a_n + \cdots$$

we shall indicate the sum of the first n term by S_n and have

$$S_1 = a_1$$
$$S_2 = a_1 + a_2$$
$$S_3 = a_1 + a_2 + a_3$$
$$\cdots \cdots \cdots \cdots \cdots \cdots$$
$$S_n = a_1 + a_2 + a_3 + \cdots + a_n$$

S_n is often called the nth partial sum. We can evaluate S_n by performing the indicated addition. If S_n is regarded as a function of n and approaches a limit S as $n \to \infty$, this limit is called the *sum* of the series and the series is said to be *convergent*. The definition may be expressed in symbolic form by writing

convergent

If $\lim_{n \to \infty} S_n = S$, then the series is convergent and has the sum S.

divergent If a series is not convergent, it is said to be *divergent*.

Example 1 The series

$$1 + \frac{1}{3} + \frac{1}{9} + \frac{1}{27} + \cdots + \frac{1}{3^{n-1}} + \cdots$$

is convergent since, by use of the formula $a(1 - r^n)/(1 - r)$ for the sum of a geometric progression, we have

$$S_n = \frac{1}{1 - \frac{1}{3}} \left[1 - \left(\frac{1}{3} \right)^n \right] = 1.5 \left[1 - \left(\frac{1}{3} \right)^n \right]$$

and $\lim_{n \to \infty} S_n = 1.5$.

Example 2   The series

$$1 + 3 + 5 + \cdots + (2n - 1) + \cdots$$

is divergent since S_n can be made greater than any chosen number by taking n sufficiently large.

Example 3   The series

$$1 - 1 + 1 - 1 + \cdots + (-1)^{n-1} + \cdots$$

is divergent since the value of S_n alternates between 1 and zero.

26.3   A necessary condition for convergence

If

$$S_1, S_2, S_3, \ldots, S_n, \ldots \tag{1}$$

has a limit, then

$$S_m, S_{m+1}, \ldots, S_{m+n}, \ldots \tag{2}$$

also has one since (2) is obtained by omitting the first $m - 1$ terms of (1). As a matter of fact, (1) and (2) have the same limit. Thus,

$$\lim_{n \to \infty} S_n = \lim_{n \to \infty} S_{m+n} \tag{3}$$

If S_n is the nth partial sum of the convergent infinite series Σa_k, then

$$a_n = S_n - S_{n-1}$$

and $$\lim_{n \to \infty} a_n = \lim_{n \to \infty} (S_n - S_{n-1}) = \lim_{n \to \infty} S_n - \lim_{n \to \infty} S_{n-1} = 0$$

Consequently, we know that if

$$a_1 + a_2 + a_3 + \cdots + a_n + \cdots$$

is convergent, then

$$\lim_{n \to \infty} a_n = 0 \tag{26.1}$$

necessary condition for convergence   This statement is sometimes expressed in the form: *a necessary condition for Σa_n to be convergent is that* $\lim\limits_{n \to \infty} a_n = 0$.

The condition (26.1) is not sufficient to assure convergence.

26.4   Geometric and harmonic series

An infinite series of the form

$$\sum_{k=1}^{\infty} ar^{k-1} = a + ar + ar^2 + \cdots + ar^{n-1} + \cdots$$

is called a *geometric series*. By use of the identity

$$1 - r^n = (1 - r)(1 + r + r^2 + \cdots + r^{n-1})$$

we can find that

$$S_n = a \sum_{k=1}^{n} r^{k-1} = a \frac{1 - r^n}{1 - r} = \frac{a}{1 - r} - \frac{ar^n}{1 - r} \qquad r \neq 1$$

Now, since

$$\lim_{n \to \infty} r^n = 0 \qquad \text{for } |r| < 1$$

we have

$$\lim_{n \to \infty} S_n = \frac{a}{1 - r} - \lim_{n \to \infty} \frac{ar^n}{1 - r} = \frac{a}{1 - r}$$

for $|r| < 1$. Consequently, a geometric series with $|r| < 1$ is convergent and has the sum $a/(1 - r)$. If $|r| \geq 1$, then the series

$$a + ar + ar^2 + \cdots + ar^{n-1} + \cdots$$

is divergent since it does not satisfy the necessary condition for convergence.

We shall now consider the *harmonic series*

harmonic series

$$1 + \frac{1}{2} + \frac{1}{3} + \frac{1}{4} + \cdots + \frac{1}{n} + \cdots$$

We notice that the first and second terms are each equal to or greater than ½ and that

$$\tfrac{1}{3} + \tfrac{1}{4} > \tfrac{1}{4} + \tfrac{1}{4} = \tfrac{1}{2}$$

$$\tfrac{1}{5} + \tfrac{1}{6} + \tfrac{1}{7} + \tfrac{1}{8} > \tfrac{1}{8} + \tfrac{1}{8} + \tfrac{1}{8} + \tfrac{1}{8} = \tfrac{1}{2}$$

$$\tfrac{1}{9} + \cdots + \tfrac{1}{16} > \tfrac{1}{16}(8) = \tfrac{1}{2}$$

$$\tfrac{1}{17} + \cdots + \tfrac{1}{32} > \tfrac{1}{32}(16) = \tfrac{1}{2}$$

By continuing this process, we can get ½ as many times as we please. Consequently, by adding a sufficiently large number of terms of the series, we can get a sum larger than any number selected in advance. Therefore, *the harmonic series is divergent.*

We shall give a second proof of the divergence of the harmonic series. If an infinite series Σa_k with S_n as nth partial sum is convergent with S as the sum, then for every $\varepsilon > 0$, there exists a number N such that

$$|S - S_n| < \varepsilon \qquad \text{for every } n > N$$

Therefore,

$$|S_n - S_m| \leq |S_n - S| + |S - S_m| \leq 2\varepsilon \qquad \text{for } n, m > N \qquad (1)$$

Consequently, we know that the difference between any two partial sums S_m and S_n is arbitrarily small if $n > N$ and $m > N$. This is the basis for the second proof of the divergence of the harmonic series, and it can be used at times to show the divergence of other series. If we use (1) with $m = 2n$ in connection with the harmonic series, we have

$$S_{2n} - S_n = \frac{1}{n+1} + \frac{1}{n+2} + \cdots + \frac{1}{2n} > \frac{1}{2n} + \cdots + \frac{1}{2n}$$

$$= n\left(\frac{1}{2n}\right) = \frac{1}{2}$$

hence, $S_{2n} - S_n > \frac{1}{2}$ for $n > 1$. Consequently, the harmonic series is divergent since for a convergent series $S_{2n} - S_n$ must be arbitrarily small and hence less than $\frac{1}{2}$.

EXERCISE 26.1

Show that the necessary condition for convergence is satisfied by the series in each of Problems 1 to 4, and find the sum of each series.

1. $1 + \frac{2}{3} + \frac{4}{9} + \cdots + (\frac{2}{3})^{n-1} + \cdots$

2. $\frac{4}{5} + \frac{4}{25} + \frac{4}{125} + \cdots + \frac{4}{5^n} + \cdots$

3. $\frac{2}{7} + \frac{2}{7^2} + \frac{2}{7^3} + \cdots + \frac{2}{7^n} + \cdots$

4. $\frac{3}{\sqrt{2}} + \frac{3}{2} + \frac{3}{2\sqrt{2}} + \cdots + \frac{3}{\sqrt{2^n}} + \cdots$

In each of Problems 5 to 12 write the first four terms of the series.

5. $a_n = 1/(4n^2 - 1)$ 6. $a_n = (\frac{2}{3})^n$

7. $a_n = 2^n/(n+1)$ 8. $a_n = 3/(2^n - 1)$

9. $\sum_{n=1}^{\infty} \frac{2}{n!}$ 10. $\sum_{n=2}^{\infty} \frac{x^n}{n \ln n}$

11. $\sum_{n=1}^{\infty} \frac{(-1)^n n!}{\ln(n+2)}$ 12. $\sum_{n=0}^{\infty} \frac{(-1)^n \ln(n+1)}{(n+2)!}$

In each of Problems 13 to 16, find an expression for a_n under the assumption that the first term corresponds to $n = 1$.

13. $1 + 4 + 7 + 10 + \cdots$ 14. $\frac{1}{3} - \frac{2}{3^2} + \frac{3}{3^3} - \frac{4}{3^4} + \cdots$

15. $\frac{1}{2} + \frac{1 \cdot 3}{2 \cdot 4} + \frac{1 \cdot 3 \cdot 5}{2 \cdot 4 \cdot 6} + \cdots$ 16. $\frac{1}{5} + \frac{2}{8} + \frac{3}{13} + \frac{4}{20} + \cdots$

The nth partial sum of an infinite series is given in each of Problems 17 to 24. Find the first four terms of each series by making use of the fact that $a_n = S_n - S_{n-1}$ for $n > 1$, state whether the series is convergent or divergent, and give the sum if convergent.

17 $S_n = \dfrac{n}{2n - 1}$

18 $S_n = \dfrac{2n}{3n + 2}$

19 $S_n = \dfrac{n + 2}{n^2}$

20 $S_n = \dfrac{n^2 - 1}{n + 3}$

21 $S_n = 3^n$

22 $S_n = 3^{-n}$

23 $S_n = \ln(n + 5)$

24 $S_n = \dfrac{\ln(n + 1)}{n}$

Show that the series in each of Problems 25 to 32 is divergent.

25 $2 - 2 + 2 - 2 + \cdots + (-1)^{n+1}\, 2 + \cdots$

26 $1 + 2 + 3 + 4 + \cdots + n + \cdots$

27 $\frac{4}{3} + \frac{16}{9} + \frac{64}{27} + \cdots + (\frac{4}{3})^n + \cdots$

28 $\dfrac{1}{2} + \dfrac{2}{3} + \dfrac{3}{4} + \cdots + \dfrac{n}{n + 1} + \cdots$

29 $\displaystyle\sum_{n=1}^{\infty} \dfrac{n}{3n + 1}$

30 $\displaystyle\sum_{n=1}^{\infty} \dfrac{2n - 1}{3n + 2}$

31 $\displaystyle\sum_{n=1}^{\infty} \dfrac{n}{n + 91}$

32 $\displaystyle\sum_{n=1}^{\infty} \dfrac{n!}{(n - 1)(n + 1)}$

26.5   Two theorems on limits

We shall now give two theorems on limits that are useful in establishing the convergence of some infinite series. The first of these theorems is:

a theorem on limits

If a function $S_n = f(n)$ never decreases as n increases but remains less than or equal to some constant K, then $\lim\limits_{n \to \infty} S_n$ exists and is less than or equal to K.

We shall now prove this theorem. The set of numbers S_n has an upper bound K by assumption; hence, the set has a least upper bound S. Therefore, $S - \varepsilon$ is not an upper bound of the set S_n for all $\varepsilon > 0$. Consequently, for some N that depends on ε, we have $S_N > S - \varepsilon$; however, $S_n \geq S_N$ for every $n > N$. Accordingly, $S_n > S - \varepsilon$ and

$$0 \leq S - S_n < \varepsilon \qquad \text{for all } n > N$$

and we have shown that

$$\lim_{n \to \infty} S_n = S$$

where S is less than or equal to K.

The second theorem can be proved similarly and is given without proof. It is:

a limit theorem

If a function $S_n = f(n)$ never increases as n increases but remains greater than or equal to some constant K, then $\lim_{n\to\infty} S_n$ exists and is greater than or equal to K.

26.6   Altering terms of a series

We shall consider two infinite series Σa_k and Σb_k with partial sums S_n and T_n, which differ only in their first m terms. Consequently,

$$S_n - T_n = S_m - T_m \qquad \text{and} \qquad S_n = S_m - T_m + T_n$$

for $n \geq m$ since the two series are identical after the first m terms. Hence,

$$\lim_{n\to\infty} S_n = S_m - T_m + \lim_{n\to\infty} T_n$$

and both limits exist if either does. Therefore, *both series converge or both diverge; furthermore, if they converge their sums differ by* $S_m - T_m$.

We shall now consider Σa_k and $\Sigma c a_k$, $c \neq 0$. If S_n is the sum of the first n terms of Σa_k, then cS_n is the sum of the first n terms of $\Sigma c a_k$. Furthermore,

$$\lim_{n\to\infty} cS_n = c \lim_{n\to\infty} S_n$$

and both limits exist if either does. Therefore, *if $\lim_{k\to\infty} \Sigma a_k$ exists, so does $\lim_{k\to\infty} \Sigma c a_k$ and $\lim_{k\to\infty} \Sigma c a_k = c \lim_{k\to\infty} \Sigma a_k$.*

Example 1   The geometric series

$$1 + \frac{1}{2} + \frac{1}{4} + \cdots + \frac{1}{2^{n-1}} + \cdots \tag{1}$$

converges and has the sum $1/(1 - \frac{1}{2}) = 2$; consequently, if we replace the first and second terms by 3 and 2, respectively, the new series

$$3 + 2 + \frac{1}{4} + \cdots + \frac{1}{2^{n-1}} \tag{2}$$

converges and has a sum that differs from the sum (1) by $3 + 2 - (1 + \frac{1}{2}) = 3.5$. Therefore, the sum (2) is 5.5.

Example 2   If we multiply each term of (1) by 3, we get a convergent series whose sum is 3 times the sum of (2), hence, is 6.

26.7   The integral test

We shall now give a test for convergence that makes use of an improper integral and deals with a series in which all of the terms have positive values. The test is called the *integral test* and is:

If f is a continuous, positive-valued, decreasing function in $[1, \infty)$ *and if*

integral test

$$\sum_{k=1}^{\infty} a_k = f(1) + f(2) + \cdots + f(n) + \cdots$$

then Σa_k *converges or diverges according as the improper integral*

$$\int_1^{\infty} f(x)\, dx$$

exists or fails to exist.

We shall use Figs. 26.1 and 26.2 in proving this theorem. Each figure is constructed in part by laying off ordinates $a_1 = f(1)$, $a_2 = f(2)$, $\ldots$, $a_n = f(n)$, $\ldots$ at the points corresponding to $x = 1, 2, \ldots, n, \ldots$ and then drawing a smooth curve through their extremities. We then complete the construction of rectangles of heights $f(2)$, $f(3)$, $\ldots$, $f(n)$, $\ldots$ in Fig. 26.1 and of heights $f(1)$, $f(2)$, $\ldots$, $f(n-1)$, $\ldots$ in Fig. 26.2. From the figures we see that

$$f(2) + f(3) + \cdots + f(n)$$

$$= \sum_{k=2}^{n} f(k) \le \int_1^n f(x)\, dx \le f(1) + \cdots + f(n-1) = \sum_{k=1}^{n-1} f(k) \quad (1)$$

since the integral is equal to the area under the curve.

If the improper integral $\int_1^{\infty} f(x)\, dx$ exists, then

$$\sum_{k=2}^{n} f(k) \le \int_1^n f(x)\, dx < \int_1^{\infty} f(x)\, dx$$

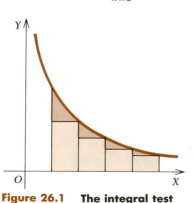

Figure 26.1   The integral test

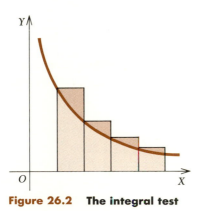

Figure 26.2   The integral test

and the infinite series $\Sigma a_k = \Sigma f(k)$ converges by the first theorem of Sec. 26.5. Furthermore, if the improper integral is infinite, then the series $\Sigma f(k)$ diverges by (1).

Example 1   Test $\displaystyle\sum_{k=1}^{\infty} \frac{1}{e^k}$ for convergence by use of the integral test.

Solution: The integral test is applicable since $a_k = f(k) = e^{-k}$ is a continuous, positive-valued, decreasing function. Furthermore,

$$\int_1^{\infty} e^{-x}\, dx = -e^{-x}\Big|_1^{\infty} = 0 + \frac{1}{e} = \frac{1}{e}$$

Since the improper integral has a definite finite value, the series is convergent.

Example 2   Use the integral test to determine whether the harmonic series $\displaystyle\sum_{k=1}^{n} \frac{1}{k}$ is convergent or divergent.

Solution: The integral test is applicable since $a_k = f(k) = 1/k$ is a continuous, positive-valued, decreasing function. Furthermore,

$$\int_1^{\infty} \frac{dx}{x} = \ln x\Big|_1^{\infty} = \infty$$

and the given harmonic series diverges.

26.8   The hyperharmonic or k series

We shall now consider a series which is often used in determining whether another series is convergent or divergent. It is

$$\sum_{n=1}^{\infty} \frac{1}{r^k} = \frac{1}{1^k} + \frac{1}{2^k} + \frac{1}{3^k} + \cdots + \frac{1}{n^k} + \cdots$$

and is called the *k series* or *hyperharmonic series*. Since the general term is a continuous, positive, decreasing function, we can apply the integral test. Thus,

$$\int_1^{\infty} \frac{dx}{x^k} = \left. \frac{x^{-k+1}}{-k+1} \right|_1^{\infty} \qquad k \neq 1$$

$$= \begin{cases} 0 - \dfrac{1}{-k+1} & \text{for } k > 1 \\[2ex] \infty - \dfrac{1}{-k+1} & \text{for } k < 1 \end{cases}$$

convergence and divergence of the k series

Consequently, *the k series converges for $k > 1$ and diverges for $k < 1$*; furthermore, *it diverges for $k = 1$* since in that case it is a harmonic series.

26.9   The comparison test

It is often possible to establish the convergence or divergence of a series by comparing it with a series which is known to be convergent or divergent. For this purpose, we shall state and prove three theorems.

first comparison test

The first one states that *a series of positive terms is convergent if each of its terms is less than or equal to the corresponding term of a series that is known to be convergent.*

In order to prove this theorem, we shall represent the known convergent series by

$$\sum_{k=1}^{\infty} c_k = c_1 + c_2 + c_3 + \cdots + c_n + \cdots \qquad (1)$$

and the series that is being tested by

$$\sum_{k=1}^{\infty} t_k = t_1 + t_2 + t_3 + \cdots + t_n + \cdots \qquad (2)$$

Furthermore, we shall let C be the sum of the series (1) and let C_n and T_n be the sum of the first n terms of (1) and of (2), respectively. Therefore,

$$T_n \leq C_n < C$$

for all values of n since each term of (2) is less than or equal to the corresponding term of (1). Consequently, $\lim_{n \to \infty} T_n$ exists and is less than or equal to C by the first theorem in Sec. 26.5.

The second of the three theorems used in determining whether a series is convergent or divergent by comparing it with another series states that *a series of positive terms is divergent if each of its terms is greater than or equal to the corresponding term of a series that is known to be divergent.*

second comparison test

We shall leave the proof of this theorem as a problem in Exercise 26.2.

These comparison theorems tell whether a series is convergent or divergent only when it is *term by term* less than a known convergent series or *term by term* larger than a known divergent series.

Example 1

Test the series

$$\frac{1}{1\cdot 2\cdot 3} + \frac{1}{2\cdot 3\cdot 4} + \frac{1}{3\cdot 4\cdot 5} + \cdots + \frac{1}{n(n+1)(n+2)} + \cdots$$

$$= \sum_{k=1}^{\infty} \frac{1}{k(k+1)(k+2)}$$

for convergence by use of a comparison test.

Solution: We shall compare the given series with the known convergent series $\Sigma \dfrac{1}{k^3}$. The nth terms of the series being tested and the known convergent series are

$$t_n = \frac{1}{n(n+1)(n+2)} \qquad \text{and} \qquad c_n = \frac{1}{n^3}$$

We must now see if $t_n < c_n$ for all n; hence, we must see if

$$\frac{1}{n(n+1)(n+2)} < \frac{1}{n^3} \tag{3}$$

Multiplying each member by $n^3(n+1)(n+2)$, we find that (3) becomes

$$n^2 < (n+1)(n+2) = n^2 + 3n + 2$$

This is a true statement; hence, (3) is true and the given series is convergent.

Example 2

Determine whether $\displaystyle\sum_{n=1}^{\infty} \frac{1}{\sqrt{n}}$ is convergent or divergent by use of a comparison test.

Solution: Since this is a k series with $k = \frac{1}{2} < 1$, we know the series is divergent but shall show that it is by use of a comparison test. We shall compare it with the harmonic series. We shall show that $t_n = 1/\sqrt{n}$ is greater than $d_n = 1/n$. If we put

$$\frac{1}{\sqrt{n}} > \frac{1}{n} \tag{4}$$

and raise each member to the second power, we have

$$\frac{1}{n} > \frac{1}{n^2}$$

This is true for all $n > 1$; hence, (4) is true for all $n > 1$ and the given series is divergent.

another comparison test

We shall now state and prove a third comparison theorem. It states that *if $c_1, c_2, \ldots, c_n, \ldots$ is a sequence of positive numbers with*

$$\lim_{k \to \infty} c_k = c \qquad c > 0$$

then the two series of positive terms

$$\sum_{k=1}^{\infty} a_k \qquad \text{and} \qquad \sum_{k=1}^{\infty} c_k a_k$$

both converge or both diverge.

We begin the proof by observing that since $\lim_{n \to \infty} c_n = c > 0$, there exists an integer N such that

$$\frac{c}{3} < c_k < \frac{4c}{3}$$

for all integers $k > N$. Consequently,

$$\frac{c}{3} a_k < c_k a_k < \frac{4c}{3} a_k$$

for all integers $k > N$. Therefore, if the series Σa_k converges so does the series $\Sigma c_k a_k$ since it is less $\Sigma \dfrac{4c a_k}{3}$; furthermore, if $\Sigma c_k a_k$ converges so does Σa_k since it is less than $\dfrac{3}{c} \Sigma c_k a_k$. This completes the part of the proof relative to convergence. The proof relative to divergence is similar and will not be given.

EXERCISE 26.2

Test the series in each of Problems 1 to 12 for convergence by use of the integral test.

1 $\dfrac{1}{\sqrt{2}} + \dfrac{2}{\sqrt{5}} + \dfrac{3}{\sqrt{10}} + \cdots + \dfrac{n}{\sqrt{n^2 + 1}} + \cdots$

2 $\dfrac{1}{5^3} + \dfrac{1}{7^3} + \dfrac{1}{9^3} + \cdots + \dfrac{1}{(2n + 3)^3} + \cdots$

3 $\dfrac{1}{2} + \dfrac{2}{5} + \dfrac{3}{10} + \cdots + \dfrac{n}{n^2 + 1} + \cdots$

4 $\dfrac{1}{2} + \dfrac{1}{5} + \dfrac{1}{10} + \cdots + \dfrac{1}{n^2 + 1} + \cdots$

5 $\displaystyle\sum_{n=2}^{\infty} \dfrac{n}{(n^2 - 1)^3}$ **6** $\displaystyle\sum_{n=1}^{\infty} \dfrac{n^3}{n^4 + 9}$

7 $\displaystyle\sum_{n=3}^{\infty} \dfrac{1}{\sqrt{n^2 - 4}}$ **8** $\displaystyle\sum_{n=1}^{\infty} \dfrac{3}{(3n + 2)^2}$

9 $\displaystyle\sum_{n=1}^{\infty} ne^{-n}$ **10** $\displaystyle\sum_{n=1}^{\infty} \dfrac{\ln n}{n}$

11 $\displaystyle\sum_{n=1}^{\infty} n\left(\dfrac{2}{3}\right)^n$ **12** $\displaystyle\sum_{n=1}^{\infty} \dfrac{n}{n^4 + 1}$

13   Can we conclude that a given series is convergent if each of its terms is smaller than the corresponding term of a divergent series?
14   If each term of a given series is larger than the corresponding term of a convergent series, can we conclude that the given series is divergent?
15   Show by use of the integral test that the series

$$\frac{1}{4} + \frac{1}{7} + \frac{1}{10} + \cdots + \frac{1}{3n + 1} + \cdots$$

is divergent and then by use of the comparison test show that $\displaystyle\sum_{n=1}^{\infty} \dfrac{1}{\sqrt{3n + 1}}$ is divergent.
16   Prove the second comparison test of Sec. 26.9.

Determine whether the series in Problems 17 to 28 are convergent or divergent.

17 $\displaystyle\sum_{n=1}^{\infty} \dfrac{1}{n^n}$ **18** $\displaystyle\sum_{n=1}^{\infty} \dfrac{1}{n(n + 1)}$

19 $\displaystyle\sum_{n=1}^{\infty} \dfrac{1}{(n + 1)(n + 3)}$ **20** $\displaystyle\sum_{n=1}^{3} \dfrac{n + 2}{(n + 3)3^n}$

21 $\displaystyle\sum_{n=2}^{\infty} \dfrac{1}{\ln n}$ **22** $\displaystyle\sum_{n=2}^{\infty} \dfrac{1}{n \ln n}$

23 $\displaystyle\sum_{n=3}^{\infty} \dfrac{\ln n}{n^2}$ **24** $\displaystyle\sum_{n=1}^{\infty} \dfrac{n}{\ln (n + 1)}$

25 $\displaystyle\sum_{n=1}^{\infty} \dfrac{1}{\sqrt{(n + 1)(n + 2)}}$ **26** $\displaystyle\sum_{n=1}^{\infty} \dfrac{n}{e^n}$

27 $\displaystyle\sum_{n=1}^{\infty} \dfrac{1 + \cos n}{n^2}$ **28** $\displaystyle\sum_{n=1}^{\infty} \dfrac{n^2}{e^n}$

26.10   The ratio test

We shall now discuss another test that is used in determining whether a series is convergent or divergent. It is called the *ratio test* and states that

If, for the series of positive terms

ratio
test
$$a_1 + a_2 + a_3 + \cdots + a_n + \cdots = \sum_{n=1}^{\infty} a_n$$

$\lim\limits_{n \to \infty} (a_{n+1}/a_n)$ *exists and is equal to L, then the series is convergent if L < 1, is divergent if L > 1, and may be convergent or divergent if L = 1. If* $\lim\limits_{n \to \infty} (a_{n+1}/a_n)$ *fails to exist, the series is divergent.*

To prove this theorem, we assume that $L < 1$ and let r be any number such that $L < r < 1$. Now since the limit of a_{n+1}/a_n as n approaches infinity is L, there exists a number k such that for $n > k$, the value of the ratio is less than r. Thus, we have

$$a_{k+1} < ra_k$$
$$a_{k+2} < ra_{k+1} < r^2 a_k$$
$$a_{k+3} < ra_{k+2} < r^2 a_{k+1} < r^3 a_k$$

and, in general, $a_{k+N} < r^N a_k$ for every positive integer N. Consequently, the series

$$a_{k+1} + a_{k+2} + a_{k+3} + \cdots$$

is convergent since each of its terms is less than that of the convergent geometric series

$$ra_k + r^2 a_k + r^3 a_k + \cdots$$

with $r < 1$. Therefore

$$\sum_{n=1}^{\infty} a_n = \sum_{n=1}^{k} a_n + \sum_{n=k+1}^{\infty} a_n$$

is convergent since $\sum\limits_{n=1}^{k} a_n$ is a finite number.

The proof for the case of divergence can be obtained from the fact that if $L > 1$, there exists a number k such that

$$\frac{a_{n+1}}{a_n} > 1$$

for all integral $n > k$. Therefore, $a_{n+1} > a_n$ for $n > k$, and the necessary condition for convergence is not satisfied. The argument will not be given for the case in which $\lim\limits_{n \to \infty} (a_{n+1}/a_n)$ fails to exist.

We still must justify the part of the theorem that treats the situation in which $L = 1$. To do this, we consider the hyperharmonic or k series $\Sigma \dfrac{1}{n^k}$. The ratio test gives

$$\lim_{n \to \infty} \frac{a_{k+1}}{a_k} = \lim_{n \to \infty} \frac{1/(n+1)^k}{1/n^k} = \lim_{n \to \infty} \left(\frac{n}{n+1}\right)^k = 1$$

and we know that this series converges for $k > 1$ and diverges for $k \le 1$. Hence, the ratio test is inconclusive for $L = 1$.

Example 1   Apply the ratio test to

$$\frac{1}{2} + \frac{2}{2^2} + \frac{3}{2^3} + \cdots + \frac{n}{2^n} + \cdots$$

Solution: Since $a_n = n/2^n$, we have

$$\lim_{n \to \infty} \frac{a_{n+1}}{a_n} = \lim_{n \to \infty} \frac{(n+1)/2^{n+1}}{n/2^n} = \lim_{n \to \infty} \frac{1}{2}\frac{n+1}{n} = \frac{1}{2}$$

Consequently, the given series converges.

Example 2   Apply the ratio test to the series

$$\frac{1}{1!} + \frac{2^2}{2!} + \frac{3^3}{3!} + \cdots + \frac{n^n}{n!} + \cdots$$

Solution: Since $a_n = n^n/n!$, it follows that

$$\frac{a_{n+1}}{a_n} = \frac{(n+1)^{n+1}/(n+1)!}{n^n/n!} = \frac{(n+1)(n+1)^n n!}{n^n(n+1)n!}$$

$$= \left(\frac{n+1}{n}\right)^n = \left(1 + \frac{1}{n}\right)^n$$

Consequently,

$$\lim_{n \to \infty} \frac{a_{n+1}}{a_n} = \lim_{n \to \infty} \left(1 + \frac{1}{n}\right)^n = e > 1$$

and the series diverges.

EXERCISE 26.3

Test the following series for convergence by use of the ratio test. If the ratio is 1, use another test.

1    $\frac{3}{4} + 2(\frac{3}{4})^2 + 3(\frac{3}{4})^3 + \cdots + n(\frac{3}{4})^n + \cdots$

2    $\dfrac{1}{2^1} + \dfrac{1}{2^2} + \dfrac{1}{2^3} + \cdots + \dfrac{1}{2^n} + \cdots$

3 $\dfrac{1}{2} + \dfrac{3}{5} + \dfrac{9}{10} + \cdots + \dfrac{3^{n-1}}{n^2 + 1} + \cdots$

4 $1 + \dfrac{3}{2!} + \dfrac{3^2}{3!} + \cdots + \dfrac{3^{n-1}}{n!} + \cdots$

5 $1 + \dfrac{1}{6} + \dfrac{2!}{6^2} + \cdots + \dfrac{n!}{6^n} + \cdots$

6 $\dfrac{1^2}{1!} + \dfrac{2^2}{2!} + \dfrac{3^2}{3!} + \cdots + \dfrac{n^2}{n!} + \cdots$

7 $1 + \dfrac{2!}{1 \cdot 3} + \dfrac{3!}{1 \cdot 3 \cdot 5} + \cdots + \dfrac{n!}{1 \cdot 3 \cdot 5 \cdots (2n - 1)} + \cdots$

8 $1 + 2 + \dfrac{2^2}{2!} + \cdots + \dfrac{2^{n-1}}{(n - 1)!} + \cdots$

9 $\dfrac{7}{1!} + \dfrac{7^2}{2!} + \dfrac{7^3}{3!} + \cdots + \dfrac{7^n}{n!} + \cdots$

10 $\dfrac{3}{1^3} + \dfrac{4}{2^3} + \dfrac{5}{3^3} + \cdots + \dfrac{n + 2}{n^3} + \cdots$

11 $\dfrac{1 \cdot 2 \cdot 3}{4} + \dfrac{2 \cdot 3 \cdot 4}{4^2} + \dfrac{3 \cdot 4 \cdot 5}{4^3} + \cdots + \dfrac{n(n + 1)(n + 2)}{4^n} + \cdots$

12 $\dfrac{1 \cdot 2}{2^3} + \dfrac{2 \cdot 3}{3^3} + \dfrac{3 \cdot 4}{4^3} + \cdots + \dfrac{n(n + 1)}{(n + 1)^3} + \cdots$

13 $\dfrac{4}{1 \cdot 2 \cdot 3} + \dfrac{4^2}{2 \cdot 3 \cdot 4} + \dfrac{4^3}{3 \cdot 4 \cdot 5} + \cdots + \dfrac{4^n}{n(n + 1)(n + 2)} + \cdots$

14 $\dfrac{1}{2 \cdot 3} + \dfrac{1 \cdot 4}{2 \cdot 4 \cdot 5} + \dfrac{1 \cdot 4 \cdot 7}{2 \cdot 4 \cdot 6 \cdot 7} + \cdots + \dfrac{1 \cdot 4 \cdots (3n - 2)}{2^n(n!)(2n + 1)} + \cdots$

15 $\dfrac{1 + 3}{1} + \dfrac{1 + 3\sqrt{2}}{2} + \dfrac{1 + 3\sqrt{3}}{3} + \cdots + \dfrac{1 + 3\sqrt{n}}{n} + \cdots$

16 $\dfrac{5}{6} + \dfrac{9}{27} + \dfrac{13}{62} + \cdots + \dfrac{1 + 4n}{7n^2 - 1} + \cdots$

17 $\displaystyle\sum_{n=1}^{\infty} \dfrac{n}{e^{2n-1}}$ **18** $\displaystyle\sum_{n=1}^{\infty} \dfrac{\pi}{2^n - 3n}$

19 $\displaystyle\sum_{n=1}^{\infty} \dfrac{n!}{1 \cdot 3 \cdot 5 \cdots (2n - 1)}$ **20** $\displaystyle\sum_{n=1}^{\infty} \dfrac{n!}{n^n}$

21 $\displaystyle\sum_{n=1}^{\infty} \dfrac{(2n - 1)(2n + 1)}{3^n}$ **22** $\displaystyle\sum_{n=1}^{\infty} \dfrac{3^n}{n(n + 1)}$

23 $\displaystyle\sum_{n=2}^{\infty} \dfrac{\ln n}{n}$ **24** $\displaystyle\sum_{n=2}^{\infty} \dfrac{1}{\sqrt{n^2 - 1}}$

26.11   Alternating series

alternating series   A series is called an *alternating series* if its terms are alternately positive and negative. Thus, if $a_1, a_2, a_3, \ldots, a_n, \ldots$ are all positive numbers, then the series

$$a_1 - a_2 + a_3 - a_4 + \cdots + (-1)^{n+1}a_n + \cdots = \sum_{k=1}^{\infty} (-1)^{k+1}a_k \quad (1)$$

is an alternating series.

a sufficient condition   A *sufficient condition* for the convergence of an alternating series is given by the statement that *the alternating series (1) is convergent if each term is numerically less than or equal to the preceding one and if* $\lim_{n\to\infty} a_n = 0$.

We shall now prove this statement. The partial sums of (1) with an even number of terms are

$$S_{2n} = (a_1 - a_2) + (a_3 - a_4) + \cdots + (a_{2n-1} - a_{2n})$$

where $a_1 - a_2 \geq 0$, $a_3 - a_4 \geq 0, \ldots$, $a_{2n-1} - a_{2n} \geq 0$. Consequently,

$$0 \leq S_2 \leq S_4 \leq \cdots \leq S_{2n} \leq \cdots$$

However, we can also write

$$S_{2n} = a_1 - (a_2 - a_3) - (a_4 - a_5) - \cdots - (a_{2n-2} - a_{2n-1}) - a_{2n}$$

where $a_2 - a_3 \geq 0$, $a_4 - a_5 \geq 0, \ldots$, $a_{2n-2} - a_{2n-1} \geq 0$, $a_{2n} \geq 0$. Therefore, for every integer n, $S_{2n} \leq a_1$ and, by the first theorem of Sec. 26.5,

$$\lim_{n\to\infty} S_{2n}$$

exists and is equal to or less than a_1.

Finally, since $S_{2n+1} = S_{2n} + a_{2n+1}$, it follows that

$$\lim_{n\to\infty} S_{2n+1} = \lim_{n\to\infty} (S_{2n} + a_{2n+1}) = \lim_{n\to\infty} S_{2n}$$

since $\lim_{n\to\infty} a_{2n+1} = 0$. We now have shown that

$$\lim_{n\to\infty} S_n = \lim_{n\to\infty} S_{2n+1} = \lim_{n\to\infty} S_{2n} = S$$

where $S \leq a_1$. We have shown not only that under the stated hypotheses, (1) converges but that the sum is not greater than a_1.

We shall now obtain an approximation to the error committed by using the sum S_n of the first n terms of (1) in place of the sum S of the series. To do this, we begin by noticing that if (1) is an alternating series with $a_{n+1} \leq a_n$ for all n and $\lim_{n\to\infty} a_n = 0$, then the remainder

$$R_n = (-1)^n(a_{n+1} - a_{n+2} + a_{n+3} - \cdots)$$

of the series after the first n terms is also an alternating series that satisfies the same conditions. Since

$$|R_n| = a_{n+1} - a_{n+2} + \cdots$$

it follows that

$$|R_n| \leq a_{n+1}$$

and since $S = S_n + R_n$, we have $S - S_n = R_n$. Hence,

$$|S - S_n| = |R_n| \leq a_{n+1}$$

error in S_n Therefore, *the error made by using S_n as an approximation to the sum S of (1) is less than or equal to the first term omitted.*

Example   The series

$$1 - \frac{1}{2} + \frac{1}{3} - \frac{1}{4} + \cdots + (-1)^{n+1}\frac{1}{n} + \cdots \tag{2}$$

has $a_{n+1} \leq a_n$ since

$$\frac{1}{n+1} - \frac{1}{n} = \frac{-1}{(n+1)n} < 0$$

Furthermore, $$\lim_{n \to \infty} a_n = \lim_{n \to \infty} \frac{1}{n} = 0$$

Consequently, (2) is convergent. The error made by using the sum of the first four terms as an approximation to the sum of the series is less than or equal to ⅕ since that is the first term omitted.

26.12   Absolute convergence

Corresponding to each series Σa_k, there is the series of absolute values

$$\sum_{k=1}^{\infty}|a_k| = |a_1| + |a_2| + \cdots + |a_n| + \cdots$$

absolute convergence The series Σa_k is said to *converge absolutely* if the series of absolute values, $\Sigma|a_k|$ converges.

We shall now prove that *if a series is absolutely convergent, it is convergent.*

In proving this statement, we shall make use of three infinite series. They are Σa_n, $\Sigma|a_n|$, and $\Sigma(a_n + |a_n|)$, and we shall designate their nth partial sums by S_n, T_n, and U_n, respectively. We then have

$$S_n + T_n = U_n$$

Since Σa_n may contain some negative terms, it follows that $0 \leq a_n + |a_n| \leq 2|a_n|$. Consequently, $0 \leq U_n \leq 2T_n \leq 2T$ for all integral values of n where T is the value of $\Sigma|a_n|$. The sequence of partial sums U_n satisfies the relation $U_1 \leq U_2 \leq \cdots \leq U_n \leq \cdots \leq 2T$ and, by the first theorem of Sec. 26.5,

$$\lim_{n \to \infty} U_n = U \leq 2T$$

Furthermore, $\lim_{n \to \infty} S_n = \lim_{n \to \infty} U_n - \lim_{n \to \infty} T_n = U - T$

and the series Σa_n is convergent with sum $S = U - T \leq T$.

The converse of the above theorem is not true. In proof of this, we offer the alternating series

$$1 - \frac{1}{2} + \frac{1}{3} - \frac{1}{4} + \cdots + (-1)^{n+1}\frac{1}{n} + \cdots$$

which is convergent whereas the corresponding series of absolute values is the divergent harmonic series. A series that is convergent but *conditional* not absolutely convergent is said to be *conditionally convergent*. *convergence*

In applying the ratio test of Sec. 26.10 to series that contain both positive and negative terms, we must replace a_{n+1}/a_n by $|a_{n+1}/a_n|$ and "convergent" by "absolutely convergent." The ratio test then is: If $\lim_{n \to \infty} |a_{n+1}/a_n| = L$, then the series is absolutely convergent if $L < 1$ and divergent if $L > 1$. The test fails if $L = 1$. If the limit fails to exist, the series is divergent.

EXERCISE 26.4

Test the series in each of Problems 1 to 20 for convergence and absolute convergence.

1 $\dfrac{1}{2} - \dfrac{1}{5} + \dfrac{1}{10} - \dfrac{1}{17} + \cdots + \dfrac{(-1)^{n+1}}{n^2 + 1} + \cdots$

2 $\dfrac{1}{e} - \dfrac{4}{e^2} + \dfrac{9}{e^3} - \dfrac{16}{e^4} + \cdots + \dfrac{(-1)^{n-1}n^2}{e^n} + \cdots$

3 $\dfrac{-1}{5} + \dfrac{4}{12} - \dfrac{9}{31} + \cdots + \dfrac{(-1)^n n^2}{n^3 + 4} + \cdots$

4 $\dfrac{1}{\sqrt{5}} - \dfrac{1}{2\sqrt{2}} + \dfrac{1}{\sqrt{13}} - \dfrac{1}{2\sqrt{5}} + \cdots + \dfrac{(-1)^{n+1}}{\sqrt{n^2 + 4}} + \cdots$

5 $-1 + \dfrac{1}{2} - \dfrac{2}{2^2} + \cdots + \dfrac{(-1)^n n}{2^n} + \cdots$

6 $1 - \dfrac{1}{4} + \dfrac{2}{4^2} - \dfrac{3}{4^3} + \cdots + \dfrac{(-1)^{n-1}n}{4^n} + \cdots$

7 $\dfrac{1}{3} - \dfrac{2}{5} + \dfrac{3}{7} - \dfrac{4}{9} + \cdots + \dfrac{n(-1)^{n+1}}{2n+1} + \cdots$

8 $-\dfrac{1}{\sqrt{1\cdot 2}} + \dfrac{1}{\sqrt{2\cdot 3}} - \dfrac{1}{\sqrt{3\cdot 4}} + \cdots + \dfrac{(-1)^n 1}{\sqrt{n(n+1)}}$

9 $\displaystyle\sum_{n=1}^{\infty} \dfrac{(-1)^n}{(2n-1)!}$ **10** $\displaystyle\sum_{n=1}^{\infty} \dfrac{(-1)^{n+1}n^6}{e^n}$

11 $\displaystyle\sum_{n=0}^{\infty} (-1)^{n+1}\dfrac{1}{(n+2)^2}$ **12** $\displaystyle\sum_{n=1}^{\infty} \dfrac{(-1)^n}{(2n-1)2n}$

13 $\displaystyle\sum_{n=1}^{\infty} \dfrac{1}{n}\left(\dfrac{3}{2}\right)^n (-1)^n$ **14** $\displaystyle\sum_{n=1}^{\infty} (-1)^{n+1}\dfrac{1+\sqrt{n}}{n}$

15 $\displaystyle\sum_{n=1}^{\infty} \dfrac{(-1)^n 1\cdot 4\cdots(3n-2)}{2^n(n!)(2n+1)}$ **16** $\displaystyle\sum_{n=1}^{\infty} (-1)^n(\sqrt{n+1} - \sqrt{n-1})$

17 $\displaystyle\sum_{n=1}^{\infty} \dfrac{(-1)^{n+1}n}{2n-1}$ **18** $\displaystyle\sum_{n=1}^{\infty} \dfrac{(-1)^n}{2n-1}$

19 $\displaystyle\sum_{n=1}^{\infty} (-1)^{n+1}\dfrac{1}{n\sqrt{n}}$ **20** $\displaystyle\sum_{n=1}^{\infty} (-1)^n\dfrac{3^{n-1}}{n!}$

What is the largest possible value of the error made by using the specified number of terms as an approximation to the sum of the series in each of Problems 21 to 24?

21 $\displaystyle\sum_{n=1}^{\infty} \dfrac{(-1)^n}{n^2+1}$, four terms   **22** $\displaystyle\sum_{n=1}^{\infty} \dfrac{(-1)^n n}{4^n}$, five terms

23 $\displaystyle\sum_{n=1}^{\infty} \dfrac{(-1)^{n+1}2}{2n-1}$, eight terms   **24** $\displaystyle\sum_{n=1}^{\infty} \dfrac{(-1)^{n-1}n}{(n+2)!}$, six terms

26.13   Power series

So far in this chapter we have studied series made up of constant terms. We shall now consider series whose terms are functions of a variable x. The series

$$x + 2x^2 + 3x^3 + 4x^4 + \cdots + nx^n + \cdots$$

and

$$1 + (x+3) + \dfrac{(x+3)^2}{2!} + \dfrac{(x+3)^3}{3!} + \cdots + \dfrac{(x+3)^{n-1}}{(n-1)!} + \cdots$$

are series whose terms are functions of a variable x. The first is
*power
series* a power series in x, and the second is a power series in $x + 3$.
In general

$$a_0 + a_1(x - c) + a_2(x - c)^2 + \cdots + a_n(x - c)^n$$

$$+ \cdots = \sum_{n=0}^{\infty} a_n(x - c)^n$$

is called a *power series in $x - c$* provided that each coefficient of a
power of $x - c$ is a constant.

Power series are convergent for some values of x and divergent for
others. If a power series is convergent for $a \leq x \leq b$, this interval is
*interval
of
convergence* called the *interval of convergence*. It may be an open interval, a
closed interval, or a half-open interval. Our next task will be the use
of the ratio test to determine the open interval of convergence. After
that is done, the two series of constant terms corresponding to the
end points must be tested to see if the interval of convergence is
open, half-open, or closed.

Example 1   Find the interval of convergence of

$$1 + \frac{2x}{3} + \frac{3x^2}{9} + \frac{4x^3}{27} + \cdots + \frac{nx^{n-1}}{3^{n-1}} + \cdots$$

Solution: In this series, $a_n = nx^{n-1}/3^{n-1}$; hence

$$\left| \frac{a_{n+1}}{a_n} \right| = \left| \frac{(n + 1)x^n}{3^n} \frac{3^{n-1}}{nx^{n-1}} \right| = \left| \frac{(n + 1)x}{3n} \right|$$

Consequently, $\lim\limits_{n \to \infty} \left| \dfrac{a_{n+1}}{a_n} \right| = \lim\limits_{n \to \infty} \left| \dfrac{(n + 1)x}{3n} \right| = \left| \dfrac{x}{3} \right|$

and the series converges if

$$\left| \frac{x}{3} \right| < 1$$

that is, if x is in $(-3, 3)$. We must now see if the two series obtained
by replacing x by 3 and by -3 are convergent. If $x = 3$, the given
series becomes

$$1 + 2 + 3 + \cdots + n + \cdots$$

It is divergent since it does not satisfy the necessary condition for
convergence. If $x = -3$, the given series becomes the divergent series

$$1 - 2 + 3 - 4 + \cdots + (-1)^{n+1}n + \cdots$$

Therefore, the given series converges for $-3 < x < 3$.

Example 2   Find the interval of convergence of

$$\sum_{n=1}^{\infty} \frac{(-1)^{n+1}(x-1)^n}{2^n n}$$

Solution: Since $a_n = (-1)^{n+1}(x-1)^n/2^n n$ for the given series, it follows that

$$\left|\frac{a_{n+1}}{a_n}\right| = \left|\frac{(x-1)^{n+1}}{2^{n+1}(n+1)} \frac{2^n n}{(x-1)^n}\right| = \left|\frac{(x-1)n}{2(n+1)}\right|$$

and $\displaystyle\lim_{n\to\infty}\left|\frac{a_{n+1}}{a_n}\right| = \left|\frac{x-1}{2}\right|$

Therefore, the series converges if

$$\left|\frac{x-1}{2}\right| < 1$$

that is, if $|x-1| < 2$. Now, solving for x, we find that the series converges if x is in $(-1, 3)$. We must check the series obtained by replacing x by 3 and by -1. If $x = 3$, the given series becomes

$$1 - \frac{1}{2} + \frac{1}{3} - \frac{1}{4} + \cdots + (-1)^{n+1}\frac{1}{n} + \cdots$$

It is convergent since it is an alternating series with $a_{n+1} < a_n$ and $\displaystyle\lim_{n\to\infty} a_n = 0$. Furthermore, if $x = -1$, the given series becomes

$$-\left(1 + \frac{1}{2} + \frac{1}{3} + \cdots + \frac{1}{n} + \cdots\right)$$

It is divergent since it is the negative of the harmonic series. Consequently, the interval of convergence of the given series is $(-1, 3]$.

EXERCISE 26.5

Find the interval of convergence of the power series in each of Problems 1 to 24.

1. $1 + x + x^2 + \cdots + x^{n-1} + \cdots$

2. $1 - x + x^2 - x^3 + \cdots + (-1)^{n+1}x^{n-1} + \cdots$

3. $1 + x + 2x^2 + \cdots + (n-1)x^{n-1} + \cdots$

4. $\dfrac{1}{2} + \dfrac{x}{2^2} + \dfrac{x^2}{2^3} + \cdots + \dfrac{x^{n+1}}{2^n} + \cdots$

5. $1 + \dfrac{2x}{5} + \dfrac{3x^2}{5^2} + \cdots + \dfrac{nx^{n-1}}{5^{n-1}} + \cdots$

6. $1 + 8x + 8^2x^2 + \cdots + 8^{n-1}x^{n-1} + \cdots$

7 $\quad \dfrac{x}{2} + \dfrac{x^2}{2 \cdot 3} + \dfrac{x^3}{2^2 \cdot 4} + \cdots + \dfrac{x^n}{2^{n-1}(n+1)} + \cdots$

8 $\quad \dfrac{x}{2 \cdot 1} + \dfrac{x^2}{2^2 \cdot 2^2} + \dfrac{x^3}{2^3 \cdot 3^2} + \cdots + \dfrac{x^n}{2^n \cdot n^2} + \cdots$

9 $\quad \dfrac{x}{1!} + \dfrac{x^2}{2!} + \cdots + \dfrac{x^n}{n!} + \cdots$

10 $\quad x - \dfrac{x^3}{3!} + \dfrac{x^5}{5!} - \cdots + \dfrac{(-1)^{n+1}x^{2n-1}}{(2n-1)!} + \cdots$

11 $\quad \dfrac{1}{x} + \dfrac{2!}{x^2} + \dfrac{3!}{x^3} + \cdots + \dfrac{n!}{x^n} + \cdots$

12 $\quad 3^3 \dfrac{1+1^2}{3!} x^3 + 3^4 \dfrac{1+2^2}{4!} x^4 + \cdots + 3^{n+2} \dfrac{1+n^2}{(n+2)!} x^{n+2} + \cdots$

13 $\quad 1 + \dfrac{2}{x} + \dfrac{4}{2!x^2} + \cdots + \dfrac{2^{n-1}}{(n-1)!x^{n-1}} + \cdots$

14 $\quad (x-1) + \dfrac{(x-1)^2}{2 \cdot 2!} + \dfrac{(x-1)^3}{3 \cdot 3!} + \cdots + \dfrac{(x-1)^n}{n \cdot n!} + \cdots$

15 $\quad x + 2!x^2 + 3!x^3 + \cdots + n!x^n + \cdots$

16 $\quad \dfrac{x}{2} + \dfrac{2x^2}{3 \cdot 2} + \dfrac{3x^3}{4 \cdot 2^2} + \cdots + \dfrac{nx^n}{(n+1) \cdot 2^{n-1}} + \cdots$

17 $\quad \displaystyle\sum_{n=1}^{\infty} \dfrac{(x+2)^{n-1}}{n^2}$ $\qquad$ 18 $\quad \displaystyle\sum_{n=1}^{\infty} \dfrac{(x-2)^n}{n}$

19 $\quad \displaystyle\sum_{n=1}^{\infty} \dfrac{(x-2)^n}{(2n-1)(2n)}$ $\qquad$ 20 $\quad \displaystyle\sum_{n=1}^{\infty} \dfrac{1}{n}\left(\dfrac{1}{x}\right)^n$

21 $\quad \displaystyle\sum_{n=1}^{\infty} \dfrac{1 \cdot 3 \cdots (2n-1)}{2^n n!} x^n$ $\qquad$ 22 $\quad \displaystyle\sum_{n=1}^{\infty} \dfrac{2 \cdot 3 \cdots (n+1)}{1 \cdot 2 \cdots n}(x-1)^n$

23 $\quad \displaystyle\sum_{n=1}^{\infty} \dfrac{(x-3)^n}{n \cdot 3^n}$ $\qquad$ 24 $\quad \displaystyle\sum_{n=1}^{\infty} \dfrac{n(x+1)^n}{4^n}$

26.14   Taylor's series

In this section we shall find what the coefficient C_n of $(x - a)$ must be if $\Sigma C_n(x - a)^n$ converges and represents $f(x)$ in an interval $(a - r, a + r)$. In Sec. 26.16, we find the conditions under which $\Sigma C_n(x - a)^n$ converges and represents the function. Specifically, we shall now assume that

$$f(x) = C_0 + C_1(x - a) + C_2(x - a)^2 + \cdots + C_n(x - a)^n + \cdots \quad (1)$$

converges and represents the function in the open interval $a - r <$

differentiating
series

$x < a + r$, and we shall determine the coefficients. If we put $x = a$ in (1), we get $f(a) = C_0$. To find the other coefficients, we shall make use of the fact that if a power series converges and represents $f(x)$ in an interval, then the series obtained by differentiating this series term by term converges in the same interval, except possibly at the end points, and represents the derivative of the function.° Therefore to compute the values of the coefficients $C_1, C_2, C_3, \ldots, C_r, \ldots$, we shall obtain $f'(x), f''(x), f'''(x)$, and a formula for $(d^r/dx^r)f(x) = f^{(r)}(x)$. Since it is assumed that

$$f(x) = \sum_{n=0}^{\infty} C_n(x - a)^n = C_0 + \sum_{n=1}^{\infty} C_n(x - a)^n$$

it follows that

$$f'(x) = \sum_{n=1}^{\infty} nC_n(x - a)^{n-1} = C_1 + \sum_{n=2}^{\infty} nC_n(x - 1)^{n-1}$$

$$f''(x) = \sum_{n=2}^{\infty} n(n - 1)C_n(x - a)^{n-2} = 2 \cdot 1 C_2 + \sum_{n=3}^{\infty} n(n - 1)C_n(x - a)^{n-2}$$

$$f'''(x) = \sum_{n=3}^{\infty} n(n - 1)(n - 2)C_n(x - a)^{n-3}$$

$$= 3 \cdot 2 \cdot 1 C_3 + \sum_{n=4}^{\infty} (n)(n - 1)(n - 2)C_n(x - a)^{n-3}$$

We now call attention to the fact that the factors that involve n in the first summation in each of the above derivatives are $n(n - 1)(n - 2) \cdots (n - r + 1)$ where r is the order of the derivative, and that the exponent of $x - a$ is $n - r$. It can be proved by use of mathematical induction that

$$f^{(r)}(x) = \sum_{n=r}^{\infty} n(n - 1)(n - 2) \cdots (n - r + 1)C_n(x - a)^{n-r}$$

$$= r!C_r + \sum_{n=r+1}^{\infty} n(n - 1)(n - 2) \cdots (n - r + 1)C_n(x - a)^{n-r} \tag{2}$$

We now employ (2) to determine the values of C_n, $n = 1, 2, 3, \ldots$. If we replace x by a in (2), we have

$$f^{(r)}(a) = r!C_r + \sum_{n=r+1}^{\infty} n(n - 1)(n - 2) \cdots (n - r + 1)C_n(a - a)^{n-r}$$

Furthermore, since $n - r \geq 1$ if $n \geq r + 1$, it follows that $(a - a)^{n-r} =$

° A proof of this statement can be found in Woods, "Advanced Calculus," sec. 23, Ginn, Boston, 1934.

0. Consequently, $f^{(r)}(a) = r!C_r$ and it follows that $C_r = f^{(r)}(a)/r!$ Thus $C_1 = f'(a)$, $C_2 = f''(a)/2!$, $C_3 = f'''(a)/3!$, and so on. Therefore,

$$f(x) = f(a) + f'(a)(x - a) + \frac{f''(a)}{2!}(x - a)^2 + \cdots$$

$$+ \frac{f^{(r)}(a)}{r!}(x - a)^r + \cdots \quad (26.1)$$

Taylor's series   This is called the *Taylor's series* for $f(x)$ in powers of $x - a$ and is sometimes called the expansion about $x = a$. The interval of convergence can be found in any valid expansion.

Example 1   Expand $f(x) = \sin x$ in powers of $x - \pi/4$.

Solution: We shall begin by finding that

$$f(x) = \sin x \qquad f\left(\frac{\pi}{4}\right) = \frac{1}{\sqrt{2}}$$

$$f'(x) = \cos x \qquad f'\left(\frac{\pi}{4}\right) = \frac{1}{\sqrt{2}}$$

$$f''(x) = -\sin x \qquad f''\left(\frac{\pi}{4}\right) = -\frac{1}{\sqrt{2}}$$

$$f'''(x) = -\cos x \qquad f'''\left(\frac{\pi}{4}\right) = -\frac{1}{\sqrt{2}}$$

It is not necessary to find other derivatives since if the orders of two derivatives of $\sin x$ differ by an integral multiple of four, the derivatives are equal. If we now substitute in (26.1), we find that

$$\sin x = \sin\frac{\pi}{4} + \cos\frac{\pi}{4}\left(x - \frac{\pi}{4}\right) - \frac{\sin(\pi/4)(x - \pi/4)^2}{2!}$$

$$- \frac{\cos(\pi/4)(x - \pi/4)^3}{3!} + \cdots$$

$$= \frac{1}{\sqrt{2}}\left[1 + \left(x - \frac{\pi}{4}\right) - \frac{(x - \pi/4)^2}{2!} - \frac{(x - \pi/4)^3}{3!}\right.$$

$$\left. + \frac{(x - \pi/4)^4}{4!} + \cdots\right]$$

If $a = 0$ in Taylor's series, we have

$$f(x) = f(0) + f'(0)x + \frac{f''(0)x^2}{2!} + \frac{f'''(0)x^3}{3!} + \cdots + \frac{f^{(n)}(0)x^n}{n!} + \cdots$$

$$(26.2)$$

Maclaurin's series   This is called *Maclaurin's series* and enables us to express $f(x)$ in terms of powers of x.

Example 2   Expand $f(x) = (1 + x)^{1/2}$ in powers of x by use of Maclaurin's series.

Solution: We begin by finding derivatives of $f(x)$ and evaluating each for $x = 0$. Thus, we get

$$f(x) = (1 + x)^{1/2} \qquad\qquad\qquad f(0) = 1$$

$$f'(x) = \frac{1}{2}(1 + x)^{-1/2} \qquad\qquad f'(0) = \frac{1}{2}$$

$$f''(x) = \frac{1}{2}\left(-\frac{1}{2}\right)(1 + x)^{-3/2} \qquad\qquad f''(0) = \frac{-1}{2^2}$$

$$f'''(x) = \frac{1}{2}\left(-\frac{1}{2}\right)\left(-\frac{3}{2}\right)(1 + x)^{-5/2} \qquad f'''(0) = \frac{1 \cdot 3}{2^3}$$

$$f^{(4)}(x) = \frac{1}{2}\left(-\frac{1}{2}\right)\left(-\frac{3}{2}\right)\left(-\frac{5}{2}\right)(1 + x)^{-7/2} \qquad f^{(4)}(0) = \frac{-1 \cdot 3 \cdot 5}{2^4}$$

. . . .

$$f^{(n)}(x) = \frac{(-1)^{n+1}(1)(3)(5) \cdots (2n - 3)}{2^n}(1 + x)^{-(2n-1)/2} \qquad n \geq 2$$

$$f^{(n)}(0) = \frac{(-1)^{n+1}(1)(3) \cdots (2n - 3)}{2^n} \qquad n \geq 2$$

Therefore, $(1 + x)^{1/2} = 1 + \dfrac{x}{2} + \displaystyle\sum_{n=2}^{\infty} \dfrac{(-1)^{n+1}(1)(3) \cdots (2n - 3)}{2^n n!} x^n$

The reader should have no trouble in showing that the same terms are obtained by use of the binomial expansion.

To find the interval of convergence, we shall use the ratio test. Thus,

$$\left|\frac{a_{n+1}}{a_n}\right| = \left|\frac{\dfrac{(1)(3)(5) \cdots (2n - 3)(2n - 1)}{(n + 1)!2^{n+1}} x^{n+1}}{\dfrac{(1)(3)(5) \cdots (2n - 3)}{n!2^n} x^n}\right|$$

$$= \left|\frac{(2n - 1)x}{(n + 1) \cdot 2}\right| \rightarrow |x| \qquad \text{as } n \rightarrow \infty$$

Consequently the series converges for $|x| < 1$, that is, for x in $(-1, 1)$, and it may converge for $x = \pm 1$ so far as is determined by the ratio test. It can be shown by other tests that the series diverges for $x = \pm 1$.

EXERCISE 26.6

Verify the expansions in Problems 1 to 4, and show that the series converge for all values of x.

1 $e^x = 1 + x + \dfrac{x^2}{2!} + \cdots + \dfrac{x^n}{n!} + \cdots$

2 $\sin x = x - \dfrac{x^3}{3!} + \dfrac{x^5}{5!} - \cdots + (-1)^n \dfrac{x^{2n+1}}{(2n+1)!} + \cdots$

3 $\cos x = 1 - \dfrac{x^2}{2!} + \dfrac{x^4}{4!} - \cdots + (-1)^n \dfrac{x^{2n}}{(2n)!} + \cdots$

4 $\cosh x = 1 + \dfrac{x^2}{2!} + \dfrac{x^4}{4!} + \cdots + \dfrac{x^{2n}}{(2n)!} + \cdots$

Verify the expansions in Problems 5 to 7, and show that the series are convergent in $(-1, 1)$.

5 $\ln(1 + x) = x - \dfrac{x^2}{2} + \dfrac{x^3}{3} - \dfrac{x^4}{4} + \cdots + (-1)^{n+1} \dfrac{x^n}{n} + \cdots$

6 $\ln(1 - x) = -x - \dfrac{x^2}{2} - \dfrac{x^3}{3} - \dfrac{x^4}{4} - \cdots - \dfrac{x^n}{n} - \cdots$

7 $\text{Arctan } x = x - \dfrac{x^3}{3} + \dfrac{x^5}{5} - \dfrac{x^7}{7} + \cdots + (-1)^{n+1} \dfrac{x^{2n-1}}{2n-1} + \cdots$

8   By use of the series given in Problems 5 and 6, show that
$$\ln\frac{1+x}{1-x} = 2\left(x + \frac{x^3}{3} + \frac{x^5}{5} + \cdots + \frac{x^{2n-1}}{2n-1} + \cdots\right)$$

9   Expand $\sin x$ in a Taylor's series with $a = \pi/6$.
10   Expand $\cos x$ in a Taylor's series with $a = \pi/4$.
11   Expand $\ln x$ in powers of $x - 2$.
12   Expand $\sqrt{x}$ in powers of $x - 1$.
13   Expand e^x in powers of $x - 1$.
14   Expand e^{-x} in powers of $x + 1$.
15   Expand $1/x$ in a Taylor's series about $x = 1$.
16   Expand $\sinh x$ in a Taylor's series about $x = 1$.
17   Use the expansion of e^y to obtain the Maclaurin series for e^{-x^2}. What is the interval of convergence?
18   Compute $\sin 9° = \sin(\pi/20)$, by use of the first two terms of the series in Problem 2.
19   Compute $\cos 20° = \cos(\pi/9)$ by use of the first three terms of the series in Problem 3.
20   Compute $\sqrt{e}$ by use of the first two terms of the series in Problem 1.
21   Show by use of Problem 2 that as x approaches zero, $\sin x$ and x differ by terms of the third and higher degree.
22   Show by use of Problem 3 that as x approaches zero, $\cos x$ and $(1 - x^2/2)$ differ by terms of the fourth and higher degree.
23   By use of the series in Problem 2, show that $\sin(-x) = -\sin x$.
24   By use of the series in Problem 3, show that $\cos(-x) = \cos x$.

In each of the following problems find the first three terms of the Maclaurin series for the given function.

25    $\tan x$ 26 $\sec x$

27    $\ln (\sec x + \tan x)$ 28 $\ln \sec x$

29    $\ln \cos x$ 30 $x^2 e^{-x}$

31    $e^x \sin x$ 32 $x \ln (1 + x)$

26.15   Operations with power series

Operations that we can perform in dealing with a finite number of terms are not necessarily permissible where infinite series are involved. Thus, changing the order of the terms in a conditionally convergent series may change its sum or cause it to be divergent. We shall now state without proof several theorems concerning operations on infinite series.*

theorems on operations on series

1.   *Two power series in x may be added term by term for values of x for which both series converge.*

2.   *Two power series*

$$f(x) = a_0 + a_1 x + a_2 x^2 + \cdots$$

and $$g(x) = b_0 + b_1 x + b_2 x^2 + \cdots$$

may be multiplied together for all values of x for which both series are absolutely convergent and

$$f(x)g(x) = a_0 b_0 + (a_0 b_1 + a_1 b_0)x + (a_0 b_2 + a_1 b_1 + a_2 b_0)x^2$$
$$+ (a_0 b_3 + a_1 b_2 + a_2 b_1 + a_3 b_0)x^3 + \cdots$$

3.   *A power series in x may be differentiated term by term with respect to x for all values of x inside its interval of convergence. We thus have*

$$f'(x) = a_1 + 2a_2 x + 3a_3 x^2 + 4a_4 x^3 + \cdots$$

4.   *A power series in x may be integrated term by term with respect to x provided that the limits of integration are within the interval of convergence of the original series.*

Example 1   If we make use of the fact that

$$\sin x = x - \frac{x^3}{3!} + \frac{x^5}{5!} - \cdots$$

---

* The proofs can be found in many advanced calculus books. See Woods, "Advanced Calculus," secs. 25 and 23, Ginn, Boston, 1934.

and
$$\cos x = 1 - \frac{x^2}{2!} + \frac{x^4}{4!} - \cdots$$

we find that
$$\sin x + \cos x = 1 + x - \frac{x^2}{2!} - \frac{x^3}{3!} + \frac{x^4}{4!} + \cdots$$

Example 2 If we use those two expansions again, we have
$$\sin x \cos x = x - x^3 \left(\frac{1}{3!} + \frac{1}{2!} \right) + x^5 \left(\frac{1}{5!} + \frac{1}{4!} + \frac{1}{2!3!} \right) + \cdots$$
$$= x - \frac{2x^3}{3} + \frac{2}{15} x^5 - \cdots$$

The reader may be interested in showing that the above three terms are the first three terms of the Maclaurin expansion of ½ sin 2x.

Example 3 If we differentiate the series given above for sin x, we get
$$1 - \frac{x^2}{2!} + \frac{x^4}{4!} - \cdots$$

as should be the case since $D_x \sin x = \cos x$.

Example 4 Since
$$e^{-x^2} = 1 - x^2 + \frac{x^4}{2!} - \frac{x^6}{3!} + \cdots$$

converges for all x, it follows that
$$\int_0^{0.5} e^{-x^2} \, dx = \int_0^{0.5} \left(1 - x^2 + \frac{x^4}{2!} - \frac{x^6}{3!} + \cdots \right) dx$$
$$= x - \frac{x^3}{3} + \frac{x^5}{5 \cdot 2!} - \frac{x^7}{7 \cdot 3!} + \cdots \Big|_0^{0.5} = 0.4613$$

26.16 Taylor's series with a remainder

We shall be interested in determining the remainder R_n of the Taylor's series for $f(x)$ after n terms. For this purpose, we shall assume that $f(x)$ and its first n derivatives are continuous in an interval $[a, b]$ and let K be determined by

determination of R_n
$$f(b) - f(a) - \frac{f'(a)}{1!} (b - a) - \frac{f''(a)}{2!} (b - a)^2 - \cdots$$
$$- \frac{f^{(n-1)}(a)}{(n-1)!} (b - a)^{n-1} - \frac{K}{n!} (b - a)^n = 0 \quad (1)$$

We now consider the function

$$F(x) = f(b) - f(x) - \frac{f'(x)}{1!}(b - x) - \frac{f''(x)}{2!}(b - x)^2 - \cdots$$

$$- \frac{f^{(n-1)}(x)}{(n-1)!}(b - x)^{n-1} - \frac{K}{n!}(b - x)^n$$

This function satisfies all conditions of Rolle's theorem; hence, there exists a value x_1 of x in (a, b) for which $F'(x_1) = 0$.

If we differentiate $F(x)$, we find that all terms except the last two add to zero in pairs; hence,

$$F'(x_1) = \left[-\frac{f^{(n)}(x_1)}{(n-1)!} + \frac{K}{(n-1)!} \right](b - x_1)^{n-1} = 0$$

Consequently, $K = f^{(n)}(x_1)$ $a < x_1 < b$

If we now substitute this for K in (1) and solve for $f(b)$, we get

$$f(b) = f(a) + \frac{f'(a)}{1!}(b - a) + \frac{f''(a)}{2!}(b - a)^2 + \cdots$$

$$+ \frac{f^{(n-1)}(a)}{(n-1)!}(b - a)^{n-1} + \frac{f^{(n)}(x_1)}{n!}(b - a)^n$$

with x_1 in (a, b). If we replace b by any x in (a, b), we have

$$f(x) = f(a) + \frac{f'(a)}{1!}(x - a) + \cdots + \frac{f^{(n-1)}(a)}{(n-1)!}(x - a)^{n-1} + R_n \quad (26.3)$$

Taylor's series with a remainder   *where the remainder R_n after n terms is given by*

$$R_n = \frac{f^{(n)}(x_1)}{n!}(x - a)^n \qquad a < x_1 < x \qquad (26.4)$$

Therefore, the Taylor's series for $f(x)$ converges to $f(x)$ in $(a - r, a + r)$ if all derivatives exist and if R_n approaches zero as n approaches infinity. The formula for R_n given by (26.4) is known as Lagrange's form for the remainder.

26.17   Computation by use of Taylor's series

Since the first n terms of a Taylor or Maclaurin series is a polynomial, either is very useful for calculation of approximate values of functions for specified values of the variable; furthermore, the remainder R_n enables us to determine the accuracy of the calculation. If the series happens to be an alternating one, then we can make use of Sec. 26.11 in checking the accuracy of the result.

Example   By use of the Lagrange form for R_n, we shall determine the number of terms that must be used in order to have sin α correct to four decimal places if sin x is expanded in powers of $x - \pi/4$ and α does not differ from $\pi/4$ by more than 0.1 radian.

Solution: Since all derivatives of sin x are either $\pm$sin x or $\pm$cos x and the numerically largest value of each is 1, we know that

$$|f^{(n)}(x_1)| \leq 1$$

Therefore,

$$R_n < \frac{1}{n!}\left(\alpha - \frac{\pi}{4}\right)^n \leq \frac{1}{n!}(0.1)^n$$

since $|\alpha - \pi/4| \leq 1$.

Suppose sin $\alpha = N + R_n$ where N is the number yielded by the first n terms of the Taylor expansion. If N is the approximate value of sin α to four decimal places, then $|R_n| < 5 \times 10^{-5}$. Hence we shall determine n by trial and error so that

$$\frac{(0.1)^n}{n!} < \frac{1}{20,000}$$

If $n = 3$,

$$\frac{(0.1)^n}{n!} = \frac{1}{6,000}$$

If $n = 4$,

$$\frac{(0.1)^n}{n!} = \frac{1}{240,000}$$

Hence we must use four terms of the expansion.

EXERCISE 26.7

1   Use the Maclaurin series for e^x and e^{-x} to find the series for $(e^x + e^{-x})/2 = \cosh x$.

2   Use the Maclaurin series for e^x and e^{-x} to find the series for $(e^x - e^{-x})/2 = \sinh x$.

3   Show that $\cos x = (e^{ix} + e^{-ix})/2$ by use of Maclaurin series for e^{ix}, e^{-ix}, and $\cos x$.

4   Show that $\sin x = (e^{ix} - e^{-ix})/2i$ by use of Maclaurin series for e^{ix}, e^{-ix}, and $\sin x$.

5   Obtain the first five terms of the Maclaurin series for $e^x \cos x$ by multiplying the series for e^x and $\cos x$. Check by expanding $e^x \cos x$ directly. What is the interval of convergence?

6   Express xe^x in powers of $x - 1$.

7   Express $x \ln x$ in powers of $x - 1$.

8 Express $e^x \sin x$ in powers of x. What is the interval of convergence?

9 Differentiate the Maclaurin series for e^x, and explain the result.

10 Differentiate the Maclaurin series for $\sin x$, and explain the result.

11 Differentiate the Maclaurin series for $\sinh x$ and thereby get a series for $\cosh x$.

12 Differentiate the Maclaurin series for $\ln (1 + x)$ and thereby get a series for $1/(1 + x)$.

13 Expand $1/\sqrt{1 - x^2}$ by the binomial theorem. By integration obtain the Maclaurin series for Arcsin x from this.

14 Express $1/(1 - x^2)$ in powers of x by division. From this result, obtain by integration the series for the function $\ln [(1 + x)/(1 - x)]$.

15 Evaluate $\int_0^x (e^{-x^2}/2) \, dx$ by use of the Maclaurin series.

16 Find a series expansion for Arctan x by integrating each member of

$$\frac{1}{1 + t^2} = 1 - t^2 + t^4 - t^6 + \cdots$$

from 0 to x.

17 Sketch the curve $y = (e^x - e^{-x})/2x$, and compute to three decimal places the area under it in the interval from $x = 0$ to $x = 1$.

18 Sketch the curve $y = (e^x - 1)/x$, and compute to three decimal places the area under it in the interval from $x = 0$ to $x = 0.5$.

19 Sketch the curve $y = e^x \ln (x + 1)$, and compute the area bounded by the curve, $x = 0$, $x = 0.5$, and $y = 0$ by using the first three terms of the expansion.

20 Sketch the curve $y = [\ln (1 + x)]/x$ and compute to three decimal places the area under it from $x = 0$ to $x = 0.5$.

21 How many terms of the Maclaurin series for $\cos x$ must be used to get $\cos 0.1$ correct to five decimal places?

22 How many terms of the expansion of $\sin x$ about $\pi/4$ must be used to get $\sin (4\pi/15)$ correct to four decimal places?

23 How many terms of the Maclaurin series for $\sqrt{1 + x}$ must be used to get $\sqrt{1.01}$ correct to six decimal places?

24 How many terms of the Maclaurin series for e^x must be used to get e correct to three decimal places?

27

Differential equations

27.1 Introductory concepts

An equation that involves derivatives of the dependent variable with respect to the independent variable or differentials of both variables is called an ordinary *differential equation*. We shall consider some ordinary differential equations. The equations $y' + 3x = 4$, $y'' + 5y' + 2y = x^2$, $(y')^2 + 2y' = \cos x$, and $(x + 1)\,dy + (3x - 2)\,dx = 0$ are differential equations. The order of the highest ordered derivative in a differential equation is called the *order* of the differential equation. The exponent of the derivative of highest order is called the *degree* of the differential equation. Thus, the differential equation $(y''')^2 + 2x(y'')^4 - y' = \tan x$ is of the order 3 and degree 2 since the order of the highest ordered derivative is 3 and it occurs with the exponent 2.

Any relation, between the variables that occur in the differential equation, that satisfies the equation is called a *solution* of the equation. If a solution of an equation of order n involves n arbitrary constants, it is called the *general solution*. Any solution that is obtained from the general solution by assigning values to the arbitrary constants is called a *particular solution*.

differential equation

order

degree

solution:

general

particular

Example

Show that $y = A \cos 2x + B \sin 2x$ is the general solution of $y'' + 4y = 0$, and find a particular solution.

Solution: Since $y = A \cos 2x + B \sin 2x$ contains two arbitrary constants, it is the general solution of the given second-order differential equation if it is a solution. We can see that it is a solution by differentiating twice and thereby obtaining

$$y'' = -4A \cos 2x - 4B \sin 2x$$

Hence, $y = A \cos 2x + B \sin 2x$ is a solution since $y'' + 4y = 0$. We get a particular solution by assigning values to the arbitrary constants A and B. For instance, letting $A = 0$ and $B = 1$, we see that $y = \sin 2x$ is a particular solution of the given differential equation.

27.2   Variables separable

If a differential equation is of the first order and the first degree, it can be put in the form

$$M(x, y)\, dx + N(x, y)\, dy = 0 \qquad (1)$$

variables separable   If M and N are such that we can write the equation in the form $f(x)\, dx + g(y)\, dy = 0$, we say that the *variables are separable*. If we can integrate $f(x)\, dx$ and $g(y)\, dy$, the solution of (1) can be obtained by so doing and setting the sum of the integrals equal to a constant since $f(x)\, dx + g(y)\, dy = 0$ is another form of (1).

Example

Find the general solution of $(1 + y^2)\, dx - 2y \sqrt{1 - x^2}\, dy = 0$.

Solution: We can separate the variables by dividing by $\sqrt{1 - x^2}(1 + y^2)$. Thus, we obtain

$$\frac{dx}{\sqrt{1 - x^2}} - \frac{2y\, dy}{1 + y^2} = 0$$

Now integrating, we see that

$$\text{Arcsin } x - \ln|1 + y^2| = k$$

is the general solution of the given equation.

EXERCISE 27.1

Give the order and the degree of the differential equation in each of Problems 1 to 4.

   1   $y''' + 2y'' - y^7 = 2x$ 2 $(y')^2 + y = \sin x$

3 $\left(\dfrac{d^2y}{dx^2}\right)^3 + \left(\dfrac{dy}{dx}\right)^5 + 2y = x$

4 $\left(\dfrac{d^4y}{dx^4}\right)^3 + \left(\dfrac{dy}{dx}\right)^6 = \tan x$

5 Show that $y = e^{2x}$ is a solution of $y' - y = e^{2x}$.

6 Show that $y = \tan x$ is a solution of $y'' = 2y \sec^2 x$.

7 Show that $y = ax^2 + bx$ is the general solution of $x^2 y'' - 2xy' + 2y = 0$.

8 Show that $y = A \cos 2x$ is the general solution of $y' + 2y \tan 2x = 0$.

Solve each differential equation given in Problems 9 to 20 by separation of the variables.

9 $x \, dy + y \, dx = 0$

10 $x^2 \, dy + 2xy \, dx = 0$

11 $y^3 \, dx + (x - 1) \, dy = 0$ 12 $(1 + y) \, dx = (1 + x) \, dy$

13 $y \, dx + (1 + x^2) \, dy = 0$

14 $2xy \, dx + (1 + x^2) \, dy = 0$

15 $\sqrt{1 - y^2} \, dx + (1 + x^2) \, dy = 0$

16 $e^y \sec x \, dx + \cos x \, dy = 0$

17 $dy/dx = x(1 + y)/y(1 - x)$

18 $dy/dx + x^2(1 + y^2)/(x - 1) = 0$

19 $dy/dx + \sec x \tan y = 0$

20 $L \, di/dt + Ri = E$

27.3   Exact differential equations

A differential equation of the form

$$M \, dx + N \, dy = 0 \qquad (1)$$

where M and N are functions of x and y is said to be *exact* if there is a function $f(x, y)$ such that

$$df \equiv M \, dx + N \, dy \qquad (2)$$

definition of an exact equation   This requires that $M \, dx + N \, dy$ be the differential of some function of x and y. Equation (1) then has the solution

$$f(x, y) = c \qquad (3)$$

We shall now derive a test for use in determining whether an equation is exact. We have

$$df = \frac{\partial f}{\partial x} \, dx + \frac{\partial f}{\partial y} \, dy$$

which is identical with the left member of (1), as required by (2), if and only if

$$\frac{\partial f}{\partial x} \equiv M \qquad \text{and} \qquad \frac{\partial f}{\partial y} \equiv N \tag{4}$$

since dx and dy vary independently. Now, we make use of the fact that $f_{yx} \equiv f_{xy}$ and, from (4), get

$$\frac{\partial^2 f}{\partial y \, \partial x} \equiv \frac{\partial M}{\partial y} \qquad \text{and} \qquad \frac{\partial^2 f}{\partial x \, \partial y} \equiv \frac{\partial N}{\partial x}$$

Consequently,
$$\frac{\partial M}{\partial y} \equiv \frac{\partial N}{\partial x} \tag{5}$$

a necessary condition is a necessary condition for exactness.

We shall now show that (5) is also a sufficient condition for exactness of (1) by finding a function $f(x, y)$ which satisfies both of Eqs. (4). The general solution of $f_x = M$ is found by integration to be

$$f = \int_a^x M(x, y) \, \partial x + K(y) \tag{6}$$

for a any convenient constant and K an arbitrary function of y. The arbitrary function of y is introduced instead of the usual arbitrary constant since M is a function of x and y and we are integrating with respect to x while y is held constant.

We now substitute the value of f given by (6) into the second equation in (4) and have

$$\frac{\partial}{\partial y} \left[\int_a^x M(x, y) \, \partial x + K(y) \right] = N \tag{7}$$

Now
$$\frac{\partial}{\partial y} \left[\int_a^x M \, \partial x + K \right] = \int_a^x \frac{\partial M}{\partial y} \, \partial x + \frac{dK(y)}{dy}$$

$$= \int_a^x \frac{\partial N}{\partial x} \, \partial x + \frac{dK(y)}{dy} \qquad \text{\textit{by use of (5)}}$$

$$= N(x, y) - N(a, y) + \frac{dK(y)}{dy}$$

However,
$$\frac{\partial}{\partial y} \left[\int_a^x M \, \partial x + K \right] = N(x, y) \qquad \text{\textit{by use of (7)}}$$

Consequently,
$$\frac{dK(y)}{dy} = N(a, y)$$

and
$$K(y) = \int N(a, y) \, dy$$

Therefore, by use of (6),

$$f = \int_a^x M(x, y)\, \partial x + \int N(a, y)\, dy$$

which satisfies both of Eqs. (4).

a necessary
and sufficient
condition

We can summarize this discussion by saying that

M dx + N dy = 0 is exact if and only if

$$\frac{\partial M}{\partial y} \equiv \frac{\partial N}{\partial x}$$

and the solution is

solution of
an exact
equation

$$\int_a^x M(x, y)\, \partial x + \int N(a, y)\, dy = C \qquad (27.1)$$

where a is any convenient constant and C is an arbitrary constant.

Example

Test $(3x^2y - y)\, dx + (x^3 - x + 2y)\, dy = 0$ for exactness, and find the solution if it is exact.

Solution: Since

$$M = 3x^2\, y - y \qquad \text{and} \qquad N = x^3 - x + 2y$$

we have $\dfrac{\partial M}{\partial y} = 3x^2 - 1$ \qquad and \qquad $\dfrac{\partial N}{\partial x} = 3x^2 - 1$

Consequently, the equation is exact and we can find the solution by use of (27.1). Thus, choosing $a = 0$, we see that the solution is

$$\int_0^x (3x^2y - y)\, \partial x + \int (0^3 - 0 + 2y)\, dy = C$$

$$x^3y - xy \Big|_0^x + y^2 = C \qquad \textit{integrating}$$

$$x^3y - xy + y^2 = C$$

27.4   Integrating factors

If the differential equation

$$M(x, y)\, dx + N(x, y)\, dy = 0 \qquad (1)$$

is not exact and the equation

$$I(x, y)M(x, y)\, dx + I(x, y)N(x, y)\, dy = 0$$

integrating
factor

is exact, we say that $I(x, y)$ is an *integrating factor* of (1). There is no general method for finding an integrating factor, but a familiarity with differentiation formulas will sometimes help in determining them.

Example   Show that

$$2y \, dx + x \, dy = 0 \tag{2}$$

is not an exact differential equation and that $I(x, y) = x$ is an integrating factor.

Solution: Since $M = 2y$ and $N = x$, it follows that $M_y = 2$ and $N_x = 1$. Consequently, $M_y \neq N_x$ and the equation is not exact. To show that $I(x, y) = x$ is an integrating factor, we must show that

$$x(2y \, dx + x \, dy) = 0 \tag{3}$$

is exact. For this equation, $M = 2xy$ and $N = x^2$; hence, $M_y = 2x = N_x$ and (2) is exact. Therefore, $I(x, y) = x$ is an integrating factor of (1).

EXERCISE 27.2

Show that the equation in each of Problems 1 to 12 is exact, and find its general solution.

1   $(2x + y) \, dx + (x + y) \, dy = 0$
2   $(3x - y) \, dx - (x - y) \, dy = 0$
3   $(x + 2y - 5) \, dx + (2x - y + 1) \, dy = 0$
4   $(2x - 3y - 2) \, dx + (-3x + 2y - 1) \, dy = 0$
5   $(y^3 + 1/x) \, dx + 3xy^2 \, dy = 0$
6   $(y^2 - 1) \, dx + (2xy + 1/y) \, dy = 0$
7   $(e^y + 1) \, dx + (xe^y - y) \, dy = 0$
8   $(ye^x + y) \, dx + (e^x + x) \, dy = 0$
9   $(e^x + 1) \, dx + (e^y + 3) \, dy = 0$
10   $(\cos x + y) \, dx + (2y + x) \, dy = 0$
11   $\tan y \, dx + x \sec^2 y \, dy = 0$
12   $(y/x) \, dx + \ln x \, dy = 0$

In each of Problems 13 to 16, show that the equation is not exact and that the given $I(x, y)$ is an integrating factor; then, find the general solution.

13   $2y \, dx + x \, dy = 0, \, I(x, y) = x$
14   $y \, dx - x \, dy = 0, \, I(x, y) = 1/xy$
15   $y \, dx + x \ln x \, dy = 0, \, I(x, y) = 1/x$
16   $xe^x \, dx + (xe^x/y + x/y) \, dy = 0, \, I(x, y) = y/x$

Show that the equations in Problems 17 to 24 are not exact; find an integrating factor and the general solution of each.

17 $(x + y)\,dx + dy = 0$ 18 $y\,dx + (1 + y)\,dy = 0$

19 $x\,dy - y\,dx = 0$ 20 $2y\,dx - x\,dy = 0$

21 $y\,dx - x\,dy + xy^2\,dx = 0$ 22 $x\,dy - y\,dx = x\,dx$

23 $2x\,dy + y\,dx = dy/y$ 24 $y^2\,dx + 2xy\,dy = dx$

27.5   Homogeneous equations

In this section, we shall study equations of the form $M(x, y)\,dx + N(x, y)\,dy = 0$ where $M(kx, ky) = k^n M(x, y)$ and $N(kx, ky) = k^n N(x, y)$. *homogeneous* Such an equation is called a *homogeneous differential equation;* *equation* furthermore M and N are homogeneous with n as degree of homogeneity. The equation

$$(x^2 - xy)\,dx + (2xy + y^2)\,dy = 0$$

is homogeneous since

$$M(kx, ky) = (kx)^2 - kxky = k^2(x^2 - xy) = k^2 M(x, y)$$

and $N(kx, ky) = 2kxky + (ky)^2 = k^2(2xy + y^2) = k^2 N(x, y)$

In the equation

$$(x - y)\,dx + xy\,dy = 0$$

$$M(kx, ky) = k(x - y) = kM(x, y)$$

and $N(kx, ky) = k^2 xy = k^2 N(x, y)$

Consequently, the degree of homogeneity of M and that of N are not equal. Therefore, the equation is not a homogeneous differential equation.

We shall show presently that the substitution $y = vx$ reduces a homogeneous differential equation to one in which the variables are separable but shall first work an example.

Example   Reduce the homogeneous equation

$$(x^2 + y^2)\,dx + xy\,dy = 0 \tag{1}$$

to one in which the variables are separable by using $y = vx$, solve the equation, and express the answer in terms of x and y.

Solution: If

$$y = vx$$

then $dy = v\,dx + x\,dv$

and the given equation becomes

$$x^2(1 + v^2)\,dx + x^2v(v\,dx + x\,dv) = 0$$

$$(1 + v^2)\,dx + v(v\,dx + x\,dv) = 0 \qquad \text{\textit{dividing by} } x^2$$

$$(1 + 2v^2)\,dx + vx\,dv = 0 \qquad \text{\textit{collecting coefficients}}$$

$$\frac{dx}{x} + \frac{v\,dv}{1 + 2v^2} = 0 \qquad \text{\textit{separating the variables}}$$

Consequently, the solution in terms of x and v is

$$\ln x + \tfrac{1}{4} \ln (1 + 2v^2) = k$$

$$4 \ln x + \ln (1 + 2v^2) = 4k$$

$$\ln x^4(1 + 2v^2) = 4k$$

Hence, using each member as an exponent of e, we have

$$x^4(1 + 2v^2) = e^{4k} = C$$

and, replacing v by y/x, we get

$$x^4 \left(1 + \frac{2y^2}{x^2}\right) = C$$

The form of the solution can be modified by removing parentheses. Thus, we find that

$$x^4 + 2x^2y^2 = C$$

is the general solution of (1).

*separating
the variables*   The reason why the substitution $y = vx$ in $M\,dx + N\,dy = 0$ yields an equation in which the variables are separable becomes apparent if we make use of the fact that $M(x, y)$ and $N(x, y)$ are homogeneous and of the same degree. Thus,

$$M(x, y) = M(x, vx) \qquad \text{for } y = vx$$

$$= x^n M(1, v)$$

and $\qquad\qquad N(x, y) = N(x, vx) = x^n N(1, v)$

Hence, $\qquad\qquad M(x, y)\,dx + N(x, y)\,dy = 0$

becomes $\qquad x^n M(1, v)\,dx + x^n N(1, v)(v\,dx + x\,dv) = 0$

Therefore, dividing through by x^n and collecting coefficients, we have

$$[M(1, v) + vN(1, v)]\,dx + xN(1, v)\,dv = 0$$

Consequently, dividing by $x[M(1, v) + vN(1, v)]$, we get

$$\frac{dx}{x} + \frac{N(1, v)\, dv}{M(1, v) + vN(1, v)} = 0$$

and the variables are separated.

EXERCISE 27.3

Solve the equations in Problems 1 to 12.

1 $(x + y)\, dx - x\, dy = 0$ 2 $(x + 2y)dx + (y + 2x)\, dy = 0$
3 $(2x - y)\, dx + (x - 2y)\, dy = 0$
4 $(x - y)\, dx + (x + y)\, dy = 0$
5 $(x^2 - y^2)\, dx + 2xy\, dy = 0$ 6 $2xy\, dx - (xy + x^2)\, dy = 0$
7 $(2y^2 - 3xy + x^2)\, dx + x(x - 2y)\, dy = 0$
8 $(xy - y^2)\, dx - x^2\, dy = 0$ 9 $(x^3 - y^3)\, dx + xy^2\, dy = 0$
10 $(x^3 + y^3)\, dx - xy^2\, dy = 0$
11 $(12x^2y - 4y^3)\, dx + x(3y^2 - 6x^2)\, dy = 0$
12 $(3x + xe^{y/x} - ye^{y/x})\, dx + xe^{y/x}\, dy = 0$

13 Show that the substitutions $x = x_1 + h$ and $y = y_1 + k$ reduce

$$(a_1x + b_1y + c_1)\, dx - (a_2x + b_2y + c_2)\, dy = 0$$

to a homogeneous equation if (h, k) is the intersection of the lines determined by $a_1x + b_1y + c_1 = 0$ and $a_2x + b_2y + c_2 = 0$.

Make use of Problem 13 to reduce each of the following equations to a homogeneous equation, solve the homogeneous equation, and then find the solution of the given equation. Only the solution of the given equation is given in the answers.

14 $(2x + y - 4)\, dx + (x - 3y + 5)\, dy = 0$
15 $(3x + 2y + 1)\, dx - (2x - 3y + 5)\, dy = 0$
16 $(x - 2y + 8)\, dx + (y - 3)\, dy = 0$

27.6 First-order linear equations

The differential equation

$$\frac{dy}{dx} + yP(x) = Q(x) \tag{1}$$

linear
first order

is a *linear first-order differential equation* since only y and its first derivative enter into the equation and they are only to the first power. As indicated in (1), P and Q are functions of x alone. We shall now determine an integrating factor $I(x)$ for this equation. If $I(x)$ is an integrating factor of (1), then

$$I\,dy + (yP - Q)I\,dx = 0 \tag{2}$$

is exact. Consequently,

$$\frac{\partial(yP - Q)I}{\partial y} = \frac{\partial I}{\partial x}$$

and

$$PI = \frac{dI}{dx} \tag{3}$$

since P, Q, and I are functions of x alone. Now, separating the variables in (3), we have

$$P\,dx = \frac{dI}{I}$$

Consequently,

$$\int P\,dx = \ln I$$

and

$$I = e^{\int P\,dx}$$

is an integrating factor of the linear equation

$$y' + yP(x) = Q(x)$$

Therefore,

$$e^{\int P\,dx}[y' + yP(x)] = Q(x)e^{\int P\,dx} \tag{4}$$

is exact; furthermore, its solution is

$$ye^{\int P\,dx} = \int Qe^{\int P\,dx}\,dx \tag{5}$$

Example   Solve the linear equation

$$y' + \frac{y}{x} = 2 \tag{6}$$

Solution: In this problem, $P(x) = 1/x$; hence,

$$\int P\,dx = \int \frac{dx}{x} = \ln x$$

and

$$e^{\int P\,dx} = e^{\ln x} = x$$

is an integrating factor of (1). Therefore,

$$xy' + y = 2x$$

is an exact equation. Its solution is

$$yx = \int 2x\,dx$$

or

$$yx = x^2 + c$$

by use of (5).

27.7   The Bernoulli equation

The equation

$$y' + yP(x) = y^n Q(x) \tag{1}$$

Bernoulli equation is similar to a linear equation; it is called the *Bernoulli equation* in honor of James Bernoulli (1654–1705). We shall show how to transform it into a linear equation. In fact, we shall show that it reduces to a linear equation if we let

$$y = z^{1/(1-n)} \qquad n \neq 1$$

For this expression for y, we have

$$\frac{dy}{dx} = \frac{1}{1-n} z^{1/(1-n)-1} \frac{dz}{dx} = \frac{1}{1-n} z^{n/(1-n)} \frac{dz}{dx}$$

and

$$y^n = (z^{1/1-n})^n = z^{n/(1-n)}$$

Consequently, (1) becomes

$$\frac{1}{1-n} z^{n/(1-n)} \frac{dz}{dx} + z^{1/(1-n)} P = z^{n/(1-n)} Q$$

and, multiplying by $(1-n) z^{-n/(1-n)}$, we have the linear equation

$$\frac{dz}{dx} + (1-n)zP = (1-n)Q$$

Consequently, *the Bernoulli equation is reduced to a linear equation by the substitution of $z^{1/(1-n)}$ for y.*

Example   Transform the Bernoulli equation

$$y' + \frac{y}{x} = x^2 y^3 \tag{2}$$

into a linear equation and solve it.

Solution: In this equation, $y^n = y^3$. Hence, $n = 3$, $1 - n = -2$, and we make the substitution $y = z^{-1/2}$. Consequently,

$$\frac{dy}{dx} = -\frac{z^{-3/2}}{2} \frac{dz}{dx}$$

and the given equation becomes

$$-\frac{z^{-3/2}}{2} \frac{dz}{dx} + \frac{z^{-1/2}}{x} = x^2 z^{-3/2}$$

Now, multiplying by $-2z^{3/2}$, we have the linear equation

$$\frac{dz}{dx} + z\left(-\frac{2}{x}\right) = -2x^2 \tag{3}$$

For this equation, $P(x) = -2/x$,

$$\int P \, dx = -2 \int \frac{dx}{x} = -2 \ln x = \ln x^{-2}$$

Therefore, $e^{\int P \, dx} = e^{\ln x^{-2}} = x^{-2}$

is an integrating factor of (3); hence,

$$x^{-2}z' - \frac{2z}{x^3} = -2$$

is an exact equation. Its solution is

$$zx^{-2} = -2x + c \qquad (4)$$

and can be transformed into a solution of (2). This is done by replacing z with y^{-2}, obtained from the substitution $y = z^{-1/2}$ that was used to change (2) into (3). Thus, (4) becomes

$$y^{-2}x^{-2} = -2x + c$$

and, multiplying by x^2y^2, we have

$$1 = x^2y^2(c - 2x)$$

as the solution of (2).

EXERCISE 27.4

Solve the linear first-order differential equation given in each of Problems 1 to 16.

1   $y' + y = 2$ 2 $y' + y/x = 4x^2$
3   $xy' + y = 2x - 1$ 4 $y' + 2y = 6$
5   $y' - y/x = xe^x$ 6 $y' - 2y/x = x - x^2e^x$
7   $y' + 2y/x = 3e^{3x}/x^2$ 8 $y' + 3y/x = \sin x/x^3$
9   $xy' - 2y = x^3 \cos x$ 10 $xy' - y = x^2 \cot x$
11   $xy' + 2y = (\sin x)/x$ 12 $xy' + 3y = (\sec^2 x)/x^2$
13   $y' + y \cot x = 2x \csc x$ 14 $y' + y \cos x = \cos x$
15   $y' \cos x - y \sin x = 2xe^{x^2}$ 16 $y' + 2y/x = x^{-3}$

Solve each of the following Bernoulli equations.

17   $y' - 2y/x = 4xy^2$ 18 $y' + y/x = xy^2$
19   $y' + y = y^3e^{2x} \sin x$ 20 $y' + 2y/x = 5/x^2y^2$

27.8   Linear equations of order greater than one

The earlier sections of this chapter have been devoted to the study of certain types of first-order differential equations. We shall now see how to solve an equation that is of the first degree in y and its

*linear
equation*

derivatives. Such an equation is called a *linear equation* of order n where n is the order of the highest ordered derivative that enters into the equation. The equation is of the form

$$a_0\frac{d^ny}{dx^n} + a_1\frac{d^{n-1}y}{dx^{n-1}} + \cdots + a_{n-1}\frac{dy}{dx} + a_ny = f(x)$$

where each coefficient a_i may be a function of x. We shall, however, consider only the case in which each coefficient is a constant.

We shall study the case in which the right-hand member of the equation is zero in the next section and reserve the case in which it is a function of x for a later section.

27.9 Right-hand member zero

*reduced
equation*

If the right-hand member is zero, we refer to the equation as the *reduced equation*. We shall now find how to determine m so that $y = e^{mx}$ is a solution of the reduced equation

*a procedure
for solving*

$$a_0\frac{d^ny}{dx^n} + a_1\frac{d^{n-1}y}{dx^{n-1}} + \cdots + a_{n-1}\frac{dy}{dx} + a_n y = 0 \qquad (1)$$

If we substitute $y = e^{mx}$ and the derivatives

$$\frac{de^{mx}}{dx} = me^{mx}, \quad \frac{d^2e^{mx}}{dx^2} = m^2e^{mx}, \quad \ldots, \quad \frac{d^ne^{mx}}{dx^n} = m^ne^{mx}$$

in (1) and divide by e^{mx}, we have

$$a_0m^n + a_1m^{n-1} + \cdots + a_{n-1}m + a^n = 0 \qquad (2)$$

Consequently, e^{mx} is a solution of (1) if and only if m is a root of the *auxiliary equation* (2).

*auxiliary
equation*

If we make use of the facts that:

(i) if e^{rx} is a solution of (1), then Ae^{rx} is also one
(ii) the sum of two solutions of (1) is also a solution, then we conclude that *if the solutions of (2) are $r_1, r_2, \ldots, r_n$ and if $C_1, C_2, \ldots, C_n$ are arbitrary constants, then*

$$y = C_1e^{r_1x} + C_2e^{r_2x} + \cdots + C_ne^{r_nx}$$

*the general
solution*

is the general solution of (1) provided that (2) has no multiple roots. The proofs of (i) and (ii) will be given as problems in Exercise 27.5.

Example 1 Solve

$$\frac{d^3y}{dx^3} - 3\frac{d^2y}{dy^2} - \frac{dy}{dx} + 3y = 0$$

Solution: If we replace y by e^{mx}, we find that the auxiliary equation is

$$m^3 - 3m^2 - m + 3 = 0$$

Its solutions are $m = -1, 1, 3$; hence,

$$y = C_1 e^{-x} + C_2 e^x + C_3 e^{3x}$$

is the general solution.

The case of equal roots. If the roots r_1 and r_2 of the auxiliary equation are equal, then the corresponding part $C_1 e^{r_1 x} + C_2 e^{r_2 x}$ of the general solution becomes $(C_1 + C_2)e^{r_1 x} = Ae^{r_1 x}$. Hence,

$$y = C_1 e^{r_1 x} + C_2 e^{r_1 x} + C_3 e^{r_3 x} + \cdots + C_n e^{r_n x}$$
$$= Ae^{r_1 x} + C_3 e^{r_3 x} + \cdots + C_n e^{r_n x}$$

contains only $n - 1$ arbitrary constants and is not the general solution. It can be shown that if $r_2 = r_1$, then $Bxe^{r_1}x$ is a solution of the equation. Therefore,

$$y_1 = Ae^{r_1 x} + Bxe^{r_1 x}$$

is the portion of the general solution that corresponds to the pair of equal roots. In Example 2 we shall verify the above statements for the equation $d^3 y/dx^3 - 3(d^2 y/dx^2) + 4y = 0$. It can also be shown that if $r_1 = r_2 = r_3 = \cdots = r_k$, then

$$y_1 = (C_1 + C_2 x + C_3 x^2 + \cdots + C_k x^{k-1})e^{r_1 x}$$

is the corresponding portion of the general solution.

Example 2   The auxiliary equation for $d^3 y/dx^3 - 3(d^2 y/dx^2) + 4y = 0$ is $m^3 - 3m^2 + 4 = 0$ and the roots are 2, 2, and -1. Show that each of $y = Ae^{2x}$, $y = Bxe^{2x}$, and $y = Ae^{2x} + Bxe^{2x}$ satisfy the equation.

Solution: We shall first obtain the first, second, and third derivatives of Ae^{2x} and Bxe^{2x} as indicated below.

y	Ae^{2x}	Bxe^{2x}
$\dfrac{dy}{dx}$	$2Ae^{2x}$	$Be^{2x} + 2Bxe^{2x}$
$\dfrac{d^2 y}{dx^2}$	$4Ae^{2x}$	$4Be^{2x} + 4Bxe^{2x}$
$\dfrac{d^3 y}{dx^3}$	$8Ae^{2x}$	$12Be^{2x} + 8Bxe^{2x}$

Now we apply the operators d^3/dx^3, $-3(d^2/dx^2)$, and 4 to $Ae^{2x} + Bxe^{2x}$ and obtain

$$\frac{d^3}{dx^3}[Ae^{2x} + (Bxe^{2x})] = \quad 8Ae^{2x} + \quad (12Be^{2x} + \ 8Bxe^{2x})$$

$$-3\frac{d^2}{dx^2}[Ae^{2x} + (Bxe^{2x})] = -12Ae^{2x} + (-12Be^{2x} - 12Bxe^{2x})$$

$$4[Ae^{2x} + (Bxe^{2x})] = \quad 4Ae^{2x} \quad\quad\quad + (4Bxe^{2x})$$

The first terms at the right of the equality signs are obtained from Ae^{2x} and those in parentheses from Bxe^{2x}. Now since the sum of the terms in each of the three columns on the right is zero, it follows that each of Ae^{2x}, Bxe^{2x}, and $Ae^{2x} + Bxe^{2x}$ satisfy the given differential equation.

Example 3 Solve $\quad \dfrac{d^4y}{dx^4} - 7\dfrac{d^3y}{dx^3} + 18\dfrac{d^2y}{dx^2} - 20\dfrac{dy}{dx} + 8 = 0$

Solution: If we substitute e^{mx} for y, we find that

$$m^4 - 7m^3 + 18m^2 - 20m + 8 = 0$$

is the auxiliary equation. Its solutions are $m = 2, 2, 2, 1$; hence, the general solution is

$$y = (C_1 + C_2x + C_3x^2)e^{2x} + C_4e^x$$

The case of complex roots. If two of the roots of the auxiliary equation of a linear differential equation of order $n \geq 2$ with right-hand member zero are $a \pm bi$, then the corresponding part of the general solution is

$$y = Ae^{(a+bi)x} + Be^{(a-bi)x} = e^{ax}(Ae^{bix} + Be^{-bix}) \qquad (3)$$

The form of this part of the solution can be changed by making use of the identities

$$e^{bix} = \cos bx + i \sin bx$$

and $$e^{-bix} = \cos bx - i \sin bx$$

Thus, (3) becomes

$$y = e^{ax}[A(\cos bx + i \sin bx) + B(\cos bx - i \sin bx)]$$
$$= e^{ax}[(A + B) \cos bx + (A - B)i \sin bx]$$
$$= e^{ax}(C \cos bx + E \sin bx)$$

where $C = A + B$ and $E = (A - B)i$ are arbitrary constants.

Example 4 Solve $y''' - 7y'' + 17y' - 15y = 0$.

Solution: If we replace y by e^{mx}, the auxiliary equation becomes

$$m^3 - 7m^2 + 17m - 15 = 0$$

The roots of this equation are $m = 3, 2 + i, 2 - i$. Consequently, the general solution of the given differential equation is

$$y = C_1 e^{3x} + C_2 e^{(2+i)x} + C_3 e^{(2-i)x} = C_1 e^{3x} + e^{2x}(C_2 e^{ix} + C_3 e^{-ix})$$
$$= C_1 e^{3x} + e^{2x}[C_2(\cos x + i \sin x) + C_3(\cos x - i \sin x)]$$
$$= C_1 e^{3x} + e^{2x}(A \cos x + B \sin x)$$

EXERCISE 27.5

Find the general solution of the differential equation in each of Problems 1 to 20.

1 $y'' - 3y' + 2y = 0$ 2 $y'' - 2y' - 3y = 0$
3 $y'' + y' - 6y = 0$ 4 $y'' - 3y' - 10y = 0$
5 $y''' - 6y'' + 11y' - 6y = 0$
6 $y''' - 7y' + 6y = 0$
7 $y''' - 5y'' + 6y' = 0$ 8 $y''' + 3y'' - 6y' - 8y = 0$
9 $y'' - 6y' + 9y = 0$ 10 $y'' + 2y' + y = 0$
11 $y''' - 4y'' + 5y' - 2y = 0$
12 $y''' + 3y'' - 4y = 0$
13 $y''' - y'' + 9y' - 9y = 0$ 14 $y''' - 4y'' + 6y' - 4y = 0$
15 $y'''' - y''' - y'' - y' - 2y = 0$
16 $y'''' - 4y''' + 4y'' + 4y' - 5y = 0$
17 $y'''' - 4y''' + 7y'' - 6y' + 2y = 0$
18 $y'''' - 2y''' + 6y'' + 22y' + 13y = 0$
19 $y'''' - 4y''' + 11y'' - 14y' + 10y = 0$
20 $y'''' - 4y''' + 9y'' - 16y' + 20y = 0$

21 Show that if e^{rx} is a solution of $y''' + ay'' + by' + cy = 0$, then $A e^{rx}$ is also one for A any constant.
22 Show that if $e^{r_1 x}$ and $e^{r_2 x}$ are solutions of $y''' + ay'' + by' + cy = 0$, then so is their sum.
23 Show that Axe^{rx} is a solution of $y'' - 2ry' + r^2 y = 0$.
24 Show that $A e^{rx} + Bxe^{rx}$ is the general solution of $ay'' + by' + cy = 0$ if the auxiliary equation has r as a double root.

27.10 Right-hand member a function of x

If we have the general solution $G(x)$ of the reduced equation

$$a_0 \frac{d^n y}{dx^n} + a_1 \frac{d^{n-1} y}{dx^{n-1}} + \cdots + a_{n-1} \frac{dy}{dx} + a_n y = 0 \qquad (1)$$

and *any* solution $A(x)$ of

$$a_0 \frac{d^n y}{dx^n} + a_1 \frac{d^{n-1} y}{dx^{n-1}} + \cdots + a_{n-1} \frac{dy}{dx} + a_n y = f(x) \qquad (2)$$

then $y = G(x) + A(x)$ is the general solution of (2) since it contains the required n arbitrary constants and satisfies Eq. (2). Consequently, in order to be able to find the general solution of (2), we need only learn how to get any solution of (2) since we saw in the last section how to find the general solution $G(x)$ of (1).°

a trial solution　If the right-hand member of (2) contains only terms that have a finite number of distinct derivatives,† we can often find a particular solution of (2) by using as a "trial solution" the sum of the terms obtained by multiplying each term in $f(x)$ and each distinct derivative of the terms of $f(x)$ by an undetermined constant. We then substitute this "trial solution" in (2) and determine the constants by equating coefficients of like terms.

Example　Find the general solution of $y'' + 3y' + 2y = x^2 + e^x$.

Solution:　The solution of the reduced equation is $G(x) = C_1 e^{-x} + C_2 e^{-2x}$. Hence, we need only find a particular solution of the given equation in order to write the general solution of it. Since the first term of the right-hand member of the given equation is x^2, its only nonzero derivatives involve x and a number. The second term, e^x, has no derivative distinct from itself. Consequently, we shall use

$$y_p = ax^2 + bx + c + he^x$$

as a "trial solution"; hence,

$$y_p' = 2ax + b + he^x \qquad y_p'' = 2a + he^x$$

These expressions, substituted into the given equation, yield

$$2a + he^x + 3(2ax + b + he^x) + 2(ax^2 + bx + c + he^x) = x^2 + e^x$$

$$2a + 3b + 2c + (6a + 2b)x + 2ax^2 + (h + 3h + 2h)e^x = x^2 + e^x$$

$$2a + 3b + 2c + (6a + 2b)x + 2ax^2 + 6he^x = x^2 + e^x$$

Now, equating coefficients of like terms, we have

$$2a + 3b + 2c \quad = 0 \qquad \text{\textit{constants}}$$

$$6a + 2b \quad\quad = 0 \qquad \text{\textit{coefficients of} } x$$

$$2a \quad\quad\quad = 1 \qquad \text{\textit{coefficients of} } x^2$$

$$6h = 1 \qquad \text{\textit{coefficients of} } e^x$$

° A discussion of how to find a solution of (2) by various methods is beyond the scope of this book but can be found in most books on differential equations; for example, see L. R. Ford, "Differential Equations," 2d ed., McGraw-Hill Book Company, New York, 1955.

† $\sin x$, $\cos x$, e^x, and polynomials in x are such terms.

The solution of this system is $a = \frac{1}{2}$, $b = -\frac{3}{2}$, $c = \frac{7}{4}$, and $h = \frac{1}{6}$. Consequently, $\frac{1}{2}x^2 - \frac{3}{2}x + \frac{7}{4} + \frac{1}{6}e^x$ is a solution of the given equation and

$$y = C_1 e^{-x} + C_2 e^{-2x} + \frac{1}{2}x^2 - \frac{3}{2}x + \frac{7}{4} + \frac{1}{6}e^x$$

is the general solution.

EXERCISE 27.6

Find a particular solution y_p of each differential equation given in Problems 1 to 20 and the general solution y_c of the corresponding equation with right-hand member zero.

1 $y'' - 3y' + 2y = 2x - 3$
2 $y'' + 2y' - 8y = -5e^x$
3 $y'' + 5y' + 6y = 5 \cos x + 5 \sin x$
4 $y'' + 2y' - 3y = 7x - 3x^2$
5 $y''' - 7y' + 6y = 8 \sin x + 6 \cos x$
6 $y''' - 2y'' - y' + 2y = 8e^{3x}$
7 $y''' - 4y'' + y' + 6y = 6x^3 + 3x^2 - 18x + 7$
8 $y''' - 4y'' + 3y' = -2e^{2x}$
9 $y'' - 4y' + 4y = 9e^{-x}$
10 $y'' - 2y' + y = 3x^2 - 12x + 5$
11 $y''' + y'' - 5y' + 3y = 5e^{2x} + 3x - 5$
12 $y''' - 3y' - 2y = 4 \cos x + 2 \sin x + 3 + 2x$
13 $y''' - y'' + 4y' - 4y = -4x^2 + 12x - 6$
14 $y''' - 4y'' + 6y' - 4y = 15e^{-x} - 4$
15 $y'''' - 2y''' + 4y'' + 2y' - 5y = -5x^2 - 4x + 8$
16 $y'''' - 4y''' + 12y'' + 4y' - 13y = -32 \sin x - 16 \cos x$
17 $y'''' - 6y''' + 14y'' - 14y' + 5y = 40e^{-x} + 5$
18 $y'''' - 6y''' + 22y'' - 48y' + 40y = 40x^2 - 136x + 212$
19 $y'''' - 8y''' + 26y'' - 40y' + 25y = -32 \cos x$
20 $y'''' - 4y''' + 14y'' - 20y' + 25y = 25x^2 - 90x + 68$

Tables

Table I   Some Integrals

1 $\displaystyle\int \frac{x\,dx}{ax+b} = \frac{x}{a} - \frac{b}{a^2}\ln|ax+b|$

2 $\displaystyle\int \frac{x\,dx}{(ax+b)^2} = \frac{b}{a^2(ax+b)} + \frac{1}{a^2}\ln|ax+b|$

3 $\displaystyle\int \frac{x^2\,dx}{ax+b} = \frac{x^2}{2a} - \frac{bx}{a^2} + \frac{b^2}{a^3}\ln|ax+b|$

4 $\displaystyle\int \frac{x^2\,dx}{(ax+b)^2} = \frac{x}{a^2} - \frac{b^2}{a^3(ax+b)} - \frac{2b}{a^3}\ln|ax+b|$

5 $\displaystyle\int \frac{dx}{x(ax+b)} = \frac{1}{b}\ln\left|\frac{x}{ax+b}\right|$

6 $\displaystyle\int x(ax+b)^n\,dx = \frac{x(ax+b)^{n+1}}{a(n+1)} - \frac{(ax+b)^{n+2}}{a^2(n+1)(n+2)} \qquad n \neq -1,\,-2$

7 $\displaystyle\int \frac{dx}{(ax+b)(cx+d)} = \frac{1}{ad-bc}\ln\left|\frac{ax+b}{cx+d}\right|$

8 $\displaystyle\int \frac{x\,dx}{(ax+b)(cx+d)} = \frac{1}{ad-bc}\left[\frac{d}{c}\ln|cx+d| - \frac{b}{a}\ln|ax+b|\right]$

9 $\displaystyle\int \frac{dx}{x(ax+b)^2} = \frac{1}{b(ax+b)} + \frac{1}{b^2}\ln\left|\frac{x}{ax+b}\right|$

10 $\displaystyle\int x\sqrt{ax+b}\,dx = \frac{2}{a^2}\left[\frac{(ax+b)^{5/2}}{5} - \frac{b(ax+b)^{3/2}}{3}\right]$

11 $\displaystyle\int x^2\sqrt{ax+b}\,dx = \frac{2}{a^3}\left[\frac{(ax+b)^{7/2}}{7} - \frac{2b(ax+b)^{5/2}}{5} + \frac{b^2(ax+b)^{3/2}}{3}\right]$

12 $\displaystyle\int \frac{dx}{x\sqrt{ax+b}} = \frac{1}{\sqrt{b}}\ln\left|\frac{\sqrt{ax+b}-\sqrt{b}}{\sqrt{ax+b}+\sqrt{b}}\right| \qquad b>0$

$\displaystyle\qquad\qquad = \frac{2}{\sqrt{-b}}\operatorname{Arctan}\frac{\sqrt{ax+b}}{\sqrt{-b}} \qquad b<0$

13 $\displaystyle\int \frac{x\,dx}{\sqrt{ax+b}} = \frac{2(ax-2b)}{3a^2}\sqrt{ax+b}$

14 $\displaystyle\int \frac{\sqrt{ax+b}}{x}\,dx = 2\sqrt{ax+b} + b\int \frac{dx}{x\sqrt{ax+b}}$

15 $\displaystyle\int \sqrt{a^2-x^2}\,dx = \frac{x}{2}\sqrt{a^2-x^2} + \frac{a^2}{2}\operatorname{Arcsin}\frac{x}{a}$

16 $\displaystyle\int x^2\sqrt{a^2-x^2}\,dx = -\frac{x}{4}(a^2-x^2)^{3/2} + \frac{a^2}{4}\int\sqrt{a^2-x^2}\,dx$

17 $\displaystyle\int (a^2-x^2)^{3/2}\,dx = \frac{x}{4}(a^2-x^2)^{3/2} + \frac{3a^2}{4}\int\sqrt{a^2-x^2}\,dx$

18 $\displaystyle\int \frac{dx}{(a^2-x^2)^{3/2}} = \frac{x}{a^2\sqrt{a^2-x^2}}$

19 $\displaystyle\int \frac{dx}{x\sqrt{a^2-x^2}} = -\frac{1}{a}\ln\left|\frac{a+\sqrt{a^2-x^2}}{x}\right|$

Table I   Some Integrals (*continued*)

20 $\displaystyle\int \frac{dx}{x^2\sqrt{a^2-x^2}} = -\frac{\sqrt{a^2-x^2}}{a^2x}$

21 $\displaystyle\int \frac{\sqrt{a^2-x^2}}{x}\,dx = \sqrt{a^2-x^2} - a\ln\left|\frac{a+\sqrt{a^2-x^2}}{x}\right|$

22 $\displaystyle\int \sqrt{x^2\pm a^2}\,dx = \frac{x}{2}\sqrt{x^2\pm a^2} \pm \frac{a^2}{2}\ln|x+\sqrt{x^2\pm a^2}|$

23 $\displaystyle\int \frac{dx}{\sqrt{x^2\pm a^2}} = \ln|x+\sqrt{x^2\pm a^2}|$

24 $\displaystyle\int \frac{dx}{(x^2\pm a^2)^{3/2}} = \frac{\pm x}{a^2\sqrt{x^2\pm a^2}}$

25 $\displaystyle\int \frac{\sqrt{x^2\pm a^2}}{x^2}\,dx = -\frac{\sqrt{x^2\pm a^2}}{x} + \ln|x+\sqrt{x^2\pm a^2}|$

26 $\displaystyle\int \frac{dx}{x\sqrt{x^2+a^2}} = -\frac{1}{a}\ln\frac{a+\sqrt{x^2+a^2}}{|x|}$

27 $\displaystyle\int \frac{dx}{x\sqrt{x^2-a^2}} = \frac{1}{a}\,\text{Arctan}\,\frac{\sqrt{x^2-a^2}}{a}$

28 $\displaystyle\int \frac{\sqrt{x^2\pm a^2}}{x}\,dx = \sqrt{x^2\pm a^2} \pm a^2\int \frac{dx}{x\sqrt{x^2\pm a^2}}$

29 $\displaystyle\int \sin^2 x\,dx = \frac{x}{2} - \frac{\sin 2x}{4}$

30 $\displaystyle\int x\sin x\,dx = \sin x - x\cos x$

31 $\displaystyle\int x\cos x\,dx = \cos x + x\sin x$

32 $\displaystyle\int x^2\sin x\,dx = 2x\sin x - (x^2-2)\cos x$

33 $\displaystyle\int x^2\cos x\,dx = 2x\cos x + (x^2-2)\sin x$

34 $\displaystyle\int x\sin^2 x\,dx = \frac{1}{4}[x^2 + \sin^2 x - x\sin 2x]$

35 $\displaystyle\int \sin^3 x\,dx = \frac{\cos^3 x}{3} - \cos x$

36 $\displaystyle\int \sin^4 x\,dx = \frac{3x}{8} - \frac{3\sin 2x}{16} - \frac{\sin^3 x\cos x}{4}$

37 $\displaystyle\int \frac{dx}{1+k\sin x} = \frac{-1}{\sqrt{k^2-1}}\ln\left|\frac{k+\sin x+\sqrt{k^2-1}\cos x}{1+k\sin x}\right| \qquad k^2>1$

$\displaystyle\qquad\qquad\quad = \frac{1}{\sqrt{1-k^2}}\,\text{Arcsin}\,\frac{k+\sin x}{1+k\sin x} \qquad k^2<1$

38 $\displaystyle\int \sin ax\sin bx\,dx = \frac{\sin(a-b)x}{2(a-b)} - \frac{\sin(a+b)x}{2(a+b)} \qquad a^2\neq b^2$

Table I   Some Integrals (*continued*)

39 $\displaystyle\int xe^{ax}\,dx = \frac{1}{a^2}(ax - 1)e^{ax}$

40 $\displaystyle\int x^2e^{ax}\,dx = \frac{1}{a^3}(a^2x^2 - 2ax + 2)e^{ax}$

41 $\displaystyle\int e^{ax}\sin bx\,dx = \frac{1}{a^2 + b^2}(a\sin bx - b\cos bx)e^{ax}$

42 $\displaystyle\int e^{ax}\cos bx\,dx = \frac{1}{a^2 + b^2}(a\cos bx + b\sin bx)e^{ax}$

43 $\displaystyle\int \ln|x|\,dx = x\ln|x| - x$

44 $\displaystyle\int x^n \ln|x|\,dx = \frac{x^{n+1}}{n + 1}\left(\ln|x| - \frac{1}{n + 1}\right) \qquad n \neq -1$

Table II   Trigonometric Functions

Angles	Sines		Cosines		Tangents		Cotangents		Angles
	Nat.	Log.	Nat.	Log.	Nat.	Log.	Nat.	Log.	
0° 00′	.0000	∞	1.0000	0.0000	.0000	∞	∞	∞	90° 00′
10	.0029	7.4637	1.0000	0000	.0029	7.4637	343.77	2.5363	50
20	.0058	7648	1.0000	0000	.0058	7648	171.89	2352	40
30	.0087	9408	1.0000	0000	.0087	9409	114.59	0591	30
40	.0116	8.0658	.9999	0000	.0116	8.0658	85.940	1.9342	20
50	.0145	1627	.9999	0000	.0145	1627	68.750	8373	10
1° 00′	.0175	8.2419	.9998	9.9999	.0175	8.2419	57.290	1.7581	89° 00′
10	.0204	3088	.9998	9999	.0204	3089	49.104	6911	50
20	.0233	3668	.9997	9999	.0233	3669	42.964	6331	40
30	.0262	4179	.9997	9999	.0262	4181	38.188	5819	30
40	.0291	4637	.9996	9998	.0291	4638	34.368	5362	20
50	.0320	5050	.9995	9998	.0320	5053	31.242	4947	10
2° 00′	.0349	8.5428	.9994	9.9997	.0349	8.5431	28.636	1.4569	88° 00′
10	.0378	5776	.9993	9997	.0378	5779	26.432	4221	50
20	.0407	6097	.9992	9996	.0407	6101	24.542	3899	40
30	.0436	6397	.9990	9996	.0437	6401	22.904	3599	30
40	.0465	6677	.9989	9995	.0466	6682	21.470	3318	20
50	.0494	6940	.9988	9995	.0495	6945	20.206	3055	10
3° 00′	.0523	8.7188	.9986	9.9994	.0524	8.7194	19.081	1.2806	87° 00′
10	.0552	7423	.9985	9993	.0553	7429	18.075	2571	50
20	.0581	7645	.9983	9993	.0582	7652	17.169	2348	40
30	.0610	7857	.9981	9992	.0612	7865	16.350	2135	30
40	.0640	8059	.9980	9991	.0641	8067	15.605	1933	20
50	.0669	8251	.9978	9990	.0670	8261	14.924	1739	10
4° 00′	.0698	8.8436	.9976	9.9989	.0669	8.8446	14.301	1.1554	86° 00′
10	.0727	8613	.9974	9989	.0729	8624	13.727	1376	50
20	.0756	8783	.9971	9988	.0758	8795	13.197	1205	40
30	.0785	8946	.9969	9987	.0787	8960	12.706	1040	30
40	.0814	9104	.9967	9986	.0816	9118	12.251	0882	20
50	.0843	9256	.9964	9985	.0846	9272	11.826	0728	10
5° 00′	.0872	8.9403	.9962	9.9983	.0875	8.9420	11.430	1.0580	85° 00′
10	.0901	9545	.9959	9982	.0904	9563	11.059	0437	50
20	.0929	9682	.9957	9981	.0934	9701	10.712	0299	40
30	.0958	9816	.9954	9980	.0963	9836	10.385	0164	30
40	.0987	9945	.9951	9979	.0992	9966	10.078	0034	20
50	.1016	9.0070	.9948	9977	.1022	9.0093	9.7882	0.9907	10
6° 00′	.1045	9.0192	.9945	9.9976	.1051	9.0216	9.5144	0.9784	84° 00′
10	.1074	0311	.9942	9975	.1080	0336	9.2553	9664	50
20	.1103	0426	.9939	9973	.1110	0453	9.0098	9547	40
30	.1132	0539	.9936	9972	.1139	0567	8.7769	9433	30
40	.1161	0648	.9932	9971	.1169	0678	8.5555	9322	20
50	.1190	0755	.9929	9969	.1198	0786	8.3450	9214	10
7° 00′	.1219	9.0859	.9925	9.9968	.1228	9.0891	8.1443	0.9109	83° 00′
10	.1248	0961	.9922	9966	.1257	0995	7.9530	9005	50
20	.1276	1060	.9918	9964	.1287	1096	7.7704	8904	40
30	.1305	1157	.9914	9963	.1317	1194	7.5958	8806	30
40	.1334	1252	.9911	9961	.1346	1291	7.4287	8709	20
50	.1363	1345	.9907	9959	.1376	1385	7.2687	8615	10
8° 00′	.1392	9.1436	.9903	9.9958	.1405	9.1478	7.1154	0.8522	82° 00′
10	.1421	1525	.9899	9956	.1435	1569	6.9682	8431	50
20	.1449	1612	.9894	9954	.1465	1658	6.8269	8342	40
30	.1478	1697	.9890	9952	.1495	1745	6.6912	8255	30
40	.1507	1781	.9886	9950	.1524	1831	6.5606	8169	20
50	.1536	1863	.9881	9948	.1554	1915	6.4348	8085	10
9° 00′	.1564	9.1943	.9877	9.9946	.1584	9.1997	6.3138	0.8003	81° 00′
	Nat.	Log.	Nat.	Log.	Nat.	Log.	Nat.	Log.	

Angles	Cosines		Sines		Cotangents		Tangents		Angles

Table II   **Trigonometric Functions** (*continued*)

Angles	Sines		Cosines		Tangents		Cotangents		Angles
	Nat.	Log.	Nat.	Log.	Nat.	Log.	Nat.	Log.	
9° 00′	.1564	**9.1943**	.9877	**9.9946**	.1584	**9.1997**	6.3138	**0.8003**	81° 00′
10	.1593	2022	.9872	9944	.1614	2078	6.1970	7922	50
20	.1622	2100	.9868	9942	.1644	2158	6.0844	7842	40
30	.1650	2176	.9863	9940	.1673	2236	5.9758	7764	30
40	.1679	2251	.9858	9938	.1703	2313	5.8708	7687	20
50	.1708	2324	.9853	9936	.1733	2389	5.7694	7611	10
10° 00′	.1736	**9.2397**	.9848	**9.9934**	.1763	**9.2463**	5.6713	**0.7537**	80° 00′
10	.1765	2468	.9843	9931	.1793	2536	5.5764	7464	50
20	.1794	2538	.9838	9929	.1823	2609	5.4845	7391	40
30	.1822	2606	.9833	9927	.1853	2680	5.3955	7320	30
40	.1851	2674	.9827	9924	.1883	2750	5.3093	7250	20
50	.1880	2740	.9822	9922	.1914	2819	5.2257	7181	10
11° 00′	.1908	**9.2806**	.9816	**9.9919**	.1944	**9.2887**	5.1446	**0.7113**	79° 00′
10	.1937	2870	.9811	9917	.1974	2953	5.0658	7047	50
20	.1965	2934	.9805	9914	.2004	3020	4.9894	6980	40
30	.1994	2997	.9799	9912	.2035	3085	4.9152	6915	30
40	.2022	3058	.9793	9909	.2065	3149	4.8430	6851	20
50	.2051	3119	.9787	9907	.2095	3212	4.7729	6788	10
12° 00′	.2079	**9.3179**	.9781	**9.9904**	.2126	**9.3275**	4.7046	**0.6725**	78° 00′
10	.2108	3238	.9775	9901	.2156	3336	4.6382	6664	50
20	.2136	3296	.9769	9899	.2186	3397	4.5736	6603	40
30	.2164	3353	.9763	9896	.2217	3458	4.5107	6542	30
40	.2193	3410	.9757	9893	.2247	3517	4.4494	6483	20
50	.2221	3466	.9750	9890	.2278	3576	4.3897	6424	10
13° 00′	.2250	**9.3521**	.9744	**9.9887**	.2309	**9.3634**	4.3315	**0.6366**	77° 00′
10	.2278	3575	.9737	9884	.2339	3691	4.2747	6309	50
20	.2306	3629	.9730	9881	.2370	3748	4.2193	6252	40
30	.2334	3682	.9724	9878	.2401	3804	4.1653	6196	30
40	.2363	3734	.9717	9875	.2432	3859	4.1126	6141	20
50	.2391	3786	.9710	9872	.2462	3914	4.0611	6086	10
14° 00′	.2419	**9.3837**	.9703	**9.9869**	.2493	**9.3968**	4.0108	**0.6032**	76° 00′
10	.2447	3887	.9696	9866	.2524	4021	3.9617	5979	50
20	.2476	3937	.9689	9863	.2555	4074	3.9136	5926	40
30	.2504	3986	.9681	9859	.2586	4127	3.8667	5873	30
40	.2532	4035	.9674	9856	.2617	4178	3.8208	5822	20
50	.2560	4083	.9667	9853	.2648	4230	3.7760	5770	10
15° 00′	.2588	**9.4130**	.9659	**9.9849**	.2679	**9.4281**	3.7321	**0.5719**	75° 00′
10	.2616	4177	.9652	9846	.2711	4331	3.6891	5669	50
20	.2644	4223	.9644	9843	.2742	4381	3.6470	5619	40
30	.2672	4269	.9636	9839	.2773	4430	3.6059	5570	30
40	.2700	4314	.9628	9836	.2805	4479	3.5656	5521	20
50	.2728	4359	.9621	9832	.2836	4527	3.5261	5473	10
16° 00′	.2756	**9.4403**	.9613	**9.9828**	.2867	**9.4575**	3.4874	**0.5425**	74° 00′
10	.2784	4447	.9605	9825	.2899	4622	3.4495	5378	50
20	.2812	4491	.9596	9821	.2931	4669	3.4124	5331	40
30	.2840	4533	.9588	9817	.2962	4716	3.3759	5284	30
40	.2868	4576	.9580	9814	.2994	4762	3.3402	5238	20
50	.2896	4618	.9572	9810	.3026	4808	3.3052	5192	10
17° 00′	.2924	**9.4659**	.9563	**9.9806**	.3057	**9.4853**	3.2709	**0.5147**	73° 00′
10	.2952	4700	.9555	9802	.3089	4898	3.2371	5102	50
20	.2979	4741	.9546	9798	.3121	4943	3.2041	5057	40
30	.3007	4781	.9537	9794	.3153	4987	3.1716	5013	30
40	.3035	4821	.9528	9790	.3185	5031	3.1397	4969	20
50	.3062	4861	.9520	9786	.3217	5075	3.1084	4925	10
18° 00′	.3090	**9.4900**	.9511	**9.9782**	.3249	**9.5118**	3.0777	**0.4882**	72° 00′
	Nat.	Log.	Nat.	Log.	Nat.	Log.	Nat.	Log.	

Angles	Cosines		Sines		Cotangents		Tangents		Angles

Table II   Trigonometric Functions (*continued*)

Angles	Sines Nat.	Log.	Cosines Nat.	Log.	Tangents Nat.	Log.	Cotangents Nat.	Log.	Angles
18° 00′	.3090	9.4900	.9511	9.9782	.3249	9.5118	3.0777	0.4882	72° 00′
10	.3118	4939	.9502	9778	.3281	5161	3.0475	4839	50
20	.3145	4977	.9492	9774	.3314	5203	3.0178	4797	40
30	.3173	5015	.9483	9770	.3346	5245	2.9887	4755	30
40	.3201	5052	.9474	9765	.3378	5287	2.9600	4713	20
50	.3228	5090	.9465	9761	.3411	5329	2.9319	4671	10
19° 00′	.3256	9.5126	.9455	9.9757	.3443	9.5370	2.9042	0.4630	71° 00′
10	.3283	5163	.9446	9752	.3476	5411	2.8770	4589	50
20	.3311	5199	.9436	9748	.3508	5451	2.8502	4549	40
30	.3338	5235	.9426	9743	.3541	5491	2.8239	4509	30
40	.3365	5270	.9417	9739	.3574	5531	2.7980	4469	20
50	.3393	5306	.9407	9734	.3607	5571	2.7725	4429	10
20° 00′	.3420	9.5341	.9397	9.9730	.3640	9.5611	2.7475	0.4389	70° 00′
10	.3448	5375	.9387	9725	.3673	5650	2.7228	4350	50
20	.3475	5409	.9377	9721	.3706	5689	2.6985	4311	40
30	.3502	5443	.9367	9716	.3739	5727	2.6746	4273	30
40	.3529	5477	.9356	9711	.3772	5766	2.6511	4234	20
50	.3557	5510	.9346	9706	.3805	5804	2.6279	4196	10
21° 00′	.3584	9.5543	.9336	9.9702	.3839	9.5842	2.6051	0.4158	69° 00′
10	.3611	5576	.9325	9697	.3872	5879	2.5826	4121	50
20	.3638	5609	.9315	9692	.3906	5917	2.5605	4083	40
30	.3665	5641	.9304	9687	.3939	5954	2.5386	4046	30
40	.3692	5673	.9293	9682	.3973	5991	2.5172	4009	20
50	.3719	5704	.9283	9677	.4006	6028	2.4960	3972	10
22° 00′	.3746	9.5736	.9272	9.9672	.4040	9.6064	2.4751	0.3936	68° 00′
10	.3773	5767	.9261	9667	.4074	6100	2.4545	3900	50
20	.3800	5798	.9250	9661	.4108	6136	2.4342	3864	40
30	.3827	5828	.9239	9656	.4142	6172	2.4142	3828	30
40	.3854	5859	.9228	9651	.4176	6208	2.3945	3792	20
50	.3881	5889	.9216	9646	.4210	6243	2.3750	3757	10
23° 00′	.3907	9.5919	.9205	9.9640	.4245	9.6279	2.3559	0.3721	67° 00′
10	.3934	5948	.9194	9635	.4279	6314	2.3369	3686	50
20	.3961	5978	.9182	9629	.4314	6348	2.3183	3652	˙40
30	.3987	6007	.9171	9624	.4348	6383	2.2998	3617	30
40	.4014	6036	.9159	9618	.4383	6417	2.2817	3583	20
50	.4041	6065	.9147	9613	.4417	6452	2.2637	3548	10
24° 00′	.4067	9.6093	.9135	9.9607	.4452	9.6486	2.2460	0.3514	66° 00′
10	.4094	6121	.9124	9602	.4487	6520	2.2286	3480	50
20	.4120	6149	.9112	9596	.4522	6553	2.2113	3447	40
30	.4147	6177	.9100	9590	.4557	6587	2.1943	3413	30
40	.4173	6205	.9088	9584	.4592	6620	2.1775	3380	20
50	.4200	6232	.9075	9579	.4628	6654	2.1609	3346	10
25° 00′	.4226	9.6259	.9063	9.9573	.4663	9.6687	2.1445	0.3313	65° 00′
10	.4253	6286	.9051	9567	.4699	6720	2.1283	3280	50
20	.4279	6313	.9038	9561	.4734	6752	2.1123	3248	40
30	.4305	6340	.9026	9555	.4770	6785	2.0965	3215	30
40	.4331	6366	.9013	9549	.4806	6817	2.0809	3183	20
50	.4358	6392	.9001	9543	.4841	6850	2.0655	3150	10
26° 00′	.4384	9.6418	.8988	9.9537	.4877	9.6882	2.0503	0.3118	64° 00′
10	.4410	6444	.8975	9530	.4913	6914	2.0353	3086	50
20	.4436	6470	.8962	9524	.4950	6946	2.0204	3054	40
30	.4462	6495	.8949	9518	.4986	6977	2.0057	3023	30
40	.4488	6521	.8936	9512	.5022	7009	1.9912	2991	20
50	.4514	6546	.8923	9505	.5059	7040	1.9768	2960	10
27° 00′	.4540	9.6570	.8910	9.9499	.5095	9.7072	1.9626	0.2928	63° 00′
	Nat.	Log.	Nat.	Log.	Nat.	Log.	Nat.	Log.	

Angles	Cosines		Sines		Cotangents		Tangents		Angles

Table II   Trigonometric Functions (*continued*)

Angles	Sines		Cosines		Tangents		Cotangents		Angles
	Nat.	Log.	Nat.	Log.	Nat.	Log.	Nat.	Log.	
27° 00′	.4540	9.6570	.8910	9.9499	.5095	9.7072	1.9626	0.2928	63° 00′
10	.4566	6595	.8897	9492	.5132	7103	1.9486	2897	50
20	.4592	6620	.8884	9486	.5169	7134	1.9347	2866	40
30	.4617	6644	.8870	9479	.5206	7165	1.9210	2835	30
40	.4643	6668	.8857	9473	.5243	7196	1.9074	2804	20
50	.4669	6692	.8843	9466	.5280	7226	1.8940	2774	10
28° 00′	.4695	9.6716	.8829	9.9459	.5317	9.7257	1.8807	0.2743	62° 00′
10	.4720	6740	.8816	9453	.5354	7287	1.8676	2713	50
20	.4746	6763	.8802	9446	.5392	7317	1.8546	2683	40
30	.4772	6787	.8788	9439	.5430	7348	1.8418	2652	30
40	.4797	6810	.8774	9432	.5467	7378	1.8291	2622	20
50	.4823	6833	.8760	9425	.5505	7408	1.8165	2592	10
29° 00′	.4848	9.6856	.8746	9.9418	.5543	9.7438	1.8040	0.2562	61° 00′
10	.4874	6878	.8732	9411	.5581	7467	1.7917	2533	50
20	.4899	6901	.8718	9404	.5619	7497	1.7796	2503	40
30	.4924	6923	.8704	9397	.5658	7526	1.7675	2474	30
40	.4950	6946	.8689	9390	.5696	7556	1.7556	2444	20
50	.4975	6968	.8675	9383	.5735	7585	1.7437	2415	10
30° 00′	.5000	9.6990	.8660	9.9375	.5774	9.7614	1.7321	0.2386	60° 00′
10	.5025	7012	.8646	9368	.5812	7644	1.7205	2356	50
20	.5050	7033	.8631	9361	.5851	7673	1.7090	2327	40
30	.5075	7055	.8616	9353	.5890	7701	1.6977	2299	30
40	.5100	7076	.8601	9346	.5930	7730	1.6864	2270	20
50	.5125	7097	.8587	9338	.5969	7759	1.6753	2241	10
31° 00′	.5150	9.7118	.8572	9.9331	.6009	9.7788	1.6643	0.2212	59° 00′
10	.5175	7139	.8557	9323	.6048	7816	1.6534	2184	50
20	.5200	7160	.8542	9315	.6088	7845	1.6426	2155	40
30	.5225	7181	.8526	9308	.6128	7873	1.6319	2127	30
40	.5250	7201	.8511	9300	.6168	7902	1.6212	2098	20
50	.5275	7222	.8496	9292	.6208	7930	1.6107	2070	10
32° 00′	.5299	9.7242	.8480	9.9284	.6249	9.7958	1.6003	0.2042	58° 00′
10	.5324	7262	.8465	9276	.6289	7986	1.5900	2014	50
20	.5348	7282	.8450	9268	.6330	8014	1.5798	1986	40
30	.5373	7302	.8434	9260	.6371	8042	1.5697	1958	30
40	.5398	7322	.8418	9252	.6412	8070	1.5597	1930	20
50	.5422	7342	.8403	9244	.6453	8097	1.5497	1903	10
33° 00′	.5446	9.7361	.8387	9.9236	.6494	9.8125	1.5399	0.1875	57° 00′
10	.5471	7380	.8371	9228	.6536	8153	1.5301	1847	50
20	.5495	7400	.8355	9219	.6577	8180	1.5204	1820	40
30	.5519	7419	.8339	9211	.6619	8208	1.5108	1792	30
40	.5544	7438	.8323	9203	.6661	8235	1.5013	1765	20
50	.5568	7457	.8307	9194	.6703	8263	1.4919	1737	10
34° 00′	.5592	9.7476	.8290	9.9186	.6745	9.8290	1.4826	0.1710	56° 00′
10	.5616	7494	.8274	9177	.6787	8317	1.4733	1683	50
20	.5640	7513	.8258	9169	.6830	8344	1.4641	1656	40
30	.5664	7531	.8241	9160	.6873	8371	1.4550	1629	30
40	.5688	7550	.8225	9151	.6916	8398	1.4460	1602	20
50	.5712	7568	.8208	9142	.6959	8425	1.4370	1575	10
35° 00′	.5736	9.7586	.8192	9.9134	.7002	9.8452	1.4281	0.1548	55° 00′
10	.5760	7604	.8175	9125	.7046	8479	1.4193	1521	50
20	.5783	7622	.8158	9116	.7089	8506	1.4106	1494	40
30	.5807	7640	.8141	9107	.7133	8533	1.4019	1467	30
40	.5831	7657	.8124	9098	.7177	8559	1.3934	1441	20
50	.5854	7675	.8107	9089	.7221	8586	1.3848	1414	10
36° 00′	.5878	9.7692	.8090	9.9080	.7265	9.8613	1.3764	0.1387	54° 00′
	Nat.	Log.	Nat.	Log.	Nat.	Log.	Nat.	Log.	

Angles	Cosines	Sines	Cotangents	Tangents	Angles

Table II   Trigonometric Functions (*continued*)

Angles	Sines		Cosines		Tangents		Cotangents		Angles
	Nat.	Log.	Nat.	Log.	Nat.	Log.	Nat.	Log.	
36° 00′	.5878	9.7692	.8090	9.9080	.7265	9.8613	1.3764	0.1387	54° 00′
10	.5901	7710	.8073	9070	.7310	8639	1.3680	1361	50
20	.5925	7727	.8056	9061	.7355	8666	1.3597	1334	40
30	.5948	7744	.8039	9052	.7400	8692	1.3514	1308	30
40	.5972	7761	.8021	9042	.7445	8718	1.3432	1282	20
50	.5995	7778	.8004	9033	.7490	8745	1.3351	1255	10
37° 00′	.6018	9.7795	.7986	9.9023	.7536	9.8771	1.3270	0.1229	53° 00′
10	.6041	7811	.7969	9014	.7581	8797	1.3190	1203	50
20	.6065	7828	.7951	9004	.7627	8824	1.3111	1176	40
30	.6088	7844	.7934	8995	.7673	8850	1.3032	1150	30
40	.6111	7861	.7916	8985	.7720	8876	1.2954	1124	20
50	.6134	7877	.7898	8975	.7766	8902	1.2876	1098	10
38° 00′	.6157	9.7893	.7880	9.8965	.7813	9.8928	1.2790	0.1072	52° 00′
10	.6180	7910	.7862	8955	.7860	8954	1.2723	1046	50
20	.6202	7926	.7844	8945	.7907	8980	1.2647	1020	40
30	.6225	7941	.7826	8935	.7954	9006	1.2572	0994	30
40	.6248	7957	.7808	8925	.8002	9032	1.2497	0968	20
50	.6271	7973	.7790	8915	.8050	9058	1.2423	0942	10
39° 00′	.6293	9.7989	.7771	9.8905	.8098	9.9084	1.2349	0.0916	51° 00′
10	.6316	8004	.7753	8895	.8146	9110	1.2276	0890	50
20	.6338	8020	.7735	8884	.8195	9135	1.2203	0865	40
30	.6361	8035	.7716	8874	.8243	9161	1.2131	0839	30
40	.6383	8050	.7698	8864	.8292	9187	1.2059	0813	20
50	.6406	8066	.7679	8853	.8342	9212	1.1988	0788	10
40° 00′	.6428	9.8081	.7660	9.8843	.8391	9.9238	1.1918	0.0762	50° 00′
10	.6450	8096	.7642	8832	.8441	9264	1.1847	0736	50
20	.6472	8111	.7623	8821	.8491	9289	1.1778	0711	40
30	.6494	8125	.7604	8810	.8541	9315	1.1708	0685	30
40	.6517	8140	.7585	8800	.8591	9341	1.1640	0659	20
50	.6539	8155	.7566	8789	.8642	9366	1.1571	0634	10
41° 00′	.6561	9.8169	.7547	9.8778	.8693	9.9392	1.1504	0.0608	49° 00′
10	.6583	8184	.7528	8767	.8744	9417	1.1436	0583	50
20	.6604	8198	.7509	8756	.8796	9443	1.1369	0557	40
30	.6626	8213	.7490	8745	.8847	9468	1.1303	0532	30
40	.6648	8227	.7470	8733	.8899	9494	1.1237	0506	20
50	.6670	8241	.7451	8722	.8952	9519	1.1171	0481	10
42° 00′	.6691	9.8255	.7431	9.8711	.9004	9.9544	1.1106	0.0456	48° 00′
10	.6713	8269	.7412	8699	.9057	9570	1.1041	0430	50
20	.6734	8283	.7392	8688	.9110	9595	1.0977	0405	40
30	.6756	8297	.7373	8676	.9163	9621	1.0913	0379	30
40	.6777	8311	.7353	8665	.9217	9646	1.0850	0354	20
50	.6799	8324	.7333	8653	.9271	9671	1.0786	0329	10
43° 00′	.6820	9.8338	.7314	9.8641	.9325	9.9697	1.0724	0.0303	47° 00′
10	.6841	8351	.7294	8629	.9380	9722	1.0661	0278	50
20	.6862	8365	.7274	8618	.9435	9747	1.0599	0253	40
30	.6884	8378	.7254	8606	.9490	9772	1.0538	0228	30
40	.6905	8391	.7234	8594	.9545	9798	1.0477	0202	20
50	.6926	8405	.7214	8582	.9601	9823	1.0416	0177	10
44° 00′	.6947	9.8418	.7193	9.8569	.9657	9.9848	1.0355	0.0152	46° 00′
10	.6967	8431	.7173	8557	.9713	9874	1.0295	0126	50
20	.6988	8444	.7153	8545	.9770	9899	1.0235	0101	40
30	.7009	8457	.7133	8532	.9827	9924	1.0176	0076	30
40	.7030	8469	.7112	8520	.9884	9949	1.0117	0051	20
50	.7050	8482	.7092	8507	.9942	9975	1.0058	0025	10
45° 00′	.7071	9.8495	.7071	9.8495	1.0000	0.0000	1.0000	0.0000	45° 00′
	Nat.	Log.	Nat.	Log.	Nat.	Log.	Nat.	Log.	

Angles	Cosines		Sines		Cotangents		Tangents		Angles

Table III   Powers and Roots

No.	Sq.	Sq. root	Cube	Cube root	No.	Sq.	Sq. root	Cube	Cube root
1	1	1.000	1	1.000	51	2,601	7.141	132,651	3.708
2	4	1.414	8	1.260	52	2,704	7.211	140,608	3.733
3	9	1.732	27	1.442	53	2,809	7.280	148,877	3.756
4	16	2.000	64	1.587	54	2,916	7.348	157,464	3.780
5	25	2.236	125	1.710	55	3,025	7.416	166,375	3.803
6	36	2.449	216	1.817	56	3,136	7.483	175,616	3.826
7	49	2.646	343	1.913	57	3,249	7.550	185,193	3.849
8	64	2.828	512	2.000	58	3,364	7.616	195,112	3.871
9	81	3.000	729	2.080	59	3,481	7.681	205,379	3.893
10	100	3.162	1,000	2.154	60	3,600	7.746	216,000	3.915
11	121	3.317	1,331	2.224	61	3,721	7.810	226,981	3.936
12	144	3.464	1,728	2.289	62	3,844	7.874	238,328	3.958
13	169	3.606	2,197	2.351	63	3,969	7.937	250,047	3.979
14	196	3.742	2,744	2.410	64	4,096	8.000	262,144	4.000
15	225	3.873	3,375	2.466	65	4,225	8.062	274,625	4.021
16	256	4.000	4,096	2.520	66	4,356	8.124	287,496	4.041
17	289	4.123	4,913	2.571	67	4,489	8.185	300,763	4.062
18	324	4.243	5,832	2.621	68	4,624	8.246	314,432	4.082
19	361	4.359	6,859	2.668	69	4,761	8.307	328,509	4.102
20	400	4.472	8,000	2.714	70	4,900	8.367	343,000	4.121
21	441	4.583	9,261	2.759	71	5,041	8.426	357,911	4.141
22	484	4.690	10,648	2.802	72	5,184	8.485	373,248	4.160
23	529	4.796	12,167	2.844	73	5,329	8.544	389,017	4.179
24	576	4.899	13,824	2.884	74	5,476	8.602	405,224	4.198
25	625	5.000	15,625	2.924	75	5,625	8.660	421,875	4.217
26	676	5.099	17,576	2.962	76	5,776	8.718	438,976	4.236
27	729	5.196	19,683	3.000	77	5,929	8.775	456,533	4.254
28	784	5.291	21,952	3.037	78	6,084	8.832	474,552	4.273
29	841	5.385	24,389	3.072	79	6,241	8.888	493,039	4.291
30	900	5.477	27,000	3.107	80	6,400	8.944	512,000	4.309
31	961	5.568	29,791	3.141	81	6,561	9.000	531,441	4.327
32	1,024	5.657	32,768	3.175	82	6,724	9.055	551,368	4.344
33	1,089	5.745	35,937	3.208	83	6,889	9.110	571,787	4.362
34	1,156	5.831	39,304	3.240	84	7,056	9.165	592,704	4.380
35	1,225	5.916	42,875	3.271	85	7,225	9.220	614,125	4.397
36	1,296	6.000	46,656	3.302	86	7,396	9.274	636,056	4.414
37	1,369	6.083	50,653	3.332	87	7,569	9.327	658,503	4.431
38	1,444	6.164	54,872	3.362	88	7,744	9.381	681,472	4.448
39	1,521	6.245	59,319	3.391	89	7,921	9.434	704,969	4.465
40	1,600	6.325	64,000	3.420	90	8,100	9.487	729,000	4.481
41	1,681	6.403	68,921	3.448	91	8,281	9.539	753,571	4.498
42	1,764	6.481	74,088	3.476	92	8.464	9.592	778,688	4.514
43	1,849	6.557	79,507	3.503	93	8,649	9.644	804,357	4.531
44	1,936	6.633	85,184	3.530	94	8,836	9.695	830,584	4.547
45	2,025	6.708	91,125	3.557	95	9,025	9.747	857,375	4.563
46	2,116	6.782	97,336	3.583	96	9,216	9.798	884,736	4.579
47	2,209	6.856	103,823	3.609	97	9,409	9.849	912,673	4.595
48	2,304	6.928	110,592	3.634	98	9,604	9.899	941,192	4.610
49	2,401	7.000	117,649	3.659	99	9,801	9.950	970,299	4.626
50	2,500	7.071	125,000	3.684	100	10,000	10.000	1,000,000	4.642

Table IV Natural Logarithms

	.00	.01	.02	.03	.04	.05	.06	.07	.08	.09
1.0	0.0000	0.0100	0.0198	0.0296	0.0392	0.0488	0.0583	0.0677	0.0770	0.0862
1.1	0.0953	0.1044	0.1133	0.1222	0.1310	0.1398	0.1484	0.1570	0.1655	0.1740
1.2	0.1823	0.1906	0.1989	0.2070	0.2151	0.2231	0.2311	0.2390	0.2469	0.2546
1.3	0.2624	0.2700	0.2776	0.2852	0.2927	0.3001	0.3075	0.3148	0.3221	0.3293
1.4	0.3365	0.3436	0.3507	0.3577	0.3646	0.3716	0.3784	0.3853	0.3920	0.3988
1.5	0.4055	0.4121	0.4187	0.4253	0.4318	0.4383	0.4447	0.4511	0.4574	0.4637
1.6	0.4700	0.4762	0.4824	0.4886	0.4947	0.5008	0.5068	0.5128	0.5188	0.5247
1.7	0.5306	0.5365	0.5423	0.5481	0.5539	0.5596	0.5653	0.5710	0.5766	0.5822
1.8	0.5878	0.5933	0.5988	0.6043	0.6098	0.6152	0.6206	0.6259	0.6313	0.6366
1.9	0.6419	0.6471	0.6523	0.6575	0.6627	0.6678	0.6729	0.6780	0.6831	0.6881
2.0	0.6932	0.6981	0.7031	0.7080	0.7129	0.7178	0.7227	0.7275	0.7324	0.7372
2.1	0.7419	0.7467	0.7514	0.7561	0.7608	0.7655	0.7701	0.7747	0.7793	0.7839
2.2	0.7885	0.7930	0.7975	0.8020	0.8065	0.8109	0.8154	0.8198	0.8242	0.8286
2.3	0.8329	0.8373	0.8416	0.8459	0.8502	0.8544	0.8587	0.8629	0.8671	0.8713
2.4	0.8755	0.8796	0.8838	0.8879	0.8920	0.8961	0.9002	0.9042	0.9083	0.9123
2.5	0.9163	0.9203	0.9243	0.9282	0.9322	0.9361	0.9400	0.9439	0.9478	0.9517
2.6	0.9555	0.9594	0.9632	0.9670	0.9708	0.9746	0.9783	0.9821	0.9858	0.9895
2.7	0.9933	0.9969	1.0006	1.0043	1.0080	1.0116	1.0152	1.0188	1.0225	1.0260
2.8	1.0296	1.0332	1.0367	1.0403	1.0438	1.0473	1.0508	1.0543	1.0578	1.0613
2.9	1.0647	1.0682	1.0716	1.0750	1.0784	1.0818	1.0852	1.0886	1.0919	1.0953
3.0	1.0986	1.1019	1.1053	1.1086	1.1119	1.1151	1.1184	1.1217	1.1249	1.1282
3.1	1.1314	1.1346	1.1378	1.1410	1.1442	1.1474	1.1506	1.1537	1.1569	1.1600
3.2	1.1632	1.1663	1.1694	1.1725	1.1756	1.1787	1.1817	1.1848	1.1878	1.1909
3.3	1.1939	1.1969	1.2000	1.2030	1.2060	1.2090	1.2119	1.2149	1.2179	1.2208
3.4	1.2238	1.2267	1.2296	1.2326	1.2355	1.2384	1.2413	1.2442	1.2470	1.2499
3.5	1.2528	1.2556	1.2585	1.2613	1.2641	1.2669	1.2698	1.2726	1.2754	1.2782
3.6	1.2809	1.2837	1.2865	1.2892	1.2920	1.2947	1.2975	1.3002	1.3029	1.3056
3.7	1.3083	1.3110	1.3137	1.3164	1.3191	1.3218	1.3244	1.3271	1.3297	1.3324
3.8	1.3350	1.3376	1.3403	1.3429	1.3455	1.3481	1.3507	1.3533	1.3558	1.3584
3.9	1.3610	1.3635	1.3661	1.3686	1.3712	1.3737	1.3762	1.3788	1.3813	1.3838
4.0	1.3863	1.3888	1.3913	1.3938	1.3962	1.3987	1.4012	1.4036	1.4061	1.4085
4.1	1.4110	1.4134	1.4159	1.4183	1.4207	1.4231	1.4255	1.4279	1.4303	1.4327
4.2	1.4351	1.4375	1.4398	1.4422	1.4446	1.4469	1.4493	1.4516	1.4540	1.4563
4.3	1.4586	1.4609	1.4633	1.4656	1.4679	1.4702	1.4725	1.4748	1.4771	1.4793
4.4	1.4816	1.4839	1.4861	1.4884	1.4907	1.4929	1.4951	1.4974	1.4996	1.5019
4.5	1.5041	1.5063	1.5085	1.5107	1.5129	1.5151	1.5173	1.5195	1.5217	1.5239
4.6	1.5261	1.5282	1.5304	1.5326	1.5347	1.5369	1.5390	1.5412	1.5433	1.5454
4.7	1.5476	1.5497	1.5518	1.5539	1.5560	1.5581	1.5602	1.5623	1.5644	1.5665
4.8	1.5686	1.5707	1.5728	1.5748	1.5769	1.5790	1.5810	1.5831	1.5851	1.5872
4.9	1.5892	1.5913	1.5933	1.5953	1.5974	1.5994	1.6014	1.6034	1.6054	1.6074
5.0	1.6094	1.6114	1.6134	1.6154	1.6174	1.6194	1.6214	1.6233	1.6253	1.6273
5.1	1.6292	1.6312	1.6332	1.6351	1.6371	1.6390	1.6409	1.6429	1.6448	1.6467
5.2	1.6487	1.6506	1.6525	1.6544	1.6563	1.6582	1.6601	1.6620	1.6639	1.6658
5.3	1.6677	1.6696	1.6715	1.6734	1.6752	1.6771	1.6790	1.6808	1.6827	1.6845
5.4	1.6864	1.6882	1.6901	1.6919	1.6938	1.6956	1.6974	1.6993	1.7011	1.7029

Table IV   Natural Logarithms (*continued*)

	.00	.01	.02	.03	.04	.05	.06	.07	.08	.09
5.5	1.7047	1.7066	1.7084	1.7102	1.7120	1.7138	1.7156	1.7174	1.7192	1.7210
5.6	1.7228	1.7246	1.7263	1.7281	1.7299	1.7317	1.7334	1.7352	1.7370	1.7387
5.7	1.7405	1.7422	1.7440	1.7457	1.7475	1.7492	1.7509	1.7527	1.7544	1.7561
5.8	1.7579	1.7596	1.7613	1.7630	1.7647	1.7664	1.7681	1.7699	1.7716	1.7733
5.9	1.7750	1.7766	1.7783	1.7800	1.7817	1.7834	1.7851	1.7868	1.7884	1.7901
6.0	1.7918	1.7934	1.7951	1.7967	1.7984	1.8001	1.8017	1.8034	1.8050	1.8066
6.1	1.8083	1.8099	1.8116	1.8132	1.8148	1.8165	1.8181	1.8197	1.8213	1.8229
6.2	1.8245	1.8262	1.8278	1.8294	1.8310	1.8326	1.8342	1.8358	1.8374	1.8390
6.3	1.8405	1.8421	1.8437	1.8453	1.8469	1.8485	1.8500	1.8516	1.8532	1.8547
6.4	1.8563	1.8579	1.8594	1.8610	1.8625	1.8641	1.8656	1.8672	1.8687	1.8703
6.5	1.8718	1.8733	1.8749	1.8764	1.8779	1.8795	1.8810	1.8825	1.8840	1.8856
6.6	1.8871	1.8886	1.8901	1.8916	1.8931	1.8946	1.8961	1.8976	1.8991	1.9006
6.7	1.9021	1.9036	1.9051	1.9066	1.9081	1.9095	1.9110	1.9125	1.9140	1.9155
6.8	1.9169	1.9184	1.9199	1.9213	1.9228	1.9242	1.9257	1.9272	1.9286	1.9301
6.9	1.9315	1.9330	1.9344	1.9359	1.9373	1.9387	1.9402	1.9416	1.9430	1.9445
7.0	1.9459	1.9473	1.9488	1.9502	1.9516	1.9530	1.9544	1.9559	1.9573	1.9587
7.1	1.9601	1.9615	1.9629	1.9643	1.9657	1.9671	1.9685	1.9699	1.9713	1.9727
7.2	1.9741	1.9755	1.9769	1.9782	1.9796	1.9810	1.9824	1.9838	1.9851	1.9865
7.3	1.9879	1.9892	1.9906	1.9920	1.9933	1.9947	1.9961	1.9974	1.9988	2.0001
7.4	2.0015	2.0028	2.0042	2.0055	2.0069	2.0082	2.0096	2.0109	2.0122	2.0136
7.5	2.0149	2.0162	2.0176	2.0189	2.0202	2.0215	2.0229	2.0242	2.0255	2.0268
7.6	2.0281	2.0295	2.0308	2.0321	2.0334	2.0347	2.0360	2.0373	2.0386	2.0399
7.7	2.0412	2.0425	2.0438	2.0451	2.0464	2.0477	2.0490	2.0503	2.0516	2.0528
7.8	2.0541	2.0554	2.0567	2.0580	2.0592	2.0605	2.0618	2.0631	2.0643	2.0656
7.9	2.0669	2.0681	2.0694	2.0707	2.0719	2.0732	2.0744	2.0757	2.0769	2.0782
8.0	2.0794	2.0807	2.0819	2.0832	2.0844	2.0857	2.0869	2.0882	2.0894	2.0906
8.1	2.0919	2.0931	2.0943	2.0956	2.0968	2.0980	2.0992	2.1005	2.1017	2.1029
8.2	2.1041	2.1054	2.1066	2.1078	2.1090	2.1102	2.1114	2.1126	2.1138	2.1150
8.3	2.1163	2.1175	2.1187	2.1199	2.1211	2.1223	2.1235	2.1247	2.1259	2.1270
8.4	2.1282	2.1294	2.1306	2.1318	2.1330	2.1342	2.1353	2.1365	2.1377	2.1389
8.5	2.1401	2.1412	2.1424	2.1436	2.1448	2.1459	2.1471	2.1483	2.1494	2.1506
8.6	2.1518	2.1529	2.1541	2.1552	2.1564	2.1576	2.1587	2.1599	2.1610	2.1622
8.7	2.1633	2.1645	2.1656	2.1668	2.1679	2.1691	2.1702	2.1713	2.1725	2.1736
8.8	2.1748	2.1759	2.1770	2.1782	2.1793	2.1804	2.1815	2.1827	2.1838	2.1849
8.9	2.1861	2.1872	2.1883	2.1894	2.1905	2.1917	2.1928	2.1939	2.1950	2.1961
9.0	2.1972	2.1983	2.1994	2.2006	2.2017	2.2028	2.2039	2.2050	2.2061	2.2072
9.1	2.2083	2.2094	2.2105	2.2116	2.2127	2.2138	2.2148	2.2159	2.2170	2.2181
9.2	2.2192	2.2203	2.2214	2.2225	2.2235	2.2246	2.2257	2.2268	2.2279	2.2289
9.3	2.2300	2.2311	2.2322	2.2332	2.2343	2.2354	2.2364	2.2375	2.2386	2.2396
9.4	2.2407	2.2418	2.2428	2.2439	2.2450	2.2460	2.2471	2.2481	2.2492	2.2502
9.5	2.2513	2.2523	2.2534	2.2544	2.2555	2.2565	2.2576	2.2586	2.2597	2.2607
9.6	2.2618	2.2628	2.2638	2.2649	2.2659	2.2670	2.2680	2.2690	2.2701	2.2711
9.7	2.2721	2.2732	2.2742	2.2752	2.2762	2.2773	2.2783	2.2793	2.2803	2.2814
9.8	2.2824	2.2834	2.2844	2.2854	2.2865	2.2875	2.2885	2.2895	2.2905	2.2915
9.9	2.2925	2.2935	2.2946	2.2956	2.2966	2.2976	2.2986	2.2996	2.3006	2.3016

Table V   Exponential and Hyperbolic Functions

x	e^x	e^{-x}	$\sinh x$	$\cosh x$	$\tanh x$
.00	1.000	1.000	.000	1.000	.000
.01	1.010	.990	.010	1.000	.010
.02	1.020	.980	.020	1.000	.020
.03	1.030	.970	.030	1.000	.030
.04	1.041	.961	.040	1.001	.040
.05	1.051	.951	.050	1.001	.050
.06	1.062	.942	.060	1.002	.060
.07	1.073	.932	.070	1.002	.070
.08	1.083	.923	.080	1.003	.080
.09	1.094	.914	.090	1.004	.090
.1	1.105	.905	.100	1.005	.100
.2	1.221	.819	.201	1.020	.197
.3	1.350	.741	.305	1.045	.291
.4	1.492	.670	.411	1.081	.380
.5	1.649	.607	.521	1.128	.462
.6	1.822	.549	.637	1.185	.537
.7	2.014	.497	.759	1.255	.604
.8	2.226	.449	.888	1.337	.664
.9	2.460	.407	1.027	1.433	.716
1.0	2.718	.368	1.175	1.543	.762
1.1	3.004	.333	1.336	1.669	.800
1.2	3.320	.301	1.509	1.811	.834
1.3	3.669	.273	1.698	1.971	.862
1.4	4.055	.247	1.904	2.151	.885
1.5	4.482	.223	2.129	2.352	.905
1.6	4.953	.202	2.376	2.577	.922
1.7	5.474	.183	2.646	2.828	.935
1.8	6.050	.165	2.942	3.107	.947
1.9	6.686	.150	3.268	3.418	.956
2.0	7.389	.135	3.627	3.762	.964
2.1	8.166	.122	4.022	4.144	.970
2.2	9.025	.111	4.457	4.568	.976
2.3	9.974	.100	4.937	5.037	.980
2.4	11.023	.091	5.466	5.557	.984
2.5	12.182	.082	6.050	6.132	.987
2.6	13.464	.074	6.695	6.769	.989
2.7	14.880	.067	7.406	7.473	.991
2.8	16.445	.061	8.192	8.253	.993
2.9	18.174	.055	9.060	9.115	.994
3.0	20.086	.050	10.018	10.068	.995
3.1	22.20	.045	11.08	11.12	.996
3.2	24.53	.041	12.25	12.29	.997
3.3	27.11	.037	13.54	13.57	.997
3.4	29.96	.033	14.97	15.00	.998
3.5	33.12	.030	16.54	16.57	.998
3.6	36.60	.027	18.29	18.31	.999
3.7	40.45	.025	20.21	20.24	.999
3.8	44.70	.022	22.34	22.36	.999
3.9	49.40	.020	24.69	24.71	.999
4.0	54.60	.018	27.29	27.31	.999
4.1	60.34	.017	30.16	30.18	.999
4.2	66.69	.015	33.34	33.35	1.000
4.3	73.70	.014	36.84	36.86	1.000
4.4	81.45	.012	40.72	40.73	1.000
4.5	90.02	.011	45.00	45.01	1.000
4.6	99.48	.010	49.74	49.75	1.000
4.7	109.95	.0090	54.97	54.98	1.000
4.8	121.51	.0082	60.75	60.76	1.000
4.9	134.29	.0074	67.14	67.15	1.000
5.0	148.41	.0067	74.20	74.21	1.000
6.0	403.4	.0025	201.7		1.000
7.0	1096.6	.00091	548.3		1.000
8.0	2981.0	.00034	1490.5		1.000
9.0	8103.1	.00012	4051.5		1.000
10.0	22026.5	.000045	11013.2		1.000

Answers

9   Domain and range are both all nonnegative real numbers.
10   Domain is $x \geq 0.5$. Range is $y \geq 0$.
11   Domain is $x \leq 2$. Range is $y \geq 0$.
13   $x/3 + 2$, all real x, all real y
14   $y = \sqrt{x - 1}$, $x \geq 1$, $y \geq 0$
15   $y = x^2 - 3$, $x \geq 0$, $y \geq -3$
17   $F = 50 + 40(d - 1)$ 18 $C = 5 + 2.75t$
19   $B = \$3 + \$0.30(t - 6)$ where t is the number of thousand gallons used.
21   Yes   22   Yes   23   Yes
25   $V = (6 - 2x)(8 - 2x)x$, all x such that $0 \leq x \leq 3$
26   $V = 320(1 - x)^2 x$, all x such that $0 \leq x \leq 1$
27   Area $= x\sqrt{36 - x^2}$, all x such that $0 \leq x \leq 6$

Exercise 2.1, page 21

1   $y - 4, 4 - y$ 2 $4 - x$
3   $-3 - y$ 7   Yes
13   $(4 + 5\sqrt{3}, -1 + 5\sqrt{3}), (4 - 5\sqrt{3}, -1 - 5\sqrt{3})$
14   $(4\tfrac{3}{5}, 0)$ 15 -2.9 17 $2, -6$
18   9.6 19 22 21 $(0, -3)$
22   $(2\tfrac{9}{7}, \tfrac{3}{7})$ 23 $(1, -2)$ 25 $(16, 13)$
26   $(39, -2)$ 27 $(1, 20)$ 29 $(5, 2)$
30   $(1, 1)$ 31 $(3.5, 1)$ 33 $(-3, -2), (0, -1)$
34   $(5, -0.5), (2, 2), (-1, 4.5)$ 35 $(2, \tfrac{2}{3})$

Exercise 2.2, page 27

1   $\sqrt{3}$ 2 -1 3 $120°, 300°$
5   $-\tfrac{4}{3}$ 6 -2 7 1
9   $-\tfrac{1}{3}$ 10 $\tfrac{5}{2}$ 11 $\tfrac{1}{8}$
17   $\text{Tan}^{-1}(-5)$ 18 0 19 $\text{Tan}^{-1}(-\tfrac{3}{8})$
21   $38°27'$ 22 $54°28'$ 23 $37°8'$

Exercise 2.3, page 34

1   $5x - 2y = 11$ 2 $x + y = 5$ 3 $5x + 2y = 22$
5   $2x - y = -10$ 6 $3x + y = 26$ 7 $2x - 3y = 15$
9   $y = -3x + 5$ 10 $y = 5x - 2$ 11 $y = -2x/7 - 1$
13   $y = x$ 14 $y = \sqrt{3}x + 1.5$ 15 $\sqrt{3}y = -x - 2$
17   $2x = y$ 18 $y = 3x - 10$ 19 $y = 2x + 17$
21   $y = -0.4x + 2$ 22 $y = 0.75x - 3$ 23 $y = 3x - 4.5$
25   9 26 2 29 -1
30   -3 31 -2 33 $x + 8y = 0$
34   $29x - 24y - 28 = 0$ 35 $319x + 33y + 425 = 0$

Exercise 2.4, page 37

1   $(x - 2)^2 + (y - 5)^2 = 3^2$ 2 $(x - 1)^2 + (y - 4)^2 = 5^2$

3   $(x - 4)^2 + (y + 2)^2 = 6^2$ 5 $x^2 + y^2 - 4x + 2y = 20$

6   $x^2 + y^2 + 6x - 10y = -5$ 7 $x^2 + y^2 + 8x + 6y = 0$

9   $x^2 + y^2 - 10x - 8y = -40$ 10 $x^2 + y^2 - 4x + 12y = -39$

11   $x^2 + y^2 + 6x - 10y = -34$ 13 $(x - 4)^2 + (y + 1)^2 = 20$

14   $(x - 4.5)^2 + (y + 2)^2 = 8.5^2$ 15 $(x - 3.5)^2 + (y - 2)^2 = 2.5^2$

17   $3, (2, 1)$ 18 $0, (4, 5)$

19   $7, (-3, 2)$ 21 $2.5, (0.5, 1.5)$

22   $1.5, (-1.5, 2.5)$ 23 $\frac{4}{3}, (2, -\frac{4}{3})$

25   $k > -5, k = -5, k < -5$ 26 $k > -25, k = -25, k < -25$

27   $k > -55, k = -55, k < -55$

31   $(1 - k^2)(x^2 + y^2) - 2(a - ck^2)x - 2(b - dk^2)y = k^2(c^2 + d^2) - a^2 - b^2$

Exercise 2.5, page 40

1   $(x - 3)^2 + (y - 6)^2 = 5^2$ 2 $(x + 2)^2 + (y - 5)^2 = 13^2$

3   $(x + 2)^2 + (y - 1)^2 = 5^2$ 5 $(x - 1)^2 + (y - 2)^2 = 5^2$

6   $(x - 4)^2 + (y + 3)^2 = 5^2$ 7 $(x - 4)^2 + (y + 3)^2 = 10^2$

9   $(x - 1)^2 + (y - 2)^2 = 5^2$ 10 $(x + 8)^2 + (y - 9)^2 = 13^2$

11   $(x - 5)^2 + (y - 3)^2 = 5^2, (x - 1)^2 + (y + 1)^2 = 1$

13   $(x - 2)^2 + (y - 1)^2 = 5^2, (x - 6)^2 + (y + 27)^2 = 25^2$

14   $(x + 3)^2 + (y - 1)^2 = 5^2, (x + {}^{1,475}\!/_{27})^2 + (y - {}^{139}\!/_{27})^2 = ({}^{1,189}\!/_{81})^2$

15   $(x - 3)^2 + (y + 1)^2 = 5^2$ 17 $(x - 2)^2 + (y - 5)^2 = 5^2$

18   $(x - 8)^2 + (y + 9)^2 = 13^2$ 19 $(x + 5)^2 + (y - 4)^2 = 17^2$

21   $(x - h)^2 + y^2 = h^2$ 22 $(x - h)^2 + (y - h)^2 = 2h^2$

23   $(x - h)^2 + (y + h)^2 = h^2$

26   $x^2 + y^2 + 2x - 4y - 4 + k(x^2 + y^2 - 6x + 2y - 6) = 0, 4x - 3y + 1 = 0$

27   $(x - h)^2 + (y - 3h + 4)^2 = 25$

29   $k = 1, x^2 + y^2 - 2x - y - 5 = 0, k = -{}^{11}\!/_{39}, (x + {}^{18}\!/_{7})^2 + (y - {}^{25}\!/_{7})^2 = {}^{25}\!/_{4}$

30   $k = 1, x^2 + y^2 - 2x - y - 5 = 0$

31   $k = 0, x^2 + y^2 + 2x - 4y - 4 = 0$

Exercise 2.6, page 46

1   None; $x = 0, y = 0$; origin   2 $y = -3; x = 3, y = 0$; none

3   $y = 2; x = -1.5, y = 0$; none   5 $x = -3; x = 0, y = \frac{1}{2}$; none

6   $x = 2.5; x = 0, y = -\frac{2}{3}$; none   7 $x = -2, y = -\frac{2}{3}; x = \frac{3}{2}, y = \frac{1}{2}$; none

9   $y = \frac{8}{3}; y = 0$; Y axis   10 $y = 2.5; y = 0$; Y axis   11 $x = 0, y = 0; y = 0$; origin

13   $x = 0, y = 0; x = \pm 2, y = 0$; origin   14 $x = 0, y = 0; x = \pm\sqrt{5}, y = 0$; origin

15   $x = 0, y = 0; x = \pm 3/\sqrt{2}, y = 0$; origin

17   $x = 0, y = 0; x = -2, y = 0$; none   18 $x = 0, y = 0; x = 1, y = 0$; none

19   $y = -\frac{3}{2}; x = \pm 2, y = 0$; Y axis

21   $x = \pm 3, y = -\frac{9}{5}; x = \pm\sqrt{5}, y = -1$; Y axis

22   $y = -1; x = \pm 2, y = 1$; Y axis   23 $x = \pm 2, y = -1; y = 1$; Y axis

25	$x = 0, y = 0; x = \pm\sqrt{3}$; origin	26	$x = 0, y = 0; x = 2$; none
27	$x = 0, y = 0; x = -2$; none	29	$x = 4.5, y = -1.5; x = 2, y = -\frac{2}{3}$; none
30	$x = \frac{8}{3}, y = -2; x = 2, y = -\frac{3}{2}$; none	31	$x = 2, y = 4; x = -1, y = -2$; none
33	$y = 1; x = -2$; none	34	$x = \pm2; x = 0$; origin
35	$y = \frac{7}{3}; y = 2$; none	37	$x = 0, y = 0; x = \pm2, y = 0$; origin
38	$x = -2, y = 6; y = 0$; none	39	$x = -\frac{3}{4}, y = 3; y = 0$; none

Exercise 3.1, page 55

1	2, 2			2	$-6, -6$
3	$-4, 6$			5	25, yes, yes
6	19, yes, yes			7	21, yes, yes
9	No, this limit does not exist.			10	No, this limit does not exist.
11	6	13	6	14	12
15	$-\frac{1}{4}$	17	8, 8, yes	18	1.5, 1.5, yes, no
19	$-1, -1$, yes, no	21	12	22	3
23	$\frac{3}{4}$	25	0	26	0
27	12	29	2	30	Limit does not exist.
31	$8x_1$	33	$2x$	34	$3x^2$
35	$-1/(2x\sqrt{x})$				

Exercise 3.2, page 60

1	It is zero.	2	It does not exist.
3	It is hk.	5	It is continuous except for $x = 0$.
6	It is continuous except for $x = 0$.		
7	It is continuous except for $x = 0$ as can be readily seen by multiplying and dividing by $2^{-1/x}$.		

Exercise 4.1, page 65

1	8, 8.5	2	27, 24	3	18.76, 17.28
5	$2\pi r$	6	$8\pi r$	7	$3x^2$
9	$12x^2$	10	$6x^2$	11	$x/4$
13	h	14	$r/2$	15	$2\pi h$
18	$\sqrt{x}/2$	19	12, 10		
21	$-1/x^2$, it indicates that the function is decreasing as x increases.				
22	$-2/(x - 1)^2$				

Exercise 4.2, page 70

1	$6x$	2	$8x + 3$	3	$2x - 5$
5	$3x^2 - 2$	6	$6x^2 + 3$	7	$4x - 7$
9	$-3/x^2$	10	$-4/(x + 2)^2$	11	$1/(2x + 1)^2$

13 $-16x/(x^2 + 4)^2$ 14 $-x\sqrt{x}/2$ 15 $-6x/(x^2 - 2)^2$

17 $-8(x^2 + 16)/(x^2 - 16)^2$ 18 $8x/(x^2 + 4)^2$ 19 $4x/(x^2 + 1)^2$

21 $12t^2$ 22 $2u - 3$ 23 $2/\sqrt{w}$

25 $\sqrt{x + 2}/2$ 26 $4t + 1$ 27 $(4 + 2t - t^2)/(t^2 + 4)^2$

29 $-1, 4$ 30 ⅔, ⅔   31 1, ½

33 $-1, -\frac{1}{16}$ 34 $-3, -\frac{1}{3}$ 35 ¾, ⁴⁄₃

37 $-\frac{3}{8}$ 38 $-\frac{2}{3}$ 39 ⅑

Exercise 4.3, page 74

1 (a) True   (b) False   2 It is increasing for $x < 5$ and decreasing for $x > 5$, 25.

5 $x > 4$ 6 $x < 2$ 7 None

9 $x > 4, x < 0$ 10 $-1 < x < 1$ 13 1; $(2, 1)$, $(-2, -1)$

14 $-\frac{4}{9}$ 15 12 17 $\sqrt{x + 2}/2, x > -2$

18 $4x^3$ 19 $V = 2x^3$, 37.5, cu ft per ft

21 128, ft per sec   22 -1.2, lb per sq in.   23 $0 < t < \frac{16}{3}$

Exercise 5.1, page 80

1 $2x + 3$ 2 $3x^2 + 6x$

3 $10x^4 + 14x + 5$ 5 $3x^2 - 2x^{-3}$

6 $10x - 3x^{-4}$ 7 $-4x^{-5} - 3x^{-2}$

9 $6x^2 + \frac{1}{2}\sqrt{x}$ 10 $x^{-2/3}/3 - x^{-3/2}$

11 $(6x^2 + 1)/2\sqrt{2x^3 + x}$ 13 $9\sqrt{4x^3 + 6x}\,(2x^2 + 1) - x^{-2}$

14 $2x(1 - 5x)/\sqrt{1 - 4x} + 6x^{-4}$ 15 $1/\sqrt{2x - 3} + 3/2(3x - 2)^{3/2}$

17 ²⁰⁹⁄₄   18 ⁵¹¹⁄₂

19 ¹⁷⁄₅₄   21 $-1\frac{1}{12}$

22 $-\frac{41}{96}$ 23 $-.23$

25 $0.5, x < 0.5, x > 0.5$ 26 ⅚, $x < ⅚, x > ⅚$

27 $-\frac{3}{4}; x < -\frac{3}{4}; x > -\frac{3}{4}$

29 $1, \frac{4}{3}; 1 < x < \frac{4}{3}; x < 1 \cup x > \frac{4}{3}$

30 $-2, \frac{2}{3}; -2 < x < \frac{2}{3}; x < -2 \cup x > \frac{2}{3}$

31 $-\frac{4}{3}, \frac{1}{2}; -\frac{4}{3} < x < \frac{1}{2}; x < -\frac{4}{3} \cup x > \frac{1}{2}$

Exercise 5.2, page 83

1 $9x^2 - 2x - 6$ 2 $24x^2 - 20x - 3$

3 $8x^3 - 9x^2 + 20x - 9$ 5 $2(x^2 - 3x - 5)/(2x - 3)^2$

6 $(3x^2 + 4x - 21)/(3x + 2)^2$ 7 $(3x^2 + 2x - 10)/(3x + 1)^2$

9 $2x(x^2 + 1)(x^2 + 4x + 11)$ 10 $3(x^2 - 2)^2(7x^2 + 2x - 2)$

11 $2x^2(x + 2)^2(5x - 2)(20x^2 + 19x - 6)$ 13 $(x^2 + 2)(9x^2 + 4x - 6)/(3x + 1)^2$

14 $x^2(x + 1)^2(10x^2 - 14x - 9)/(2x - 3)^2$

15 $(x^2 + 3x + 1)^2(20x^2 + 42x + 23)/(4x + 3)^2$

17 $(8x^2 - 8x + 1)/2\sqrt{x^2 - x}$ 18 $2(9x^2 - 12x + 2)/\sqrt{3x^2 - 4x}$

19 $x(3x^2 - 17)/\sqrt{x^2 - 5}$

21 $3(x^2 + 3x + 3)/(\sqrt{2x + 3})^3$

22 $-25/(x^2 - 5x)^{3/2}$

23 $-16/(3x^2 - 8x)^{3/2}$

25 $14(2x - 3)/(x + 2)^3$

26 $30(3x - 1)^2/(x + 3)^4$

27 $-44(2x + 5)^3/(3x + 2)^5$

29 $-a^2/(x^2 - a^2)^{3/2}$

30 $a^2/x^2\sqrt{x^2 - a^2}$

31 $-a^2/(x^2 - 2ax)^{3/2}$

35 0.42 unit per min

Exercise 5.3, page 87

1 $30x(3x^2 - 2)^4$

2 $4x^3(2x^2 - 1)^3(6x^2 - 1)$

3 $3x^2(x^3 - 3)^2(4x^3 - 3)$

5 $-1/(x - 1)\sqrt{x^2 - 1}$

6 $-1/(2x - 1)^{2/3}(x - 2)^{4/3}$

7 $-13/3(3x + 2)^{2/3}(2x - 3)^{4/3}$

9 $4x(x^2 + 3)$

10 $6(2x - 1)^2$

11 $(2x^2 - 6x - 3)(2x - 3)$

13 $2(x + 2)/(x^2 + 4x + 5)^2$

14 $x(-x^3 + 6x^2 - 6x + 2)/[x^3 + (x - 1)^3]^2$

15 $-(x^4 - 2x^2 + 4)(x^2 + 2)/(x^4 - 6x^2 + 4)^2$

17 $2x + 3\sqrt{x + 1}$

18 $8x(1 - x)(1 + x)(1 + x^2)/(x^4 + 1)^2$

19 $\sqrt{x + 1}\,(7\sqrt[4]{x + 1} + 12)/4$

21 6

22 $-4, 0, 12$

23 It is increasing, 3 units of distance per unit of time.

25 $-\frac{2}{5}$

26 $-a/b$

27 $-x/y, \pm x/\sqrt{a^2 - x^2}$

29 $-b^2x/a^2y, \pm bx/a\sqrt{a^2 - x^2}$

30 $-x/4y, \pm x/2\sqrt{2}\sqrt{4 - x^2}$

31 $p/y, \pm\sqrt{p/2x}$

33 $(2 - x)/y, \pm(2 - x)/\sqrt{4x - x^2}$

34 $-y^{1/2}/x^{1/2}, (x^{1/2} - a^{1/2})/x^{1/2}$

35 $4/y(4 - x)^2, \pm 4\sqrt{16 - x^2}/(4 - x)^2(4 + x)$

37 -2

38 -2

39 2

41 $-4, 1$

42 $1, 0$

43 $-\frac{5}{4}$

45 $\frac{3}{4}$

46 $1\frac{2}{5}$

47 $\frac{1}{2}$

49 -1

50 $\frac{4}{3}$

51 $-\frac{3}{8}$

53 $-\frac{4}{3}$

54 $-\frac{1}{2}$

55 $\frac{7}{3}$

57 $(2x + y)/(8y - x)$

58 $(4 - x)/(y - 12)$

59 $(3 - 2y)/(2x + 5)$

61 $(4x + 3y)/(2y - 3x)$

62 $(ay - x^2)/(y^2 - ax)$

63 $-(3x^2 + 4y^2)/4y(2x + 3y)$

65 $(y^2 - 8xy)/(4x^2 - 2xy + 6y^2)$

66 $x(2x^2 - y^2)/y(x^2 + 4y^2)$

67 $(4 - 8y - 6x)/(8x - 6y - 1)$

69 $(10xy^2 - y - 4x^3)/x(1 - 10xy)$

70 $-2(2 + y^3 + 2x^3)/(1 + 6xy^2)$

71 $(3 + 4y - 2x - 3x^2)/(1 - 4x + 3y^2)$

73 $x^2(x^2 + 12)/2y(x^2 + 4)^2$

74 $32(2 - x^2)/x^3y(x - 2)^2(x + 2)^2$

75 $-4/3y^2(x - 2)^2$

Exercise 6.1, page 93

5 $2(6x + 1)\,dx$

6 $3x(5x - 2)\,dx$

7 $(4x - 5)\,dx$

9 $(3x^2 - 2/x^2)\,dx$

10 $(5 + 6/x^3)\,dx$

11 $(8x^3 + 21/x^4)\,dx$

13 $(9x^2 + 2/\sqrt[3]{x})\,dx$

14 $(-x^{-2} + 2/\sqrt[3]{x})\,dx$

15 $(3x^{-2/5} - 4x^{-3})\,dx$

17 $(2x^2 - 1)/\sqrt{x^2 - 1}\,dx$

18 $3x(32 - 3x^2)/\sqrt{16 - x^2}\,dx$

19 $x^2(4x + 11)\sqrt[3]{x^3 + 3x^2}/(x + 3)\,dx$

21 $x(x^2 + 8)\sqrt{x^2 + 4}/(x^2 + 4)^2\,dx$

22 $(3x + 1)\sqrt{x^2 + x}/2(x^2 + x)^2\,dx$

23 $-18\sqrt{4x^2 - 9}/(4x^2 - 9)^2\,dx$

25 3

26 -2

27 -4

29 $-\frac{1}{4}$

30 $\frac{2}{81}$

31 -3

33 $\frac{1}{16}$ ft per min, $\frac{1}{32}$ ft

34 -4 ft per sec, -0.4 ft

35 -0.6 lb/sq in., 1.2 lb/sq in.

37 $-\frac{3}{4}\,dx$

38 $\frac{4}{3}, 4, -\frac{4}{3}$

39 $-\frac{3}{8}, -1.5, 1.5$

Exercise 6.2, page 100

1 $(3x + 2y)\,dy + (3y - 6)\,dx = 0, 3(2 - y)/(3x + 2y), (3x + 2y)/3(2 - y), y \neq 2$

2 $(2xy + 1)\,dy + (y^2 - 8)\,dx = 0, (8 - y^2)/(2xy + 1), (2xy + 1)/(8 - y^2), y \neq \pm 2\sqrt{2}$

3 $(5x + 6y)\,dy - (4x - 5y)\,dx = 0, (-5y + 4x)/(5x + 6y), (5x + 6y)/(-5y + 4x)$

5 $x(2y + x)\,dy + y(2x + y)\,dx = 0, -y(2x + y)/x(x + 2y), -x(x + 2y)/y(2x + y)$

6 $(x^2 - y)\,dx + (y^2 - x)\,dy = 0, (y - x^2)/(y^2 - x), (y^2 - x)/(y - x^2)$

7 $(2y - 5x^2)\,dy + (6 - 10xy)\,dx = 0, (10xy - 6)/(2y - 5x^2), (2y - 5x^2)/(10xy - 6)$

9 $2(x + 1), 6(x + 1)$ 10 $2x - 5, 2(2x - 5)$ 11 $3, 3(2t - 7)$

13 $(2x^2 - 1)/\sqrt{x^2 - 1}, (2x^2 - 1)t/(\sqrt{x^2 - 1})(\sqrt{t^2 + 2})$

14 $(8x - 3x^3)/\sqrt{4 - x^2}, (8x - 3x^3)/(\sqrt{4 - x^2})(\sqrt{2t - 1})$

15 $2/(x^2 + 2)^{3/2}, (2 - t)/(x^2 + 2)^{3/2}t^2\sqrt{t - 1}$

17 4π sq in.   18 19.2 cu in.   19 32π cu in.

21 $\frac{13}{16}, \frac{3}{4}$ 22 $-\frac{127}{216}, -\frac{1}{2}$ 23 16, 9

25 2.52   26 10.5   27 $\frac{43}{36}$

29 27.54   30 15.36   31 970,000

33 32 sq ft   34 $\frac{1}{150}$ in.   35 $10/\pi$

37 1,000 and greater   39 3 percent, 2 percent   41 $e/4$

42 $r/2$ 43 r

Exercise 7.1, page 106

1 $x^3/3 + C$ 2 $-x^{-2}/2 + C$ 3 $2x^{1/2} + C$

5 $-x^{-2} + C$ 6 $x^6/2 + C$ 7 $x^3 + C$

9 $x^3/3 + 3x^2/2 + 5x + C$ 10 $x^4/4 - x^2/2 - 4x + C$

11 $x^2/2 - 1/x + C$ 13 $(x^2 - 4x + 7)^4/4 + C$

14 $(2x^2 - 3x + 1)^3/3 + C$ 15 $(x^3 - 3x)^6/6 + C$

17 $(x^2 + 4x - 1)^5/10 + C$ 18 $2(x^3 + 6x)^3/3 + C$

19 $(x^4 + 5x^2)^6/12 + C$ 21 $6x^{5/2}/5 + C$

22 $2x^{4.5} + C$ 23 $-(a^2 - x^2)^{3/2}/3 + C$

25 $2x^{5/2}/5 - 4x^{1/2} + C$ 26 $2x^{7/2}/7 - 8x^{3/2}/3 + C$

27 $\sqrt{x^2 - 9} + C$ 29 $-1/x - 2x\sqrt{x} + 1.5x^{4/3} + C$

30 $-3/x - 2\sqrt{x} + 3x^{4/3}/4 + C$ 31 $6x^{11/6}/11 - 2/x^2 + C$

33 $x^3/3 + x^2/2 - 12x + C$ 34 $2x^3/3 + 3x^2/2 - 2x + C$

35 $x^4/2 + 5x^3/3 + 5x^2 + 4x + C$

37 $x^3/3 + x^2 + x + C_1, (x + 1)^3/3 + C_2, C_1 = C_2 - \frac{1}{3}$

38 $2x^4 - 4x^3 + 3x^2 - x + C_1, (2x - 1)^4/8 + C_2, C_1 = C_2 - \frac{1}{8}$

39 The differential $2x\,dx$ of the function that is raised to a power does not occur.

Exercise 7.2, page 111

1 $y = x^2 + 6x + C, C = 1$ 2 $y = x^2 - 7x + C, C = 0$

3 $y = 2x^2 - 4x + C, C = 1$ 5 $y = x^3 - 3x + C, C = 0$

6 $y = x^3 - 6x^2 - 3x + C, C = 32$ 7 $y = 2x^3 - 5x^2 + 7x + C, C = 5$

9 $y^2 - 2y = x^2 + 2x + 2C, C = -2$ 10 $x^2 + y^2 - 4x - 8y + 2C = 0, C = 2$

11 $x^2 + 2y^2 - x - 3y + C = 0, C = -15$

13 $y = t^2 + 3t$ 14 $y = (t + 1)^3/3 + 8$

15 $y = -1/(t + 2) + 3\frac{1}{2}$ 17 $3y = t^3, v = t^2$

18 $v = 2t^{1/2} + 3, 3y = 4t^{3/2} + 9t + 18$

19 $y = 4(t + 1)^{5/2} - 3t + 5, v = 10(t + 1)^{3/2} - 3$

21 $v = -32t + 80, y = -16t^2 + 80t + 50$

22 $y = -16t^2 - 60t + 700, v = -32t - 60$

23 $v = -32t, y = -16t^2 + h$

Exercise 8.1, page 117

5 $n(n + 1)(2n + 1)$ 6 $n(n + 1)(2n^2 + 4n + 1)$

7 $2n(n + 1)(2n - 1)$ 9 $2n(n + 1)(n - 1)$

10 $2n(n + 1)(n + 3)$ 11 $3(n - 1)n(n + 1)(n + 2)$

13 $\frac{15}{4}$ 14 33 15 99 17 $\frac{77}{60}$

18 $\frac{433}{168}$ 19 $\frac{869}{1,800}$ 21 $\frac{8}{3}$ 22 30

23 $\frac{255}{4}$ 25 $\frac{496}{3}$ 26 5 27 3

Exercise 8.2, page 122

1 15 2 9 3 16 5 $-\frac{32}{15}$

6 33 7 $-\frac{4}{3}$ 9 $\frac{348}{5}$ 10 $\frac{58}{7}$

11 -78 13 $\frac{19}{3}$ 14 $\frac{33}{5}$ 15 48

Exercise 8.3, page 128

1 6 2 48 3 19.5 5 12 6 $\frac{155}{6}$ 7 $\frac{125}{6}$

9 28 10 12 11 32 13 $\frac{97}{2}$ 14 2 15 $2\frac{2}{3}$

17 $\frac{52}{3}$ 18 $\frac{38}{3}$ 19 $\frac{112}{9}$ 21 $\pi a^2/4$ 22 $\pi ab/8$ 23 36

Exercise 8.4, page 131

1	$64/3$	2	19.5	3	6	5	$4/3$	6	4.5	7	36		
9	10	10	22.5	11	24	13	36	14	$125/6$	15	36		
17	$64/3$	18	$4/3$	19	108	21	$1/24$	22	$32/3$	23	3		
25	$71/6$	26	$2,483/12$	27	8	29	π	30	π	31	4π		
33	$22/3$	34	$32/3$	35	$71/6$								

Exercise 8.5, page 136

1	½	2	Divergent	3	⅛	5	$-1/3$
6	1	7	Divergent	9	zero	10	zero
11	Divergent	13	Divergent	14	2	15	Divergent
17	$-3/2$	18	Divergent	19	-20	21	Divergent
22	Divergent	23	zero	25	Divergent	26	Divergent
27	Divergent	29	1	30	⅛	31	Divergent

Exercise 9.1, page 141

1 $(y - 3)^2 = -8(x - 4)$

2 $(y - 2)^2 = -16(x - 6)$

3 $(x - 5)^2 = -8(y - 5)$

5 $(x - 7)^2 = -16(y - 1)$

6 $(y + 2)^2 = -32(x - 3)$

7 $(y - 4)^2 = -16(x + 3)$

9 $y^2 = -8(x - 7)$

10 $(x - 4)^2 = 16(y + 3)$

11 $(x + 2)^2 = -4(y - 4)$

13 $(y - 1)^2 = -8(x - 3)$

14 $(x - 2)^2 = 8(y - 6)$

15 $(x - 4)^2 = 16(y + 2)$

17 $(y - 1)^2 = 4(x - 2)$

18 $(y + 3)^2 = -4(x - 1)$

19 $(x - 2)^2 = 8(y - 3)$

21 $(y - 5)^2 = 8(x - 4)$

22 $(y - 1)^2 = 16(x - 3)$

23 $x^2 = 16(y - 1)$

25 2.25 in. from the vertex

26 $10\frac{2}{3}$ in.

Exercise 9.2, page 145

1 $C(1, 2)$, $F_1(4, 2)$, $F_2(-2, 2)$, $V_1(6, 2)$, $V_2(-4, 2)$, ends of focal chords at $(-2, 2 \pm 16/5)$ and $(4, 2 \pm 16/5)$

2 $C(-3, 0)$, $F_1(9, 0)$, $F_2(-15, 0)$, $V_1(10, 0)$, $V_2(-16, 0)$, ends of focal chords at $(9, \pm 25/13)$ and $(-15, \pm 25/13)$

3 $C(-1, 2)$, $F_1(-1, 6)$, $F_2(-1, -2)$, $V_1(-1, 7)$, $V_2(-1, -3)$, ends of focal chords at $(-1 \pm 16/5, 6)$ and $(-1 \pm 16/5, -2)$

5 $(x - 2)^2/5^2 + (y - 4)^2/3^2 = 1$

6 $(x - 4)^2/5^2 + (y + 1)^2/4^2 = 1$

7 $(y + 3)^2/13^2 + (x - 1)^2/12^2 = 1$

9 $(x + 3)^2/5^2 + (y + 1)^2/3^2 = 1$

10 $(y - 2)^2/17^2 + x^2/15^2 = 1$

11 $(y + 2)^2/13^2 + (x - 2)^2/12^2 = 1$

13 $(x - 1)^2/5^2 + (y + 1)^2/2^2 = 1$

14 $(y - 3)^2/7^2 + (x - 2)^2/6^2 = 1$

15 $(y + 4)^2/9^2 + (x - 3)^2/8^2 = 1$

17 $(y - 1)^2/5^2 + (x + 3)^2/5^2 = 1$

18 $(y + 3)^2/5^2 + (x - 2)^2/2^2 = 1$

19 $(x - 4)^2/4^2 + (y - 3)^2/3^2 = 1$

21 $25x^2 + 16y^2 - 100x - 64y - 236 = 0$

22   $169x^2 + 144y^2 - 1{,}014x - 576y - 22{,}239 = 0$
23   $40x^2 - 30xy + 24y^2 - 190x - 284y + 1{,}861 = 0$
25   $4x^2 + 3y^2 + 24y = 0$ 26 $5x^2 + 9y^2 - 108x + 324 = 0$
27   $x^2 + 4y^2 = 256$

Exercise 9.3, page 150

1   $y = 2 \pm 4(x - 1)/3$, $C(1, 2)$, $F_1(6, 2)$, $F_2(-4, 2)$, $V_1(4, 2)$, $V_2(-2, 2)$, ends of focal chords are at $(6, 2 \pm 16\!/\!3)$ and $(-4, 2 \pm 16\!/\!3)$
2   $y = \pm 5(x + 3)/12$, $C(-3, 0)$, $F_1(-16, 0)$, $F_2(10, 0)$, $V_1(-15, 0)$, $V_2(9, 0)$, end of focal chords are at $(-16, \pm 25\!/\!12)$ and $(10, \pm 25\!/\!12)$
3   $y = 2 \pm 4(x + 1)/3$, $C(-1, 2)$, $F_1(-1, 7)$, $F_2(-1, -3)$, $V_1(-1, 6)$, $V_2(-1, -2)$, ends of focal chords are at $(-1 \pm 9\!/\!4, 7)$ and $(-1 \pm 9\!/\!4, -3)$
5   $(x - 2)^2/4^2 - (y - 4)^2/3^2 = 1$ 6 $(x - 4)^2/3^2 - (y + 1)^2/4^2 = 1$
7   $(y + 3)^2/5^2 - (x - 1)^2/12^2 = 1$ 9 $(x + 3)^2/4^2 - (y + 1)^2/3^2 = 1$
10   $(y - 2)^2/8^2 - x^2/15^2 = 1$ 11 $(y + 2)^2/12^2 - (x - 2)^2/5^2 = 1$
13   $(x - 1)^2/21 - (y + 1)^2/2^2 = 1$ 14 $(y - 3)^2/13 - (x - 6)^2/6^2 = 1$
15   $(y + 4)^2/17 - (x - 3)^2/8^2 = 1$ 17 $(y - 1)^2/5^2 - (x + 3)^2/24 = 1$
18   $(y + 3)^2/21 - (x - 2)^2/2^2 = 1$ 19 $(x - 4)^2/4^2 - (y - 3)^2/3^2 = 1$
21   $y^2/3^2 - (x - 3)^2/7 = 1$ 22 $(y - 1)^2/2.5^2 - (x - 1)^2/2.75 = 1$
23   $48x^2 + 48xy + 28y^2 - 216x + 304y + 391 = 0$
25   $x^2 - 3y^2 - 4x + 14y + 9 = 0$ 26 $1.25x^2 + 12x - y^2 - 36 = 0$
27   $8x^2 - y^2 + 44x + 10y - 5 = 0$

Exercise 9.4, page 153

1   $(y - 2)^2 = 4(2)(x + 1)$ 2 $(y + 1)^2 = 4(-3)(x - 1)$
3   $(x + 3)^2 = 4(-1)(y - 2)$ 5 $(x + 1)^2/5^2 + (y + 3)^2/3^2 = 1$
6   $(x + 2)^2/13^2 + (y - 4)^2/5^2 = 1$ 7 $(y - 3)^2/5^2 + (x + 1)^2/3^2 = 1$
9   $(y + 2)^2/4^2 - (x - 1)^2/3^2 = 1$ 10 $(y + 1)^2/12^2 - (x - 1)^2/5^2 = 1$
11   $(x - 5)^2/3^2 - (y + 4)^2/4^2 = 1$ 13 1, parabola, $(2, 0)$
14   $1/\sqrt{2}$, ellipse, $(4, 0)$ 15 $\sqrt{2}$, hyperbola, $(6, 0)$
17   $\frac{1}{2}$, ellipse, $(-2, 0)$ 18 $\sqrt{15}/3$, hyperbola, $(3, 0)$
19   1, parabola, $(0, 1)$ 21 $\sqrt{7}/2$, hyperbola, $(0, 6)$
22   1, parabola, $(0, 10)$ 23 $\sqrt{5}/3$, ellipse, $(0, -1)$
25   13 26 -7 27 $F > 8$

Exercise 9.5, page 156

1   $x^2 + y^2 = 9$ 2 $4x^2 + y^2 = 4$ 3 $y^2 = 8x$
5   $4x^2 + 9y^2 = 36$ 6   Cannot be done   7 $4x^2 - y^2 = 4$
9   Cannot be done   10 $9x^2 - y^2 = 9$ 11 $4x^2 + 9y^2 = 36$
13   Cannot be done   14 $(4, -1)$ 15 $ax + ay + c = 0$
17   $ax + by = 0$ 18 $y = mx$
19   $x^2 + y^2 - 2(h - k)(x + y) = r^2$

Exercise 9.6, page 161

1 $7x'^2 - 3y'^2 = 6$ 2 $5x'^2 - 7y'^2 + 10 = 0$ 3 $20x'^2 - 5y'^2 - 9 = 0$
5 Hyperbola   6 Parabola   7 Hyperbola
9 Ellipse   10 Parabola   11 Ellipse
13 ½, $\sqrt{3}/2$ 14 $1/\sqrt{2}, 1/\sqrt{2}$ 15 $1/5\sqrt{2}, 7/5\sqrt{2}$
21 $9x'^2 + 4y'^2 = 36$ 22 $9x'^2 - 4y'^2 = 36$
23 $2x'^2 + 2x' + y'^2 + y' - 1 = 0, 8x''^2 + 4y''^2 = 7$

Exercise 10.1, page 166

1 $y = 12x - 19, x + 12y = 62$ 2 $y = 4x - 16, x + 4y + 30 = 0$
3 $y = 2, x = 1$ 5 $3x + 8y = 25, 8x - 3y = 18$
6 $9x - 2y + 22 = 0, 2x + 9y = 14$ 7 $8x - 3y = 30, 3x + 8y = 7$
9 $2x + 3y = 12, 3x - 2y = 5$ 10 $y = 4, x = 2$
11 $20x + 9y = 8, 9x - 20y = {}^{107}\!/_3$ 13 2
14 ⅝, 5/7   15 7, ⅐   17 ⅗
18 $^{41}\!/_{12}$ 19 3   21 $^{11}\!/_{17}, {}^{44}\!/_{323}$
22 $^{35}\!/_{47}, {}^{140}\!/_{43}$ 23 $^{54}\!/_{361}, {}^{54}\!/_{49}$ 25 $^{18}\!/_{31}$
26 6/7   27 $^{57}\!/_{26}$ 29 $5x - y = 9$
30 $6x - y = 30$ 31 $3x + y = -10$ 37 $3x - 5y \mp 25 = 0$
38 ⅝, -15 41 $a = 1, b = -2, c = 4$
42 $a = 2, b = -6, c = 5$ 43 $a = 1, b = -2, c = -4, d = -5$

Exercise 10.2, page 172

1 (a) 56 ft per sec, -8 ft per sec; (b) 225 ft; (c) 7.5 sec
2 (a) 96 ft per sec, 32 ft per sec; (b) 400 ft; (c) 10 sec
3 $80\sqrt{6}$ ft per sec   5 0, -2, -2, 0, 4, 10; -3, -1, 1, 3, 5, 7
6 6, 2, 3, 11, 18; -2, 0, 1, 3, 4   7 4, 8, $6\sqrt{3} - 2$; 3; 1, 3 $- 2\sqrt{3}$
10 8.9 ft per sec   11 0.8 mph   13 32 sq in. per min
14 36π sq in. per min   15 150 cc per sec
17 3.6 ft per sec   18 $^{64}\!/_{100}\pi$ in. per min
19 24π cu ft per min   21 ⅝ ft per sec
22 $(17 - 9\sqrt{2})100/\sqrt{13 - 6\sqrt{2}}$ 23 24 knots
25 $^1\!/_{12}\pi$ ft per min   26 Decreasing at 4.2 lb per sq ft per min
27 $^{150}\!/_{17}$ ft per min

Exercise 10.3, page 181

1 max., 3   2 min., -2
3 min., $-b/2a$ for $a > 0$; max., $-b/2a$ for $a < 0$
5 max., 0; min., 4   6 max., 1; min., 3
7 max., -2; min., 0   9 max., 1; min., 2

10	max., -1; min., 3	11	No max. or min.
13	min., 3	14	min., -1
15	min., -2, 1; max., 0	17	min., 2
18	max., -2; min., 2	19	min., -2
21	max., 0	22	max., 0
23	max., 1	25	max., 4
26	max., -4; min., 0	27	min., -2; max., 2
29	max., 0	30	min., -3
31	min., 0	33	min., -3
34	max., -2; min., 8	35	max., 2; min., 6

37 max., $-2\sqrt{2}$; min., $2\sqrt{2}$; min., -4; max., 4

38 min., -3; max., 3; max., $-3\sqrt{2}$; min., $3\sqrt{2}$

39 max., -4; min., 0; max., 4; max., $-2\sqrt{6}$; min., $2\sqrt{6}$

41	min., 4	42	min., ½
43	max., 0	45	min., 2; max., 1; max, 4
46	min., 3; max., 2; max., 5	47	min., -1; max., 3

Exercise 10.4, page 185

1	12.5	2	12.5	3	$^{50}\!/_{3}$, $^{25}\!/_{3}$
5	110 ft on each side	6	110 ft, 220 ft	7	2,250 cc
9	2	10	$h = 2r$	11	$h = r$
13	base $= (2\sqrt{3}/3)$ altitude	14	Midpoint	15	The radius

17 Radius $= \sqrt{6}r/3$; height $= 2\sqrt{3}r/3$ 18 Altitude $= h/3$, radius $= 2a/3$

19	$a = 2, b = -4$	22	$h = 4r$	23	12 in.

25 Height of the rectangle equal to $(4 + \pi)/4$ times the radius of the semicircle

26	$(6 + \sqrt{3})P/33$	27	6 in.

29 Height of the cylinder equal to the diameter of the hemisphere

30	$3\sqrt[4]{108}$ in.	31	12 in.	33	170	34	100
35	$3.25			37	$3s/40 + 120/s$, 40 mph		
38	$9x^2 + 16y^2 = 1,152$			41	$4/\pi$		

42 $(9, 8)$ is nearest, $(-15, 0)$ is farthest   43 Shortest distance is $\sqrt{261}/2$, longest is 22

45 $\pi L/(4 + \pi)$ for the circle, $4L/(4 + \pi)$ for the square; L for the circle, zero for the square

46	$y = 2x + 5$	47	$8x - 3y + 18 = 0$

Exercise 11.1, page 192

1	$4x - 3, 4$	2	$-2x + 1, -2$
3	$12x^2 + 10x, 24x + 10$	5	$2 + 1/x^2, -2/x^3$
6	$2x + 4/x^3, 2 - 12/x^4$	7	$3x^2 + 3/x^4, 6x - 12/x^4$
9	$1/(x + 1)^2, -2/(x + 1)^3$	10	$7/(2x + 1)^2, -28/(2x + 1)^3$

11 $(x^2 + 1)/(x^2 - 1)^2, 2x(x^2 + 3)/(x^2 - 1)^3$

13 $x/\sqrt{x^2 + 1}, 1/(x^2 + 1)^{3/2}$ 14 $(1 - x)/\sqrt{2x - x^2}, -1/(2x - x^2)^{3/2}$

15 $(x^2 + 1)/(x^3 + 3x)^{2/3}, 2(x^2 - 1)/(x^3 + 3x)^{5/3}$

17	$-p/2xy$	18	$-r^2/y^3$	19	$-b^4/a^2y^3$

21 $2(y - x - 8)/(x - 8)^2$ 22 $2(y - 1)/(1 + x)^2$ 23 $-16/(y + 4)^3$
25 $a^{1/2}/2x^{3/2}$ 26 $a^{2/3}/3x^{4/3}y^{1/3}$ 27 $a^{1/2}/2x^{3/2}$
29 $1, -\frac{1}{4}$ 30 $\frac{5}{12}, -\frac{169}{1,728}$ 31 0, 0
33 $-3r^2x/y^5$ 34 $-3b^6x/a^4y^5$ 35 $3b^6x/a^4y^5$
37 $(-1)^n 6(n!)x^{-n-1}$ 38 $(-1)^n(n + 1)!x^{-n-2}$
39 $[m!/(m - n)!]x^{m-n}, m > n; n!, m = n; 0, m < n$
42 $uv'' + 2u'v' + u''v$ 43 $uv''' + 3u'v'' + 3u''v' + u'''v$

Exercise 11.2, page 197

1 max., -2; min., 2; infl., 0 2 max., $\frac{1}{3}$; min., 0; infl., $\frac{1}{6}$
3 max., -1; min., 3; infl., 1 5 max., 2; min., 4; infl., 3
6 infl., 1 7 max., 1; min., 3; infl., 2
9 min., 0 10 max., ± 2; min., 0; infl., $\pm 2/\sqrt{3}$
11 max., 1; infl., 0, 2; min., $1 \pm \sqrt{3}$ 13 min., $-\frac{4}{3}$; infl., -2
14 max., -3; min., 3; infl., 0 15 max., -2; min., 2; infl., 0
17 max., 0; infl., ± 1
18 max., 0 for $a > 0$; min., 0 for $a < 0$; infl., $\pm 2a/\sqrt{3}$
19 infl., $3\sqrt[3]{2}$, 0, -3 21 min., 0; infl., ± 2
22 max., 0; infl., ± 1 23 min., 0; infl., $\pm 2\sqrt{2}$
25 min., -3; infl., -5.5 26 max., $\frac{5}{3}$; infl., 3
27 max., $-\sqrt{3}$; min., $\sqrt{3}$; infl., 0, ± 3 31 $b = 0$
33 $a = 1, b = -1, c = -5, d = 2$ 37 6 ft, -4 ft per min, 2 ft per min per min
38 28 ft, 24 ft per min, 6 ft per min per min
39 128 ft, 32 ft per sec, -28 ft per sec per sec, $\frac{10}{3} < t < 10$
41 0.3 ft per sec, $-\frac{9}{500}$ ft per sec per sec 42 $\frac{1}{6}$ ft per sec, $-\frac{1}{216}$ ft per sec per sec

Exercise 12.1, page 205

9 $(-1 - \sqrt{3})/4$ 10 $-2/\sqrt{3}$ 11 $(2\sqrt{2} - 1)/2$
17 $\cos A = \frac{12}{13}, \tan A = \frac{5}{12}, \cot A = \frac{12}{5}, \sec A = \frac{13}{12}, \csc A = \frac{13}{5}$
18 $\sin A = -\frac{4}{5}, \tan A = -\frac{4}{3}, \cot A = -\frac{3}{4}, \sec A = \frac{5}{3}, \csc A = -\frac{5}{4}$
19 $\sin A = \frac{4}{5}, \cos A = -\frac{3}{5}, \cot A = -\frac{3}{4}, \sec A = -\frac{5}{3}, \csc A = \frac{5}{4}$
21 $\frac{7}{25}, \frac{24}{25}, -\frac{24}{25}, 3$ 22 $-\frac{63}{65}, -\frac{56}{65}, \frac{120}{169}, \frac{2}{3}$
23 $\frac{240}{289}, -1, -\frac{240}{289}, -\frac{1}{4}$ 33 $\pi/3, \pi/2, 5\pi/3$
34 $\pi/6, 11\pi/16, \pi/12, 13\pi/12, 7\pi/12, 19\pi/12$
35 $0, \pi$

Exercise 12.2, page 211

1 1 2 -2 3 -1
5 -2 6 $\frac{1}{2}$ 7 2
9 0 10 $\frac{3}{5}$ 11 $\frac{1}{2}$
13 ∞ 14 1 15 1

17 $-12 \sin 4x$ **18** $10 \cos 5x$ **19** $\cos x + 2 \sin 2x$

21 $6\pi \sin^2 \pi x \cos \pi x$ **22** $-4 \sin 2x \cos 2x$ **23** 0

25 $-8x \sin 2x + 4 \cos 2x$ **26** $3\pi x \cos \pi x + 3 \sin \pi x$

27 $2x \cos x (\cos x - x \sin x)$ **29** $2/(\cos 2x - 1)$

30 $-2 \cos x/(1 + \sin x)^2$ **31** $1/(1 + \cos x)$

33 $10k^2 \cos 2kx$ **34** $-4a^2 \cos 2ax$

35 $-2(\sin x + 2 \cos 2x)$ **37** $-x \sin x + 2 \cos x$

38 $[(2 - x^2) \cos x + 2x \sin x]/x^3$ **39** $\sin x/(1 - \cos x)^2$

41 Arctan $2\sqrt{2}$ at each intersection

42 Arctan $3\sqrt{3}$ for $\pi/6$ and for $x = 5\pi/6$, zero for $x = 3\pi/2$

43 $\pi/3$ at each intersection

45 min. $x = 0, \pi$; max. $x = \pi/2, 3\pi/2$; infl. $x = \pi/4, 3\pi/4, 5\pi/4, 7\pi/4$

46 min. $x = \pi/2, 3\pi/2$; max. $x = 0, \pi$; infl. $x = \pi/4, 3\pi/4, 5\pi/4, 7\pi/4$

47 min. $x = 3\pi/2$; max. $x = \pi/2$; infl. $x = 0, \pi$, Arcsin $(\sqrt{6}/3)$, π + Arcsin $(\sqrt{6}/3)$

49 min. $x = 5\pi/4$; max. $x = \pi/4$, infl. $x = 3\pi/4, 7\pi/4$

50 min. $x = 5\pi/6$; max. $x = \pi/6$; infl. $x = \pi/2, 3\pi/2$, π + Arcsin ¼, 2π − Arcsin ¼

51 min. $x = 7\pi/6$; max. $x = \pi/2, 11\pi/6$; infl. $x = \pi/12, 5\pi/12, 13\pi/12, 17\pi/12$

53 $[-4, 4]$ **55** $\pi/4$ **57** $4\pi a^3 \sqrt{3}/9$ **58** $4a/3$

59 $4a/3$

Exercise 12.3, page 214

1 $3 \sec^2 3x$ **2** $-5 \csc^2 5x$ **3** $2 \sec 2x \tan 2x$

5 $-3x^2 \csc (x^3 - 1) \cot (x^3 - 1)$ **6** $(2x - 3) \sec^2 (2x - 3)$

7 $-(2x - 1) \csc^2 (x^2 - x)$ **9** $6x \sec^3 (x^2 - 3) \tan (x^2 - 3)$

10 $[-(2x + 4)/\sqrt{x^2 + 4x}] \csc^2 \sqrt{x^2 + 4x} \cot \sqrt{x^2 + 4x}$

11 $[2(x - 1)/\sqrt{x^2 - 2x}] \tan \sqrt{x^2 - 2x} \sec^2 \sqrt{x^2 - 2x}$

13 $[-(4x^3 - 1)/\sqrt{x^4 - x}] \cot \sqrt{x^4 - x} \csc^2 \sqrt{x^4 - x} + 2$

14 $9x^2 \sec^3 (x^3 + 3) \tan (x^3 + 3) + 1$ **15** $-6 \csc^2 3x \cot 3x + 5$

17 $2x(2x - 1) \tan (x^2 - x) \sec^2 (x^2 - x) + \tan^2 (x^2 - x)$

18 $2x \cot^2 (2x - 5)[\cot (2x - 5) - 3x \csc^2 (2x - 5)]$

19 $(2x \sec^2 \sqrt{x^2 + 6x}/\sqrt{x^2 + 6x})[\sqrt{x^2 + 6x} + x(x + 3) \tan \sqrt{x^2 + 6x}]$

21 $-\csc^4 x/(1 - \cot^2 x)^2$ **22** $2 \sec x/(\sec x - \tan x)^2$

23 $6 \sec^2 x/(4 \tan x + 1)^2$ **25** $\sec^2 (x/2) \tan (x/2)$

26 $2 \sec^2 (x/2)[\sec^2 (x/2) + 2 \tan^2 (x/2)]$ **27** $8 \csc^2 2x \cot 2x$

29 $2 \sec^2 x (\sec^2 x + 2 \tan^2 x)$ **30** $4 \sec 2x (\sec^2 2x + \tan^2 2x)$

31 $12 \csc^3 2x (\csc^2 2x + 3 \cot^2 2x)$

33 Arctan ⅓ for $x = 0$, Arctan 3 for $x = \pi$, Arctan ¾ for $x = \pi/3, 5\pi/3$

34 Arctan ¹²⁄₃₁ for $x = \pi/6, 7\pi/6, 5\pi/6, 11\pi/6$

35 Arctan 3 for $x = \pi/4, 3\pi/4$

Exercise 12.4, page 220

1 $1/\sqrt{5}$ **2** 0.8 **3** $-$²⁴⁄₇   **5** Zero

6 ¹³⁄₉   **7** 1 **9** ⅕   **10** ⅙, -1

11 $\pm 1/\sqrt{5 - 2\sqrt{3}}$ 13 $-1/\sqrt{4 - x^2}$ 14 $3/\sqrt{1 - 9x^2}$

15 $4/(4 + x^2)$ 17 $2/x\sqrt{x^4 - 1}$ 18 $1/\sqrt{1 - x^2}$

19 $1/\sqrt{a^2 - x^2}$ 21 $1/x\sqrt{2x - 1}$ 22 $x/|x|(1 + x^2)$

23 $1/\sqrt{1 - x^2}$ 25 $1/(16 + x^2)$ 26 $-1/(1 + x^2)$

27 $-1/|x + 1|\sqrt{2x + 1}$ 29 $\sqrt{a - x}/\sqrt{a + x}$ 30 $\sqrt{x^2 - 4}/x^2$

31 $2\sqrt{1 - x^2}$ 33 $x^2/(1 + x^2) + 2x \text{ Arctan } x$

34 $\sqrt{x^2 - 4}/x$ 35 $-4x^3/\sqrt{1 - x^2} + 8 \text{ Arccos } x$

37 $x/(4 - x^2)^{3/2}$ 38 $375x/(1 - 25x^2)^{-3/2}$

39 $-16x/(1 + 4x^2)^2$ 41 $-2x/(1 + x^2)^2$

42 $-x/(1 - x^2)^{3/2}$ 43 $2(1 - 3x^4)/x^2(x^4 - 1)^{3/2}$

45 $(0, 0)$ 47 $\pi/4$

49 $-\tfrac{3}{25}$ radian per min

50 $198/(1 + 198^2 t^2)$ radians per hr, t in hours

51 $\sqrt{14}$ ft

Exercise 13.1, page 228

1 x 2 $1/x$ 3 $1/x$ 5 $2x^2$

6 x^4 7 xe^x 9 x 10 $x^2 - x$

11 $x^2 + \ln x$ 17 $3\tfrac{9}{16}$ 18 $28\tfrac{1}{8}$ 19 $\tfrac{5}{8}$

21 $\tfrac{1}{8}$ 22 $\tfrac{8}{9}$ 23 $64\tfrac{2}{27}$ 29 5.00

30 3.55 31 0.79

33 $A = 0.6, k = 1.76$ 34 $A = 0.15, k = -0.520$

35 $A = 0.4, k = -1.6479$ 45 $3.6217, 7.9621$

46 $8.3754 - 10, 3.9801 - 10$ 47 $2.1541, 8.6377$

49 $9.25, 0.0662$ 50 $6.83, 0.286$ 51 $75.1, 0.0228$

Exercise 13.2, page 233

1 $2/x$ 2 $3/x$ 3 $1/2x$

5 $(x + 1)/(x^2 + 2x + 3)$ 6 $(x - 2)/x(x - 4)$ 7 $(2x - 3)/(3x^2 - 9x + 1)$

9 $\dfrac{2x - 1}{x^2 - x + 1} \log_3 e$ 10 $\dfrac{2x + 1}{x^2 + x - 1} \log_5 e$ 11 $\dfrac{3}{3x - 4} \log_{10} e$

13 $\cot x$ 14 $2 \tan x$ 15 $\sec x \csc x$

17 $3(1 + \ln x)$ 18 $x^2(1 + 3 \ln x)$ 19 $2x(1 + 2 \ln x)$

21 $x \tan x + \ln \sec x$ 22 $x(-x \sec x \csc x + 2 \ln \cot x)$

23 $2(x \cot x + \ln \sin x)$ 25 $1/\sqrt{x^2 + a^2}$ 26 $(2x^2 + a^2)/x(x^2 + a^2)$

27 $(9x + 5)/x(6x + 5)$ 29 $8x/(4 + x^2)(4 - x^2)$ 30 $\csc x$

31 $6x^2/(1 + x^3)(1 - x^3)$ 33 $-12e^{-3x}$ 34 $-2xe^{-x^2/2}$

35 $\cos x \, e^{\sin x}$ 37 $2^{x^2} x \ln 2$ 38 $5^{5(x+1)} \ln 5$

39 $(2x \ln 3)3^{x^2+2}$ 41 $xe^{-x/2}(4 - x)/2$ 42 $(4 - x)e^{-x/4}$

43 $e^{-x}(1 - x)$ 45 $(1 - x \ln x)/xe^x$

46 $(2 \sin 2x + \cos 2x)/2e^x$ 47 $[\cos (x/2) - 4 \sin (x/2)]/2e^{2x}$

49 $e^x(x \ln x - 1)/x \ln^2 x$ 50 1 51 $2xe^{x^2-2}(x^2 - 1)$

53 $e^{\cos x}(\cos x - \sin^2 x)$ 54 $e^{\tan x}(\tan x + \sec^2 x \ln \sec x)$

55   $e^{\sin x}[\sec x + \cos x \ln (\sec x + \tan x)]$ 66 $-3, 2$
67   $1, 2$ 70 $-\tfrac{19}{105}$ 71 $1.44e^{0.6}$
73   $x^{\sin x - 1}(\sin x + x \cos x \ln x)$ 74 $x^{\tan x - 1}(\tan x + x \ln x \sec^2 x)$
75   $(\sin x)^{x-1}(x \cos x + \sin x \ln \sin x)$ 76 $(\tan x)^{x-1}(x \sec^2 x + \tan x \ln \tan x)$
77   $x^{\sqrt{x}}(2 + \ln x)/2\sqrt{x}$ 78 $(\ln x)^{x-1}(1 + \ln x \ln \ln x)$
79   $y \sin x \, (1 - \ln \sec x)$

Exercise 13.3, page 237

21   $\cosh x = \tfrac{17}{15}$, $\tanh x = \tfrac{8}{17}$, $\coth x = \tfrac{17}{8}$, $\operatorname{sech} x = \tfrac{15}{17}$, $\operatorname{csch} x = \tfrac{15}{8}$
22   $\sinh x = \pm\tfrac{3}{4}$, $\tanh x = \pm\tfrac{3}{5}$, $\coth x = \pm\tfrac{5}{3}$, $\operatorname{sech} x = \tfrac{4}{5}$, $\operatorname{csch} x = \pm\tfrac{4}{3}$
23   $\sinh x = -\tfrac{3}{4}$, $\cosh x = \tfrac{5}{4}$, $\coth x = -\tfrac{5}{3}$, $\operatorname{sech} x = \tfrac{4}{5}$, $\operatorname{csch} x = -\tfrac{4}{3}$
25   $\sinh x = \pm\tfrac{12}{5}$, $\cosh x = \tfrac{13}{5}$, $\tanh x = \pm\tfrac{12}{13}$, $\coth x = \pm\tfrac{13}{12}$, $\operatorname{csch} x = \pm\tfrac{5}{12}$
26   $\sinh x = \tfrac{3}{4}$, $\cosh x = \tfrac{5}{4}$, $\tanh x = \tfrac{3}{5}$, $\coth x = \tfrac{5}{3}$, $\operatorname{csch} x = \tfrac{4}{3}$
27   $\sinh x = \pm\tfrac{24}{7}$, $\tanh x = \pm\tfrac{24}{25}$, $\coth x = \pm\tfrac{25}{24}$, $\operatorname{sech} x = \tfrac{7}{25}$, $\operatorname{csch} x = \pm\tfrac{7}{24}$

Exercise 13.4, page 242

1   0.9   2   0.09   3   2.5   5   2.2
6   1.6   7   1.8   9   1.63   10   2.94
11   2.37   13 $2 \cosh 2x$ 14 $3 \sinh 3x$ 15 $\tfrac{1}{2}\operatorname{sech}^2 (x/2)$
17   $2 \tanh x \operatorname{sech}^2 x$ 18 $3 \sinh^2 x \cosh x$ 19 $2 \sinh 4x$
21   $\tanh x$ 22 $\operatorname{sech} x \operatorname{csch} x$ 23 $\coth x$
25   $e^x(2 \cosh 2x + \sinh 2x)$ 26 $e^x(2 \sinh 2x + \cosh 2x)$
27   $-e^{-2x} \operatorname{sech} x \, (\tanh x + 2)$ 29 $2/\sqrt{4x^2 + 1}$
30   $2/\sqrt{4x^2 - 1}$ 31 $3/(1 - 9x^2)$
33   $2 \sec^2 x/\sqrt{\sec^2 x + 1}$ 34 $-\csc x$ 35 $\sec x$

Exercise 14.1, page 247

1   1.5   2   1   3   4   5 $(-5 \pm 2\sqrt{7})/3$
6   $-1 \pm \sqrt{3}$ 7 $\tfrac{1}{4}$ 9   3   10   4
11   $(3 \pm \sqrt{183})/6$ 13   1   14 $(-1 \pm \sqrt{7})/3$ 15 $-1 + \sqrt{3}$
17   Not continuous for $x = 1$ 18   Not continuous for $x = \tfrac{2}{3}$ 19 $f'(3)$ does not exist
21   $f'(0)$ does not exist   22 $f'(-3)$ does not exist   23 $f'(\pm 2)$ do not exist
25   5.1   26 $\tfrac{82}{27}$ 27   0.0099

Exercise 14.2, page 249

1   $\tfrac{1}{3}$ 2 $\tfrac{5}{3}$ 3 ∞ 5 -6 6   0   7 $\tfrac{3}{4}$
9   $-\tfrac{1}{4}$ 10   1   11 $\tfrac{1}{6}$ 13   1   14 $\tfrac{1}{2}$ 15   0
17   2   18   4   19   2   21 $\tfrac{1}{2}$ 22   0   23 $\ln 4$
25   0   26 ∞ 27 $\tfrac{7}{2}$ 29 $\tfrac{1}{3}$ 30 $-\infty$ 31   9
33   0   34   0   35   0

Exercise 14.3, page 253

1	∞	2	0	3	∞	5	0	6	∞	7 0
9	1	10	$-2/\pi$	11	0	13	1	14	1	15 e^2
17	1	18	1	19	1	21	e	22	e^2	23 e^{-15}

Exercise 15.1, page 259

5 $(5, -320°), (-5, 220°), (-5, -140°)$
6 $(-4, -300°), (4, -120°), (4, 240°)$
7 $(3, 330°), (-3, 150°), (-3, -210°)$
9 $(\sqrt{2}, 45°), (\sqrt{2}, -45°)$
10 $(2, 60°), (2, -60°)$
11 $(2, 120°), (2, -120°)$

Exercise 15.3, page 268

1 $r = 2/(\cos\theta + 4\sin\theta)$ 　2 $r = 1/(3\cos\theta - 4\sin\theta)$
3 $r = 1/(2\cos\theta - 5\sin\theta)$ 　5 $r = 4\tan\theta\sec\theta$
6 $r = 6\cot\theta\csc\theta$ 　7 $r = (2\cos\theta + 4\sin\theta)/\cos^2\theta$
9 $r = 2\cos\theta$ 　10 $r = 4\cos\theta$
11 $r = 18\cos\theta/(9\cos^2\theta + \sin^2\theta)$ 　13 $r = 3$
14 $r = 4\sqrt{\sec 2\theta}$ 　15 $r = 4\sqrt{\csc 2\theta}$
17 $r = 3/(1 + \cos\theta)$ or $r = -3/(1 - \cos\theta)$
18 $r = 1/(1 - \sin\theta)$ or $r = -1/(1 + \sin\theta)$
19 $r = 3/(2 + \cos\theta)$ or $r = -3/(2 - \cos\theta)$
21 $x = 2$ 　22 $y = 4$ 　23 $y + 2x = 1$
25 $x^2 + y^2 - 4x = 0$ 　26 $x^2 + y^2 + 3y = 0$ 　27 $x^4 + x^2y^2 = y^2$
29 $x^2 + y^2 - x - y = 0$ 　30 $xy = x + y$
31 $x^2y^2 = x^2 + y^2$ 　33 $xy = 1$ 　34 $x^2 - y^2 = 1$
35 $x^4 - y^4 = 2xy$ 　37 $y^2 + 2x - 1 = 0$ 　38 $x^2 - 2y - 1 = 0$
39 $8x^2 - y^2 + 12x + 4 = 0$

Exercise 15.4, page 273

1 A straight line making an angle of $70°$ with the polar axis
2 A straight line making an angle of $-3\pi/4$ with the polar axis
3 A circle of radius 6 with the center at the pole
5 A circle of diameter 18 with the center at $(9, 0°)$
6 A circle of diameter 20 with the center at $(10, 90°)$
7 A circle of diameter 8 with the center at $(4, 180°)$
9 A parabola opening to the left with the focus at the pole; $e = 1, p = 6$
10 A parabola opening downward with the focus at the pole; $e = 1, p = 4$
11 A parabola opening to the right with the focus at the pole; $e = 1, p = 7$
13 An ellipse with the right focus at the pole; $e = \frac{1}{2}, p = 12$
14 An ellipse with lower focus at the pole; $e = \frac{2}{3}, p = 18$
15 An ellipse with the upper focus at the pole; $e = \frac{3}{4}, p = 3$
17 A hyperbola with the left focus at the pole; $e = 2, p = 4$

18   A hyperbola with the right focus at the pole; $e = 4$, $p = 6$
19   A hyperbola with the lower focus at the pole; $e = \frac{5}{2}$, $p = 2$
21   $r \cos(\theta - 30°) = 5$ 22 $r \cos(\theta - 315°) = 2$ 23 $r = 7$
25   $r = 8 \cos \theta$ 26 $r = 18 \sin \theta$ 27 $r = -14 \cos \theta$
29   $r = 5/(1 + \cos \theta)$ 30 $r = 8/(1 - \cos \theta)$ 31 $r = 6/(1 + \sin \theta)$
33   $r = 21/(5 + 3 \cos \theta)$ 34 $r = 45/(4 + 3 \sin \theta)$ 35 $r = 12/(3 - 2 \cos \theta)$
37   $r = 18/(1 - 2 \sin \theta)$ 38 $r = 60/(3 - 5 \cos \theta)$ 39 $r = 24/(2 + 3 \sin \theta)$

Exercise 15.5, page 275

1   $(2, 90°)$ 2 $(\sqrt{3}, 30°)$ 3   No solution
5   $(\sqrt{3}/2, 30°), (-\sqrt{3}/2, 120°), (\sqrt{3}/2, 210°), (-\sqrt{3}/2, 300°)$
6   $(\frac{1}{2}, 45°), (\frac{1}{2}, 135°), (\frac{1}{2}, 225°), (\frac{1}{2}, 315°)$
7   $(\sqrt{2}, 135°), (-\sqrt{2}, 315°)$ 9 $(a, 0), (-a, 180°)$
10   $(a, 90°), (-a, 270°)$ 11 $(2, 45°), (-2, 135°), (2, 225°), (-2, 315°)$
13   $(0, 0), (3, 90°), (0, \pi)$ 14 $(2, 0), (0, 90°), (-2, 270°)$
15   $(0, 0), (\sqrt{3}/2, 60°), (0, 180°), (-\sqrt{3}/2, 300°)$
17   $(2, 30°), (2, 150°)$ 18 $(\frac{1}{2}, 120°), (-1, 180°), (\frac{1}{2}, 240°)$
19   $(2, 180°)$ 21 $(1.5, 120°)$ 22 $(0, 180°), (\frac{1}{2}, 120°)$
23   $(1, 0), (1, 180°)$

Exercise 15.6, page 280

1   $x = 3t, y = t^2 - \frac{1}{3}$ 2 $x = 2/t, y = 3 + t$
3   $x = (3t - 3)/t, y = t - 3$ 5 $x = 5 \cos \theta, y = 5 \sin \theta$
6   $x = 7 \sec \theta, y = 7 \tan \theta$ 7 $x = 3 \cos \theta, y = 2 \sin \theta$
9   $x = 1 + 3 \sin \theta, y = 3 \cos \theta$ 10 $x = 4 \sec \theta, y = 3 + 4 \tan \theta$
11   $x = 2 + 3 \cos \theta, y = -1 + 5 \sin \theta$ 13 $x = 2 \sec \theta, y = 2 \tan^2 \theta$
14   $x = -\tan^2 \theta, y = 2 + \sec \theta$ 15 $x = 2 + 3 \sec \theta, y = (9 \tan^2 \theta)/5$
17   $x = t^2 + 2t, y = t^3 + 2t^2$ 18 $x = at^2/(t^2 - 1), y = at^3/(t^2 - 1)$
19   $x = t^3 - t, y = t^4 - t^2$ 21 $x = 6 \sin \theta \cos \theta, y = 6 \sin^2 \theta$
22   $x = \cos \theta (1 + 3 \cos \theta), y = \sin \theta (1 + 3 \cos \theta)$
23   $x = 2 \cos \theta/(1 + 3 \sin \theta), y = 2 \sin \theta/(1 + 3 \sin \theta)$
33   $x = 4t, y = 7t$ 34 $x = vt, y = -gt^2/2$
35   $x = 12 \cos \theta, y = 4 \sin \theta$

Exercise 15.7, page 286

The answers below are the rectangular forms of the equations.
1   $x = 2y - 2$ 2 $2x = 2 + y$ 3 $x - y + 6 = 0$
5   $x - 2y = 6$ 6 $x - y + 4 = 0$ 7 $2x - 3y = 7$
9   $x = (y + 1)^2$ 10 $x = 2y^2 - 4y + 1$ 11 $3y^2 - 6y - 4x - 5 = 0$
13   $x + 2y = 4$ 14 $2x - 3y = 6$ 15 $x - 3y = 3$
17   $x^2 + y^2 = 4$ 18 $x^2 - y^2 = 9$ 19 $4x^2 + 9y^2 = 36$

21 $(x - 2)^2 + (y - 1)^2 = 1$

22 $(x - 3)^2 - (y - 2)^2 = 1$

23 $[(x - 1)/3]^2 + (y - 2)^2 = 1$

25 $xy = 1$

26 $y(x - 3) = -1$

27 $(x - 5)(y - 2) = -1$

29 $xy - y^2 = x$

30 $xy + 2x^2 = 2y$

31 $x^2 + y^2 = y$

33 $x^{2/3} + y^{2/3} = 4$

34 $x^{1/2} + y^{1/2} = 1$

35 $x^2 + y^2 = 20$

Exercise 16.1, page 291

1 $-8/t^2, 32/t^3$

2 $3(t + 1), 3/2(t - 1)$

3 $3(t^2 + 1)/2t, 3(t^2 - 1)/4t^3$

5 $-(\cot t)/2, -(\csc^3 t)/8$

6 $-(5 \cot t)/3, (-5 \csc^3 t)/9$

7 $-\cos t, -\frac{1}{6}$

9 $-\frac{1}{2}, 0$

10 $-2 \tan t, (2 \sec^4 t \csc t)/3$

11 $-1/(2 \sin^3 t \cos t), (3 \cos^2 t - \sin^2 t)/(12 \sin^5 t \cos^3 t)$

13 $\tan 2\theta, 1/(3 \cos^3 2\theta)$

14 $-\cot 2\theta, -(\csc^3 2\theta)/2$

15 $-(\cos \theta + \cos 2\theta)/(\sin \theta + \sin 2\theta), -(1 + \cos \theta)/(\sin \theta + \sin 2\theta)^3$

17 $y = x - 3, y = -x + 3$

18 $y = x - 1, y = -x + 1$

19 $y = 0, x = 0$

21 No max., min., or inflection points

22 max., at $(0, 1)$

23 max., at $(0, 1)$

Exercise 16.2, page 294

1 $\sqrt{2}$

2 1

3 1

5 $\sqrt{5}/2$

6 1

7 $\sqrt{10}$

9 $\sqrt{5}/2$

10 $\frac{5}{4}$

11 $\sqrt{13}$

13 $\sqrt{1 + \cos^2 x_1}$

14 $\sqrt{1 + \sin^2 x_1}$

15 $\sqrt{3}$

17 $\sqrt{2}$

18 $\sqrt{1 + e^2}$

19 $\sqrt{1 + \cosh^2 x_1}$

21 $\sqrt{1 + x_1^2}/x_1$

22 1

23 $\sqrt{10}/3$

25 1

26 3

27 $2\sqrt{2}$

29 2

30 $\sqrt{6}/2$

31 4

33 3

34 1

35 $4\sqrt{2}$

Exercise 16.3, page 299

K and R are reciprocals, only K is given.

1 $2\sqrt{5}/25$

2 2

3 0

5 $2\sqrt{10}/25$

6 $6\sqrt{37}/1{,}369$

7 8

9 $\frac{3}{4}$

10 $\sqrt{13}/13$

11 $\sqrt{10}/80$

13 2

14 0

15 $\sqrt{2}$

17 $4\sqrt{5}/25$

18 $\frac{1}{2}$

19 2

21 $\sqrt{5}/50$

22 $\frac{3}{16}$

23 $\sqrt{2}/16$

25 $\frac{2}{3}$

26 2

27 $\sqrt{2}/16$

29 $x = 4, K = 4$

30 $x = 0, K = \frac{1}{2}$

31 $x = 1/e, K = e$

33 $x = 3\pi/2, K = 3$

34 $x = 5\pi/6, K = 3\sqrt{3}/2$

35 $x = \pi/4, K = 4\sqrt{5}/25$

37 $x = -\ln \sqrt{2}, K = 2\sqrt{3}/9$

38 $x = 1, K = 1/\sqrt{2}$

39 $x = \pi, K = 1$

41 $(x + 4)^2 + (y - 3.5)^2 = 31.25$

42 $(x + \frac{2}{3})^2 + (y - \frac{5}{6})^2 = {}^{125}\!/_{36}$

43 $(x + 2)^2 + (y - 3)^2 = 8$

Exercise 16.4, page 305

1 $4\mathbf{i} + 3\mathbf{j}$, 5   2 $5\mathbf{i} + 12\mathbf{j}$, 13   3 $24\mathbf{i} + 7\mathbf{j}$, 25   5 $7\mathbf{i} + 2\mathbf{j}$

6 $-2\mathbf{i} + 7\mathbf{j}$ 7 $11\mathbf{i} - 4\mathbf{j}$ 9 $(\mathbf{i}\sqrt{3} + \mathbf{j})/2$ 10 $[\mathbf{i}(-\sqrt{3}) + \mathbf{j})]/2$

11 $\mathbf{i}(-1/\sqrt{2}) - \mathbf{j}1/\sqrt{2}$ 13 2, 30°   14 $\sqrt{2}$, 315°

15 5, arctan ⁴⁄₃   17 $(3\mathbf{i} - 4\mathbf{j})/5$ 18 $(-12\mathbf{i} + 5\mathbf{j})/13$

19 $\mathbf{i}4/\sqrt{41} + \mathbf{j}5/\sqrt{41}$ 21 $\mathbf{i}1/\sqrt{10} + \mathbf{j}3/\sqrt{10}$ 22 $\mathbf{i}1/\sqrt{26} + \mathbf{j}5/\sqrt{26}$

23 $\mathbf{i}2/\sqrt{5} + \mathbf{j}1/\sqrt{5}$ 25 -7 26 -21

27 11   29 arccos $7/(17\sqrt{2})$ 30 arccos $34/5\sqrt{53}$

31 arccos $(-{}^{15}\!/_{17})$ 37 Parallel   38 Parallel   39 Perpendicular

Exercise 16.5, page 311

1 $\mathbf{v} = \mathbf{i}2t + \mathbf{j}3t^2$, $\mathbf{a} = \mathbf{i}2 + \mathbf{j}6t$; $\mathbf{v} = 0$, $\mathbf{a} = 2\mathbf{i}$

2 $\mathbf{v} = \mathbf{i} + \mathbf{j}2t$, $\mathbf{a} = 2\mathbf{j}$; $\mathbf{v} = \mathbf{i} + 2\mathbf{j}$, $\mathbf{a} = 2\mathbf{j}$

3 $\mathbf{v} = \mathbf{i}(-3\cos^2 t \sin t) + \mathbf{j}(3\sin^2 t \cos t)$, $\mathbf{a} = \mathbf{i}(3\cos t)(2\sin^2 t - \cos^2 t) + \mathbf{j}(3\sin t)$
$(2\cos^2 t - \sin^2 t)$; $\mathbf{v} = \mathbf{i}(-3/2\sqrt{2}) + \mathbf{j}3/2\sqrt{2}$, $\mathbf{a} = \mathbf{i}3/2\sqrt{2} + \mathbf{j}3/2\sqrt{2}$

5 $\mathbf{v} = \mathbf{i} \sec^2 t + \mathbf{j} \sec t \tan t$, $\mathbf{a} = \mathbf{i}\, 2\sec^2 t \tan t + \mathbf{j}(\sec^3 t + \sec t \tan^2 t)$; $\mathbf{v} = \mathbf{i}$, $\mathbf{a} = \mathbf{j}$

6 $\mathbf{v} = \mathbf{i}1/(t + 2) + \mathbf{j}$, $\mathbf{a} = \mathbf{i}[-1/(t + 2)^2]$; $\mathbf{v} = \mathbf{i} + \mathbf{j}$, $\mathbf{a} = -\mathbf{i}$

7 $\mathbf{v} = \mathbf{i} \sinh t + \mathbf{j} \cosh t$, $\mathbf{a} = \mathbf{i} \cosh t + \mathbf{j} \sinh t$; $\mathbf{v} = \mathbf{j}$, $\mathbf{a} = \mathbf{i}$

9 $\mathbf{v} = \mathbf{i}\,a\cos t - \mathbf{j}\,a\csc t \cot t$, $\mathbf{a} = -\mathbf{i}\,a\sin t + \mathbf{j}a(\cot^2 t + \csc^2 t)\csc t$; $\mathbf{v} = 0$, $\mathbf{a} = \mathbf{i}\,a - \mathbf{j}a$

10 $\mathbf{v} = \mathbf{i}\,a \sec t \tan t + \mathbf{j}\,b \sec^2 t$, $\mathbf{a} = \mathbf{i}\,a(\sec^3 t + \sec t \tan^2 t) + \mathbf{j}\,2b \sec^2 t \tan t$;
$\mathbf{v} = \mathbf{i}a2\sqrt{3} + \mathbf{j}4b$, $\mathbf{a} = \mathbf{i}14a + \mathbf{j}8\sqrt{3}b$

11 $\mathbf{v} = \mathbf{i}3\cos t - \mathbf{j}6\sin 2t$, $\mathbf{a} = -\mathbf{i}3\sin t - \mathbf{j}12\cos 2t$; $\mathbf{v} = -3\mathbf{i}$, $\mathbf{a} = -12\mathbf{j}$

13 $\mathbf{T} = \mathbf{i}/\sqrt{5} + 2\mathbf{j}/\sqrt{5}$, $\mathbf{N} = -2\mathbf{i}/\sqrt{5} + \mathbf{j}/\sqrt{5}$

14 $\mathbf{T} = \mathbf{i}(2t/\sqrt{4t^2 + 1}) + \mathbf{j}(1/\sqrt{4t^2 + 1})$, $\mathbf{N} = \mathbf{i}(-1/\sqrt{4t^2 + 1}) + \mathbf{j}(2t/\sqrt{4t^2 + 1})$

15 $\mathbf{T} = \mathbf{i}[(t^2 - 1)/\sqrt{2}\,\sqrt{t^4 + 1}] + \mathbf{j}[(t^2 + 1)/\sqrt{2}\,\sqrt{t^4 + 1}]$,
$\mathbf{N} = -\mathbf{i}[(1 + t^2)/\sqrt{2}\,\sqrt{t^4 + 1}] + \mathbf{j}[(-1 + t^2)/\sqrt{2}\,\sqrt{t^4 + 1}]$

17 $\mathbf{T} = \mathbf{i}$, $\mathbf{N} = \mathbf{j}$

18 $\mathbf{T} = \mathbf{i}/\sqrt{10} + 3\mathbf{j}/\sqrt{10}$, $\mathbf{N} = -3\mathbf{i}/\sqrt{10} + \mathbf{j}/\sqrt{10}$

19 $\mathbf{T} = \mathbf{j}$, $\mathbf{N} = -\mathbf{i}$

21 $\mathbf{T} = -\mathbf{i}/\sqrt{2} + \mathbf{j}/\sqrt{2}$, $\mathbf{N} = -\mathbf{i}/\sqrt{2} - \mathbf{j}/\sqrt{2}$

22 $\mathbf{T} = \mathbf{i}/\sqrt{2} - \mathbf{j}/\sqrt{2}$, $\mathbf{N} = \mathbf{i}/\sqrt{2} + \mathbf{j}/\sqrt{2}$

23 $\mathbf{T} = 3\mathbf{i}/\sqrt{10} + \mathbf{j}/\sqrt{10}$, $\mathbf{N} = -\mathbf{i}/\sqrt{10} + 3\mathbf{j}/\sqrt{10}$

Exercise 16.6, page 314

1 $a_T = 4$, $a_N = 32$ 2 $a_T = 1$, $a_N = 16$

3 $a_T = 0$, $a_N = 8$ 5 $a_T = 0$, $a_N = 4$

6 $a_T = 0$, $a_N = 5$ 7 $a_T = 8/\sqrt{17}$, $a_N = 15/\sqrt{17}$

9 $a_T = 0$, $a_N = 2/(t^2 + 1)$ 10 $a_T = -1/e^2$, $a_N = \sqrt{2e^2 + 1}/e^3$

11 $a_T = \sinh 2t/\sqrt{\cosh 2t}$, $a_N = \sqrt{\cosh^2 2t - \sinh^2 2t}/\sqrt{\cosh 2t}$

17 $\mathbf{v} = \mathbf{u}_r(-4\sin\theta) + \mathbf{u}_\theta 2r$, $\mathbf{a} = \mathbf{u}_r 4(-2\cos\theta - r) + \mathbf{u}_\theta(-16\sin\theta)$

18 $\mathbf{v} = \mathbf{u}_r(16t\cos 2\theta) + \mathbf{u}_\theta(2rt)$, $\mathbf{a} = 4\mathbf{u}_r(4\cos 2\theta - 16t^2\sin 2\theta - rt^2) + 2\mathbf{u}_\theta(r + 32t^2\cos 2\theta)$

19 $\mathbf{v} = \mathbf{u}_r 6\sin\theta + \mathbf{u}_\theta 3r$, $\mathbf{a} = \mathbf{u}_r 9(2\cos - r) + \mathbf{u}_\theta 36\sin\theta$

Exercise 17.1, page 317

1 $\ln (2x - 1) + c$
2 $\ln (x^2 - 3x) + c$
3 $\ln 9$
5 $\frac{1}{2} \ln (x^2 - 2x) + c$
6 $\frac{1}{3} \ln 17 - \ln 2$
7 $\frac{1}{2} \ln (x^2 + 4x) + c$
9 $\ln 1.5$
10 $\ln (2 - \cos x) + c$
11 $\ln \sin x + c$
13 $\ln (e^x + 3) + c$
14 $\ln (3^x + 2) + c$
15 $\frac{1}{2} \ln (e^{x^2} + 1) + c$
17 $e^{2x}/2 + c$
18 $e^x - e^{-x} + c$
19 $e^{x^3}/3 + c$
21 $e^{\tan x} + c$
22 $e^{(\sin x + \cos x)^2}/2 + c$
23 $e^{\sec x} + c$
25 $(3^{\pi/2} - 1)/\ln 3$
26 $(15/\ln 4)$
27 $(-2/\ln 7)7^{1/x^2} + c$
29 $\ln (\sec x + \tan x) + c$
30 $-\ln (\sec x - \tan x) + c$
31 $-\ln (\csc x + \cot x) + c$
33 $\frac{1}{2} \ln^2 \sin x + c$
34 $-\frac{1}{2} \ln^2 \cos x + c$
35 $\ln (\ln x) + c$

Exercise 17.2, page 319

1 $\frac{1}{4} \sin 4x + c$
2 $2 \sin (x/2) + c$
3 $\frac{1}{3} \sin 3x + c$
5 $-(3/\pi) \cos \pi x + c$
6 $-\frac{1}{2} \cos 2x + c$
7 $\pi/2 - 1$
9 $\frac{1}{3} \tan (3x + 1)$
10 1
11 $\frac{1}{2}$
13 $-\frac{1}{2} \cot (2x - 1) + c$
14 $-\frac{1}{2} \cot 2x + c$
15 $-\frac{1}{2} \cot 2x + c$
17 0
18 $\frac{1}{3} \sec 3x + c$
19 $-\frac{1}{2} \csc 2x + c$
21 $(1/\pi) \ln (\sec \pi x + \tan \pi x) + c$
22 $\frac{1}{3} \ln [\sec (3x + 2) + \tan (3x + 2)] + c$
23 $2 \ln (\sec x + \tan x) + c$
25 $-\frac{1}{2} \ln [\csc (2x - 1) + \cot (2x - 1)] + c$
26 $\ln (\sqrt{2} + 1)$
27 $-\frac{1}{2} \ln (\csc x^2 + \cot x^2) + c$
29 $-\ln \cos x + c$
30 $-\ln (1 - \tan^2 x) + c$
31 $\ln \sqrt{2}$
33 $\frac{1}{2}$
34 $\frac{1}{3}$
35 $4 - \pi/6$
37 $\ln 2$
38 $\sqrt{2} - 1$
39 $\pi/3 + \ln 2$

Exercise 17.3, page 322

1 $-\frac{1}{8} \cos 6x - \frac{1}{4} \cos 4x + c$
2 $-\frac{3}{4} \cos 4x + \frac{3}{2} \cos 2x + c$
3 $-\frac{1}{6} \cos 6x - x \sin 2 + c$
5 $\frac{1}{3} \sin 3x + \sin x + c$
6 $\frac{1}{2} \sin 2x + 0.1 \sin 10x + c$
7 $2x + \frac{1}{3} \sin 6x + c$
9 $-\frac{1}{2} \sin 6x + \frac{3}{4} \sin 4x + c$
10 $-2x \cos 4 + 0.5 \sin (4x + 2) + c$
11 $-x + \frac{1}{4} \sin 4x + c$
13 $\frac{1}{2} \sinh 2x + c$
14 $-\sinh (3 - x) + c$
15 $-\frac{1}{3} \cosh (1 - 3x) + c$
17 $\frac{1}{2} \ln \cosh 2x + c$
18 $2 \ln \cosh x^{1/2} + c$
19 $\ln \sinh \sqrt{x^2 + 1} + c$
21 $\frac{1}{2} \tanh (2x + 1) + c$
22 $(1/m) \tanh mx + c$
23 $(-1/n) \coth nx + c$
25 $-\frac{1}{2} \operatorname{sech} x^2 + c$
26 $-3 \operatorname{sech} (3x + 2) + c$
27 $-2 \operatorname{csch} 2x + c$
29 $\frac{1}{12} \cosh^4 3x + c$
30 $\frac{2}{3} (\sinh x)^{3/2} + c$
31 $\frac{1}{8} \tanh^4 2x + c$
33 $-\frac{1}{5} \operatorname{sech}^5 x + c$
34 $-\frac{1}{4} \operatorname{csch}^4 x + c$
35 $\frac{1}{4} \cosh^2 2x + c$

Exercise 17.4, page 325

1 $-\cos x + \frac{1}{3}\cos^3 x + c$ **2** $-\cos x + \cos^3 x - \frac{3}{5}\cos^5 x + \frac{1}{7}\cos^7 x + c$

3 $-\frac{1}{2}\cos 2x + \frac{2}{3}\cos^3 2x - \frac{1}{5}\cos^5 2x + c$ **5** $\sin x - 3\sin^3 x + \frac{3}{5}\sin^5 x - \frac{1}{7}\sin^7 x + c$

6 $\frac{1}{3}\sin 3x - \frac{1}{9}\sin^3 3x + c$ **7** $\frac{1}{2}\sin 2x - \frac{1}{3}\sin^3 2x + \frac{1}{10}\sin^5 2x + c$

9 $x/2 + \frac{1}{4}\sin 2x + c$ **10** $3x/4 - \frac{1}{4}\sin 2x + \frac{1}{32}\sin 4x + c$

11 $3x/16 - \frac{1}{4}\sin 2x + \frac{3}{64}\sin 4x + \frac{1}{24}\sin^3 2x + c$

13 $-\frac{1}{6}\cos^6 x + \frac{1}{8}\cos^8 x + c$ **14** $\frac{1}{8}\sin^8 x - \frac{1}{10}\sin^{10} x + c$

15 $\frac{1}{12}\sin^6 2x - \frac{1}{8}\sin^8 2x + \frac{1}{20}\sin^{10} 2x + c$

17 $\frac{1}{3}\sin^3 x - \frac{1}{5}\sin^5 x + c$ **18** $\frac{1}{5}\sin^5 x - \frac{1}{7}\sin^7 x + c$

19 $\frac{1}{7}\cos^7 x - \frac{1}{5}\cos^5 x + c$ **21** $(12x - 3\sin 4x + 4\sin^3 2x + c)/192$

22 $(120x - 128\sin^3 2x - 24\sin 4x - 3\sin 8x + c)/3{,}072$

23 $(12x - 3\sin 4x - 4\sin^3 2x + c)/384$

25 $(4x - \sin 4x + c)/32$ **26** $(24x - 8\sin 4x + \sin 8x + c)/1{,}024$

27 $\frac{1}{2}\sec^2 x + c$ **29** $\frac{1}{2}\tan^2 x + \ln\cos x + c$

30 $(-\cot^4 x - 2\cot^2 x + 4\ln\sin x + c)/4$

31 $-\csc x + c$ **33** $x - \sin^2 x + c$

34 $(12x + 8\sin 2x + \sin 4x + c)/8$

35 $(24x - 6\sin 2x - 32\cos^3 x + 48\cos x + 3\sin 4x + c)/24$

Exercise 17.5, page 328

1 $\tan x - x + c$ **2** $(\tan^3 x - 3\tan x + 3x + c)/3$

3 $(-\cot^3 2x + 3\cot 2x + 6x + c)/6$ **5** $(\cot^2 x + 2\ln\sin x + c)/(-2)$

6 $(-\cot^4 2x + 2\cot^2 2x + 4\ln\sin 2x + c)/8$

7 $(\tan^4 x - 2\tan^2 x - 4\ln\cos x + c)/4$

9 1 **10** $\sqrt{3}/2 - 0.5$

11 $(-15\cot 2x - 10\cot^3 2x - 3\cot^5 2x + c)/30$

13 $\frac{1}{3}$ **14** $(5\tan^3 x + 3\tan^5 x + c)/15$

15 $(-9\cot^7 x - 7\cot^9 x + c)/63$ **17** $(3\tan^4 x + 2\tan^6 x + c)/12$

18 $(10\tan^6 x + 15\tan^8 x + 6\tan^{10} x + c)/60$

19 $(-5\cot^8 2x - 4\cot^{10} 2x + c)/80$ **21** $(-2\csc^6 x + 3\csc^4 x + c)/12$

22 $(-6\csc^{10} x + 15\csc^8 x - 10\csc^6 x + c)/60$

23 $(4\sec^{10} 2x - 15\sec^8 2x + 20\sec^6 2x - 10\sec^4 2x + c)/80$

25 $(2\sqrt{2} - 1)/3$ **26** $\frac{4}{3}$

27 $(-3\csc^5 x + 5\csc^3 x + c)/15$ **29** $\sec^2 x + c$

30 $-\frac{1}{3}\cot^3 x + c$

31 $2\sqrt{\tan x}\,(77\tan x + 66\tan^3 x + 21\tan^5 x)/231 + c$

Exercise 17.6, page 331

1   17.4 g   **2** $\ln 216/\ln 2$ hr = 7.76 hr   **3**   23.1 yr

6   87 percent   **7**   9.4 gal, 7.8 gal   **9**   72.7

10   45°   **11**   21.68 hr   **13**   After 2.30 hr

14 $i = 20e^{-50t}$, $i = 20/e^{100}$ **15**   \$1,000

Exercise 18.1, page 336

1 $(x^2 + 9)^{3/2}/3 + c$ 2 $(x^2 + 4)^{3/2}/3 + c$ 3 $(x^2 + 16)^{1/2} + c$

5 $-(x^2 + 25)^{1/2}/x + c$ 6 $-9\sqrt{x^2 + 9}/x^3 + c$

7 $\ln |\sqrt{x^2 + 16} + x| - \sqrt{x^2 + 16}/x + c$

9 $32 \text{ Arcsin } (x/4) - x\sqrt{16 - x^2}(8 - x^2)/2 + c$

10 $-(9 - x^2)^{3/2}(x^2 + 6) + c$

11 $81 \text{ Arcsin } (2x/3) - 2x\sqrt{9 - 4x^2}(9 - 8x^2) + c$

13 $-\sqrt{16 - x^2}/16x + c$ 14 $-\sqrt{25 - 4x^2}(25 + 2x^2)/24 + c$

15 $-\frac{1}{2} \ln |(2 + \sqrt{4 - x^2})/(x)| + c$ 17 $\sqrt{x^2 - 1} - \text{Arcsec } x + c$

18 $\sqrt{x^2 - 9} - 3 \text{ Arcsec } (x/3) + c$

19 $\frac{1}{20} [\text{Arcsec } (2x/5) - 5\sqrt{4x^2 - 25}/4x^2] + c$

21 $(x^2 - 1)^{3/2}(8 - 3x^2)/15 + c$ 22 $\frac{1}{4} [\text{Arcsec } (x/2) - 2\sqrt{x^2 - 4}/x^2] + c$

23 $(4x^2 - 9)^{3/2}(2x^2 + 3)/40 + c$ 25 $\frac{1}{2} \text{ Arcsec } (x/2) + c$

26 $\text{Arcsec } (x/3) + c$ 27 $\sqrt{4x^2 - 9}/9x + C$

29 $\ln |\sqrt{x^2 + a^2} + x| - x/\sqrt{x^2 + a^2} + C$

30 $\ln |\sqrt{x^2 + 4} + x| + C$ 31 $\sqrt{x^2 + 1} + C$

33 $-(9 - x^2)^{5/2}/5 + C$ 34 $-(9 - 4x^2)^{5/2}(10x^2 + 9)/280 + C$

35 $x/\sqrt{a^2 - x^2} - \text{Arcsin } (x/a) + C$

Exercise 18.2, page 338

5 $\text{Arcsin } (x/3) + C$ 6 $\frac{1}{6} \text{ Arctan } (3x/2) + C$

7 $\frac{1}{16} \ln |(x^2 - 4)/(x^2 + 4)| + C$ 9 $\frac{1}{2} \text{ Arctan } (e^x/2) + C$

10 $\frac{1}{12} \ln |(x^2 + 1)/(x^2 + 7)| + C$ 11 $\ln |x + 1 + \sqrt{x^2 + 2x + 7}| + C$

13 $\frac{1}{4} \ln |(2x - 3)/(2x + 1)| + C$ 14 $\ln |2x - 3 + \sqrt{4x^2 - 12x + 13}| + C$

15 $(1/b) \text{ Arcsin } (bx/a) + C$ 17 $\ln |4x - 3 + \sqrt{16x^2 - 24x - 7}| + C$

18 $\frac{1}{2} \text{ Arcsin } \dfrac{2x + 1}{3} + C$ 19 $\frac{1}{2} \text{ Arctan } [(x + 1)/2] + C$

21 $2\sqrt{9x^2 - 12x} + 3 \ln |3x - 2 + \sqrt{9x^2 - 12x}| + C$

22 $\text{Arcsin } (x - 1) + C$ 23 $\frac{1}{2} \text{ Arctan } [(2x - 1)/2] + C$

25 $\frac{1}{6} \text{ Arctan } [(x^2 + 2)/3] + C$

26 $\ln |4x^2 - 4x - 3| + \frac{3}{8} \ln |(2x - 3)/(2x + 1)| + C$

27 $\sqrt{4x^2 - 12x + 5} + 3.5 \ln |2x - 3 + \sqrt{4x^2 - 12x + 5}| + C$

29 $\ln |x^2 + 4x + 8| + \text{Arctan } [(x + 2)/3] + C$

30 $\frac{1}{2} \ln |x^2 + 4x - 5| + \frac{1}{6} \ln |(x - 1)/(x + 5)| + C$

31 $4\sqrt{x^2 - 3x + 2} + 13 \ln |x - 1.5 + \sqrt{x^2 - 3x + 2}| + C$

33 $\frac{1}{18} \ln |(3x - 4)/(3x + 2)| + C$ 34 $\frac{1}{3} \ln |3x + 1 + \sqrt{9x^2 + 6x + 10}| + C$

35 $(1/\ln 2) \text{ Arcsin } (2^x/3) + C$

37 $\sqrt{x^2 - 5x + 4} - 3.5 \ln |x - 2.5 + \sqrt{x^2 - 5x + 4}| + C$

38 $-\frac{2}{9} \sqrt{3 - 6x - 9x^2} - \frac{23}{9} \text{ Arctan } [(3x + 1)/2] + C$

39 $\frac{3}{2} \ln |x^2 + x + 1| + (7/\sqrt{3}) \text{ Arctan } [(2x + 1)/\sqrt{3}] + C$

Exercise 18.3, page 343

1 $4\sqrt{1+2x}(1+2x)(3x-1)/15 + c$ 2 $-8\sqrt{1-2x}(1-2x)(1+3x)/15 + c$

3 $-2\sqrt{2-x}(2-x)(14+3x)/15 + c$ 5 $2\sqrt{3x-2}(6x+17)/27 + c$

6 $\sqrt{2x+1}(x-3) + c$ 7 $-2\sqrt{4-x}(9+x) + c$

9 $3\sqrt[3]{x+2}(x+2)(2x-3)/14 + c$ 10 $\sqrt[3]{3x+2}(3x+2)(12x-41)/28 + c$

11 $3(x-4)^{2/3}(x+6)/15 + c$ 13 $-\sqrt{4+x^2}/4x + c$

14 $-\sqrt{1-x^2}/x + c$ 15 $-\sqrt{x^4+1}/2x^2 + c$

17 $2\sqrt{x} + 3\sqrt[3]{x} + 6\sqrt[6]{x} + 6\ln(x^{1/6}-1) + c$

18 $\sqrt{x-2}/\sqrt{x} + c$ 19 $-\sqrt{4-x}/2\sqrt{x} + c$

21 $-4(1-x)^{3/2}(2+3\sqrt{x}) + c$ 22 $-8\sqrt{2}(4-x)^{3/2}(16+\sqrt{x}) + c$

23 $e^x - 3\ln(e^x+1) + c$ 25 $\tan x - \sec x + c$

26 $\dfrac{2}{\sqrt{3}}\operatorname{Arctan}\left(\dfrac{1}{\sqrt{3}}\tan\dfrac{x}{2}\right) + c$ 27 $-\ln\left(1-\tan\dfrac{x}{2}\right) + c$

29 $x + \ln|\cos x + \sin x| + C$ 30 $-\ln|1+\cos x| + C$

31 $\ln|1-\cos x| + C$

Exercise 18.4, page 346

1 $-e^{-x}(x+1) + c$ 2 $e^x(x^2-2x+2) + c$ 3 $(a^x \ln a)(x-\ln a) + c$

5 $e^x(\sin x + \cos x)/2 + c$ 6 $e^{-x}(\cos x - \sin x)/2 + c$

7 $-e^{-2x}(2\sin 3x + 3\cos 3x)/13 + c$ 9 $x(\ln x - 1) + c$

10 $x^2(\ln x^2 - 1)/4 + c$ 11 $x^3(6\ln x - 1)/9 + c$ 13 $-x\cos x + \sin x + c$

14 $x\sin x + \cos x + c$

15 $(x^2/2)\sin 2x + (x/2)\cos 2x - \frac{1}{4}\sin 2x + c$

17 $x\operatorname{Arccos} x - \sqrt{1-x^2} + c$ 18 $x\operatorname{Arcsin} 2x + \sqrt{1-4x^2}/2 + c$

19 $(\operatorname{Arcsin} x)(2x^2-1)/4 + 3x\sqrt{1-x^2}/2 + c$

21 $x\operatorname{Arctan} x - \frac{1}{2}\ln(1+x^2)$ 22 $\frac{1}{2}(x^2\operatorname{Arctan} x + \operatorname{Arctan} x - x) + c$

23 $(x/2)[\sin(\ln x) - \cos(\ln x)] + c$ 25 $\frac{1}{2}\sec x\tan x + \frac{1}{2}\ln(\sec x + \tan x) + c$

26 $-\frac{1}{2}\csc x\cot x - \frac{1}{2}\ln(\csc x + \cot x) + c$

27 $x\cosh x - \sinh x + c$ 29 $-2\sqrt{x}\cos\sqrt{x} + 2\sin\sqrt{x} + c$

30 $2\sqrt{x}\sin\sqrt{x} + 2\cos\sqrt{x} + c$

31 $(\operatorname{Arcsin}\sqrt{x})(2x-1)/4 + 3\sqrt{x}\sqrt{1-x}/2 + c$

33 $\frac{1}{2}\sin x\sqrt{1+\sin^2 x} + \frac{1}{2}\ln(\sqrt{1+\sin^2 x} + \sin x) + c$

34 $\frac{1}{2}\sec x\sqrt{1+\sec^2 x} + \frac{1}{2}\ln(\sqrt{1+\sec^2 x} + \sec x) + c$

35 $x^3/3 - x\cos x + \sin x + x/2 - (\sin 2x)/4 + c$

Exercise 18.5, page 352

1 $x + 3\ln|x-3| + C$ 2 $2x + 7\ln|x-1| + C$

3 $-2x/3 + \frac{11}{9}\ln|3x-5| + C$ 5 $1.5\ln|x-2| - 0.5\ln|x+4| + C$

6 $5\ln|x+3| - 2\ln|x+2| + C$ 7 $x + 3\frac{1}{3}\ln|x-5| - \frac{7}{3}\ln|x-2| + C$

9 $3\ln|x| + 4\ln|x-1| - \ln|x-3| + C$

10 $0.5 \ln |x(x - 2)| - \ln |x - 1| + C$

11 $4 \ln |x + 2| - 2 \ln |x - 3| - \ln |x - 1| + C$

13 $x + \frac{1}{3} \ln |x + 1| + \frac{2}{3} \ln |x - 2| - 2 \ln |x + 2| + C$

14 $\ln |x| + 2 \ln |x - 1| - 0.5 \ln |2x - 3| + x^2/2 + 2x + C$

15 $x^2 - x + 7 \ln |x - 2| - 3 \ln |2x - 1| - \ln |x + 1| + C$

17 $-1/(x - 1) + 2 \ln |x - 1| - 4 \ln |x - 3| + C$

18 $\frac{11}{32} \ln |(x - 6)/(x - 2)| - 1/4(x + 2) + C$

19 $(\ln |x - 2| - \ln |x + 2| + 4/x)/16 + C$

21 $\ln |\sec \theta + \tan \theta| + C$ 22 $-\ln |\csc \theta + \cot \theta| + C$

23 $0.5 \sec \theta \tan \theta + 0.5 \ln |\sec \theta + \tan \theta| + C$

25 $\ln |x - 4| - \ln |x - 3| + 1/(x - 3) + C$

26 $\frac{5}{2} \ln |x - 1| - 6/(x - 1) - 1\frac{1}{2} \ln |x - 3| + C$

27 $0.5 \ln |2x - 1| + 2 \ln |x - 1| + 3 \ln |x + 1| + C$

29 $-\ln |x| + \frac{3}{2} \ln |x^2 + x + 1| + \dfrac{11}{\sqrt{3}} \operatorname{Arctan} \dfrac{2x + 1}{\sqrt{3}} + C$

30 $3 \ln |x - 4| - \ln |x^2 + 2x + 3| - \dfrac{3}{\sqrt{2}} \operatorname{Arctan} \dfrac{x + 1}{\sqrt{2}} + C$

31 $\dfrac{1}{12} \ln |x - 2| - \dfrac{1}{24} \ln |x^2 + 2x + 4| - \dfrac{1}{4\sqrt{3}} \operatorname{Arctan} \dfrac{x + 1}{\sqrt{3}} + C$

33 $\ln |x - 1| + 2 \ln |x - 4| + \dfrac{3}{2} \ln |x^2 + x + 4| - \dfrac{\sqrt{15}}{3} \operatorname{Arctan} \dfrac{2x + 1}{\sqrt{15}} + C$

34 $-\ln |x| - 3/x + (7/\sqrt{5}) \operatorname{Arctan} (x/\sqrt{5}) + C$

35 $\frac{1}{32} \ln |(x - 2)/(x + 2)| - \frac{1}{16} \operatorname{Arctan} (x/2) + C$

37 $\left[\ln |x^2 + 4| + \operatorname{Arctan} \dfrac{x}{2} + \ln |x^2 + 2x + 5| + \operatorname{Arctan} \dfrac{x + 1}{2} \right]/2 + C$

38 $\ln |x^2 + x + 3| + \operatorname{Arctan} \dfrac{x + 1}{2} + C$

39 $\ln |x|| x^2 - x + 1| + 1/(x^2 - x + 1) + C$

Exercise 18.6, page 354

1	Divergent	2	2	3	1	5	1
6	Divergent	7	$\ln 2$	9	0	10	$\pi/2$
11	0	13	Divergent	14	3	15	Divergent
17	$3\frac{2}{3}$	18	$\pi/2$	19	$\pi a^2/4$	21	$2\sqrt{6}\pi$
22	12π	23	πa^2				

Exercise 19.1, page 358

1	$e^2 - e$	2	$(e - 1)/2$	3	$2e^3$	5	1
6	$\frac{1}{2}$	7	$e^2(3e^2 - 1)/4$	9	$\frac{4}{3}$	10	$\pi/2$
11	2π	13	2			14	$(\sqrt{2} + \sqrt{2} \ln \sqrt{2} - 1)/\sqrt{2}$
15	$\ln 4 - \ln 3$	17	$\ln 81$	18	$\frac{8}{3}$	19	4
21	$6\frac{4}{3}$	22	$25\pi/4$			23	$12.5[\sqrt{2} - \ln (\sqrt{2} + 1)]$

25 $5\pi/3$ 26 $\ln 16$ 27 $15\!/\!4 \ln 15\!/\!7$ 29 $3\frac{2}{3}$
30 3   31 18   33 $1.5 \ln 3 - \frac{2}{3}$ 34 $6 \ln 9 - 16\frac{2}{3}$
35 $2.25 - 3 \ln 4$ 37 $(e-1)^2/e$ 38 e 39 $2\sqrt{2}$

Exercise 19.2, page 362

3   First   5 4π 6 2π 7 $\pi a^2/2$
9   4   10 16   11 4   13 $19\pi/2$
14 $9\pi/4$ 15 $3\pi a^2/2$ 17 $\pi - 3\sqrt{3}/2$ 18 a^2
19 $2a^2(\pi/3 - \sqrt{3}/4)$ 21 $4(\sqrt{3} - \pi/3)$ 22 $\frac{4}{3}$
23 8π

Exercise 19.3, page 368

1   39π 2 $18\sqrt{3}\pi$ 3 $\pi(e-2)$ 5 8π
6   $1.6\sqrt{2}\pi$ 7 104π 9 $500\pi/3$ 10 $64\pi(2-\sqrt{2})/3$
11 162π 13 $7{,}936\pi/5$ 14 $4\pi a^2 b/3$ 15 $4\pi ab^2/3$
17 $\pi^2/2$ 18 $\pi(\pi - 2)$ 19 $\pi(4-\pi)/4$ 21 $\pi/2$
22 $128\pi/9$ 23 $32\pi a^3/105$

Exercise 19.4, page 372

1   $4\sqrt{2}\pi/5$ 2 $4\pi/21$ 3 $(e^8-1)\pi/4e^8$ 5 $32\pi a^3/105$
6   $4\pi ab^2/3$ 7 $2\pi^2 a^2 b$ 9 $4\pi a^3/3$ 10 $\pi(4-\pi)/4$
11 16π 13 $\pi(e^2-1)/2$ 14 $2\pi(e^2-1)/e$ 15 $2\pi^2 r^3$
17 100π 18 4π 19 $76\pi/5$ 21 20π
22 $5\pi^2 a^3$ 23 $6\pi^3 a^3$ 25 $\pi(2\pi - 3\sqrt{3})$ 26 $\pi(4\pi + 3\sqrt{3})/3$
27 $8\pi \ln 2$

Exercise 19.5, page 375

1   $6(9-h)^2/27$, 54 cu in.   2 $6(14-z)^2/49$, 112 cu in.
3   21.87 cu ft   5 $2{,}000\frac{2}{3}$ 6 $288\sqrt{3}$ 7 162π
9   $4\pi p^3$ 10 $4\sqrt{3}p^3$ 11 $16ab^2/3$ 13 $\frac{8}{15}$
14 $625\pi/12$ 15 $\sqrt{3}\pi/16$ 17 $16a^3/3$ 18 18 cu ft
19 $176\pi/3$

Exercise 19.6, page 381

1   $3\sqrt{2}$ 2 $14\frac{1}{3}$ 3 $16(19\sqrt{19} - 1)/27$

5   $\ln(2 + \sqrt{3})$ 6 $3 + \ln 2$ 7 $\ln \dfrac{e^b - e^{-b}}{e^a - e^{-a}}$

9   $8(10\sqrt{10} - 1)/27$ 10 $\ln(\sqrt{2} + 1) - \ln\sqrt{3} + 2 - \sqrt{2}$

11   $6a$ 13 $8a$ 14 3π

15   $\dfrac{4}{a}\displaystyle\int_0^a \sqrt{\dfrac{a^4 - (a^2 - b^2)x^2}{a^2 - x^2}}\,dx$ 17 $6a$

18   $8a$ 19 $2\sqrt{2}$ 21 $13\pi/3$

22   49π sq in.   23 $12\pi a^2/5$

25   $\pi[\sqrt{2} + \ln(\sqrt{2} + 1)]$ 26 $\pi[\sqrt{2} + \ln(\sqrt{2} + 1)]$

27   18π 29 $12\pi a^2/5$ 30 $64\pi a^3/3$ 31 $2\pi a^2$

33   $4\pi a^2$ 34 $32\pi a^2/3$ 35 $4\pi^2 a^2$

Exercise 19.7, page 384

1   2,000 lb   2   2,500 lb   3   1,000 lb

5   10,000 lb   6   800/3 lb   7 400π lb

9   $800(3\pi - 4)$ lb, $800(3\pi + 4)$ lb   10 $3{,}750\pi$ lb

11   1,600/3 lb   13   1,031.25 lb   14   1,218.75 lb

15   1,687.5 lb   17   1,031.25 lb   18   500/3 lb

19   40,000 lb

Exercise 19.8, page 388

1   30 in.-lb   2   80 ft-lb   3   33.75 ft-lb

5   $\pi/3$ 6   28   7 $24/\pi$

9   1,356 ft-lb   10   10,800 ft-lb   11   12,400 ft-lb

13   $40{,}000\pi$ ft-lb   14 $45{,}000\pi$ ft-lb   15 $60{,}000\pi$ ft-lb

17   $88{,}400\pi$ ft-lb   18 $4{,}250\pi/3$ ft-lb   19 $416\pi/3$ ft-lb

Exercise 20.1, page 392

1   112.5, $335\frac{1}{3}$ 2   68, 64   3   22.6, $6\frac{1}{3}$ 5   0.98, 1

6   0.71, ln 2   7   0.78, 0.74   9   18.1   10   23.68

11   3.06   13   1.18   14   1.86   15   0.88

17   1.91   18   58.5   19   11.2

Exercise 20.2, page 398

1   16.90   2   20.23   3   1.74   5   2.54

6   1.19   7   0.14   9   17.33, $5\frac{2}{3}$ 10 $17\frac{2}{3}$, $17\frac{2}{3}$

11   $57\frac{2}{3}$, $57\frac{2}{3}$ 14 $16r^2/3$ 15 32π 17   20.6

18   $110\frac{1}{3}$ 19 $9\frac{2}{3}$ 21   1.1, $\frac{1}{60}$ 22   56, 0

23   0.5000, 0.00005

Exercise 21.1, page 406

1   $(8/3, 0)$
2   $(2.4, 1.5)$
3   $(-20/3\pi, 20/3\pi)$
5   $(9/5, 9/5)$
6   $(0.8, 2)$
7   $(c/3a, c/3b)$
9   $(92/75, 8/5)$
10   $(68/105, 814/735)$
11   $(a/5, a/5)$
13   $[1/(e-1), (e+1)/4]$
14   $[(\pi-2)/2, \pi/8]$
15   $(\pi/2, \pi/8)$
17   $(\pi a, 5a/6)$
18   $(3p/5, 0)$
19   $\left(\dfrac{4a}{3\pi}, \dfrac{4a}{3\pi}\right)$

21   $\bar{x} = \dfrac{3[18\sqrt{5} - \ln(2+\sqrt{5})]}{4[2\sqrt{5} + \ln(\sqrt{5}+2)]}, \bar{y} = \dfrac{4(5\sqrt{5}-1)}{2\sqrt{5} + \ln(2+\sqrt{5})}$

22   $(2/\pi, -2/\pi)$
23   $(0, 4a/3)$
26   $2\pi^2 r^2 R$

Exercise 21.2, page 412

In Problems 1 to 8 the numbers are the values of I_x, I_y, k_x, and k_y respectively.

1   $3\tfrac{2}{3}, 8, \sqrt{8/3}, \sqrt{2}$
2   $32, 8/3, 2\sqrt{2}, \sqrt{2/3}$
3   $500/3, 2{,}500/7, \sqrt{5}, 5\sqrt{21}/7$
5   $4{,}096/105, 128/15, 8\sqrt{70}/35, 2\sqrt{6}/3$
6   $2^9/15, 3\tfrac{2}{3}, 8\sqrt{30}/15, 2\sqrt{10}/5$
7   $(e^3-1)/9e^3, (2e-5)/e, \sqrt{e^2+e+1}/3e^2, \sqrt{(2e-5)/(e-1)}$
9   $16 - 4\pi$
10   $256/5$
11   $bh^3/12$
13   $4/9$
14   $0.5 - \ln 2$
15   $[\sqrt{2} + \ln(1+\sqrt{2})]/6$
17   $1/20$
18   $2/7$
19   $(e^3 - 3e^2 - 3e - 7)/36$
21   $1{,}696a^3/105$
22   $[(1+e^2)^{1.5} - 2\sqrt{2}]/3$
23   $1.5a^3$
25   $17\pi a^4/4$
26   $37\pi ab^3/4$
27   $5e - 9e^{-1}$

Exercise 21.3, page 417

1   $\pi a^6/4b^2, 3a/4$
2   $32\pi/3, 5/3$
3   $96\pi, 5$
5   $\pi a^2 b^2/4, 3a/8$
6   $32\pi a^4/3, 4a/3$
7   $9\pi a^2 b^2/4, 27a/16$
9   $\pi(3 - e^2)/4e^2, (3 - e^2)/2(e^2 - 1)$
10   $(\pi^2 + 4)\pi/16, (\pi^2 + 4)/4\pi$
11   $\pi(9\ln^2 3 - 9\ln 3 + 4)/2, (9\ln^2 3 - 9\ln 3 + 4)/[2(3\ln^2 3 - 6\ln 3 + 4)]$
13   $\pi a^4/40, 21a/128$
14   $\pi a^4/4, 3a/8$
15   $9\pi a^4/4, 27a/16$
17   $52\pi/3, \sqrt{39}/3$
18   $64\pi/3, 2\sqrt{6}/3$
19   $256\pi/7, 2\sqrt{35}/7$
21   $4\pi(3 - e), \sqrt{2(3 - e)}$
22   $\pi(\pi^3 - 24\pi + 48)/4, 0.5\sqrt{(\pi^3 - 24\pi + 48)/(\pi - 2)}$
23   $\pi(3e^4 + 1)/8, 0.5\sqrt{(3e^4 + 1)/(e^2 + 1)}$

Exercise 22.1, page 422

1   $7, 6/7, 3/7, 2/7$
2   $5, -4/5, 0, 3/5$
3   $\sqrt{61}, 6/\sqrt{61}, 4/\sqrt{61}, -3/\sqrt{61}$
5   $\sqrt{21}, -1/\sqrt{21}, -4/\sqrt{21}, -2/\sqrt{21}$
6   $3\sqrt{6}, -1/\sqrt{6}, -2/\sqrt{6}, -1/\sqrt{6}$
7   $\sqrt{17}, 3/\sqrt{17}, 2/\sqrt{17}, -2/\sqrt{17}$
13   $\pm 2/3$
14   $\pm 1/2$
15   0
17   $(0, 0, 1.5)$

18   $4 \pm \sqrt{15}$ 19 $-2 \pm 2\sqrt{2}$ 21 $\pi/3$ 22 $\pi/2$

23   $3\pi/4$ 25 $\frac{2}{7}$ 26 $-\frac{2}{7}$ 27 $-\frac{1}{3}$

Exercise 22.2, page 430

1   -12 2 -11 3 0

5   $-42\mathbf{i} + 21\mathbf{j} + 14\mathbf{k}$ 6 $-33\mathbf{i} - \mathbf{j} + 9\mathbf{k}$ 7 $-35\mathbf{i} + 4\mathbf{j} - 28\mathbf{k}$

9   $3\sqrt{6}$ 10 $\sqrt{465}$ 11 $4\sqrt{91}$

13   $(8\mathbf{i} + \mathbf{j} - 22\mathbf{k})/3\sqrt{61}$ 14 $(13\mathbf{i} - 9\mathbf{j} + 20\mathbf{k})/5\sqrt{26}$

15   $(-37\mathbf{i} + 22\mathbf{j} - 23\mathbf{k})/\sqrt{2,382}$ 17 $\pi/2$ 18 $2\pi/3$

19   0 21   Arccos $(-9/\sqrt{318})$ 22   Arccos $(11/\sqrt{178})$

23   Arccos $(-11/\sqrt{627})$ 25 $\mathbf{v} = 4\mathbf{i} + \mathbf{j} + 3\mathbf{k}, \mathbf{a} = 2\mathbf{i}$

26   $\mathbf{v} = 2\mathbf{i} + \mathbf{k}, \mathbf{a} = -\mathbf{j} + \mathbf{k}$ 27 $\mathbf{v} = e\mathbf{i} + \mathbf{j} + 3\mathbf{k}, \mathbf{a} = e\mathbf{i} - \mathbf{j} + 2\mathbf{k}$

Exercise 22.3, page 434

1   $\pi/2$ 2   Arccos $(-\frac{4}{9})$

3   Arccos $(9/2\sqrt{39})$ 5 $(1, 2, 3)$

6   $(-1, 1, -3)$ 7 $(6, 4, 1)$

9   $19x + y - 13z + 2 = 0$ 10 $2x - 3y + z + 2 = 0$

11   $2x + 6y - z + 5 = 0$ 13 $(x - 2)/(-1) = (y - 4)/5 = (z - 5)/4$

14   $(x - 7)/2 = (y - 6)/3 = (z + 2)/(-3)$ 15 $(x - 3)/(-2) = y/1 = (z + 5)/5$

17   $6x + 2y - 3z - 3 = 0$ 18 $2x - 5y + \sqrt{7}z = 8 - \sqrt{7}$

19   $7x + 5y - z - 20 = 0$ 21 $4x + 5y - 3z = 50$

22   $3x - 6y + 2z - 49 = 0$ 23 $2x - 2y + z = 1$

26   $3x/7 - 2y/7 + 6z/7 = 3$ 27 $2x/7 + 6y/7 + 3z/7 = 5$

30   7 31 7

Exercise 23.2, page 442

1   Circular   2   Circular   3   Circular   5   Parabolic

6   Parabolic   7   Parabolic   9   Elliptic   10   Elliptic

11   Hyperbolic   13 $(x - 2)^2 + (y - 1)^2 + (z - 3)^2 = 16$

14   $(x - 3)^2 + (y - 5)^2 + (z + 2)^2 = 49$ 15 $(x + 4)^2 + y^2 + (z - 6)^2 = 49$

17   $(x - 1)^2 + (y + 2)^2 + (z + 4)^2 = 9^2$ 18 $(x + 2)^2 + (y - 3)^2 + (z - 5)^2 = 7^2$

19   $(x + 1)^2 + (y - 4)^2 + (z + 3)^2 = 26$ 21 $(x - 1)^2 + (y - 2)^2 + (z - 3)^2 = 4^2$

22   $(x - 2)^2 + (y + 3)^2 + (z + 4)^2 = 5^2$ 23 $(x + 3)^2 + (y - 1)^2 + (z - 4)^2 = 6^2$

Exercise 23.3, page 446

5   Ellipsoid, $x^2/3^2 + y^2/2^2 + z^2/4^2 = 1$ 6   Ellipsoid, $x^2/4^2 + y^2/3^2 + z^2/1^2 = 1$

7   Ellipsoid, $x^2/5^2 + y^2/2^2 + z^2/3^2 = 1$

9   Elliptic hyperboloid of one sheet, $x^2/2^2 - y^2/1^2 + z^2/4^2 = 1$

10 Elliptic hyperboloid of one sheet, $x^2/3^2 + y^2/1^2 - z^2/2^2 = 1$
11 Elliptic hyperboloid of one sheet, $x^2/2^2 + y^2/3^2 - z^2/5^2 = 1$
13 Elliptic hyperboloid of two sheets, $x^2/2^2 - y^2/1^2 - z^2/2^2 = 1$
14 Elliptic hyperboloid of two sheets, $x^2/3^2 - y^2/6^2 - z^2/2^2 = 1$
15 Elliptic hyperboloid of two sheets, $-x^2/4^2 - y^2/1^2 + z^2/3^2 = 1$
17 Elliptic cone, $x^2/1^2 + y^2/2^2 - z^2/3^2 = 0$
18 Elliptic cone, $x^2/5^2 + y^2/2^2 - z^2/4^2 = 0$
19 Elliptic cone, $x^2/9^2 - y^2/3^2 + z^2/2^2 = 0$
21 $x'^2/3^2 + y'^2/2^2 + z'^2/5^2 = 1$ 22 $x'^2/2^2 - y'^2/3^2 + z'^2/2^2 = 1$
23 $x'^2/5^2 - y'^2/2^2 - z'^2/3^2 = 1$ 25 $x'^2/2^2 + y'^2/3^2 - z'^2/2^2 = 1$
26 $x'^2/3^2 - y'^2/1^2 - z'^2/2^2 = 1$ 27 $x'^2/3^2 + y'^2/3^2 - z'^2/2^2 = 0$

Exercise 24.1, page 457

1 $x(3x - 8y^2), 8y(-x^2 + 2)$ 2 $x(3x + 10y), 5x^2$
3 $4y^2/(y - 4x)^2, y(y - 8x)/(y - 4x)^2$ 5 $2x/(x^2 + y^2), 2y/(x^2 + y^2)$
6 $-y/(x^2 + y^2), x/(x^2 + y^2)$ 7 $-xy \sin xy + \cos xy, -x^2 \sin xy$
9 $e^y + ye^x, xe^y + e^x$ 10 $e^{x-y} - 2xy, -e^{x-y} - x^2$
11 $\sin y - y \sin x, x \cos y + \cos x$ 13 12, 3
14 $\sqrt{2}, 3\sqrt{2}$ 15 11, 8, -1 17 $-x/z, -y/z$ 18 $-y/2z, -x/2z$
19 $x/4z, y/9z$ 21 $-2xz/(x^2 + 8z), 8/(x^2 + 8z)$
22 $(xy - 2)/2z, x^2/4z$ 23 $(y - 2xz)/(x^2 + 2yz), (x - z^2)/(x^2 + 2yz)$
25 $-\frac{2}{3}, -\frac{1}{2}$ 26 10, 12   27 $\pm 1.5, \pm \frac{1}{2}$

Exercise 24.2, page 463

1 $z - 1 = 2(x - 1) + 6(y - 3), (x - 1)/2 = (y - 3)/6 = (z - 1)/(-1)$
2 $z - 4 = 8(x - 2) + 2(y - 1), (x - 2)/8 = (y - 1)/2 = (z - 1)/(-1)$
3 $z - 1 = 6(x - 3) - 8(y - 2), (x - 3)/6 = (y - 2)/(-8) = (z - 1)/(-1)$
5 $z + 3 = 2(x - 6) + 1(y - 3), (x - 6)/2 = (y - 3)/1 = (z + 3)/(-1)$
6 $z - 3 = -1.5(x - 2) - 0.75(y - 4), (x - 2)/1.5 = (y - 4)/0.75 = (z - 3)/1$
7 $z - 3 = 8(x - 2) + 4(y - 2), (x - 2)/8 = (y - 2)/4 = (z - 3)/(-1)$
9 $dz = 2(x - y)\, dx - 2(x - 3y)\, dy$ 10 $dz = (2x\, dx - dy)/2\sqrt{x^2 - y}$
11 $dz = dx/x + dy/y$ 13 $du = z(y - 1)\, dx + xz\, dy + x(y - 1)\, dz$
14 $du = e^x\, dx - dy/y - dz/z$ 15 $dw = dx/x - \sin y\, dy - e^z dz$
17 3.98   18 63.36   19 92.0 cc   21 18.5 cu in
22 319   23 2.26 sq ft

Exercise 24.3, page 468

1 $2x - y + (-x + 2y)3$ 2 $2x - 3y^2/x$
3 $\cos x - (\sin y)e^x$ 5 $5x$ 6 $\cos(x + y) - 2 \sin x$
7 $-2e^{-t}/x + 2e^t/y$ 9 $u_r = 3(x + y), u_s = x - y$
10 $u_r = (-y + sx)/(x^2 + y^2), u_s = (-y + rx)/(x^2 + y^2)$
11 $u_r = (-s \sin xy)(2ry + xe^{rs}), u_s = (-r \sin xy)(ry + xe^{rs})$

13   $du/dt = (y - z)\cos t - (x + z)\sin t + 2(y - x)\cos 2t$
14   $du/dt = -2x\sin t - 2y\cos t + 4z\cos 2t$
15   $du/dt = 2t/y + (-x + y^2)/y^2 - 1$
17   $u_r = (y + z)s + 2x - y - z$, $u_s = (y - z)r - y + z$
18   $u_r = -2ryz/(r^2 - s^2)^2 + x(y + z)$, $u_s = 2syz/(r^2 - s^2) - xy + xz - 2$
19   $u_r = 2x\cos s + 2ys\cos r + 2zs$, $u_s = -2xr\sin s + 2y\sin r + 2zr$
25   $\sqrt{2}/2$ 26 $(5 - \sqrt{3})/2$ 27 $-\sqrt{2}/2$ 29 $37\sqrt{2}/6$
30   $-\sqrt{3}/2$ 31 $2\sqrt{3}$ 33 $2\sqrt{2}, 5\pi/4$ 34 $2\sqrt{2}, 7\pi/4$
35   $1, 7\pi/4$ 37 $0.627°$ per in.

Exercise 24.4, page 472

1   $-y/(x + 3y^2)$ 2 $-y/(x - 5)$ 3 $(2xy + 1)/(1 - x^2)$
5   $x(2 - 3x)/4y^3$ 6 $-(x^3 + y)/(x + y)$ 7 $x(3x + 2y)/(2y - x^2)$
9   $-x/9z, -y/9z$ 10 $(2x + y)/(2z - y), (x + z)/(2z - y)$
11   $(x^2 - y)/(z^2 - 3y), (x + 3z)/(z^2 - 3y)$
13   $(3x^2 - u^2)/(2ux - yz)$ 14 $-(y + 2vt)/(x^2 + t^2)$
15   $(2xy - s)/(y - 2zs)$ 17 $2y^2, 4xy, 2x^2$ 18 $-.2, 2y, 2x + 6y$
19   $-4y/(x + y)^3, 2(x - y)/(x + y)^3, 4x/(x + y)^3$
21   $-x\cos(x - y) - 2\sin(x - y), -x\cos(x - y) + \sin(x - y), -x\cos(x - y)$
22   $-4y\sin(2x - y), 2y\sin(2x - y) + 2\cos(2x - y), -y\sin(2x - y) - 2\cos(2x - y)$
23   $0, 1/y, -x/y^2$

Exercise 24.5, page 476

1   None   2   None   3   min. at $(0, 0)$ 5   min. at $(1, 1)$
6   min. at $(1, 1)$ 7   max. at $(-1, -1)$ 9   None
10   No information   11   max. at $(\pi/3, -\pi/6)$ 13 $N/3, N/3, N/3$
14   $2^{3/4}$ 15   Cube   17   Cube   19 $8abc$

Exercise 25.1, page 483

1   2   2   24   3   11   5 $17\frac{}{30}$
6   $104\frac{}{3}$ 7 $14\frac{}{3}$ 9   36   10   18
11   4   13 $(\pi + 3)/48$ 14 $(3\pi - 8)/8$ 15 $\pi/2$
17   1   18   24   19   1   21 $\int_0^1\int_{y^{1/2}}^{y^{1/3}} dx\, dy = \frac{1}{12}$
22   $\int_{-1}^1\int_0^{\sqrt{y+1}} dx\, dy = 4\sqrt{2}/3$ 23 $\int_0^1\int_0^{\sqrt{1-x^2}} dy\, dx = \pi/4$

Exercise 25.2, page 489

1   $\frac{4}{3}$ 2 $\frac{2}{3}$ 3 $\frac{3}{4}$ 5 $125\frac{}{24}$
6   12   7 $10\ln 5 - 8$ 9 πab 10 $37\frac{}{324}$

11 $\frac{1}{12}$ **13** π **14** $1.5\pi a^2$ **15** 4.5π

17 $\frac{1}{3}$ **18** $4(\sqrt{3} - \pi/3)$ **19** $5\pi/4$ **21** $\pi a^4/4$

22 $\frac{2}{9}$ **23** $\pi - 2$ **25** $2\int_0^{\pi/3}\int_{1+\cos\theta}^{3\cos\theta} r^3 \sin^2\theta \, dr \, d\theta$

26 $2\int_0^{\pi/3}\int_0^{1+\cos\theta} r^3 \sin^2\theta \, dr \, d\theta + 2\int_{\pi/3}^{\pi/2}\int_0^{3\cos\theta} r^3 \sin^2\theta \, dr \, d\theta$

27 $\int_0^{\pi/3}\int_0^{1-\cos\theta} r^3 \cos^2\theta \, dr \, d\theta + \int_{\pi/3}^{\pi/2}\int_0^{\cos\theta} r^3 \cos^2\theta \, dr \, d\theta$

30 $3\pi a^4/2$ **31** $\frac{44}{105}$ **33** $a^4/8$ **34** 576

35 $\frac{1}{8}$

Exercise 25.3, page 495

1 18   **2** $17\pi/2$ **3** $\frac{1}{12}$ **5** $4\pi a^3/3$

6 $\sqrt{2}/15$ **7** $9(3\pi - 4)/4$ **9** 2π **10** $1.9\sqrt{6}$

11 $2ma^3/3$ **13** 4   **14** 2π **15** $\frac{1}{2}$

17 $\int_0^1\int_0^{\sqrt{1-x^2}}\int_0^{9-x^2-y^2} dz \, dy \, dx$ **18** $\int_0^{2a}\int_0^x\int_0^{(x^2+y^2)/4a} dz \, dy \, dx$

19 $\int_0^{1/2}\int_{2x^2}^{1/2}\int_0^{1-2y} dz \, dy \, dx$ **21** $\int_0^6\int_0^{6-x}\int_0^{(6-x-y)/2} (y^2 + z^2) \, dz \, dy \, dx$

22 $\int_0^1\int_{\sqrt{x}}^1\int_0^{1-y} (x^2 + z^2) \, dz \, dy \, dx$ **23** $\int_0^{\sqrt{2}/2}\int_{x^2}^{1/2}\int_0^{1-2y} (y^2 + z^2) \, dz \, dy \, dx$

25 $\int_0^6\int_0^{6-x}\int_0^{(6-x-y)/2} x^2 \, dz \, dy \, dx$ **26** $\int_0^1\int_{\sqrt{x}}^1\int_0^{1-y} y^2 \, dz \, dy \, dx$

27 $\int_0^{\sqrt{2}/2}\int_{x^2}^{1/2}\int_0^{1-2y} z^2 dz \, dy \, dx$

Exercise 25.4, page 504

1 $r^2 + z^2 = a^2$, $\rho = a$ **2** $r^2 = 2z$, $\rho = 2\cot\phi\csc\phi$

3 $r^2 \cos 2\theta = a^2$, $\rho^2 \sin^2\phi \cos 2\theta = a^2$ **5** $x^2 + y^2 + z^2 = a^2$, $\rho = a$

6 $x^2 + y^2 = 2z$, $\rho = 2\cot\phi\csc\phi$ **7** $x^2 - y^2 = a^2$, $\rho^2 \sin^2\phi \cos 2\theta = a^2$

9 $x^2 + y^2 + z^2 = a^2$, $r^2 + z^2 = a^2$ **10** $x^2 - y^2 = a^2$, $r^2 \cos 2\theta = a^2$

11 $2xy = z$, $r^2 \sin 2\theta = z$ **13** 9π

14 67.5π **15** 18π **17** 8π **18** 24π

19 $112\pi/3$ **21** $8(3\pi - 4)/9$ **22** $8\pi(4 - \sqrt{2})/3$

23 $4\pi/3$ **25** $4a^3(3\pi - 4)/9$ **26** $224\pi/3$ **27** 288 cu in.

29 $ma^2/2$ **30** $2ma^2/5$ **31** $m(3a^2 + 4b^2)/4$

Exercise 25.5, page 508

1 $4a^2$ **2** $7\sqrt{2}\pi/6$ **3** $10\pi/3$ **5** $8a^2$

6 $(2\sqrt{2} - 1)\pi$ **7** $13\pi/12$ **9** $7\pi/6$ **10** $2\pi ah$

11 16π

Exercise 26.1, page 513

1 3   2 1   3 $\frac{1}{3}$ 5 $\frac{1}{3}$, $\frac{1}{15}$, $\frac{1}{35}$, $\frac{1}{63}$

6 $\frac{2}{3}$, $\frac{4}{9}$, $\frac{8}{27}$, $\frac{16}{81}$ 7 1, $\frac{4}{3}$, 2, $\frac{16}{5}$ 9 2, 1, $\frac{1}{3}$, $\frac{1}{12}$

10 $x^2/(2\ln 2)$, $x^3/(3\ln 3)$, $x^4/(4\ln 4)$, $x^5/(5\ln 5)$

11 $-1/\ln 3$, $2/\ln 4$, $-6/\ln 5$, $24/\ln 6$ 13 $3n - 2$

14 $(-1)^{n+1}n/3^n$ 15 $\dfrac{1 \cdot 3 \cdot 5 \cdots (2n - 1)}{2 \cdot 4 \cdot 6 \cdots 2n}$

17 1, $-\frac{1}{3}$, $-\frac{1}{15}$, $-\frac{1}{35}$; convergent, $\frac{1}{2}$ 18 $\frac{2}{3}$, $\frac{1}{10}$, $\frac{1}{22}$, $\frac{2}{77}$; convergent, $\frac{2}{3}$

19 3, -2, $-\frac{4}{9}$, $-\frac{13}{72}$; convergent, zero   21 3, 6, 18, 54; divergent

22 $\frac{1}{3}$, $-\frac{2}{9}$, $-\frac{2}{27}$, $-\frac{2}{81}$; convergent, zero

23 $\ln 6$, $\ln 7 - \ln 6$, $\ln 8 - \ln 7$, $\ln 9 - \ln 8$; divergent

Exercise 26.2, page 520

1 Divergent	2 Convergent	3 Divergent	5 Convergent
6 Divergent	7 Divergent	9 Convergent	10 Divergent
11 Convergent	13 No	14 No	17 Convergent
18 Convergent	19 Convergent	21 Divergent	22 Divergent
23 Convergent	25 Convergent	26 Convergent	27 Convergent

Exercise 26.3, page 523

1 Convergent	2 Convergent	3 Divergent	5 Divergent
6 Convergent	7 Convergent	9 Convergent	10 Convergent
11 Convergent	13 Divergent	14 Divergent	15 Divergent
17 Convergent	18 Convergent	19 Convergent	21 Convergent
22 Divergent	23 Divergent		

Exercise 26.4, page 527

1 Absolutely convergent	2 Absolutely convergent	3 Conditionally convergent
5 Absolutely convergent	6 Absolutely convergent	7 Divergent
9 Absolutely convergent	10 Absolutely convergent	11 Absolutely convergent
13 Divergent	14 Conditionally convergent	15 Divergent
17 Divergent	18 Conditionally convergent	19 Absolutely convergent
21 $\frac{1}{26}$	22 $\frac{3}{2}(4^5)$	23 $\frac{2}{17}$

Exercise 26.5, page 530

1 $(-1, 1)$	2 $(-1, 1)$	3 $(-1, 1)$	5 $(-5, 5)$
6 $(-\frac{1}{8}, \frac{1}{8})$	7 $[-2, 2]$	9 All finite x	10 All finite x
11 None	13 $x \neq 0$	14 All finite x	15 $x = 0$
17 $[-3, -1]$	18 $[1, 3)$	19 $[1, 3]$	21 $[-1, 1]$
22 $(0, 2)$	23 $[0, 6)$		

Exercise 26.6, page 534

9 $\frac{1}{2}[1 + \sqrt{3}(x - \pi/6) - (x - \pi/6)^2/2! - \sqrt{3}(x - \pi/6)^3/3! + \cdots]$

10 $(\sqrt{2}/2)[1 - (x - \pi/4) - (x - \pi/4)^2/2! + (x - \pi/4)^3/3! + \cdots]$

11 $\ln 2 + (x - 2)/2 - (x - 2)^2/8 + (x - 2)^3/24 - \cdots$

13 $e[1 + (x - 1) + (x - 1)^2/2! + (x - 1)^3/3! + \cdots]$

14 $e[1 - (x + 1) + (x + 1)^2/2! - (x + 1)^3/3! + \cdots]$

15 $1 - (x - 1) + (x - 1)^2 - (x - 1)^3 + \cdots$

17 $1 - x^2 + x^4/2! - x^6/3! + \cdots + (-1)^n x^{2n}/n! + \cdots$, all values

18 0.15643

19 0.93970

25 $x + x^3/3 + 2x^5/15$

26 $1 + x^2/2 + 5x^4/24$

27 $x + x^3/6 + x^5/24$

29 $-x^2/2 - x^4/12 - x^6/45$

30 $x^2 - x^3 + x^4/2$

31 $x + x^2 + x^3/3$

Exercise 26.7, page 539

1 $\cosh x = 1 + x^2/2! + x^4/4! + x^6/6! + \cdots$

2 $\sinh x = x + x^3/3! + x^5/5! + x^7/7! + \cdots$

5 $1 + x - x^3/3 - x^4/6 - x^5/40$, all values of x

6 $e[1 + 2(x - 1) + 3(x - 1)^2/2 + 2(x - 1)^3/3 + 5(x - 1)^4/24 + \cdots]$

7 $(x - 1) + (x - 1)^2/2 - (x - 1)^3/6 + (x - 1)^4/12 - (x - 1)^5/20 + \cdots$

13 $x + \dfrac{x^3}{2 \cdot 3} + \dfrac{1 \cdot 3x^5}{2 \cdot 4 \cdot 5} + \dfrac{1 \cdot 3 \cdot 5x^7}{2 \cdot 4 \cdot 6 \cdot 7} + \dfrac{1 \cdot 3 \cdot 5 \cdot 7x^9}{2 \cdot 4 \cdot 6 \cdot 8 \cdot 9} + \cdots$

14 $2(x + x^3/3 + x^5/5 + x^7/7 + x^9/9 + \cdots)$

15 $x - \dfrac{x^3}{2 \cdot 3} + \dfrac{x^5}{2!2^2 \cdot 5} - \dfrac{x^7}{3!2^3 \cdot 7} + \dfrac{x^9}{4!2^4 \cdot 9} - \cdots$

17 1.057

18 0.574

19 $\frac{7}{48}$

21 Three

22 Four

23 Four

Exercise 27.1, page 542

1 Third, first

2 First, second

3 Second, third

9 $xy = c$

10 $x^2y = c$

11 $1 - 2y^2 \ln |x - 1| = Ky^2$

13 $\text{Arctan } x + \ln y = c$

14 $y(1 + x^2) = c$

15 $\text{Arctan } x + \text{Arcsin } y = c$

17 $c + x + y = \ln |(1 + y)/(1 - x)|$

18 $x^2 + 2x + 2 \ln |x - 1| + 2 \text{ Arctan } y = c$

19 $\ln |\sec x + \tan x| \sin y = c$

Exercise 27.2, page 546

1 $2x^2 + 2xy + y^2 = c$

2 $3x^2 - 2xy + y^2 = c$

3 $x^2 + 4xy - 10x - y^2 + 2y = c$

5 $xy^3 + \ln |x| = c$

6 $xy^2 - x + \ln |y| = c$

7 $2xe^y + 2x - y^2 = c$

9 $e^x + x + e^y + 3y = c$

10 $\sin x + xy + y^2 = c$

11 $x \tan y = c$

13 $x^2y = c$

14 $x = cy$

15 $y \ln |x| = c$

17 $e^x, e^x(x - 1 + y) = c$

18 $e^{x+y}, ye^{x+y} = c$

19 $1/x^2, y = cx$

21 $1/y^2, 2x + x^2y = 2cy$

22 $1/x^2, y = x \ln |x| + cx$

23 $y, xy^2 = y + c$

Exercise 27.3, page 549

1 $y = x \ln |x| - cx$ 2 $x^2 + 4xy + y^2 = c$ 3 $(x + y)^3(x - y) = k$

5 $x^2 + y^2 = kx$ 6 $(y - x)^2 = cy$ 7 $y = cx - x \ln |x|$

9 $y^3 = 3x^3(c - \ln |x|)$ 10 $y^3 = 3x^3(\ln |x| - c)$ 11 $ky(y^2 - 6x^2) = x^4$

14 $2(x - 1)^2 + 2(x - 1)(y - 2) - 3(y - 2)^2 = c$

15 $\ln |x^2 + y^2 + 2x - 2y + 2|^{3/2} = c + 2 \operatorname{Arctan} (y - 1)/(x + 1)$

Exercise 27.4, page 552

1 $y = 2 + ce^{-x}$ 2 $xy = x^4 + c$ 3 $xy = x^2 - x + c$

5 $y = xe^x + cx$ 6 $y = x^2(\ln |x| - e^x + c)$

7 $x^2y = e^{3x} + c$ 9 $y = x^2(\sin x + c)$ 10 $y = x(\ln |\sin x| + c)$

11 $x^2y = -\cos x + c$ 13 $y \sin x = x^2 + c$ 14 $y = 1 + ce^{-\sin x}$

15 $y \cos x = e^x + c$ 17 $x^2 = y(c - x^4)$ 18 $1 = xy(c - x)$

19 $e^{2x}y^2(2 \cos x + c) = 1$

Exercise 27.5, page 556

1 $y = c_1e^{2x} + c_2e^x$ 2 $y = c_1e^{3x} + c_2e^{-x}$ 3 $y = c_1e^{2x} + c_2e^{-3x}$

5 $y = c_1e^x + c_2e^{2x} + c_3e^{3x}$ 6 $y = c_1e^x + c_2e^{2x} + c_3e^{-3x}$

7 $y = c_1 + c_2e^{2x} + c_3e^{3x}$ 9 $y = (c_1 + c_2x)e^{3x}$

10 $y = (c_1 + c_2x)e^{-x}$ 11 $y = c_1e^x + c_2xe^x + c_3e^{2x}$

13 $y = c_1e^x + c_2 \cos 3x + c_3 \sin 3x$ 14 $y = c_1e^{2x} + e^x(c_2 \cos x + c_3 \sin x)$

15 $y = c_1 \cos x + c_2 \sin x + c_3e^{2x} + c_4e^{-x}$

17 $y = (c_1 \cos x + c_2 \sin x + c_3 + c_4x)e^x$

18 $y = e^{2x}(c_1 \cos 3x + c_2 \sin 3x) + (c_1 + c_2x)e^{-x}$

19 $y = e^x(c_1 \cos x + c_2 \sin x + c_3 \cos 2x + c_4 \sin 2x)$

Exercise 27.6, page 558

1 $y_p = x, y_c = ae^x + be^{2x}$ 2 $y_p = e^x, y_c = ae^{2x} + be^{-4x}$

3 $y_p = \sin x, y_c = ae^{-2x} + be^{-3x}$ 5 $y_p = \cos x, y_c = ae^x + be^{2x} + ce^{-3x}$

6 $y_p = e^{3x}, y_c = ae^x + be^{-x} + ce^{2x}$ 7 $y_p = x^3 + x, y_c = ae^{2x} + be^{3x} + ce^{-x}$

9 $y_p = e^{-x}, y_c = ae^{2x} + bxe^{2x}$

10 $y_p = 3x^2 - 1, y_c = ae^x + bxe^x$

11 $y_p = x + e^{2x}, y_c = ae^x + bxe^x + ce^{-3x}$

13 $y_p = x^2 - x, y_c = a \cos 2x + b \sin 2x + ce^x$

14 $y_p = 1 - e^{-x}, y_c = e^x(a \cos x + b \sin x + ce^x)$

15 $y_p = x^2, y_c = e^x(a \cos 2x + b \sin 2x + c) + he^{-x}$

17 $y_p = e^{-x} + 1, y_c = e^{2x}(a \cos x + b \sin x) + ce^x + hxe^x$

18 $y_p = x^2 - x + 3, y_c = e^x(a \cos 3x + b \sin 3x) + ce^{2x} + hxe^{2x}$

19 $y_p = \sin x, y_c = e^{2x}(a \cos x + b \sin x + cx \cos x + hx \sin x)$

Index